RECENT ADVANCES IN ENVIRONMENTAL SCIENCE

RECENT ADVANCES
IN
ENVIRONMENTAL SCIENCE

Editor

Dr. K.G. Hiremath

Head Department of Botany
Dhempe College of Arts & Science
Miramar, Panaji
Goa–403 001

DISCOVERY PUBLISHING HOUSE
NEW DELHI-110002

First Published-2003
Reprinted: 2014
ISBN 978-81-7141-679-0

Published by:

DISCOVERY PUBLISHING HOUSE
4831/24, Ansari Road, Prahlad Street,
Darya Ganj, New Delhi-110002 (India)
Phone: 23279245 • Fax: 91-11-23253475
E-mail:dphtemp@indiatimes.com

Printed at:

Dynamic Printers, Delhi

Dedicated to

My Father

Who has Always

Inspired Me to be

Sincere and Hardworking

FOREWORD

Protection of the environment is the responsibility of every citizen of India. In fact, the concern about conservation of nature is growing all over the world. This is largely due to the efforts of World Commission on Environment and Development which enunciated the idea of "Sustainable Development" and the U N Conference on Environment and Development (UNCED) or the "Earth Summit" held at Rio de Janeiro in June 1992 which was signed by 154 countries. It emphasized the need for "Conservation of Biological Diversity".

With this background, it is important to describe the state of environment in various regions. The present book largely deals with the environmental problems of Goa and its adjoining areas. In addition to these, it includes some articles of a general nature covering other regions.

In the terrestrial environment, deforestation is a serious problem in India and efforts are being made to carry out afforestation programmes in the Himalayas. Conversation of wetlands and protection of rivers from pollution are the other important aspects to be considered. In the marine environment, protection of estuaries and coastal regions from pollution and other man-made changes are important areas to be looked into. Similarly, mangroves of both east and west coasts of India require a lot of conservation measures. In the Goa region, mining waste has created the depletion of benthic fauna.

The book entitled "Recent Advances in Environmental Science" edited by Dr. K. G. Hiremath includes a lot of important papers dealing with different aspects of the environment.

I am sure the book will be well received and it will be an important addition to all libraries. It will serve as a teaching aid and would be very useful for research workers pursuing studies in the field of environmental science.

S. Z. Qasim
Society for Indian Ocean Studies
1, Aruna Asaf Ali Marg,
New Delhi–110067.

PREFACE

Man has to increase and widen his knowledge about environment which needs concomitant action to safeguard the biosphere. The threat to the environment is largely due to: industrialisation and unprecedented growth in human population. During the past few decades overexploitation of resources, deforestation and indiscriminate destruction of wild life have added to the impact. Morever, the environmental problems are by nature complex, as they include a series of intricate feedback processes. The changes brought about by industrialization, mechanization, new political and world orders can rival the forces of nature.

The tropics are the storehouses of millions of species of flora and fauna and unfortunately the same regions have high human population. The World Conservation Strategy and the report of World Commission on Environment and Development have stressed the need to conserve habitats. These habitats are so complex that their understanding requires an interdisciplinary approach involving input from Botany, Zoology, Geology, Forestry, Wild life management, Physics, Chemistry, Microbiology, Biotechnology, Space Technology,etc. Ecologists find it difficult to keep pace with developments in all environmental aspects which are dependent on changes in other sciences. The motivation to write this book is to give glimpses of developments occurring in a wide range of subjects related to the environment. The book shares the vision of unity of all sciences to ameliorate the environmental problems.

The book has its origin in the need faced by the editor when trying to locate a book with comprehensive treatment of wide range of topics. A major purpose, of editing this book, therefore, is to serve the needs of students, teachers and researchers of remote areas where library fecilities are either limited or inadequate. It is hoped, therefore, that, this humble attempt will help them to identify problem-oriented research programmes. However, there seems to be no end to emerging areas in environmental science.

I wish to thank the contributors who are well known in their respective field of studies. I extend my gratitude, to those who gave so generously of their time and expertise towards making this volume a reality. I offer some advice while reading this book .

(a) Select and read those chapters which you like to use for your work.

(b) Skip lightly the vignettes and have a taste of flavour of the entire book.

(c) Be patient with deliberate redundancy.

(d) Some chapters need to be revisited after reading another chapter.

I take this opportunity to thank Dr. S. Z. Qasim, former member of the planning Commission for kindly writing the Foreword for this book. I take this opportunity to thank Dr. A. G. Untawale for his constant encouragement and guidance. Thanks are also due to Mr. Devanand Kavlekar for the help rendered in computer work.

Dr. K.G. Hiremath

CONTENTS

MARINE ENVIRONMENT

LIST OF CONTRIBUTORS

1. Amminedu E, Department of Geo-Engineering, Andhra University, Visakhapatnam-530003 A.P. India
2. Ansari Z. A, National Institute of Oceanography, Dona Paula, Goa–403004, India
3. Dr. Basavarajappa S, Dept. of Applied Zoology, Kuvempu University, B. R. Project, Karnatak-577115 India.
4. Bhosle N. B. National Institute of Oceanography Dona Paula, Goa–403004 India.
5. Borges Sonali D. Department of Zoology Goa University, Goa–403206, India.
6. Borkar Manoj, Department of Zoology, Carmel College for Women, Nuvem, Goa–403604, India.
7. Bukhari Mehtab, Department of Botany Goa University, Taleigao Plateau, Goa–403206, India.
8. Chaturvedi Neera, Marine and Water Resource Group, Remote Sensing Applications Area, Space Applications Centre (ISRO) Ahmedabad–380015 India.
9. Chauhan O. S. National Institute of Oceanography Dona Paula Goa-403004 India.
10. Chittibabu C. V. Department of Medicinal Botany, Government Siddha Medical College, Chennai–600106, Tamil Nadu India.
11. Dhar R. L. National Geophysical Research Institute, Uppal Road, Hyderabad–500007 India.
12. D'Souza Fraddry National Institute of Oceanography Dona Paula, Goa–403206 India.
13. De Jaysankar National Institute of Oceanography Dona Paula, Goa–403004 India.
14. Dubey S. K. Department of Microbiology, Goa University, Taleigao Plateau, Goa–403206 India.
15. Furtado Ramila, National Institute of Oceanography Dona Paula, Goa–403004 India.
16. Gangolli S. Department of Microbiology, Goa University, Taleigao Plateau, Goa–403206.
17. Gaonkar Uday Department of Botany Goa University Taleigao Plateau Goa–403206 India.

18. Gavali Jitendra G. Ecology Lab, Department of Botany Faculty of Science, M S University of Baroda Vadodara–390002 India.

19. Gavali (Saha) Deepa Ecology Lab, Department of Botany Faculty of Science, M S University of Baroda Vadodara–390002 India.

20. Gurunadharao V.V.S. National Geophysical Research Institute, Uppal Road, Hyderabad–500007 India.

21. Hegde Ganesh R. P. G. Department of Botany Karnatak University Dharwad–580003 India.

22. Holmes David S. Department of Microbiology, Goa University, Taleigao Plateau Goa–403206 India.

23. Hosmani Shankar P. Department of Studies in Botany University of Mysore Manasagangotri, Mysore–570006 India.

24. Ingole B. S. National Institute of Oceanography Dona Paula Goa–403004 India.

25. Iyer S. R. National Institute of Oceanography Dona Paula Goa–403004 India.

26. Jaiswal Varsha Department of Botany Goa University Taleigao Plateau Goa–403206.

27. Khade Sharda W., Department of Botany Goa University, Goa–403206 India.

28. Khanna Shalini, Department of Chemistry Faculty of Science, Dayalbagh Educational Institute Dayalbagh, Agra–282005 India.

29. Mathew Thomas Sada, Vasco-Goa, India.

30. Modassir Yasmin Department of Zoology Dhempe College of Arts & Science Miramar, Goa–403001 India.

31. Monika Ram D., Department of Botany University of Delhi, Delhi–110007 India.

32. Nandakishore P., EPTRI, Survey No. 91/4, Gachibowli, Hyderabad–500019 India.

33. Partha S., Department of Studies in Botany University of Mysore Manasagangotri Mysore–570006 India.

34. Parthasarathy N., Salim Ali School of Ecology and Environmental Sciences, Pondicherry University, Pondicherry–605014 India.

35. Patil Neelam A., Botany Department Tuljaram Chaturchand College, Baramati (Maharashtra) 413102 India

36. Prakash B. A., National Geophysical Research Institute, Uppal Road, Hyderabad–500007 India.

37. Puranik S. C., Department of Studies in Geology Karnatak University Dharwad–580003 India.

38. Ramaiah N., National Institute of Oceanography Dona Paula Goa–403004 India.

39. Rodrigues B. F., Department of Botany Goa University, Goa–403206 India.

40. Rao G. R. Department of Microbilogy, Goa University, Goa–403206 India.

41. Sharma Rahul, National Institute of Oceanography, Dona Paula, Goa–403004 India.

42. Sankaran S., EPTRI, Survey No. 91/4, Gachibowli, Hyderabad–500019 India.

43. Sathe P. V., National Institute of Oceanography, Dona Paula, Goa–403 004 India.

44. Shanbhag A. B., Department of Zoology, Goa University, Goa–403206 India.

45. Sivadas Sanitha, Department of Zoology, Dhempe College of Arts and Science, Miramar Goa-403 001 India.

46. Soni Prafulla, Ecology and Environment Division, Forest Research Institute, Dehradun–248006 India.

47. Srinivasa Reddy K., EPTRI, Survey No. 91/4, Gachibowli, Hyderabad–500019, India.

48. Srivastava M. M., Department of Chemistry, Faculty of Science, Dayalbagh Educational Institute, Dayalbagh, Agra–282005, India.

49. Srivastava Shalini, Department of Chemistry, Faculty of Science, Dayalbagh Educational Institute, Dayalbagh, Agra–282005, India.

50. Subbarao N., Department of Geology, Andhra University, Visakhapatnam–530003 India.

51. Untawale A. G., 48, Nirmiti, Sagar Society, Dona Paula-Goa, 403004 India.

52. Uttangi J.C., Mission Compound, Dharwad, Karnataka–580001, India.

53. Vasudeva Rao P. V. H., Department of Geo-Engineering, Andhra University, Visakhapatnam–530003 A.P. India

54. Veena Chandra, Extension Division, Forest Research Institute, Dehradun–248006, India.

55. Venkateswara S. Rao, Department of Geology, Andhra University, Visakhapatnam–530003, A. P. India.

56. Walia Renu, Department of Zoology, Goa University, Goa–403206, India.

1

Introduction

*K.G. Hiremath

Soil, water and air are in constant state of change that affect living organisms. The population growth in developing countries and consumer demand in developed countries has hastened the flux of matter, much faster than any system can tolerate. Biogeochemical cycles are playing a subordinate role to man. Developing countries are unfortunately treading the path of developed countries instead of learning from them. In 1990s the environmental problems were noticed locally. These initial disturbances have got amplified through various feedback processes and have reached ominous dimensions which warrants dispassionate look at the problems. Man has realized that influence of man on biosphere and vice-versa are inescapable. Greenhouse effect and ozone depletion are some of the examples. Green house gases, albeit in tiny concentrations are influencing 'Earth's climate'. Never before the countries were so alarmed about the deteriorating environments as today.

Although technological changes are a source of many environmental problems. They can also offer solutions. Technology has to be viewed as prerequisite for successful conservation. The environmental problems are getting their due attention which is apparent from large scale multidisciplinary research that is being carried out throughout the world. The purpose of this section on terrestrial environment is to give a broad understanding of new emerging areas of environmental science. This section has articles covering a wide range of topics. The cohesiveness may be assessed by the readers.

The global biodiversity is changing at an unprecedented rate. Understanding of floral diversity is essential to assess the impact and suggest measures to optimize productivity and rehabilitate the disturbed forest. A comprehensive paper on "Plant diversity in undisturbed and human impacted sites of tropical evergreen forest in Kolli Hills. Eastern Ghats India" gives

* Head, Department of Botany Dhempe College of Arts & Science, Miramar, Panaji, Goa, India 403001

glimpses of impact of human interference on floral biodiversity. Chittibabu and Parthasarathy have carried out a detailed survey of trees and nontree species to assess the impact.

A paper on "The Saga of Sahyadri : Societal and environmental issues of Konkan and Goa" by Dr. Sathe and Chauhan explores the stresses caused by man on western ghats. The authors cover a wide range of issues such as physiography, deforestation, flooding, mangroves, sand dunes, coastal regulation zone and bring about recommendations. India has a rich repository of medicinal plants. Ethnomedicines have properties not only to cure but also to prevent the diseases. The paper by Dr. Hegde "Medicinal Plants of India – Status and Prospects" suggests various ways to cultivate medicinal plants and document our traditional knowledge of medicinal plants. Attempts to patent medicinal palnts and their products by western countries can only be counteracted by scientific documentation of our ethnomedicine. A paper on "Ethnobotany and tropical forest conservation" by Veena Chandra deals with a brief history of Ethnobotany and discusses its role in conservation.

Heavy metal pollution is of great ecological significance due to toxicity and accumulation behavior of such metals in the ecosystem. Paradoxically forests are areas of rich mineral resources. Thus mining activities have led to destruction of forests, loss of fertile land, pollution of water bodies and increase in health hazards. Recolonization of degraded areas can be achieved by three different ways : physical, chemical and vegetative methods. Dr. Soni's paper on "Mining scenario in India – Environmental Problems and Prospects" gives the status of mining leases environmental impacts and assessment. The paper suggests to generate background information and reviews the technological developments towards restoration of mine wastes. The reclamation measures involving physical, chemical and biological methods have been dealt in detail.

The role of micro-organisms in reclaiming degraded areas has received little attention as compared to correction of nutritional deficiency, toxicity and moisture deficiency. The paper by Dr. Rodrigues et.al. explores the prospects of using arbuscular mycorrhizae. It reviews the information available about mycorrhizae and suggests the methods to use them in reclamation of iron ore mine wastes.

The decision to use or not to use a chemical fertilizer or pesticide is an ecological decision. The green revolution although brought development in agriculture sector, the practices such as indiscriminate use of chemical fertilizer and pesticides has its adverse effect on natural resources like air, soil and water. The overuse and dangerous handling of pesticides are widespread. Many chemicals banned by developed counties are commonly marketed in developing countries.Pesticides misuse has set a treadmill of overspray, evoloution of pest resistance and more spraying. India is the third largest consumer of pesticides in the world and highest among South Asian countries. The use of pesticides has become an indispensable tool in view of the loss caused by harmful organisms which amounts to one third of agricultural loss. The article on "A holistic biotechnological approach in farming environment for quality agriproducts in the 21st century" by Dr. Basavarajappa gives a general overview of soil nutrients, fertilizers, organic manure, vermiculture, biofertilizers, biopesticides. The paper proposes a prudent use of conventional and modern biotechnological methods to increase healthy agricultural products.

The pesticides have undergone biological magnification and significantly high levels are found in food. A paper on "Ecofriendly pesticides – An emerging trends" by Dr. Srivastava et.al. reviews the status of pesticide consumption., their effect, pattern of pesticide consumption etc. The paper highlights new strategies to protect plants and control pests. An array of bio-pesticides can be obtained from a wide diversity of flora and fauna available in India. The authors put forth a challenge to identify, modify and commercially produce the biopesticides.

The use of chemical fertilizers tremendously increase the oxygen demand of the soil. Roots, soil fauna and soil microorganisms use oxygen and hence oxygen content of soil is an important factor which governs the agricultural products. A paper by Dr. Patil on 'Oxygen supplementation through soil: A boon to agriculture in new millennium" gives information, procedure and suggestions to use oxygenated peptone to supplement soil oxygen. This case study carried out on a variety of cereals, pulses, vegetables, floriculture and fruit crops has shown improvement in yield due to increase in oxygen content, micorrhizal activity and nutrient availability.

Nitrogen is another element required by the plant growth and the role of blue green algae has been widely recognised. A case study on "Blue green algae from soils of Karnataka" by Dr. Hosmani & Partha has assessed the diversity of blue green algae in relation to physicochemical parameters. The paper discusses in detail the relationship between chemical fertilizers, natural fertilizers and blue green algae. The study concludes that soil manured with chemical and natural fertilizers may interact to produce antagonistic effect on the growth of blue green algae.

Air pollutants interact with both biotic and abiotic components of ecosystem to create a complex environmental problem. Moreover, the air pollutants travel long distances depending on wind velocity and direction. Green house effect, ozone depletion and smog are some of the complex effects of air pollution. The paper on "Plant and air pollution" by Gavali and Gavali has reviewed the available information on impact of air pollution on plants. Vegetation as a tool to monitor air pollution has been brought forth . The paper is organized to provide an overview of impact of air pollution on morphology, anatomy, physiology, and phenology of plants. It also discusss the bioaccmulation capacity of plants.

High frequency of land falling of cyclones cause irrepairable loss to Indian economy every year. Coastal Andhra Pradesh experiences cyclones which lead to loss of property and crops. A systematic survey and study of six coastal districts of Andhrapradesh bring forth the need to develop coastal shelterbelt of plants for effective eco-conservancy for ameliorating the consequences of cyclones. Dr. Vasudev Rao et.al. have discussed these aspects in their paper on "Shelter-cum-bioenergy belt along the coast of Andhra Pradesh, India"

Solid waste management is an international mission since pollutants do not honour any border. An integrated approach involving recovery, recycling, improved collection, thermal treatment, biodegradation and land filling is required. The choice of methods/combination of methods depends on local conditions. Dr. Gurunadha Rao et.al. in their paper "Solid waste disposal practices : Implications on ground water pollution" approach the problem by an integrated approach involving hydrogeological, geophysical and numerical modeling. The wastes contain a variety of biodegradable and non-biodegradable substances, many of them are toxic and cause ground water pollution. The paper reviews the problem associated with land filling and drainage problems associated with it.

Environmental impact assessment (EIA) before any major industrial, domestic, irrigational project has become a necessity. Countries all over the world have realized its importance rather than studying the impact later. The geophysical surveys were carried out to assess the impact of the proposed alumina plant in Orissa, India. The paper on "Environmental Impact Assessment Studies of ground water regime around a proposed Alumina plant in Doragarh, Raygada district, Orissa" by Dr. Gurunadha Rao et.al. characterizes the hydrogeological system and gives a groundwater flow and mass transport model. The predictions of the model showed that the contaminants take more than 50 years to cause contamination of a stream. The principles, concepts and models have relevance to any EIA involving groundwater contamination.

Urbanization has created a larger demand for water for industries and domestic purpose. Bulk consumers exploit ground water resources which has led to depletion of ground water. The sewage water has further compounded the problem by polluting ground water. A case study on "Groundwater environs of Hubli Dharwad municipal corporation area, Karnatak" by Dr. Puranik bring forth the problems of groundwater pollution. Such hydrogeological studies are needed in planning the cities for expansion and disposal of sewage water.

Biotechnological potential of microorganisms in agriculture, industry and medicine is a frontier area which has brought revolutionary changes. Several bioluminescent biosensors have become an important tool in biotech industries, health care products, food processing and environmental monitoring. Microorganisms especially bacteria can be used as biosensors to detect and/or measure very low level of chemicals or analytes. Biotechnological methods are less expensive as compared to many traditional analytical methods such as atomic absorption spectrometry, atomic emission spectrometry. Mass spectrometry etc. Bacteria possess metal resistance determinants which make them suitable biosensors. Dr. Dubey et. al. in their paper "Detection and measurement of ultra trace levels of mercury and copper by genetically engineered microbial sensors" have described the ability of microbial sensors to detect toxic metals in contaminated environmental samples. The paper concludes that bacterial biosensors can distinguish bioavailable component in the total chemical concentration which is vital for toxicity studies and biomediation. A paper by Ramaiah in marine environment section also discusses this aspect.

Crocodiles form an important component of aquatic ecosystem. A case study on "Ecological history, contemporary status and conservation future of marsh crocodiles *Crocodylus palustris* in Goa" by Dr. Borkar reviews the history and present status of crocodiles. Conservation strategies have been suggested to protect crocodiles in Goa which may hold good for other regions also. The authors stresses the need to highlight the ecological role of crocodiles to general public.

Wetlands harboured the ancient civilizations and have been a source of water, food and recreation. A paper on "Carambolim lake, an ideal wetland of Goa" by Dr. Shanbhg et.al. focuses on the importance of wetland and threat posed by mankind. The paper highlights the seasonality of a variety of flora and fauna, ranging from microscopic plankton and benthos to macrophytes and avifauna. The study is a model which has wide applicability.

Wild life is valuable resource of every country which is affected by unrestricted and selfish activities of man. According to red data book of ICUN (International Union for Conservation of Nature & Natural Resources) about 245 species of animals (including 139 species of birds) have already become extinct and nearly 600 more species are heading towards extinction. Although there are protected areas, their management is wanting in its quality. The base line data about flora and fauna is essential to manage these areas. A paper on"Ornitho-botanical aspects of Anashi National Park in Western Ghats (India) by Dr. Uttangi survey the avifauna and its associated plants to bring forth their interdependency. It discusses the interaction among plants, insects, bird, butterflies and animals. Many measures to conserve the habitat have been suggested.

The wide range of topics covered in this section has reminded us of the fragility of our natural systems. Although the countries are separate entities, nature reminds us of humanity's basic unity. The book shares the vision of unity of all sciences to ameliorate the environmental problems.

2

Plant Diversity in Undisturbed and Human-impacted Sites of Tropical Evergreen Forest in the Kolli Hills, Eastern Ghats, India

***C.V.Chittibabu &**
N. Parthasarathy

INTRODUCTION

Global diversity is changing at an unprecedented rate (Sala *et al.* 2000) as a complex response to several human-impacts and the latter date back to antiquity and even to pre-history. Documenting such impacts on genetic diversity of forest is a difficult matter and little quantitative data exist (Ledig, 1992). Primary forests of Asia, particularly those of the Western Ghats and the Eastern Ghats of peninsular India are disappearing at an alarming rate due to human impacts and are replaced by forests comprising inferior species or their land use pattern changed (Parthasarathy, 1999). The disappearance of tropical forests comes at a time when our knowledge on their structure and dynamics is woefully inadequate (Hubbell & Foster, 1992). Understanding plant diversity in a human-impacted forest is necessary for assessment of potential impacts, the amelioration of effects of disturbance, optimization of productivity and rehabilitation of ecosystem (Congdon & Herbohn, 1993).

Tropical forests are interesting as they excel in species diversity (Condit *et al.*, 1996), which is generated by species interaction such as competition and niche diversification (Bada 1984; Pianka, 1996) and due to high humidity and temperature (Ojo & Ola-Adams, 1996). Trees have vital roles (both structural and functional) in maintaining the equilibrium of a forest ecosystem. Tree diversity is fundamental to total rainforest biodiversity, because they provide resources and habitat structure for almost all other rainforest species (Cannon *et al.*, 1998). So, plant diversity

* Salim Ali School of Ecology & Environmental Sciences, Pondicherry University, Pondicherry-605 014, India.

inventories of tropical forests mainly focus on trees (Gentry, 1988) as against non-trees, which are stated to contribute little to the total species diversity in rain forests (Hartshorn, 1983; Whitmore, 1984). However, non-trees such as lianas also form an important structural and functional component of tropical rain forests (Grubb 1977; Whitmore 1984; Hegarty & Caballe 1991). Also, some tropical forests might possess the most herb rich plant communities on earth (Gentry & Dodson 1987), and ground herbs might prove to be a good indicator of forest succession status (Turner *et al.* 1996). To substantiate the latter statement we lack adequate data (Whitmore 1984; Gentry 1988; Poulsen 1996a; Annaselvam & Parthasarathy 1999). Hence, investigations involving the whole plot diversity are essential but are fewer and the scales of such inventory vary considerably (Whitmore *et al.*1986; Gentry & Dodson 1987; Kelly *et al.* 1994; Killeen *et al.* 1998; Parthasarathy *et al.* 2001).

Quantitative plant diversity inventories on different plant communities of Indian tropical forests are available from various forests of the Western Ghats (Sukumar *et al.* 1992; Pascal & Pelissier 1996; Ganesh *et al.* 1996; Parthasarathy & Karthikeyan 1997a; Ghate *et al.* 1998; Annaselvam & Parthasarathy 1999; Ayyappan & Parthasarathy 1999; Parthasarathy 1999; Muthuramkumar & Parthasarathy 2000; Padaki & Parthasarathy 2000; Srinivas & Parthasarathy 2000) and on the Indian Coromandel coast (Parthasarathy & Karthikeyan 1997b; Parthasarathy & Sethi 1997). Whereas, the Indian Eastern Ghats is poorly studied for these aspects except those of the Shervarayan hills (Kadavul & Parthasarathy 1999a, 2000) and of the Kalrayan hills (Kadavul & Parthasarathy 1999b, 1999c).

This paper reports the results of the plant diversity inventory of mature trees (≥ 30 cm gbh), regeneration trees (≥ 3 cm girth and < 30 cm gbh), lianas (≥ 5 cm gbh) and understory plants (shrubs, undershrubs, herbaceous climbers and herbs) in four 2-ha plots as replicates of undisturbed and human-impacted sites of tropical evergreen forest in the Kolli hills, Eastern Ghats, south India.

MATERIALS AND METHODS

Study Area

The study was conducted in a tropical evergreen forest at Kolli hills (11° 10′ – 11° 30′N; 78° 15′ - 78° 30′E) in the Eastern Ghats, south India (Fig. 1). The Kolli hills cover an area of 282.9 km^2 and support heterogeneous vegetation along the elevational gradient. Deciduous forests dominate at lower elevation < 900 m above sea level (a.s.l.). The patches of evergreen forest (locally called 'sholas'), < 30 m tall, are found between 900 m and 1300 m a.s.l. Tree branches are often festooned with epiphytic moss, ferns, lichens, aroids and orchids. Epiphylls are also abundant.

Four patches of evergreen forests, located 1-7 km apart from each other, were selected as replicate sites of undisturbed and disturbed forests. The site Perumakkai shola (PS) at 1000 m a.s.l. on a gentle slope of a valley with a seasonal stream and the site Vengodai shola (VS) at 1050 m a.s.l. on a raised slope are relatively undisturbed because they are relatively distant from human settlements. The site Kuzhivalavu shola (KS) at 1200 m a.s.l. on a steep slope and the site Mottukkadu shola (MS) at 1250 m a.s.l. on a hilltop are disturbed owing to their proximity to human habitation (Chittibabu & Parthasarathy 2000a).

The geological substrate is charnockite (hypersthene granite) associated with gneisses and varied metamorphic rocks and a thin layer of ferruginous sandy soil (Legris & Meher-Homji 1984). Climatological data (1951-1980) of the nearest station, Salem (278 m a.s.l.), 50 km from the study site, reveals a mean annual temperature of 28.3° C. By applying the lapse rate of 0.44° C per 100 m elevation (Singh *et al.* 1994), the estimated mean annual temperature of Kolli hills at 1100 m is 24.7° C and the estimated minimum temperature is 15.7° C for the coldest months (December and January). The rainfall is seasonal, the bulk of which (500 mm) falls during the southwest monsoon (June-September) and 300 mm during the northeast monsoon (October and November), totaling 1014 mm year $^{-1}$ at Salem for that period.

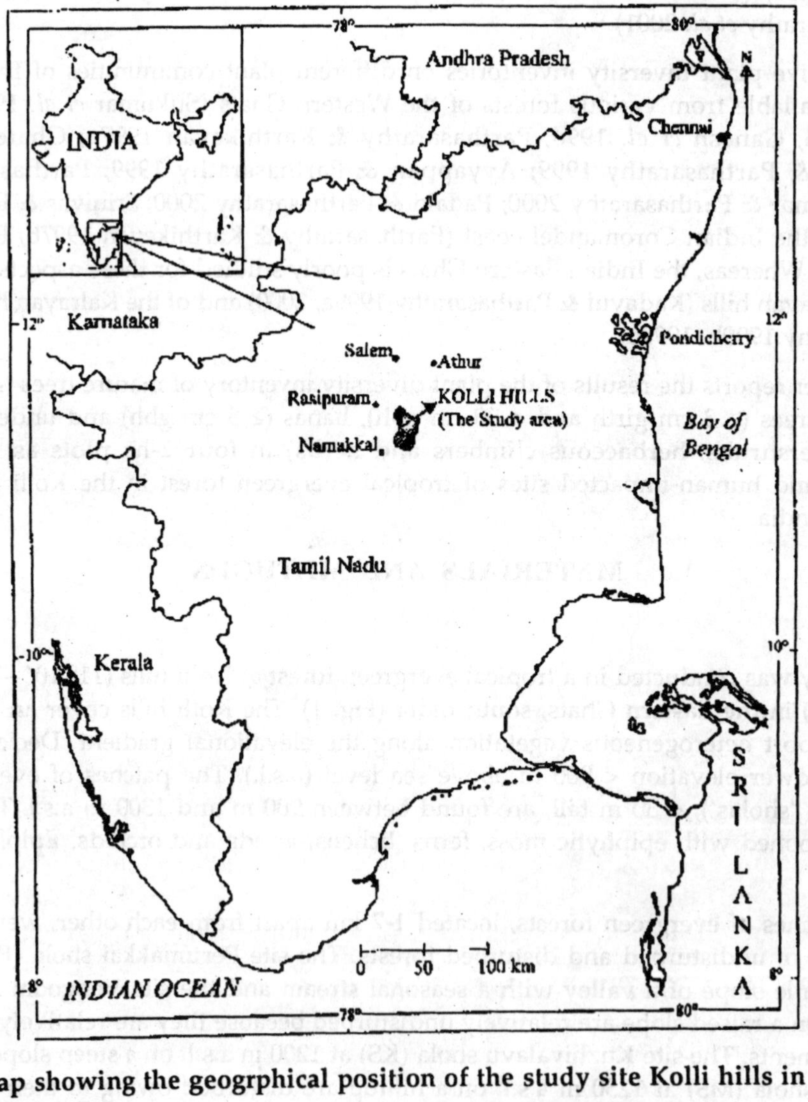

Fig.1. Map showing the geogrphical position of the study site Kolli hills in Tamil Nadu, South India.

Data Collection

Fieldwork was undertaken during May 1996 - June 1997. Four 2-ha (100 m × 200 m) permanent plots were established in each of the four sites to investigate the diversity of mature trees, regeneration trees, lianas and understory plants. All mature trees ≥ 30 cm gbh and lianas ≥ 5 cm gbh were inventoried by gridding each plot into two-hundred 10 m × 10 m quadrats. Eight 10 m × 50 m subplots, distributed one each in the center of every 100 m × 100 m (1 ha) area, were marked for regeneration studies. The regeneration plots were also subdivided into 10 m × 10 m quadrats. For understory plant inventory, two-hundred 4 m^2 (2 m × 2 m) quadrats were regularly laid in each of the 10 m × 10 m grids of 2-ha study plot.

All trees (≥ 30cm gbh) were appended with sequentially numbered aluminium tags using nails and their girth at breast height (gbh), at 1.3 m, was measured; those with buttresses were measured above the latter. In the case of trees with multiple stems, each stem was measured separately. In addition, the following were recorded: tree height (measured for trees up to 5 m tall and others were estimated), trees multistemmed, thorny trees; dead trees; and cut trees. For tree regeneration studies, all tree seedlings (3-10 cm girth) and tree saplings (10-30 cm gbh) were identified and measured for their girth.

All lianas ≥ 5 cm gbh in the study plots were identified and measured for their gbh. The understory plants were enumerated and identified in the eight-hundred 4 m^2 sample quadrats, besides noting their life-form and fruit types. Life- forms constituting the understory plants in this study are herbs (non-woody small plants < 1 m tall), undershrubs (small plants of 1 to 1.5 m tall with moderate stem thickness), shrubs (>1.5 m to 3 m tall with thick stems and branching at ground level without a distinct trunk), and herbaceous climbers. Grasses, sedges, aroids, root parasites and pteridophytes, in this study, are treated under the synusia of herbs.

Species were identified using pertinent regional floras (Gamble & Fischer 1915-1935; Nair & Henry 1983; Henry *et al.* 1987, 1989; Matthew 1991), and the field key of Pascal and Ramesh (1987), and by reference to the collections at the herbarium of Botanical Survey of India, Coimbatore (MH) and at the herbarium of School of Ecology, Pondicherry University. Plant nomenclature follows Nair and Henry (1983) and Henry *et al.* (1987, 1989).

Data Analysis

Diversity indices Shannon (H') and Simpson (D) were calculated (Magurran 1988) for all plant communities. These indices have low sensitivity to sample size and are widely used. Species-area curves for mature trees and lianas were constructed by sequential summing of the 10 m × 100 m (0.1 ha) scores. While, for regeneration trees the 10 m × 10 m (0.01) scores and for understory plants the pooled sample results obtained from the 10 m × 100 m (0.1 ha) scores were used.

Importance value index (IVI) of Curtis and McIntosh (1950), which depicts the sociological structure of a species in its totality in the community, for a total score of 300 for mature trees, and a modified IVI for a score of 200, were calculated for regeneration trees, lianas and understory plants.

RESULT AND DISCUSSION

The plant diversity of trees, lianas and understory species in the 8-ha plots totaled 156 species and 15,826 individuals (Table 1). Interestingly, the contribution of non-tree species such as lianas and understory plants (78 species, 50%) to total species richness is as much as the tree species richness (78 species, 50%) in the tropical evergreen forest sites in the Kolli hills, Eastern Ghats, India. This is notable and underlines the need for quantitative inventory on non-tree components as carried out at different scales in a few other tropical forests (Hall & Swaine 1981; Gentry & Dodson 1987; Poulsen & Balslev 1991; Poulsen & Nielsen 1995; Poulsen & Pendry 1995; Poulsen 1996b; Annaselvam & Parthasarathy 1999). Globally, whole plant diversity inventories at 0.05 to 1.5 ha scales recorded a range of 219 to ~500 species in various tropical forests (Richards 1996), while the Varagalair forest of the Anamalais, Western Ghats on a 30-ha scale harboured 431 species (Parthasarathy *et al.* 2001).

Table—1. Diversity and density of trees, lianas, understory plants and regeneration trees in the four 2-ha plots and in total 8-ha of tropical evergreen forest area in Kolli hills

Variable	*Undisturbed*		*Disturbed**		*Total*
	PS	*VS*	*KS*	*MS*	
Species richness					
Mature trees (≥ 30 cm gbh)	58	51	42	39	78
Regeneration trees (≥ 3 cm and < 30 cm gbh)	36	30	23	12	46
Lianas (≥ 5 cm gbh)	21	16	9	14	26
Understory plants	42	39	24	19	52
Total	**121**	**106**	**75**	**72**	**156**
Population density					
Mature trees (≥ 30 cm gbh)	1054	1151	651	969	3825
Regeneration trees (≥ 3 cm and < 30 cm gbh)	525	536	624	252	1937
Lianas (≥ 5 cm gbh)	113	156	25	90	384
Understory plants	2158	2027	2865	2630	9680
Total	**3850**	**3870**	**4165**	**3941**	**15826**

(*Human-impacted)

TREE DIVERSITY

Species Richness and Diversity Indices

Mature tree species diversity (≥ 30cm gbh) in the total 8-ha area totaled 78 species in 61 genera and 36 families (Table 2). Undisturbed study plots PS and VS had greater species richness (58 and 51 species) than the human-impacted sites KS and MS (42 and 39 species), indicating an average species reduction of 26% in the latter. Most diversity indices and stand density also showed a similar trend with disturbance as did species richness. Diversity of trees in regeneration stands totaled 46 species from 38 genera and 25 plant families (Table 3). Plot-wise, species richness and number of genera and families of regeneration trees were greatest (36 species, 31 genera

Table—2. Consolidated diversity details of all mature trees ³ 30 cm gbh in four 2 ha plots and in total 8-ha of tropical evergreen forest in the Kolli hills, Eastern Ghats, South India

Variable	*Undisturbed*		*Human-impacted*		*Total*
	PS	*VS*	*KS*	*MS*	
Species richness	58	51	42	39	78
(species ha^{-1} in parenthesis)	(54,45)	(46,35)	(32,27)	(35,26)	
Genus richness	48	43	35	33	61
Family richness	31	28	23	22	36
Diversity indexes:					
(a) Shannon (H')	3.296	2.791	2.870	2.376	3.34
(b) Simpson's (D)	0.056	0.111	0.080	0.190	0.06
Stand density (stems ha^{-2})	1054	1151	651	969	382
(trees ha^{-1} in parenthesis)	(580; 474)	(632; 519)	(385; 266)	(483; 486)	
Mean stand density (stems ha^{-1})	527	575.5	325.5	484.5	478
Stand basal area (m^2 ha^{-2})	106.0	93.2	103.0	46.6	348.8
Mean stand basal area (m^2ha^{-1})	53.0	46.6	51.5	23.3	43.6
Number of multi-stemmed trees	68	131	75	172	446
Number of cut trees (stumps)	-	5	33	74	107
Number of dead trees	1	2	24	22	49
Number of thorny trees	-	-	4	2	6
Number of trees buttressed	194(18%)	286(25%)	291(44%)	71(7%)	842(22%)

and 21 families) in the undisturbed plot PS and lowest (12 species, 12 genera and 10 families) in the disturbed plot MS (Table 3). Species richness of tree seedlings (28 to 35 species) and saplings (24 species) in undisturbed plots PS and VS was greater than those of the seedlings (8 to 19 species) and saplings (12 to 19 species) in disturbed plots KS and MS (Table 4).

Table—3. Consolidated details of biodiversity inventory of regeneration stands in undisturbed (PS, VS) and human-impacted (KS, MS) plots of the Kolli hills, Eastern Ghats, India.

Variable	*Undisturbed*		*Disturbed**		*Total*
	PS	*VS*	*KS*	*MS*	
Species richness	36	30	23	12	46
Number of genera	31	27	23	12	38
Number of families	21	16	17	10	25
Density	525	536	624	252	1937
Diversity index: Shannon (H')	3.29	2.86	2.50	1.80	3.14
Simpson (D)	0.04	0.07	0.10	0.25	0.06
Stand basal area (m^2)	0.47	0.43	0.42	0.44	1.75

Table—4. Vegetation structure of the forest stand comprising regeneration and adult trees in the four 2-ha plots of Kolli hills (SR – Species Richness, PD – Population Density, BA – Basal Area in m^2).

	Plot	*Tree seedlings (0.1 ha)*			*Tree saplings (0.1 ha)*			*Adult trees (2-ha)*		
		SR	*PD*	*BA*	*SR*	*PD*	*BA*	*SR*	*PD*	*BA*
Undisturbed:	Perumakkai shola (PS)	35	347	0.11	24	178	0.35	58	1054	106.0
	Vengodai shola (VS)	28	354	0.09	24	182	0.33	51	1151	93.2
Disturbed:	Kuzhivalavu shola (KS)	19	465	0.14	19	159	0.28	42	651	103.0
	Mottukkadu shola (MS)	8	61	0.03	12	191	0.41	39	969	46.6
Total		**43**	**1227**	**0.37**	**38**	**710**	**1.37**	**78**	**3825**	**348.8**

Tree species inventories at defined study sites and in minimum diameter classes give a reliable instrument to indicate the diversity level of a study site (Wattenberg and Breckle 1995). The species richness of 78 tree species in 8 ha area of the tropical evergreen forest in the Kolli hills of Eastern Ghats reflects its low diversity status. The Kolli hill sites, compared to the semi-evergreen forests of Indian Eastern Ghats, i.e., the Shervarayan hills showing 70 tree species in 4-ha (Kadavul and Parthasarathy 1999a), and the Kalarayan hills with 73 species in 4-ha (Kadavul and Parthasarathy 1999b), have more similar species richness. Also within peninsular India, species richness for the same tree girth threshold (≥ 30 cm gbh) in various sites of the Western Ghats ranged from 30 species ha^{-1} in Nelliampathy (Chandrashekara and Ramakrishnan 1994), to 57 species ha^{-1} in Mylodai and Courtallum reserve forest (Parthasarathy and Karthikeyan 1997a), 64 to 82 species ha^{-1} in the medium-elevation forest of Kalakad (Parthasarathy 2001), to 80 to 85 species ha^{-1} in high elevation evergreen forests at Kalakad (Parthasarathy 1999), 90 species on a 3.82 ha scale in Kalakad Mundanthurai Tiger reserve (Ganesh *et al.* 1996) and a range of 59 to 79 species ha^{-1} in the 30 hectares of tropical evergreen forest in Varagalaiar, Anamalais (Ayyappan and Parthasarathy 1999). Across the tropics, richness of woody species ≥ 10 cm dbh ranges from a low value of 20 species ha^{-1} in flooded Varzea forest of Rio Xingu, Brazil (Campbell *et al.* 1992), a medium-value of 137 to 168 species ha^{-1} for the four 1-ha plots of terra firme forest, Central Amazonia (Ferreira and Prance 1998) to 307 species ha^{-1} in Amazonian Ecuador (Valencia *et al.* 1994). Such a comparison (Chittibabu & Parthasarathy 2000c) places the present study site in low-diversity forest category like the adjacent Shervarayan and Kalrayan hills of the Eastern Ghats.

Species-area curves for mature trees in each of the four 2-ha contiguous plots rapidly increased up to 0.5 ha. It continued to increase much more slowly up to 1.5 ha and thereafter the species addition was little to none, with a slight spurt at the end of the curves in sites VS and MS (Fig. 2a). The tree species-area accumulation curves approach an asymptote, indicating adequate sampling for each 'shola' patch. Regeneration tree species-area curves of plots PS, VS and MS did not stabilize, while that of plot KS reached an asymptote (Fig. 2b).

Tree Density and Basal Area

In all, 3825 mature trees ≥ 30 cm gbh were enumerated in the 8-ha inventory. Stand density showed a wide variation ranging from as low as 266 trees ha^{-1} in KS to 632 trees ha^{-1} in VS, and the mean stand density was 478 trees ha^{-1}. The individual 2-ha plots varied in basal area also ranging from 23.3 m^2 ha^{-1} in the highly disturbed site MS to 53.0 m^2 ha^{-1} in the relatively undisturbed PS. The mean basal area was 43.6 m^2 ha $^{-1}$. As regard to basal area, the individual species that domineered each site include: *Alseodaphne semecarpifolia* var. *angustifolia* of Lauraceae (13.64 m^2, 12.8%) in PS; *Olea paniculata* (Oleaceae; 30.87 m^2, 33.13%) in VS; *Meliosma simplicifolia* (Sabiaceae; 17.43 m^2, 16.9%) in KS; and *Syzygium cumini* (Myrtaceae; 8.99 m^2, 19.2%) in MS.

The overall regeneration stand basal area contributed by seedlings was 0.37 m^2 (21%) and by saplings 1.37 m^2 (79%) (Table 4). Among the plots, greatest basal area contribution by saplings (0.41 m^2, 94%) and seedlings (0.14 m^2, 34%) was from the disturbed plots MS and KS respectively.

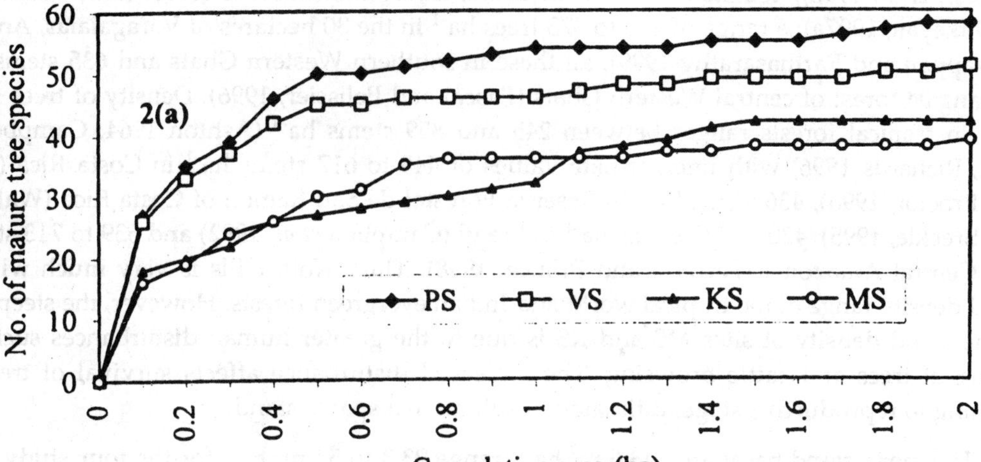

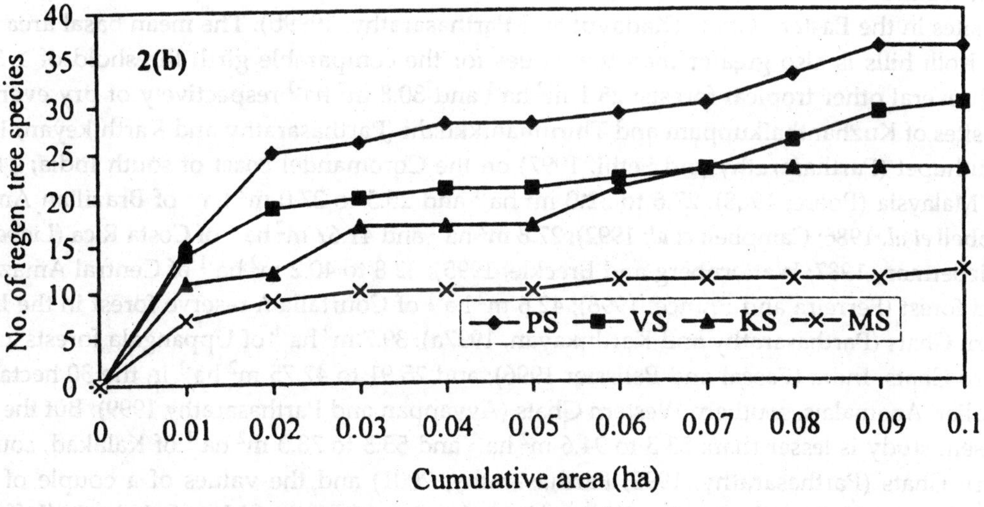

Fig.2. Species-area curves for the mature trees (2a) and regeneration trees (2b) in the four study sites of Kolli hills.

Basal area contribution by adult trees for each 2-ha plot ranged from a low value of 46.6 m^2 of the disturbed plot MS to a high value of 106.0 m^2 of the undisturbed plot PS (Table 4).

The absolute stand density of our four study sites ranged from 266 to 632 trees ha^{-1}, with a whole study area average of 478 trees ha^{-1}. The Shervarayan and Kalrayan hills of Eastern Ghats, respectively, stocked a range of 640 to 986 trees ha^{-1} (Kadavul and Parthasarathy, 1999a) and 367 to 667 trees ha^{-1} (Kadavul and Parthasarathy, 1999b). Whereas the tree densities in various tropical evergreen forests of the Western Ghats of peninsular India were: 574 to 915 stems ha^{-1} in medium-elevation forest of Kalakad (Parthasarathy 1999); 852 to 965 stems ha^{-1} in high elevation forest of Kalakad (Parthasarathy 2001); 583 stems ha^{-1} in Kalakad-Mundanthurai area (Ganesh *et al.* 1996); 482 stems ha^{-1} in Mylodai, Courtallam reserve forest (Parthasarathy and Karthikeyan, 1997a); a range of 270 to 673 trees ha^{-1} in the 30 hectares of Varagalaiar, Anamalais (Ayyappan and Parthasarathy 1999), all these in southern Western Ghats and 635 stems ha^{-1} in Uppangala forest of central Western Ghats (Pascal and Pelissier, 1996). Density of trees (^{3} 30 cm gbh) in tropical forests ranges between 245 and 859 stems ha^{-1} (Ashton 1964; Campbell *et al.* 1992; Richards 1996) with intermediate values of 448 to 617 stems ha^{-1} in Costa Rica (Heaney and Proctor, 1990), 436 stems ha^{-1} in Reserva Forestal de San Ramon of Costa Rica (Wattenberg and Breckle, 1995), 420 to 777 stems ha^{-1} in Brazil (Campbell *et al.* 1992) and 639 to 713 stems ha^{-1} in Central Amazonia (Ferreira and Prance, 1998). Thus, Kolli Hills is very much within the stand density range of the tropical world and Indian evergreen forests. However, the steep decline in the stand density of sites MS and KS is due to the greater human disturbances such as the felling of trees and cattle browsing. The increased disturbance affects survival of trees from seedling to reproductive stage, ultimately resulting in a sparse stand.

The mean stand basal area 43.6 m^2 ha^{-1} (range 23.3 to 53 m^2 ha^{-1} for the four study sites) of the Kolli hills is admittedly greater than the pantropical average of 32 m^2 ha^{-1} (Dawkins, 1959), and the values of Shervarayan hills (34.9 m^2 ha^{-1}) and Kalrayan Hills (33.7 m^2 ha^{-1}), the two closer sites in the Eastern Ghats (Kadavul and Parthasarathy, 1999b). The mean basal area value of the Kolli hills is also greater than the values for the comparable girth threshold of ≥ 30 cm gbh of several other tropical forests: 28.1 m^2 ha^{-1} and 30.8 m^2 ha^{-1} respectively of dry evergreen forest sites of Kuzhanthaikuppam and Thirumanikkuzhi (Parthasarathy and Karthikeyan, 1997b) and Puthupet (Parthasarathy and Sethi, 1997) on the Coromandel coast of south India; 24.2 m^2 $ha^{-1}$of Malaysia (Poore, 1968), 27.6 to 32.0 m^2 ha^{-1} and 25.5 to 27.0 m^2 ha^{-1} of Brazilian Amazon (Campbell *et al.* 1986; Campbell *et al;* 1992); 27.8 m^2 ha^{-1} and 41.67 m^2 ha^{-1} of Costa Rica (Lieberman and Lieberman, 1987; Watternberg and Breckle, 1995); 32.8 to 40.2 m^2 ha^{-1} of Central Amazonian upland forest (Ferreira and Prance, 1998); 42.6 m^2 ha^{-1} of Courtallam reserve forest in the Indian Western Ghats (Parthasarathy and Karthikayan, 1997a); 39.7 m^2 ha^{-1} of Uppangala forests, central Western Ghats, India (Pascal and Pelissier 1996); and 25.91 to 47.75 m^2 ha^{-1} in the 30 hectares of Varagaliar, Anamalais, southern Western Ghats (Ayyappan and Parthasarathy 1999). But the value of present study is lesser than: 53.3 to 94.6 m^2 ha^{-1} and 55.3 to 78.3 m^2 ha^{-1} of Kalakad, southern Western Ghats (Parthasarathy, 1999; Parthasarathy, 2001) and the values of a couple of other tropical forests: 47 (for alluvium) to 49.5 m^2 ha^{-1} (for slope forest) of New Caledonia (Jeffre and Veillon, 1990), and 62 m^2 ha^{-1} of Monteverde, Costa Rica (Nadkarni *et al.*, 1995).

Tree Species Dominance and IVI

The population density of mature tree species varied considerably in the four study sites ranging from one individual to 808 individuals in the 8-ha area. The population density of top 12 trees also varied greatly between the undisturbed and human-impacted plots (Table 5). Notably, sites MS and VS were dominated by single species, *Memecylon umbellatum* of Melastomataceae (39% and 26% of the stand); whereas, the other two sites, PS and KS, had a set of three species codominating the stand. *Nothopegia heyneana* (12%), *Memecylon umbellatum* (9.8%) and *Diospyros ovalifolia* (9%) shared major density of site PS, and *Meliosma simplicifolia* (16.7%), *Myristica dactyloides* (12.5%) and *Phoebe wightii* (9.6%) together dominated the forest stand of KS. Overall, in the 8-ha area *Memecylon umbellatum* alone formed 20% of the total stand density. Our study site, based on abundant species, in the tropical evergreen forest of the Eastern Ghats can be classified as *Memecylon-Phoebe-Olea* association with *Neolitsea* and *Myristica* as codominants. The dominant species in each site and their score out of 300 were: *Nothopegia heyneana* (IVI =29.0) and *Alseodaphne semecarpifolia* (28.8) in PS; *Beilschmedia gemmiflora* (60.0) and *Memecylon umbellatum* (46.6) in VS; *Olea paniculata* (46.9) and *Myristica dactyloides* (40.2) in KS; *Memecylon umbellatum* (69.8) and *Neolitsea scrobiculata* (42.6) in MS.

In the total 0.4 ha regeneration stands, four species *Memecylon umbellatum, Neolitsea scrobiculata, Myristica dactyloides* and *Phoebe wightii* dominated with an IVI score exceeding 15.

Table—5. Population density of tree species ≥ 30cm gbh encountered in each 2-ha plot and in total 8-ha of tropical evergreen forest in the Kolli hills, Eastern Ghats, south *India*

Sl. No.	Species	Family	Density (no. of trees) in site				
			PS	VS	KS	MS	Total
1.	*Memecylon umbellatum* Burm.f.	Melastomataceae	104	300	26	378	808
2.	*Phoebe wightii* Meisner	Lauraceae	6	107	63	54	230
3.	*Beilschmiedia gemmiflora* (Bl.) Kosterm.	Lauraceae	12	159	19	-	190
4.	*Neolitsea scrobiculata* (Meisner) Gamble	Lauraceae	8	19	10	149	186
5.	*Myristica dactyloides* Gaertn.	Myristicaceae	41	35	82	13	171
6.	*Mallotus philippensis* (Lam.) Muell.- Arg. var. *Philippensis*	Euphorbiaceae	13	70	46	34	163
7.	*Syzygium cumini* (L.) Skeels	Myrtaceae	12	66	2	71	151
8.	*Scolopia crenata* (Wight & Arn.) Clos var. *crenata*	Flacourtiaceae	85	5	53	-	143
9.	*Nothopegia heyneana* (Hook.f.) Gamble	Anacardiaceae	128	8		1	137
10.	*Meliosma simplicifolia* (Roxb.) Walp. subsp. *simplicifolia*	Sabiaceae	-	-	109	-	109
11.	*Diospyros ovalifolia* Wight	Ebenaceae	97	2	-	7	106
12.	*Persea macrantha* (Nees) Kosterm.	Lauraceae	10	32	33	29	104

The set of abundant tree species recorded in the regeneration stands of the Kolli hills differed from its respective overstorey dominants. In a total density of 1937 regenerating trees, *Canthium dicoccum, Maesa indica* and *Cipadessa baccifera* contributed 121, 118 and 78 stems, while in the total 3825 adult trees, the above mentioned species comprised 16, 2 and 18 stems respectively (Chittibabu and Parthasarathy 2000a). In general, the composition of regeneration stands in various tropical forests differs from their respective overstories (Busing 1994; Jones *et al.* 1994; Richards 1996; Uhl and Murphy 1981; Cao and Zhang 1997; Kadavul and Parthasarathy 2000), and some of the dominant overstorey species may not be represented in the regeneration stand (Kadavul and Parthasarathy 2000).

Although the composition of future vegetation could be predicted from the regeneration stand composition (Swaine and Hall 1988; Jayasingham and Vivekanantharaja 1994), environmental changes over a period could defy the above expectation and the regeneration stand alone may not represent the future stand (Swaine and Hall 1988). A number of factors, such as, light demanding nature (Hubbell and Foster 1987; Kadavul and Parthasarathy 2000), regeneration strategies (Grime and Hillier 1992), and predators of seeds and seedlings (Janzen 1970), besides environment (Brokaw 1987; Jurik and Pleasants 1990; Benitze-Malvido 1998) and human disturbance (Benitze-Malvido 1998; Montergnini *et al.* 1998) would contribute to the variation in species composition of regeneration and overstorey layers.

LIANA DIVERSITY

Liana Species Richness and Density

Liana diversity of the Kolli hills in the 8-ha totaled 26 species (≥ 5 cm gbh) representing 18 families and 24 genera. The range of species, genera and family richness in the four 2-ha study plots was nine to 21, nine to 19 and seven to 15, respectively. The richness of liana taxa was lower in the human-impacted sites than in undisturbed sites (Table 6). Liana species-area accumulation curves (Fig. 3a) stabilized for the two disturbed sites (KS and MS), while they did not appear to approach an asymptote for the two undisturbed sites (PS and VS). In all, 384 liana individuals with a mean value of 48 individuals ha^{-1} and a range of 25 to 156 individuals per plot were recorded. The basal area of lianas ≥ 5 cm gbh in the 8-ha was 0.23 m^2 ha^{-1}. Basal area of individual study plots was smaller in the human-impacted sites (0.02 and 0.15 m^2 ha^{-1}) than in the undisturbed sites (0.40 and 0.37 m^2 ha^{-1}).

The human-impacted sites of Kolli hills have lower species richness of lianas than the undisturbed sites. This difference is probably due to selective tree felling and cattle browsing in the human-impacted sites thus depriving lianas of their hosts. For example, in the human-impacted site KS, there was a five-fold decrease in the density of smaller (30-60 cm girth) trees (51.5 stems ha^{-1}) as compared with the undisturbed site VS (268 stems ha^{-1}) for the same girth as smaller trees were selectively cut (Chittibabu & Parthasarathy, 2000a). Human disturbance also directly reduces diversity and density of lianas since they are cut for various purposes. In the disturbed sites, thicker lianas were cut together with their support trees. Consequently, decreased species richness and density of thicker (>10 cm) lianas in the human-impacted sites (KS and MS) as compared to undisturbed sites (PS and VS) were notable.

Table—6. Biodiversity inventories of lianas ≥ 5 cm gbh in four 2-ha plots of tropical evergreen forest at Kolli hills, Eastern Ghats, south India

Variable	*Undisturbed site*		*Disturbed site*		
	PS	*VS*	*KS*	*MS*	*8-ha Total*
Species richness (per plot)	21	16	9	14	26
Species richness (ha^{-1})	(21, 17)	(16, 14)	(9, 2)	(14, 10)	
Genus richness	19	15	9	13	24
Family richness	15	14	7	12	18
Diversity index					
Shannon (H')	2.60	2.44	2.02	2.31	2.89
Simpson (D)	0.09	0.10	0.12	0.11	0.06
Density (ha^{-1})	56.5	78	12.5	45	48
Basal area (m^2 ha^{-1})	0.40	0.37	0.02	0.15	0.23

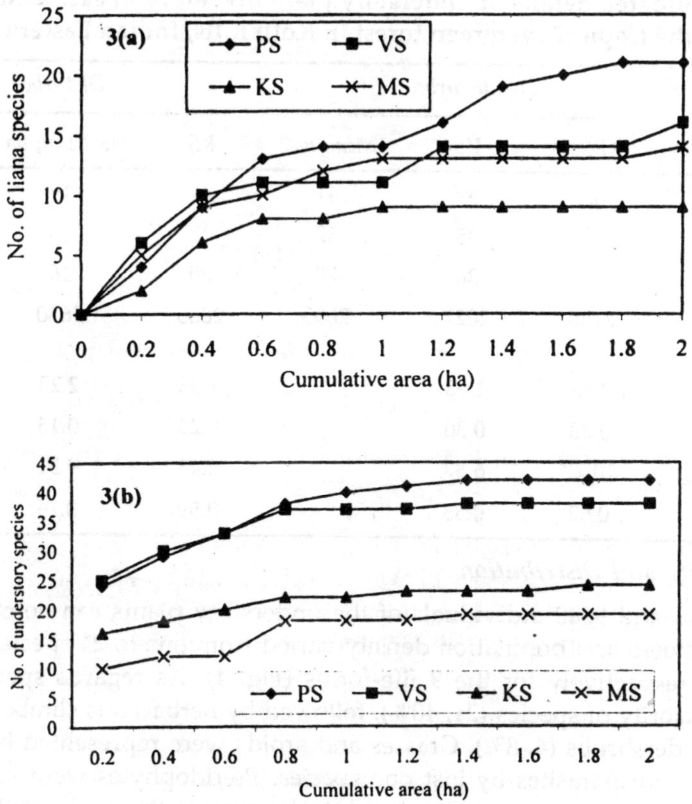

Fig.3. Species-area curves for the lianas (3a) and understory plants (3b) in the four study sites of Kolli hills.

Liana Species IVI

Of the 26 liana species, five species were common to all the four sites, four species occurred in three sites, 11 species in two sites and 6 species in one of the four sites. *Hiptage benghalensis* (Malpighiaceae) was the most abundant species (IVI = 25.0), followed by *Elaeagnus indica* (17.1), and *Gnetum ula* (16.9). Maximum gbh attained by liana species varied considerably, ranging from 5 cm for the root climber *Rhaphidophora laciniata* to 78.4 cm for the twiner *Entada pursaetha*.

UNDERSTORY PLANT DIVERSITY

Understory Richness and Diversity Indices

Understory plant diversity in the four sites of tropical evergreen forest in the Kolli hills totaled 52 species from 49 genera and 35 families (Table 7). Species richness of the individual plots varied. The undisturbed sites PS and VS had 42 and 39 species per plot respectively (41 species ha^{-1}), while the human-impacted sites KS and MS harboured 24 and 19 species (22 species ha^{-1}). Shannon (H′) and Simpson (D) indices marginally varied for the four sites (Table 7). Understory species-area accumulation curves of the four 2-ha plots stabilized at 1.2 ha to 1.4 ha scale (Fig. 3b). The asymptote of the curves suggests that the samples captured most of the species in the study plots.

Table—7. Consolidated details of understory plant inventory in each 2-ha plot and in the 8-ha of the tropical evergreen forest in Kolli hills, Indian Eastern Ghats.

Variable	*Undisturbed*			*Disturbed*			*Total*
	PS	*VS*	*Mean ha^{-1}*	*KS*	*MS*	*Mean ha^{-1}*	
Species richness	42	39	41	24	19	22	52
Number of genera	39	35	37	23	18	21	49
Number of families	32	26	29	20	16	18	35
Population density	2158	2027	2093	2865	2630	2748	9680
Diversity index							
Shannon (H′)	2.33	1.93		1.86	2.23		2.51
Simpson (D)	0.23	0.30		0.23	0.15		0.16
Hill (N_1)	10.27	6.87		6.43	9.31		12.36
Evenness (E_1)	0.62	0.53		0.59	0.76		0.64

Life-form Diversity and Distribution

The 52 species and 9680 individuals of the understory plants can be classified into 9 life-forms. Species richness and population density varied from one to 21 species and from eight to 3895 individuals respectively for the 9 life-forms (Fig. 4). As regards species richness, herbs constituted the majority of species (21, 40%), followed by herbaceous climbers (12, 23%), shrubs (6, 11.5%), and undershrubs (4, 8%). Grasses and aroids were represented by two species each, while sedges and root parasites by just one species. Pteridophytes were represented by three species. The grasses, *Cyrtococcum trigonum* in particular, contributed to the bulk of total population density of understory plants, followed by aroids, herbs and pteridophytes. The attenuated species richness (54%) of understory plants (mean 22 species ha^{-1}) in human-impacted sites (KS & MS), compared to the species richness (mean 41 species ha^{-1}) of undisturbed sites (PS & VS), is in line with the diversity of trees ≥ 30 cm gbh and lianas ≥ 5 cm gbh of the study plots in the Kolli hills (Table 1).

The trend of decreasing diversity along increasing disturbance of the study sites in the Kolli hills can be explained as in the case of tree and liana inventories (Chittibabu & Parthasarathy 2000a; Chittibabu & Parthasarathy, 2001). Disturbance at the study site includes cattle grazing and browsing and anthropogenic activities, such as collection of fuel-wood, medicinal plants and minor forest produce, and selective tree felling. Disturbed site KS has just one-fifth the tree density (52 stems ha^{-1}) in the smaller girth class (30-60 cm) of the 268 stems ha^{-1} for the same girth threshold in the undisturbed site VS. This indicates that selective tree felling has occurred in the study site (Chittibabu & Parthasarathy, 2000a). Hence, anthropogenic activities together with intraplot variation of topographic features might result in reduced plant diversity in the disturbed sites. The increase in understory plant density, particularly weeds in the disturbed sites, could be due to diaspore dispersal of a few species facilitated by cattle and human activities.

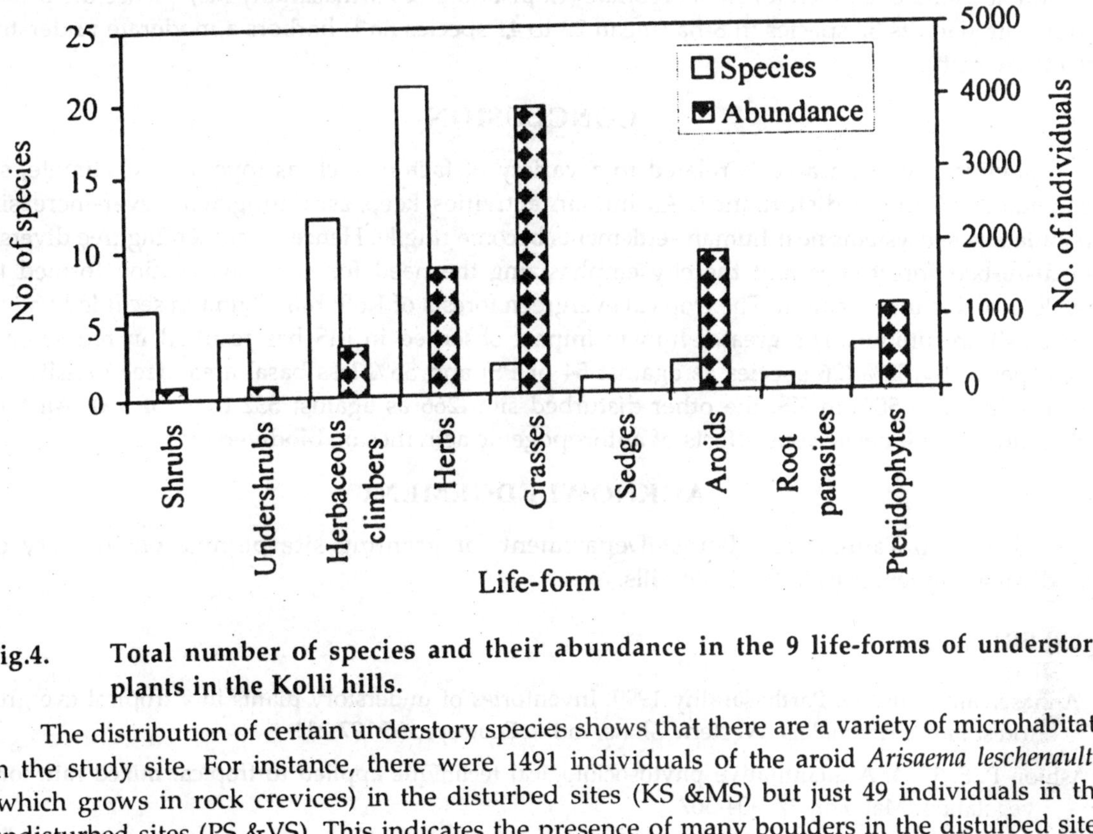

Fig.4. Total number of species and their abundance in the 9 life-forms of understory plants in the Kolli hills.

The distribution of certain understory species shows that there are a variety of microhabitats in the study site. For instance, there were 1491 individuals of the aroid *Arisaema leschenaultii* (which grows in rock crevices) in the disturbed sites (KS &MS) but just 49 individuals in the undisturbed sites (PS &VS). This indicates the presence of many boulders in the disturbed sites and this is corroborated by our field observation. On the other hand, the pteridophytes *Dryopteris concolor, Leptochilus decurrens* and *Pteris quadriaurita*, growing in moist and shady localities, were predominant (725 individuals) in the undisturbed sites compared with their density (426 individuals) in the disturbed sites. Nevertheless, the distribution of plant species could also be due to their ecological amplitude (Chittibabu & Parthasarathy, 2000a). The uniform abundance of *Cyrtococcum trigonum* could explain its environmental plasticity, being both shade and light tolerant (Chittibabu & Parthasarathy 2000b).

A comparison of understory plant diversity in the Kolli hills with those of other tropical forest inventories (Smith 1970; Hall & Swaine 1981; Gentry & Dodson 1987; Poulsen & Balslev 1991; Poulsen & Nielson 1995; Poulsen & Pendry 1995; Poulsen 1996a & b; Turner *et al.* 1996; Annaselvam & Parthasarathy, 1999) becomes increasingly difficult because of the considerable differences in plot size, quadrats sampled and synusiae included in various investigations. Yet, the literature survey (Table 4 of Annaselvam & Parthasarathy 1999) reveals that understory plant species richness ranged from 50 species ha^{-1} in moist tropical forest at Cuyabeno, Amazonian Ecuador (Poulsen & Nielsen, 1995) to 121 species ha^{-1} in 0.75 ha of evergreen forest in Brunei (Poulsen & Pendry, 1995). The Indian Western Ghats inventories yielded 155 understory plant species in 30 ha (17 to 83 species ha^{-1}) of tropical evergreen forest at Varagalaiar, Anamalais (Annaselvam & Parthasarathy, 1999), and 55 species in three 1-ha plots (21 to 29 species ha^{-1}) of tropical lowland evergreen forest at Agumbe (Gopisundar & Parthasarathy ms). Hence the present study site with its 52 species in 8-ha (mean 22 to 41 species ha^{-1}) harbors a moderate understory plant diversity.

CONCLUSION

Biodiversity of an area is related to a variety of factors such as topography, climate, soil and natural/human disturbance. As human activities keep escalating with ever-increasing population, ecosystems near human settlements become fragile. Hence, documenting tree diversity in disturbed forest sites and thereby emphasizing the need for site conservation formed the nucleus of this investigation. The tropical evergreen forests of Kolli hills found in secluded patches ('sholas') are unique. The greater human impact observed in MS has resulted in reduction of tree species by 52% (26 species as against 54 of PS) and 56% less basal area. Tree density was also reduced to 58% in KS, the other disturbed site (266 as against 632 trees of VS). All these data indicate the deleterious effects of anthropogenic activities on biodiversity.

ACKNOWLEDGEMENT

Thanks to Tamil Nadu Forest Department for granting site permission to carry out biodiversity inventories in the Kolli hills.

REFERENCES

Annaselvam J. and N. Parthasarathy 1999. Inventories of understory plants in a tropical evergreen forest in the Anamalais, Western Ghats, India. Ecotropica 5:197-211

Ashton P. S. 1964. A quantitative phytosociological technique applied to tropical mixed rain forest vegetation. Mal. For. 27: 304-307

Ayyappan N. and N. Parthasarathy 1999. Biodiversity inventory of trees in a large-scale permanent plot of tropical evergreen forest at Varagalaiar, Anamalais, Western Ghats, India. Biodiver. Conserv. 8:1533-1554

Bada S.O. 1984. Growth patterns in an untreated tropical rainforest ecosystem Ph D thesis, University of Ibadan, Nigeria

Benitze-Malvido J. 1998. Impact of forest fragmentation on seedling abundance in a tropical rain forest. Conserv. Biol. 12:380-389

Brokaw N.V.L. 1987. Gap-phase regeneration of three pioneer tree species in a tropical forest. J. Ecol. 75:9-19

Busing R.T. 1994. Canopy cover and tree regeneration in old-growth cover forests of the Appalachian mountains. Vegetatio 115:19-27

Campbell D. G., Daly D. C., Prance G. T. and U. N. Maciel 1986. Quantitative ecological inventory of terra firme and varzea tropical forest on the Rio Xingu, Braz. Ama. Britt. 38:369-393

Campbell D. G., Stone J. L. and Jr. A. Rosas 1992. A comparison of the phytosociology and dynamics of three floodplain (varzea) forest of known ages, Rio Jurua, western Brazilian Amazon. Bot. J. Linn. Soc. 108:213-237

Cannon C.H., Peart D. R. and M. Leighton 1998. Tree species diversity in commercially logged Bornean rainforest. Science 28:1366-1368

Cao M. and J. H. Zhang 1997. Tree species diversity of tropical forest vegetation in Xishuangbanna, SW China. Biodiver. Conserv. 6:995-1006

Chandrashekara U.M. and P. S. Ramakrishnan 1994. Vegetation and gap dynamics of a tropical wet evergreen forest in the Western Ghats of Kerala, India. J. Trop. Ecol. 10:337-354

Chittibabu C.V. and N. Parthasarathy 2000a. Attenuated tree species diversity in human- impacted tropical evergreen forest sites in Kolli hills, Eastern Ghats, India. Biodiver. Conserv. 9:1493-1519

Chittibabu C.V. and N. Parthasarathy 2000b. Understory plant diversity in a tropical evergreen forest in the Kolli hills, Eastern Ghats, India. Ecotropica 6:129-140

Chittibabu C.V. and N. Parthasarathy 2000c. Tree Diversity, Species Conservation and Sustainable Management of Tropical Forest: A Case Study from Indian Eastern Ghats IN Arunachalam A. and M. L. Khan (Eds) Sustainable Management of Forests-India pp111-134. International Book Distributors, Dehradun, India

Chittibabu C.V. and N. Parthasarathy 2001. Liana diversity and host relationships in a tropical evergreen forest of Indian Eastern Ghats. Ecol. Res. 16:519-529

Condit R., Hubbell S. P., Lafranke J. V., Sukumar R., Manokaran N., Foster R. B., and P. S. Ashton 1996. Species-area and species individual relationships for tropical trees: a comparison of three 50-ha plots. Journal of Ecology 84:549-562

Congdon R. A. and J. L. Herbohn 1993. Ecosystem dynamics of disturbed and undisturbed sites in north Queensland wet tropical rain forest I Floristic composition, climate and soil chemistry. J. Trop. Ecol. 9:349-363

Curtis J. T. and R..P.McIntosh 1950. The interrelations of certain analytic and synthetic phytosociological characters. Ecology 31:434-455

Dawkins H. C. 1959. The volume increment of natural tropical high-forest and limitations of improvements. Emp. For. Rev. 38:175-180

Ferreira L. V. and G. T.Prance 1998. Species richness and floristic composition in four hectares in the Jau National Park in upland forests in Central Amazonia. Biodiv. Conserv. 7:1349-1364

Gamble J. S. and C. E. C. Fischer 1915-1935. Flora of the Presidency of Madras VolsI-III London: Adlard and son Ltd

Ganesh T., Ganesan R., Soubadradevy M., Davidar P. and K. S. Bawa 1996. Assessment of plant biodiversity at a mid-elevation evergreen forest of Kalakad-Mudanthurai Tiger reserve, Western Ghats, India. Curr. Sci. 71:379-392

Gentry A. .H. 1988. Changes in plant community diversity and floristic composition on environmental and geographical gradients. Ann. Mo. Bot. Gard. 75:1-34

Gentry A. .H. and C. Dodson 1987. Contribution of non-trees to species richness of a tropical rain forest. Biotropica 19:149-156

Ghate U., Joshi N. V. and M. Gadgil 1998. On the patterns of tree diversity in the Western Ghats of India. Current Science 75:594-602

Grime J. P. and H.S. Hillier 1992. The contribution of seedling regeneration to the structure and dynamics of plant communities and larger units of landscape pp 349-363 IN Fenner. M (Ed) Seeds: The Ecology of regeneration in plant communities. C.A.B. International. Willingford

Grubb P. J. 1977. Control of forest growth and distribution on wet tropical mountains: with special reference to mineral nutrition. Ann. Rev. Ecol. Syst. 8:83-107

Hall J. B. and M. D. Swaine 1981. Distribution and ecology of vascular plants in a tropical rain forest. W. Junk Publishers, The Hague.

Hartshorn G. S. 1983. Plants pp. 118-157 IN Janzen, D.H. (Ed.) Costa Rican Natural History., The University of Chicago Press, Chicago

Heaney A. and J. Proctor 1990. Preliminary studies on forest structure and floristics on Volcan Barva, Costa Rica. Journal of Trop. Ecol. 6:307 – 320

Hegarty F.E. and G. Caballe 1991. Distribution and abundance of vines in forest communities pp.313-335 IN Putz F. E. and H.A. Mooney (Eds) The Biology of Vines, Cambridge University Press, Cambridge

Henry A. N., Kumari G. R. and V. Chitra 1987. Flora of Tamil Nadu, India Ser1 Vol 2 Botanical survey of India, Coimbatore

Henry A. N., Chitra V. and N. P. Balakrishnan 1989. Flora of Tamil Nadu, India Ser1 Vol. 3 Botanical survey of India, Coimbatore

Hubbell S. P. and R. B. Foster 1987. The spatial context of regeneration in a neotropical forest. pages 205-231 IN Crawley M., Gray A. and R. B. Foster (Eds) Colonization, succession and stability. Blackwell, Oxford.

Hubbell S. P. and R. B. Foster 1992. Short-term dynamics of a neotropical forest: why ecological research matters to tropical conservation and management. Oikos 63: 48-61

Janzen D.H. 1970. Herbivores and the number of tree species in Tropical Forests. Amer. Nat. 104: 501-528.

Jayasingam T. and S. Vivekanantharaja 1994. Vegetation survey of the Wasgomuva National Park Sri Lanka: analysis of the Wasgomuvaoya forest. Vegetatio 113: 1-8.

Jeffre T. and J. M. Veillon 1990. Etude floristique et structurale de deux forêts denses humides sur roches ultrabasiques en Nouvelle-Caledonie. *Buletin de la Museum National Histoire Naturalle,* Paris 12: 243-273

Jones E. W. 1950. Some aspects of natural regeneration in the Benin rain forest. Emp. For. Rev. 29:108-124.

Jurik T.W. and J. M. Pleasants 1990. The effects of light and plant size on fate of seedlings of the Prairie compass plant (*Siphium laciniatum* L.) Funct. Ecol. 4: 661-665.

Kadavul K. and N. Parthasarathy 1999a. Plant biodiversity and conservation of tropical semi-evergreen forest in the Shervarayan hills of Eastern Ghats, India. Biodiv. Conserv. 8: 421-439

Kadavul K. and N. Parthasarathy 1999b. Biodiversity of woody species and conservation of tropical semi-evergreen forest in Kalrayan hills, Eastern Ghats, India. Trop. Ecol. 40:77-90

Kadavul K. and N. Parthasarathy 1999c. Lianas in two tropical semi-evergreen forest sites on the Kalrayan hills, Eastern Ghats, south India. Trop. Biodiv. 6:197-208

Kadavul K. and N. Parthasarathy 2000. Regeneration Patterns of Woody Species in Tropical Semi-Evergreen Forest of Shervarayan Hills, Eastern Ghats, South India. Int. J. Ecol. Env. Sci. 26: 37-48.

Kelly D. L., Tanner E. V. J., Nic Lughadha E. M. and V. Kapos 1994. Floristics and biogeography of a rain forest in the Venezuelan Andes. J. Biogeogr. 21:421-440

Killeen T. J., Jardim A., Mamani F. and N. Rojas 1998. Diversity, composition and structure of a tropical semideciduous forest in the Chiquitania region of Santa Cruz, Bolivia. J. Trop. Ecol.14:803-827

Ledig F. T. 1992. Human impact on genetic diversity in forest ecosystems. Oikos 63: 87-108

Legris P. and V. M. Meher-Homji 1984. The Eastern Ghats: Phytogeographic aspects Indian Review of Life Sci. 4:115-136

Lieberman D. and M. Lieberman 1987. Forest growth and dynamics at La Selva, Costa Rica (1969-1982). J. Trop. Ecol. 3: 347-358

Magurran A. 1988. Ecological diversity and its measurement. Princeton University Press, New Jersey

Matthew K. M. 1991. An Excursion Flora of Central Tamilnadu, India. Oxford and IBH, New Delhi

Montergnini F., Eibl B., Szczipanski L. and R. Rioș 1998. Tree regeneration and species diversity following conventional and uniform spacing methods of selective cutting in a sub-tropical humid forest reserve. Biotropica 30: 349- 361.

Muthramkumar S. and N. Parthasarathy 2000. Alpha diversity of lianas in a tropical evergreen forest in the Anamalais, Western Ghats. Diver. Distr. 6:1-14.

Nadkarni N. M., Matelson T. J. and W. A. Haber 1995. Structural characteristics and floristic composition of a neotropical cloud forest, Monteverde, Costa Rica. J. Trop. Ecol. 11: 481-495

Nair N. C. and A. N. Henry 1983. Flora of Tamilnadu, India Ser. 1, vol I. Botanical survey of India, Coimbatore

Ojo L O and B. A. Ola-Adams 1996. Measurement of tree diversity in the Nigerian rainforest. Biodiv. Conserv. 5:1253-1270

Padaki A. and N. Parthasarathy 2000. Abundance and distribution of lianas in tropical lowland evergreen forest of Agumbe, central Western Ghats, India. Trop. Ecol. 41(2):143-154

Parthasarathy N. 1999. Tree diversity and distribution in undisturbed and human-impacted sites of tropical wet evergreen forest in southern Western Ghats, India. Biodiv. Conserv. 8: 1365-1381

Parthasarathy N. 2001. Changes in forest composition and structure in three sites of tropical evergreen forest around Sengaltheri, Western Ghats. Curr. Sci. ——

Parthasarathy N. 2001. Plant biodiversity and conservation of tropical evergreen forest in the Anamalais, Western Ghats, India pp. 557-560 IN Ganeshaiah K. N., Uma Shaanker R. and K. S. Bawa (Eds) Tropical Ecosystems: Structure, Diversity and Human Welfare. Oxford-IBH, New Delhi

Parthasarathy N. and R. Karthikeyan 1997a. Biodiversity and population density of woody species in a tropical evergreen forest in Courtallum reserve forest, Western Ghats, India. Trop. Ecol. 38: 297-306

Parthasarathy N. and R. Karthikeyan 1997b. Plant biodiversity inventory and conservation of two tropical dry evergreen forests on the Coromandel coast, south India. Biodiv. Conserv. 6: 1063-1083

Parthasarathy N. and P. Sethi 1997. Trees and liana species diversity and population structure in a tropical dry evergreen forest in south India. Trop. Ecol. 38: 19-30

Pascal J. P. and R. Pelissier 1996. Structure and floristic composition of a tropical evergreen forest in south-west India. Journal of Trop. Ecol. 12:191-214

Pascal J. P. and B. R. Ramesh 1987. A field key to the trees and lianas of the evergreen forests of the Western Ghats (India). Institut Francais de Pondichery, Travaux de la section Scientifique et Technique, Tome XXIII, Pondicherry

Pianka E. R. 1966. Latitudinal gradients in species diversity: a review of concepts. Amer. Nat. 100: 33-46

Poore M. E. D. 1968. Studies in Malaysian rainforest 1 The forest on triassic sediments in Jengka forest reserve. J. Ecol. 56: 143-196

Poulsen A. D. 1996a. The herbaceous ground flora of the Batu Apoi forest Reserve, Brunei Darussalam pp. 43-57 IN Edwards D. S., Booth W. E. and S. C. Choy (eds.) Tropical rainforest research–current issues. Kluwer Academic Publishers, Dordrecht.

Poulsen A. D. 1996b. Species-richness and density of ground herbs within a plot of lowland rain forest in north-west Borneo. J. Trop. Ecol. 12: 843-851.

Poulsen A. D., and H. Balslev 1991. Abundance and cover of ground herbs in an Amazonian rain forest. J. Veg. Sci. 2: 315-322.

Poulsen A. D. and I. H. Neilsen 1995. How many ferns are there in one hectare of tropical rain forest? Amer. Fern J. 85: 29-35.

Poulsen A. D. and C. A. Pendry 1995. Inventories of ground herbs at three altitudes on Bukit Belalong, Brunei, Borneo. Biodiv. Conserv. 4: 745-757

Richards P. W. 1996. The tropical rain forest: an ecological study. Cambridge University Press, London.

Sala O. E. *et al.* 2000. Global Biodiversity Scenarios for the Year 21000. Science 287:1770-1774

Singh S. P., Adhikari B. S. and D. B. Zobel 1994. Biomass, productivity, leaf longevity, and forest structure in the central Himalaya. Ecol. Mono. 64: 401-421.

Srinivas V. and N. Parthasarathy 2000. Comparative analysis of tree diversity and dispersion in the tropical lowland evergreen forest of Agumbe, central Western Ghats, India. Trop. Biodiv. 7:45-60

Sukumar R., Dattaraja H. S., Suresh H. S., Radhakrishnan J., Vasudeva R., Nirmala S. and N. V. Joshi 1992. Long-term monitoring of vegetation in a tropical deciduous forest in Mudumalai, southern India. Curr. Sci. 62: 608-616

Swaine M. D. and J. B. Hall 1988. The mosaic theory of forest regeneration and the determination of forest composition in Ghana. J. Trop. Ecol. 4:253-269.

Turner M. I., Tan H. T. W., and K. S. Chua 1996. Relationship between herb layer and canopy composition in a tropical rain forest successional mosaic in Singapore. J. Trop. Ecol. 12: 843-851.

Uhl C. and P. G. Murphy 1981. Composition, structure and regeneration of a tierra firm forest in the Amazon basin of Venezuela. Trop. Ecol.. 22: 219-237.

Valencia R., Balslev H. and C. G. P. Y. Mino 1994. High tree alpha-diversity in Amazonian Ecuador. Biodiv. and Conserv. 3: 21-28

Wattenberg I. and S. Breckle 1995. Tree species diversity of a premontne rain forest in the Cordillera de Tilaren, Costa Rica. Biotropica 1: 21-30

Whitmore T. C. 1984. Tropical rain forests of the Far East. Clarendon Press, Oxford.

Whitmore T. C., Peralta R. and K. Brown 1986. Total species count in a Costa Rican Tropical rain forest. J. Trop. Ecol. 1:375-378

3

The Saga of Sahyadri: Societal Environmental Issues of Konkan and Goa

***P. V. Sathe & O. S. Chauhan**

INTRODUCTION

When we think of seashore, we imagine pristine sunny beaches, forested mountains and pretty rivers meandering down the lush-green hill slopes. We also imagine people merrily flocking to the beaches every evening, wondering at the waves splashing on the shore, the fishermen tending their nets and receiving fishing boats with a happy song. Few among us realize that behind this picture of sun, sand and song lies a conflict, which is invisible to a visitor. A perennial conflict between man and forces of nature, that makes life on the coast rather harsh.

Most of this dream-like landscape consisting of sandy beaches and rocky shore with dynamic coastline is unfit for cultivation. Even fresh water is scarce on the coast due to unsuitability of shore as deep lasting aquifer. Most of the coastal rivers are saline as the tidal water from the sea mixes with the fresh river water every day. This makes the coastal rivers unsuitable either for domestic water supply or for irrigation. Coastal regions are humid, thus making summers unpleasant, especially for outdoor activities. Humid air also breeds a host of germs that affects respiratory health. During storms and cyclones, the coastal region becomes the first target for devastation.

The region along the Sahyadri ranges on India's west coast is internationally acclaimed for its sun and sand making it a heavenly abode. With the growing demand for the coastal space due to demographic burden, tourism and industrialization, the region is subjected to rapid changes. This article describes the societal environmental issues besieging the costal region on the foothills of Sahyadri hills, in particular, Konkan and Goa.

* National Institute of Oceanography, Dona Paula, Goa.

Constraints of Physiography

The Sahyadri range of mountains that runs parallel to the coast lies too close to the sea for comfort. In the Konkan coast of Maharashrta, the mountain range is placed right on the seashore making access to the shore almost impossible. Unlike east coast, the rocky headlands jetting out into the sea are a common sight in Konkan. Due to narrow coastal belt and the adjacent rising highland, there is an acute shortage of cultivable flatlands. Flatlands on mountain-tops are mainly dry, rocky and uncultivable. Neither people live nor any agriculture is undertaken on them. Flat fertile lands are found only in valleys and along river banks. There is a bitter contest among people in owning up these few pockets of flatlands and putting them to multiple uses such as agriculture, grazing, transport, recreation and commerce. Narrow unwinding lanes and congested markets where animals move freely and artisans and craftsmen setup their equipment for the day's business is a typical sight in Konkan. The national highway has connected larger interior towns but remote coastal villages remain inaccessible by roads. Even the mega Konkan railway project which was supposed to improve communication and transport in Konkan, is placed far from the seashore and does not connect major population pockets on the coast. During spells of heavy rains, and other natural calamities such as storms or earthquakes, large number of villages simply gets cut off from the civilized world. Life is surely harsh for men and animals alike with difficult terrain to tread on foot, shortage of drinking water and lack of flat fertile lands.

It is an irony of fate that while the spacious airy vast uplands on mountain tops remain uninhabited, the low-lying damp areas located on the flood plains of the rivers are congested with over population. This is because of availability of fresh water in low lands. If water can be supplied over the uplands, the vast rocky flatlands lying unutilized will be quickly developed to share the burden of demographic explosion and industrial activities. With ease of transport and communication and healthy environment, comfortable life will await people of Konkan uplands.

As we move south into Goa, the Sahyadri range is pushed eastward by about 20 to 30 km and large plains of flatlands begin to appear near the sea shore. This makes transport and communication relatively easier in Goa than in Konkan. Since the mountains lie away from the sea, the shore is easily accessible. This has been both boon and bane for Goa. Boon because this helped tourism grow giving livelihood to thousands of families. Bane because coastal regions of Goa became thickly populated creating myriads of societal, environmental and other infrastructure-related problems. Earlier, people used to live away from the shore at an elevation of about 5 to 10 meters, overlooking the agricultural lowlands towards the river banks or seashore. These lowlands also served as grazing lands for domestic animals, play-grounds for children and open vacant spaces to hold traditional feasts and festivals. A true harmony indeed!

With the advent of tourism, agriculture was given a go-bye, and the low-lying flatlands were quickly converted into residential use. Being agricultural lands, they were characterized by high water tables and their soil had poor water absorption capacity. They were prone to water logging and flooding during the wet months. This resulted into flooding of new residential areas during monsoon because these lowlands were meant for paddy cultivation in the past in which large pools of water was held for days on. Even absorption tanks of domestic drain water could not function efficiently in such agricultural lands. In the absence of a central drainage system in Goa, stagnating drain water in muddy soils behind new houses in Goa became a common sight. No wonder then, that malaria and other water borne diseases, which were unknown in Goa, took a firm foothold.

Disappearance of open fertile flatlands from Goa also had other far-reaching societal implications. The domestic animals lost their grazing lands. They turned orphan and began wandering in urban areas in search of food and settling on the highways for shelter. They became "intruders", when, in fact, they are in their own natural domains, in what used to be their grazing lands. It is we who are intruders into their territory! If animals lost their grazing lands, children too lost their playgrounds around their homes. More often than not, a child growing in Goan towns will land onto a road as soon as he steps out of his house (except those lucky few who can afford to live in planned housing complexes). Few of us are aware that most schools and colleges in Panaji have no playgrounds. What childhood memories will children carry when they spend their entire schooling lives in concrete buildings with narrow dark corridors and congested staircases! The tourism demands are so high that it is a matter of time that few open spaces left in Goan towns disappear altogether.

The Sandbar, Navigational Curse

The characteristics of a coast, by and large, result from interaction of physical forces such as waves, tides, currents and supply and availability of sand. The surface circulation along the Sahyadri ranges is by and large seasonal, but for most of the time the littoral circulation is clockwise. As a result, any waste dumped in the sea tends disperse and pollute the southern regions along the coast. Wherever rivers meet Arabian sea, this current usually erodes northern regions off the mouths, and the eroded material is transported and dumped along the southern region of the river mouths. As a result, the northern regions off the river mouths are rocky, while beaches are formed on southern side. The beaches have a tendency to grow seaward because of continuous deposition of eroded material from the north. Though, beaches become tourist attraction in south, the eroded rocky headlands in the north cannot be put to any commercial or recreational use.

During monsoon however, the entire picture reverses to the detriment of local residents who live by the side of the smaller rivers. High turbidity with increased runoff from the rivers mixes with saline waters in the estuary, and this leads to the flocculation of the clay, silt and sand in the coastal stretches of the river mouths. The strong onshore wind and wave action during monsoon prevents the silt from moving seaward and a temporary sand bar develops right at the river mouth. This sand bar blocks the river mouth completely and the river navigation closes down. All small ports along the river become non-functional. All trade, transport and commerce that the small ports support get a forced holiday during the monsoon. On the other hand, the obstructed river becomes flood-prone during the spells of heavy rain and tidal cycles. This is the case with almost all smaller rivers along the west coast. Fortunately, the sandbar gradually opens all by itself once the monsoon is over. However, by then, the cessation of the commercial activities during the wet months, already takes a toll on the economic growth of the region.

Forests, the Contractors' Paradise

Due to tropical climate and high rainfall the Sahyadri Range is blessed with many natural pockets of forests on the slopes. These forests used to serve as the natural source of firewood, fodder and timber to the local residents for their domestic needs. Besides, the forests supported other activities such as fruit and honey collection, harvesting of medicinal plants and occasional game hunting by local residents. They used forests for centuries together but it did not result in deforestation. Naturally so, because they were concerned with its conservation as their very existence depended on them. During the last two decades, the forest department in their misplaced zeal to protect forests, began preventing local people from entering forests for their traditional activities without providing them alternate source of fuel or fodder. Laws were enacted that prevented people from cutting trees from their own free-hold lands even for domestic use. Surprisingly, contractors were freely given permissions as usual to cut forests on a large scale for commercial use. This forced people to buy firewood from contactors at exorbitant prices eroding their meager resources instead of collecting firewood directly from forests.

It appears, as if, Government grows forests so that contractors can cut them for commercial use. Worse still, the new forests grown under Government programmes are all pockets of monoculture. Thousand of eucalyptus, nilgiri or cajuarina trees, all look-alike, standing in unison along hill slopes, strictly guarded by armed watchmen, waiting for a contractor to fell them down is no one's idea of environmental conservation. Thus, the state sponsored conservation schemes, deprived forests of their biodiversity and made people pay heavily for firewood and fodder from contractors.

We need social forestry to reverse this trend. The pockets of forests found around remote villages need to be handed over to the local communities for maintenance. The local communities should be encouraged to make local laws for cutting the forests for domestic use and charge the local community for forest maintenance. This experiment has succeeded well in Himachal Pradesh where "Van Rakshak Samiti's" are appointed by Gram Panchayats to look after forests. Madhya Pradesh has gone a step further in which, the right to award contracts for cutting forests is removed from Gram Panchayats and given to Gramsabha's directly. We need to learn from these examples of Local Self Governance and make similar arrangements in Konkan and Goa if forests along Sahyadri are to survive in harmony with people *and not with contractors.*

Flooding, Manmade and Natural

Flooding during monsoons is a regular feature all along the Sahyadri foothills. Surprisingly, the reasons of flooding are mostly manmade. The undulating landscape has such a well-defined surface drainage pattern that all rainwater should quickly drain into sea due to high gradient of the landform. Unfortunately, haphazard constructions carried out without regard to the drainage pattern has resulted in accumulation of waters at unexpected locations and its diversion to new channels leading to flash floods. Traditionally, people used to build ponous fences by putting up layers of uneven stones over one another. Water used to pass through such fences into natural broad pathways. Modern concrete fences block water completely and obstruct the natural flow governed by the slope gradient.

The town planning departments, municipalities and village Panchayats who give construction licenses are totally oblivious of the need to preserve the natural drainage of the landscape and thus, ignore centuries old ancestral wisdom. Visit any newly constructed residential complex on hill slopes in Goa during rains and you will invariably find the criss-cross RCC compound walls blocking the water flow in myriads of ways. One frequently treads through ankle deep water while moving through new residential complexes.

Another cause of flooding is the shallowing of rivers themselves. Unplanned massive deforestation carried out during past several decades has exposed soil overburden in the watershed of the rivers. This has supplied the rivers with silt many times higher than their carrying capacity. As a result, rivers which were once capable of absorbing the natural rainwater into their flow are now unable to carry the same during monsoon. The water then spills over and across their banks inundating fields and townships. True, this is a temporary phenomenon but it does create havoc in the region as waters enter houses, markets, schools and public places. The foot roads get submerged. Domestic animals get trapped on their way home. Remote localities of a village get cut off for days together. This brings discomfort and miseries to people every monsoon.

The only lasting solution to this annual flooding is dredging (desilting) of rivers. Dredging will make rivers deep once again put an end to the annual flooding. Besides, this will also make rivers navigable, thus offering alternate mode of transport to a large number of "transport-starved" remote villages of Konkan. There are several small ports along the shore that have closed down due to siltation. One can see large warehouses of Portuguese, Dutch and English traders on these ports indicating that they had once formed important links in marine trade through navigable waterways. Desilting can rejuvenate these ports once again. There is usually opposition to desilting from local fishing communities because it destroys the bottom dwelling shellfish thus depriving the local marginal fishing community of their livelihood. Ideally, such fishing communities should be suitably compensated for the loss of their livelihood and desilting should be carried out along selected stretches for all rivers along the western ghats.

If freshwater flooding is an annual problem in interiors, saltwater flooding is a daily problem near river mouths. Both are caused by shallowing of rivers. The shallow river mouths are simply unable to accommodate the daily incoming tidal waters in addition to their own freshwaters. During high tides, tidal water overflows the river banks, spills over the plains and inundates the landscapes. The fields and grazing lands get submerged under salt water everyday with highest intensity during spring tides. This water seeps into the soil and wells turn salty. Landforms that once supported freshwater ecology because of depth and width factor of river channels, have now turned hostile and become unfit for human habitats. As rivers are becoming shallower and shallower, more and more fertile flatlands are turning

Fig. 1. **Diwar island in Mandovi river is a classic example of the entire ecology getting submerged under salt water due to shallowing of river Mandovi**

salty year after year. As it is, the region has shortage of flat fertile lands due to the geological setting, and added to that, is the new problem of saltwater ingress that makes them uncultivable. In Goa, the problem is further accentuated by mining. One ton of ore extracted produces 3 tons of refuge, which finds its way into the nearby rivers. This speeds up shallowing of rivers. Diwar island in Mandovi river is a classic example of the entire ecology getting submerged under salt water due to shallowing of river Mandovi (Figs 1 and 2).

Fig. 2. **Diwar island in Mandovi river is a classic example of the entire ecology getting submerged under salt water due to shallowing of river Mandovi**

Mangroves, the Natural Solution

Nature has provided a solution to this problem within the coastal ecosystem. In fact, if the estuary is left to itself, the solution emerges from within. Thick rows of mangroves which bedeck the river banks are nature's first line of defense against both the evils, namely, flooding and the salt water ingress (Fig. 3). Mangrove is a unique group of plants that grow only where both saltwater and freshwater is available. River mouths and the inter-tidal zone (the place in the river upstream upto

Fig. 3. **Thick rows of mangroves which bedeck the river banks are nature's first line of defense against both the evils, namely, flooding and the salt water ingress.**

which salt water from the sea reaches during high tides) are natural habitats for mangroves since both fresh and salt waters are available here. Mangroves grow into thick forests and accumulate silt at their base. This makes a natural bund along riverbanks. This bund not only stops erosion but also prevents the salt water ingress into the habitats thus protecting the freshwater ecology. All freshwater sources such as ponds, streams and wells are restored and agriculture becomes possible right besides the estuaries where mangroves exist. Wood from mangroves is used as firewood while twigs and leaves serve as fodder for domestic animals. Mangrove habitats are good locations for fish breeding, capture fisheries and prawn culture since the refuge of the mangrove provides nutrition to the estuarine biota. Thus mangroves not only contribute to a healthy ecosystem but also to the socio-economic well being of its people.

Unfortunately, mangroves are destroyed all along the river-banks for short-term economic gains of a few people. The tourism industry requires land right on the river banks where protection of mangroves becomes secondary. Construction of houses, jetties, roads and other commercial ventures that come up right on the river banks destroys them completely. True, the particular owner of the swampy land off the river-bank benefits by cutting mangroves and setting up of some other profitable venture but the cost is borne by the entire coastal community in terms of salt water ingress, flooding, erosion and a host of other adverse effects. In fact, the saltwater ingress is such a serious problem along the west coast that the residents have abandoned the entire coastal villages. There is already a legislation protecting mangrove habitats but it has not served its purpose due to lack of awareness. Besides, there are pressure groups from within the tourism industry that work against implementation of such laws. The forest department needs to take up large scale plantation of mangroves all along the estuarine banks and later hand over their maintenance to local residents.

Sand Dunes

Long rows of majestic sand dunes as high as 3 to 10 meters exist along the beaches of Konkan and Goa. A casual visitor to the beach often ignores them because the visitor looks seaward while on the beach (rather than landward). If mangroves are nature's biological line of defense, sand dunes are nature's physical line of defense against marine invasion. What mangroves do along river banks, sand dunes do along the seashore. They break the fury of storms and protect life and property along the coast during monsoon. They support a variety of plant life, which in turn, supports an entire ecosystem around them. Vegetation on the dunes stabilizes them. Sand dunes protect the beaches by stopping erosion. It is believed that sand dunes protect the freshwater sources on the coast by stopping the saltwater ingress underground though their exact role in the process is not understood. Sand dunes are usually not destroyed by local population. In fact, local population has co-existed with sand dunes for thousands of years. Sand dunes have made it possible for people to settle very close to the seashore in Goa by protecting them from marine ingress.

During the last decade however, sand dunes have been disappearing from almost all beaches in Goa. The beach tourism in Goa in particular is responsible for destroying sand dunes. Star hotels are constructed right on the beaches by razing the sand dunes to the ground so that tourists can have unrestricted access to their sites (Fig. 4). Sometimes, roads, parking lots and hawkers' markets are built by flattening the dunes in the name of "landscaping" the sea-face by the

Fig. 4. Star hotels are constructed right on the beaches by razing the sand dunes to the ground so that tourists can have unrestricted access to the seashore from their rooms.

Government to support tourism. Once sand dunes are destroyed, all vegetation supported by it disappears rendering the beach flat and lifeless. The beach becomes erosion-prone and begins to narrow down as in case of Anjuna beach of Goa. The sand becomes lose and begins to blow straight onto the roads, in the houses and markets. Miramar beach in Goa is a classic example. The Miramar circle is always full of sand. The nearby houses keep accumulating sand all the time. This is because the new coastal road is built right on the sand dunes by razing them to the ground. Like mangroves, sand dunes are also protected by law but it is seldom implemented. It is a matter of time before sand dunes disappear completely from Goa's beaches exposing it to the wrath of the oceanic forces.

The CRZ Controversy

The word CRZ (Coastal Regulation Zone) evokes different responses from people with different background. Hoteliers, and builders consider it as a curse to their professional ambitions, and a hindrance in providing the best to the tourist, while NGO's worship the provisions under CRZ. Lawyers take great pride in stalling coastal projects, and star hotels halfway through the construction by drawing court's attention to violations committed by the owners. Initiated in eighties, debates on CRZ are usually heated and lost in the rhetoric surrounding the definitions and scope of CRZ. One rarely comes across a discussion on the real purpose behind CRZ guidelines.

One does not have to be an environmentalists to see the benefits of maintaining the CRZ free of development, which, more often than not means setting up of private commercial enterprise right on the seashore. Prime locations for a builder mean sites on the seashore, preferably on the beach itself. Starred hotels are ready to pay any amount to secure such "prime locations" for their business for they get many fold returns from such investments year after year. If this trend is not checked, every inch on the seafront will be privatized and ordinary people will be left with no place for recreation. Open and vacant places in the cities are breathing grounds of the society where children play, elders meet and feasts and festivals are held. Besides, seashore also supports a variety of life forms in the inter-tidal zone that not only adds to the biodiversity but also enhance the carbon and nitrogen cycle of the region. It stands to reason then that seashore should be kept in its pristine form for the benefit of the entire humanity and not owned up by a private individual for his profit.

These form the basic arguments in support of CRZ guidelines. In case of Konkan and Goa, there is yet another factor overlooked by environmentalists while making a case for CRZ protection. A large number of people have traditionally settled right near the seashore behind the sand dunes for centuries together because of availability of flat lands. Rest of the landscape was uneven and hilly. Most of the people are engaged in traditional occupations in which their livelihood is derived from their immediate environment such as fishing, toddy tapping, marginal farming, sand extraction, poultry, horti-culture etc. Ideally, development should mean strengthening their economies by providing them with necessary infrastructure, equipments, loans and markets for their produce. Unfortunately, hard development of the seashore means displacing these original inhabitants in the interest of some commercial venture such as a hotel or an industry that occupies their ecological space. No amount of compensation is adequate for such people to recreate their ecological space elsewhere. Certainly a fisherman cannot take his sea with him, nor a bee-keeper can take his patch of forests to his new place. As a result, such people after receiving petty cash for their lost habitats usually turn destitute in the course of time. We need CRZ guidelines to ensure that their ecological space is not encroached upon in the name of development. A 500 meter belt from the seashore and 200 meters from riverbank will ensure that people in traditional occupations can continue to earn their livelihood with dignity and recreational space for the rest of the population remains intact.

Recommendations

Several rivers have become flood-prone and un-navigable because of accumulation of silt. As years pass, they will become more and more shallow and will be unable to carry their load into the sea. This has far reaching implications. Villages that were once connected by riverine network are now cutoff because rivers are so shallow that they are simply not accessible by boats. Ironically, these are the very villages that get flooded during monsoon. A major programme of dredging of selected stretches of rivers need to be undertaken all along the coast of Konkan especially near the river mouths. The silt removed through dredging operations should be used to fill up the regions on the riverbank so that they are raised and thus protected from flooding.

Mangrove plantation should be undertaken all along the inter-tidal zone to prevent salt water ingress. This will not only restore agriculture and freshwater sources but also provide firewood and fodder to the inhabitants. Besides, this will rejuvenate the estuarine ecosystem and enhance fishery potential of the region.

There is a strong case for water harvesting in this region, specially on the hill tops and along the hill slopes. The harvested water should be used to augment the drinking water during summer and also to develop forests on the hill-tops. Afforestation of rocky hill-tops will keep the environment cool as well as recharge the ground water table around the hill slopes. This will ensure that wells do not turn dry during summer.

The rocky headlands jetting out in the sea are ideal locations for both wind energy based power plants and harnessing of solar power. These headlands are barren, rocky and bereft of any vegatation. They receive plenty of sunlight and wind. The successful windmill experiment of Devgad region of Sindhudurg district should be extended to other areas.

The low-lying flood-prone areas near the estuaries, popularly known as Khazan lands in Goa have become practically waste lands. They have immense potential for developing as aquaculture sites. The modern fish farming technology should be made available to the people living around the Khazan lands.

All open and vacant spaces in the cities and townships should be preserved and developed only as recreational spots benefiting the entire society. Conversion of agricultural land for residential use should be completely banned. All development including residential complexes should be shifted away from the seashore towards the hills. The surface drainage maps need to be prepared and made available to planning authorities down to the Panchayat level so that they can be trained to integrate them in their development planning.

Most municipalities do not have any central drainage system for domestic sewage. Each house takes care of its sewage through construction of septic and absorption tanks. A scientifically planned central drainage system for waste water needs to be constructed keeping in mind the physiography of the landscape for each city before it is too late.

The tourism focus should be changed to eco-tourism rather than beach tourism in which the forests, hills, springs, lakes, sanctuaries and other interior features of Goa become key tourist locations. This will protect both the geomorphic features of the coastline and the marine life in the inter-tidal zone.

The region produces a large variety of fruits such as mangoes, cashew nuts, coconuts, jackfruits etc. But this produce has neither brought prosperity to the growers nor is the produce any cheaper in the region. This is because the production, distribution and marketing is not organized. The growers are not assured of any price for their produce. A cooperative movement is needed for production and distribution of prime fruits of the region. It should also support scientific research on modern methods of growing, preservation and processing of these fruits.

4

Medicinal Plants of India: Status and Prospects

*Ganesh R. Hegde

INTRODUCTION

India has a rich heritage of herbal treatment for various diseases. The rich plant resource of India is a priceless asset. The traditional medicinal plant knowledge is the invaluable inherited property of Indian people. This system not only includes curative properties but also has preventive properties. The WHO (1976) defines traditional medicine as 'the sum total of all knowledge and practices, whether explicable or not, used in diagnosis, prevention and elimination of physical, mental or social imbalance and relying exclusively on practical experience and observations handed down from generation to generation, verbally or in writing'. Some system of medicine like Ayurveda, Siddha, Unani etc. have well recorded written documents. The medicinal system which does not have the written documents but is inherited by verbal and cultural means is called ethnomedicine (Hughes, 1968). It was Hahnemann, the founder of Homoeopathy, who introduced the term Allopathy for the first time to separate from other medicinal systems. Amongst the various non-ethnomedicine systems, Allopathy differs in its philosophy and methodology. Thus all the systems other than Allopathy are treated as 'alternative Medicine'. Certain specific systems such as Naturopathy, Aroma therapy, Hydrotherapy etc. are regarded as Complementary medicines (Kameswar Rao, 2000). A recent WHO study indicates that more than 80,000 higher plant species of the known 2,50,000 species on the earth, are medicinal. That is to say that nearly 30% of the world's plant species are being used for medicinal purposes.

MEDICINAL PLANTS IN INDIA

There is no authentic record of the beginning of herbal medicine practice in India. But *Rig Veda* is the oldest written document giving the clue for origin of this Science. It is discernible that Indian Hindus are the pioneers on this earth in the exploitation of medicinal plants. The

* P.G.Department of Botany Karnatak University Dharwad-580 003

Vedic Aryans learnt the use of about 100 plants. It is clear from the works of Charaka and Sushruta that Indo Aryans were familiar with large number of medicinal plants. Charaka, Sushruta and Ashtangahrudaya Samhitas or Treatise are the major source text of Ayurveda, 'the Science of life'. Ayurveda is the oldest medical science of India dating back to about 1000 BC. Ayurveda is thus considered as the Hindu system of medicine. It consisted mainly the herbal treatments. Ayurveda was respected and popularised during the period of Buddha. Following the Greek invasion, several drugs of Central Asia and Asia Minor entered the Indian land. The Hindu system of medicine had a set back during the Mohammadan invasion when Unani system was introduced and popularised. After the discovery of sea-route to India by Portuguese, the spice trade to Europe was established. This interested several Europeans and the credit of compiling uses of Indian medicinal plants goes to Von Rheede in his book called *Hortus Malabarica*. However, significant attention was given to the study of medicinal plants only after the formation of Asiatic Society of Bengal by Sir William Jones, who subsequently published The Pharmacopoeia of India (1867) which stands as an authoritative work on the native remedies of India. This was followed by several monumental works, both at national level and regional levels (Mhasker et al., 2000). However, the European invasion in India resulted in the introduction of allopathic system of medicine and thus Ayurveda and ethnomedical practice existed in the various pockets of India mostly with the traditional practitioners. Although Allopathy became popular in India the essence of herbal medicines was not completely lost (Farooqi and Sreeramu, 2001). This is evidenced by the important works such as *Indian Medicinal Plants* by Kirtikar and Basu (1918), *Glossary of Indian Medicinal Plants* by Chopra et al. (1956), *Indian Materia Medica* by Nadkarni (1954), *Indian Medicinal Plants* by Dey (1980) etc.

DISTRIBUTION OF MEDICINAL PLANTS IN INDIA

India is bestowed with varied climatic conditions and soil. Physiographycally the mainland of India is divided into three distinct zones as the Himalayas, the Indo-Gangetic plains and the Peninsular India or the Deccan. The rainfall varies from almost nil to the highest in the world. Temperature in most regions rises above 38 ^{0} C during summer. It rarely goes below 20^{0} C in the Southern regions and at few places in the northern regions it falls below 4^{0} C. The soil types of India are mainly the alluvial soils rich in lime, the fertile black soils which are rich in lime, the poor quality red soils and the lateritic soils which is rich in hydrated oxides of aluminium and iron. India is one of the 12 centres of origin and diversity of several plant species in the world. The total estimated angiosperms in India are around 18,000 species. It has the heterogeneous assemblage of species from surrounding Malaya, Tibet, China, Japan and Europe. Some are from other continents like America, Africa and Australia (Rao, 1994). However according to Nayar (1980) as many as 5000 species are endemic to the Indian subcontinent.

The richness in the biodiversity in India is obviously due to its varied climate, topography and soil types. The altitude varies from sea level to the tallest mountains of the world. The habitat range from wet Western Ghats to the dry deserts of Rajasthan. Warm and humid coastal plains in the southern Peninsular India to cold icy mountains of Himalayas in the north. There are accordingly almost all types of forests ranging from scrub/thorn forests to the tropical evergreen rainforests; coastal mangroves to the temperate and alpine forests. Champion and Seth (1968) have classified the vegetation of India into 16 major forest types. Of these the tropical moist deciduous forests are the dominant (37%) followed by tropical dry deciduous forests (28.6%). It is also estimated that 11% of the known plant species of the world are represented in India (Rao, 1994).

Nearly 70% of the medicinal plants are found in the tropical regions from the foothills of Himalayas and Eastern and Western Ghats. Less than 30% occur in temperate regions. A detailed study reveals that higher percentage of their occurrence is in moist and dry deciduous forests. Comparatively lesser percentage is in evergreens. Since screening of wild plants for their drug value is being done by various Institutions like National Laboratories, Universities and Pharmacology Institutions it is difficult to estimate the exact number of medicinal plants. Of the 18,000 angiosperms of India about 2,500 are the important sources of medicinal and aromatic chemical compounds. As per the estimation of Kapoor and Mitra (1979) about 540 plant species are being used in various herbal formulations in India. According to Farooqi and Sreeramu (2001) at least 11,00 medicinal plants are used in various systems of alternative medicines in India.

A latest habit wise analysis reveals that highest drug plants are trees (354 species) followed by herbs (344 species). The shrubs (220 species) and climbers (127 species) are less represented. The major families with medicinal properties are Fabaceae (117 species), Poaceae (89 species), Euphorbiaceae (61 species), Asteraceae (57 species), Rubiaceae (48 species), Cucurbitaceae (43 species), Convolvulaceae (36 species), Malvaceae (36 species), and Solanaceae (36 species). A very small proportion of Indian medicinal plants are from lower plants like algae, fungi, lichens, ferns etc. (Anonymous, 1997). While the percentage of algae, fungi and lichens are comparatively less, ferns are employed in Ayurveda, Unani, Homoeopathy and other traditional systems of medicine. According to Manickam (1999) at least 56 fern species are used for various diseases in different parts of India including Andaman and Nicobar Islands. As the majority of the ferns need shade and moisture, they form the ground vegetation in the tropical wet evergreen forests. However, they are also distributed on the mountain ranges of Himalayas and the hilly regions of Southern India.

PRESENT STATUS OF INDIAN MEDICINAL PLANTS

As per the authentic records there are over 6 lakh licensed and registered alternative medicine practitioners in India. They are practicing either Ayurveda, Siddha, Unani and of late Tibetian system of treatment. It is also estimated that nearly 4635 ethnic communities in India are treating human and veterinary health disorders (Anonymous, 1997). In the recent years the Alternative medicines have attracted the people world over and India with oldest heritage has taken several steps for collecting the information on herbal cure. Interestingly a large treasure of herbal knowledge is still with various tribal communities of India. Although many organisations, including NGO's, are trying to document the plants that heal, the inventory is not complete. India still has several tribal communities in various States and the real ethnomedical knowledge is with them (Bhandary, 1995). A systematic documentation of the plants used by such communities for treating various diseases is of immediate necessity. Attempts are being made throughout India to extract the knowledge and properly identify the plants. For instance in Karnataka State Maruthi *et al.* (2000) have listed 30 angiosperm species being used in the drug preparations for curing various skin diseases. The Study was undertaken in various villages of

Davanagere district and information was collected from traditional age-old practitioners. Bhandary et al. (1996) have collected information from the Gaulis and Siddhis of Uttara Kannada District. Information about more than 100 plants has been collected. The findings include more than 50 hitherto unknown medicinal uses of known medicinal plants. Gowda (1997) has collected medicinal properties of plants being used in the formulations by one of the original tribals of Uttara Kannada called the Halakki community. Harsha *et al* (2002) have documented 45 medicinal plant species used by Kunabi community people who are residing in the forests of the same district. Foundation for the Revitalisation of Local Health Traditions (FRLHT) set up at Bangalore is not only collecting the ethnomedicinal informations but is also involved in the conservation of endangered plants in the Southern India. Works on the similar lines are being taken up in the other states also (Singh and Ali, 1994). One of the main drawbacks of the ethnomedicobotany and other alternative medicine is the difficulty in proper identification of the medicinal plants. Names of such plants in Ayurveda formulations are all in Sanskrit. Authentic identification is a challenging task as the names vary from place to place. However, proper effort on this line is put by Government of India, Ministry of Health, Delhi which has documented plants used in important formulations.

Some of the disorders in humans are due to the pathogens like viruses, bacteria, fungi, protozoa etc. With the advancement in Allopathy and other modern medical sciences most of the microbial diseases are being cured or controlled. In recent years, however, some of them have caused a serious threat due to development of resistance to these drugs. A large number of plant extracts have proved their anti microbial properties. This is very much evidenced in the traditional systems of medicine, as for example *Phyllanthus fraternus* is effective in the treatment of jaundice. The database on medicinal plants prepared by Kameswar Rao (2000) lists 771 plant species having antibacterial activity, 263 species having anti fungal properties and 96 species with antiviral activity. This excludes 96 species already being used for jaundice. Amongst the various protozoan diseases, malaria and amoebic dysentery are very serious. Atleast 55 species of angiosperms are found to be effective in curing various protozon diseases. Gastro-intestinal disorder is another severe problem in Indians, especially during summer, caused chiefly by bacteria and viruses. Significant work has been done on the curative properties of plant extracts and at least 946 species are found to be effective in curing gastro-intestinal disorders. Similarly, 197 species are used in dental care, 142 species have anti-inflammatory and astringent effects, 101 species have analgesic effects, 39 species have anti diabetic properties and 37 species are found to be effective in promotion of fertility (Kameswar Rao, 2000).

Another development in the study of medicinal plants in India is the isolation and characterisation of bioactive compounds in the medicinal plants. Most of the traditional medicinal plants have been subjected to test and several commercially important compounds have been successfully isolated. National Institutions like CDRI and CIMAP in Lucknow and various University Laboratories are involved in the isolation and characterisation of useful secondary metabolites. Ravishankar and Ramachandra Rao (2000) have listed 24 such plant-derived medicinally important compounds and their sources. Plants produce the secondary metabolites, mostly as by products, in the metabolic pathway. These may be in the form of phenylpropanoids, alkaloids, terpenoids, quinines or steroids. Although many of these have been fully studies and their chemical nature is understood not all of them can be artificially synthesised. Hence plants are the only sources for them. They may be produced in larger or smaller quantities in various parts of the plants like roots, stems, leaves, flowers, fruits or seeds (Sarma, 1996). It is thus necessary to manipulate the plants for enhanced production of the secondary metabolites and it is here the biotechnology will play a major role. Micropropagation of medicinal plants has been discussed in detail by Naik (1998). It has been found that the secondary metabolites produced by the differentiated tissues *in vitro* many a times is either equivalent or more. One of the important achievements in this line is the development of plant cell culture as a source of large number of secondary metabolites which are produced independent of agro-climatic requirements. The disadvantage of such cultures is the instability of cell lines. The organ cultures like shoot culture and root culture are more stable. However, the root culture is more difficult and the growth is slower. A major break through in the research in this line is the production of hairy root cultures. A bacterium *Agrobacterium rhizogenes* is found to induce hairy roots in large number of host plants and such hairy roots have the ability to grow fast without supply of exogenous auxins. Since their growth in many cases is not light dependant their mass production in various bioreactors has been possible (Ravishankar and Ramachandra Rao, 2000).

CONSERVATION OF MEDICINAL PLANTS

It is estimated that there are over 6 lakh licensed and registered alternative medicine practitioners in India. They include experts in Ayurveda, Siddha, Unani and Tibetan system of medicine. There are nearly 4635 ethnic communities treating human and veterinary health disorders. Apart from this, a number of ayurvedic drug manufacturers are mostly dependent on the raw materials from the nature. The raw materials are roots, bark, wood, stem or the whole plant. According to FRLHT report the highest percentage of plant parts used is the roots (29.6%), followed by whole plant (16.3%), and Bark (13.5%). Fruits constitute for 10.3%, leaves 5.8%, stem 5.5%, flowers 5.2% and rhizomes 4.4%. Lowest percentage is that of wood (2.2%).

The popularity of alternative medicine has led to the depletion of our natural medicinal plant wealth in India. The process is fast increasing for several reasons. Most of the medicinal plants are collected in a destructive manner. About 95% of the collections are from wild and involve most unscientific collection. A large percentage of medicinal plants thus collected go waste due to untimely collection. Majorities of the harvesters, mostly the economically poor, are unaware of the right season and time of collection. The collectors are also not aware of the appropriate stage of the plant growth when the parts of the plant contain the optimum percentage of active principles. They also lack the knowledge of proper collection and storage and are attacked by microorganisms. Manufacturers reject greater percentages of such materials. Medicinal plant trade is also not free from the involvement of the middleman, as there is no coordination between collector and purchaser. There are no organised centres for purchase, storage and disposal of medicinal plants. Owing to habitat degradation, loss and over harvesting the medicinal plants are under various degrees of threat. As per the latest guidelines laid by IUCN around 200 species of medicinal plants have been rated as RET (Rare, Endangered and Threatened) plants. By the end of 20th century 112 plants in South India have been categorised as RET plants. Similarly status of 33 species are critical, 19 species are endangered, 16 species are vulnerable in North India (Anonymous, 1997).

There is immediate necessity of conservation and sustainable use of medicinal plants in India. Looking to the endemic features of most of the rare species it is clear that plant diversity at generic, species and ecosystem level can be conserved for years to come in the nature itself. Medicinal plants have the wide distribution and are confined to few pockets. Discovery of *Myristica* swamps in Uttara Kannada district of Karnataka by Subash Chandran (2000) confirms the existence of many such endemic areas in the unexplored areas of Indian forests, especially in the Western Ghat regions. Identification of such areas and declaring them as reserved plots to make a net work of reserves would form the ideal *in situ* gene banks. They would act as the repositories of wild germ plasm. A net work of 33 such Medicinal Plant Conservation Areas (MPCAS) has been established in South India where inventory and ecological studies are undertaken, apart from conservation and propagation of germ plasm. A proper regulation on wild collection and sustainable harvesting of medicinal plants from nature is the urgent need of present day. Restriction on the collection and trade of drug plants is also inevitable.

Another step in the conservation of medicinal plants is the establishment of herbal gardens in various parts of India to conserve regional plant species. Several Universities, NGO's, Research Foundations, Agriculture Universities and Forest Departments are actively involved in such *ex situ* conservation of medicinal plants. Botanical gardens are no exceptions for this. These are not only educative but are the invaluable repositories of plant species. They can act as the back up pools at times of need. Department of Agriculture Botany and Crop Physiology, JNK Vishwa Vidyalaya, Jabalpur (M.P.) has collected and conserved 1027 species of medicinal and aromatic plants (Tiwari, 2000). It is also growing 48 RET medicinal and aromatic plants. The ethnic park in Tumkur district of Karnataka is the abode of 345 native medicinal plants (Abdul Kareem, 2001). On the similar lines there is need to establish nurseries for supply of economically important medicinal plants for large scale cultivation in the fields. It is also necessary to encourage large-scale cultivation of medicinal plants in preferable non-agriculture lands such as wastelands. In spite of all these it is likely that few species may be lost in future. It is thus appropriate to have the seed banks or gene banks. In view of this Government of India has planned to have 3 such banks at ICAR, New Delhi, CIMAP at Lucknow and TBGRI in Thiruvanathapuram (Anonymous, 1997).

Medicinal plants are therefore now recognised as an important component of natural resources. For all practical purposes medicinal plants are on par with any other economically important plant species, whether in nature or under domestication. In a meeting held during Jan. 1998, The Indian Subcontinent Plant Specialist Group identified *Abrus precatorius, Adhatoda vasica, Centella asiatica, Costus speciosus, Gloriosa superba, Rauwolfia serpentina, Saraca asoca, Strebles asper, Tribulus terrestris and Withania somnifera* for detailed study and protection (Kameswar Rao, 2000).

CULTIVATION OF MEDICINAL PLANTS

With the current growth rate, the world population is expected to double by 2020. Apart from supply of food, proper medicines for different ailments is of great concern. Possibly due to the belief that the alternative medicines are without side effects, plant based drugs would be very popular in future. At present there are at least 121 major plant drugs, whose structure is known, but they can not be synthetically produced. Of these 45 are from tropical species. There are at least 35 major medicinal plants that can be cultivated in India. Thus commercial production of medicinal plants has a greater scope in India. The variety of agro climatic conditions in India make it a suitable place for growing medicinal plants requiring different environmental conditions. Simple plants like *Andrographis paniculata, Bacopa monnieri, Boerhaavia diffusa, Eclipta alba, Swertia chirayata, Valeriana wallichii* are few of many such plants having great demand. What is needed at present is the improved agricultural practices and proper processing technologies. The Chinese herb *Artemisia annua* is a good source of artesininin, a highly potent antimalaria drug. This plant has been already introduced in India by CIMAP. The drug is found to be very effective in killing malarial parasites which are resistant to even chloroquine phosphate. *Taxus baccata* is another such commercially important plant for the isolation of taxol, an anticancerous drug. Farooqi et al (1999) have given the information about cultivation of 29 medicinal plants. A detailed aspect of 36 medicinal plants giving information about cultivation method, manure requirement, interculture, pests and diseases, harvesting and drying, grading and yield is given by Farooqi and Sreeramu (2001).

Medicinal plants are not free from diseases, nematodes and insect pests. Selection and introduction of any species into agriculture practice is the process of domestication of the strain, which makes it more susceptible. 'Phytoplasma' a wall-less prokaryote is known to cause hundreds of plant diseases. The most common diseases are the periwinkle yellows of *Catharanthus roseus* and Sandal spike disease of *Santalum album*. Hiruki and Wang (2000) have discussed the heat therapy method developed for the inactivation of phytoplasmas in different organs of many plant species. The method involves tissue culturing of the shoot tips of required plant in MS medium at 35°C up to 5 weeks, followed by meristem culture at 25°C resulting in disease free plantlets. Obviously medicinal plants can not be sprayed with synthetic chemicals and therefore, alternative natural or herbal extracts are to be developed. The weedicide nature of *Artemisia annua* extract, nematicidal effect of *Tagetes minuta* extract, insecticide property of *Azadirachta indica* extract etc. are giving promising results. Genetically engineered transgenic insect resistant varieties would appear to be the greatest solution in this line (Sushil Kumar et al., 1997).

Loss of biodiversity and the medicinal plants, due to anthropogenic pressure, be the major concern. Inventory of vulnerable biodiversity spots has to be accelerated and should fall under the *in situ* conservation. There should be check on the wild and unscientific collections of drug plants. Creation of public awareness is an urgent necessity. Our knowledge of ethnomedicine is still not complete. There is a greater scope for the preparation of database on the regional medicinal plants in India. One of the major steps to minimise the wild collection of medicinal plants is to bring them under cultivation. During 1995-1996 India exported medicinal plant commodities worth 189 crores (Farooqi and Sreeramu, 2001). India has the greatest potential for the cultivation of medicinal plants. Because of its varied agro-climatic conditions, selected high yielding varieties can come under modern agriculture. A large number of plants are still to be screened for the properties of their bioactive compounds. Though biotechnology has entered the field of medicinal plants much has to be done in India, especially in field of hairy root cultures. Since India has the greatest heritage of medicinal plant and also good number of endemic plants, many plants can be tried on this line for commercial exploitation (Kureja et al., 1997). A lot is to be done to bridge the gap between ethnomedical knowledge and scientific knowledge.

REFERENCES

Abdul Kareem, M., Hiremath, S., Nimbalkar, S.D. and Suresh, D.R. 2001. Ethnomedicinal parks-prefectures of conservation. *Amruth*, 5(5):1-19.

Anonymous, 1997. *Medicinal Plants of India. Guidelines for National Policy and Conservation Programmes.* Conservation Science Division, Ministry of Environ. and Forests. Govt. of India, New Delhi.

Bhandary, M.J., Chandrashekar, K.R. and Kaveriappa, K.M. 1995. Medical ethnobotany of the Siddis of Uttara Kannada District, Karnataka, India.

Bhandary, M.J., Chandrashekar, K.R. and Kaveriappa, K.M. 1996. Ethnobotany of Gowlis of Uttara Kannada District, Karnataka. *J. Econ. Taxon. Bot. Add. Ser.* 12:244-249.

Champion, H.G. and Seth, S.K. 1968. *A revised survey of forest types in India.* Delhi.

Chopra, R.N., Nayar, S.L. and Chopra, I.C. 1956.*Glossary of Indian medicinal plants*.CSIR, New Delhi.

Dey, A.C. 1980. *Indian medicinal plants used in Ayurvedic preparations.* Bishen Singh Mahendra Pal Singh, Dehra Dun.

Farooqi, A.A., Khan M.M. and Vasundhara, M. 1999. *Production technology of medicinal and aromatic crops.* Natural Remedies Pvt. Ltd. Bangalore.

Farooqi, A.A. and Sreeramu, B.S. 2001. *Cultivation of Medicinal and Aromatic Crops.* Universities Press (India) Ltd., Hyderabad.

Gowda, T.S. 1997. *Uttara Kannadada Halakki Okkaligaru*.(in Kannada) Naisargika Sampanmula Samrakshana Samsthe, Bangalore.

Harsha, V.H., Hebbar, S.S., Hegde, G.R. and Shripathi, V. 2002. Ethnomedical knowledge of plants used by Kunabi tribe of Karnataka in India. *Fitoterapia* (in press)

Hiruki, C. and Wang, K. 2000. Phytoplasma disease of medicinal and aromatic plants and their heat therapy. *Proc. Ind. Phyt. Path. Soc. New Delhi, Golden Jubilee Volume*:1206-1213.

Hughes, C.C. 1968. Ethnomedicine. In *International Encyclopedia of Social Scineces.* Macmillan, New York, Vol. 10, p. 99. *J. Ethnopharmacol.* 47:149-158.

Kameswar Rao, C. 2000. *Database of Medicinal Plants.* KSST for the Deptt. Of Forests, Environ. And Ecol. Govt. of Karnataka, Bangalore. pp.458.

Kapoor, S.L. and Mitra, R. 1979. *Herbal drugs in Indian pharmaceutical industry*, Lucknow.

Kirtikar, K.R. and Basu, B.D. 1918. *Indian Medicinal Plants.* Allahabad.

Kureja, A.K., Zehra, M., Sultana, A. and Banerjee, S. 1997. Hairy roots of medicinal plants: Potential and prospects. In *Biotechnologicl Applications of Plant Tissue Culture.* Ravishankar, g.A. and Venkataraman, L.V. (Eds.) Oxford & IBH Pub. Co. Pvt. Ltd. N. Delhi. pp. 74-90.

Manickam, V.S. 1999. Medicinal ferns of India. *Amruth*, 3(1):3-9.

Maruthi, K.R., Krishna, V., Manjunatha, B.K. and Nagaraja, Y.P. 2000. Traditional medicinal plants of Davanagere District, Karnataka with reference to cure of skin diseases. *Environment and Ecology*, 18(2):441-446.

Mhasker, K.S., Blatter, E. and Cain, J.K. 2000. *Kirtikar and Basu's illustrated medicinal plants.* 3rd ed., 11 vols. Sri Satguru Publishers, Delhi.

Nadkarni, KM 1954. *Indian Materia Medica* Vol.1 & 2. Popular Prakashan, Bombay.

Naik, G.R. 1998. Micropropagation studies in medicinal and aromatic plants. In: Khan, I.A. and Khanum, A., Eds. *Role of Biotechnology in Medicinal and Aromatic Plants*. Hyderabad; Ukaz Publications. pp.50-56.

Nayar, M.P. 1980. Endemism and patterns of distribution of endemic genera (angiosperms). *J. Econ. Tax. Bot*. 1:99-110.

Rao,R.R. 1994. *Biodiversity in India (Floristic Aspects)*. Bishen Singh Mahendra Pal Singh, Dehra Dun.

Ravishankar, G.A. and Ramachandra Rao, S. 2000. Biotechnological production of phyto-pharmaceuticals. *J. Biochem. Mol. Biol. and Biophys*. 4:73-102.

Sarma, O.P. 1996. *Hill's Economic Botany*. Tata McGraw-Hill Publishing Co. Ltd. New Delhi.

Singh, V.K. and Ali, Z.A. 1994. Folk medicines in primary health care: common plants used for the treatment of fevers in India. *Fitoterapia*, 65 (1):68-74.

Subash Chandran, M.D. 2000. On the conservation of biodiversity of the Western Ghats of Karnataka. Proc. Workshop on Karnataka Biodiversity Information System, Govt. of Karnataka, Department of Forest, Ecology and Environment and K.S.C.S.T.:1-12.

Sushil Kumar, Shukla, Y.N., Lavania, U.C., Shok Sharma and Singh, A.K. 1997. Medicinal and aromatic plants: Prospects for India. *J. Medicinal and Aromatic Plant Sciences*, 19:361-365.

Tiwari, J.P. 2000. *Biodiversity of Medicinal and Aromatic Plants conserved in Herbal gardens*. Dept. Agri. Bot. Crop Physiol., JNKVV, Jabalpur.

—

5

Ethnobotany and Tropical Forest Conservation

*Veena Chandra

INTRODUCTION

The rate of extinction of species is increasing at an alarming rate. Many temperate life forms are on the verge of extinction. About half of world's forests are in the tropics and the majority of the world's threatened species inhabit these forests. Tropical region are best known for their rain forests, streamy jungles with towering trees, epiphytes, and dense under stories of smaller trees, shrubs and vines. In addition to rain forests, there are mangroves, moist forests, dry forests, and savannas. In recent times, however, they have been cut at a rapid rate. It has been estimated that 17 million ha of tropical forest are disappearing annually. Between 1985 to 1990, 210 million acres (85 million ha) of tropical forests were destroyed. Covering less than 10% of the earth's surface, these tropical forests are believed to contain 50% of the world's species (Wilson, 1985). They are the most complex species rich ecosystem in the world. For many years, concern has been expressed about the rapid and continuing loss of these forests. Removal of natural vegetation and over-cropping of trees are causing degradation in approximately 2 billion ha of these tropical forests. The unpleasant conclusion is that the human race is causing one of the first major reductions of global vascular plant diversity since the origin of life. The extinction of plants may have a more serious impact on human welfare than the disappearance of animals.

The staggering amount of species diversity concentrated in tropical regions is best demonstrated with following example. For instance, a district in the Gangetic plain has 500-600 species while even a smaller area in the northeast region of India has over 1000 species.

REASONS FOR CONSERVATION

Why should any one care whether or not species become extinct? To the layman, our way of life seems wholly dependent on a few domesticated species that appear to be thriving–wheat,

* Extension Division, Forest Research Institute, Dehradun, India-248006.

rice, corn, cattle and so on. Of what possible consequence could the extinction of some poorly known tropical species be for the progress? Several different arguments have been put forward as to why species should be protected.

1. **Aesthetic:** Basically stated, the aesthetic argument claims that all nature has an inherent beauty and must be protected as part of our global heritage. Bird watching and wild life tourism is one of the major sources of foreign exchange in some countries. Nature has certainly served as an inspiration to artists through the ages. Many plants of tropical forests find uses in homes and gardens: ferns, palms, split-leaf philodendron, marantas, bromeliads, and orchids, to name a few. The horticulture and floriculture industries are dependent on wild forms.

2. **Ethical:** The ethical argument claims that no one species has the right to wipe out another. Even aboriginal societies have taboos against the over hunting of certain species. They associate plants and animals with Gods and Goddesses as their symbols or vehicles (vahan), or favourites, and so are biological entities for various plants. Western literature is also full of such associations. These associations have one direct relevance to biodiversity, that of recalling and popularizing these mythological associations for saving these species or ecosystems from destruction. Cultural associations are witnessed with specific trees.

3. **Diversity and Stability:** The argument for diversity and stability claims that ecosystems are complex mechanisms, the continued existence of which is dependent on the presence of certain key species. This argument is particularly important in the tropics, where plant-animal interactions play a much greater role in ecosystem function than in temperate zone.

4. **Scientific Value:** This argument claims that species must be protected and studied for their scientific value even those species that have little scientific value.

5. **Utilitarian:** This claims that species must be preserved because of their utility to humankind. Man simply cannot survive without plants and animals. Our civilization depends on forests for everything from food to medicine to the oxygen we breathe. To allow dwindling species to become extinct would be shortsighted, as there are numerous cases in which such organisms have turned out to be very important.

Economic Benefits of Conservation: All forests have both economic and ecological value, but tropical forests covering less than 10 percent of the earth's land area, contain the vast majority of the world's plant and animal genetic resources. The potential utility of fauna and flora is not yet fully known or rather little known. New species are being discovered everyday and there can be little doubt that many "super species" are waiting to be discovered. When Aublet described the rubber (*Hevea*) tree in 1775, who could have predicted that its latex would be a major factor in bringing about the Industrial Revolution (Schultes 1977)? As Hugh Synge put "Plant conservation, with its basis in plant systematics and taxonomy, has tended to give equal weight to each threatened plant rather than emphasize the useful ones......... A change in emphasis is needed with the far more effort on wild plants used by people, whether or not these plants enter the monetory economy."

More emphasis should be given on the economic benefits that are gained if a species are protected and rationally utilized. This approach has particular appeal in developing countries where economic conditions are often difficult.

Animals, which comprise only a small fraction of total tropical forest biomass, are forest products of paramount importance to local people who utilize wild game as sources of food and revenue. With adequate management, sustainable productivity of tropical wildlife could become ecologically sound economic alternative in tropical countries.

Protecting Tropical Forests

In most tropical countries, "forest utilization" has simply meant "timber removal". The process of timber extraction demands economies of scale, which usually require the use of heavy machinery. Skidding and road building can cause tremendous damaging both to fragile tropical soils and to the residual stand. Felling can result in the elimination of the established seedling crop, which is required to form the new stand. Furthermore, road damage associated with logging in tropical areas can lead to increased sediment load in streams, accelerated soil erosion, decline in water quality and uniformity of flow, as well as impoverishment of essential soil micro-organisms and nutrients (Gilmour, 1971; Auchter, 1978). Tropical forests have been utilized as nonrenewable resources, usually being turned into crop or pastureland that is only good for two or three years in many areas (Guppy 1984).

Ashton and Plotkin (1982) suggested that more attention be paid to nontimber products such as medicine, fibres, oils, wax, latex and tannins. Many of the so-called secondary forest products already participate in international commerce, including rattan (*Calamus* species) from several tropical Asian countries ($1.2 billion per year). In India, latex, forage, pesticides, tannins, dyes, resins, spices, oils, medicines, and other nonwood products extracted from the forests are estimated to be worth over $ 1 billion a year (UNESCO, 1978).

Many medicines are derived from plants found in tropical rain forests. Quinine, an ancient drug used for malaria, curare, an anesthetic and muscle relaxant used in surgery, and rosy periwinkle, a treatment for Hodgkin's disease and leukemia are some of the noteworthy medicines from forests.

To protect tropical ecosystems for future generations, we must demonstrate the potential value of nontimber plant products especially the medicinal plants. The most cost-effective method for accomplishing this is through the science of ethnobotany.

Ethnobotany and Conservation

Ethno botany is defined as the study of the relationship between people and plants and most commonly refers to the study of indigenous uses of plants. In other words, it is the marriage between cultural anthropology and botany, investigating the roles of plants as medicine, nourishment, natural resources or gateways to the Gods. It is considered a relatively young field of study.

Ethno botany came into being when the earliest man observed animals eating certain plants, and he gathered and hunted for his food and for heeling his wounds, or sought cover from rains and hailstorm. This knowledge got wider use and success in experiments on humans and led to our recognized foods and medicines.

Since the stone age, (which is really a complete misnomer and should instead be called 'the plant-age', since many artifacts were fashioned from plants than from stone) human beings of all cultures and races have been incredibly innovative with regard to utilizing plant materials for their various needs. Plants offer medicines for all types of diseases, from the common cold to cancer. Phytochemical compounds have played a significant role in the manufacture of many important drugs that have saved thousands of lives over the centuries, and the hunt for new plant medicines is still on. In fact, these days it is a race against time. As the forces of development threaten habitats of rich biodiversity, scientists all over the world are scrambling to identify new plant species and learn about their traditional uses before they are lost forever.

In fact, the knowledge regarding the uses of plants is one of the primary human concerns and has been practiced by all cultures for thousands of years - its just that it wasn't called 'Ethno botany' then. In the course of nomadic roaming this knowledge was exchanged with neighboring tribes, friends and foes. Thus, the plant knowledge has been passed around the world since the beginning of time, and frequently the actual plants themselves spread along with it. Looking at the recent history, there are numerous records of ethnobotanical fieldtrips and acquisitions as well as detailed descriptions of plants and their uses dating to pre-classical times. The Arabs for example, had a keen interest in plants. On their expeditions to the East they collected information about local plant uses and brought as many new plants back as they had taken with them. The same is true for every conquering nation that has ventured into foreign and unchartered territory. The Romans, on their crusade through Europe, actively sought out local 'herbalists' and often utilized their knowledge by enlisting them for service to their troops. Plants were big business in the Old World and many of the ancient trade routes were in fact established for trade in plant-products, such as Frankincense, exotic spices or silk .

The Spanish conquistadores also took detailed records of the plants used by the indigenous people they encountered in the New World. In fact, Columbus 'discovery' of the Americas was an 'accidental', side effect of searching for a quicker way to India and the profit promising allure of the Southeast Asia. The plants brought by him and others have become indispensable items of the western diet.

Ethnobotany is thus really one of the oldest fields of human inquiry. It recently acquired recognition as a science. Scientific study is characterized by a 'subject/object' relationship in which the observer is required to detach him or herself from the observed. This really signifies a move away from actual interaction with plants, perhaps, because they have become less significant, i.e. less present in our consciousness in this technological age. They have been pushed out of the sphere of direct experience (i.e. personal relationship) into the realm of external phenomena, where they can be studied objectively and rationally, labeled and given scientific names, dissected and analyzed in the search of 'active compounds', which might subsequently be exploited as new wonder drugs.

Empirical investigation on the other hand, is the exact opposite approach. The investigator must be involved with the object of their study, fine-tune themselves to subtle impressions and effects, which scientists consider completely subjective. Traditional healers of all cultures follow this approach to learning about plants. Scientists find it baffling how these traditional healers, without the use of elaborate equipment or any formal schooling can be so adept at identifying plants, discern their uses and find effective remedies among the thousands of plants that surround them. Not only that, they manage to combine several species that may work together synergistically, but not individually, or remedies that cure when given in one dose but kill when administered in another. 'Trial and error', the scientist postulates, but that is merely an excuse for their own ignorance and inability to comprehend. The shaman would hardly be able to find remedies by sitting in his hut and postulating a hypothesis about the plants around them, nor would he waste his energy by wandering about, randomly munching potentially lethal plants or insects.

Ethnobotanists and other scientists continue to be baffled. But while they can't explain the phenomena, they make an effort to record the results-and then go and test 'their discoveries' according to an entirely different set of rules within their labs. And although Ethnobotanists don't usually look down at their native teachers who might have been their guides and shown them their sacred plants, there is nevertheless an underlying cultural superiority complex at work. While researchers might have a deep respect and admiration for their 'medicine men' they still consider the tools of modern science as superior and their own cosmology and scientific framework as 'the real reality'.

Thus, the stories and teachings that belong to each plant are considered simply as 'folklore', and disregarded or perhaps noted for the sake of curiosity in their notebook, only to be promptly forgotten. But it is precisely these anecdotes, stories, mythologies, rituals and communications with the gods that hold real keys to the character and meaning of a plant. The same source from which this knowledge is derived can also offer clues to the nature of a patient's disease, within a spiritual context. However, the task of modern medicine is more akin to the process of fixing a car than to counseling a soul. Ethnobotanists are rarely healers. They are just in the business of prospecting for potentially useful plants. They don't need to know anything about healing and don't have to share the shaman's concern for the balance or imbalance of a patients soul. This doesn't mean to say that scientists should throw away their instruments and quit their scientific method of inquiry, rather, it is a question of using a balanced approach between analytical and synergistic modes of thinking in order to understand plants as well as people as total beings - and not just as collections of interesting compounds or physiological biomachines. However, understanding what the plant-lore conveys is quite another matter. It requires a study of the cosmology and beliefs of the people who tell the stories and the ability to go beyond the boundaries of so-called 'solid realities' - more a job for anthropologists than botanists or chemists. There are few rare individuals who have learned to combine all three skills and understand ethnobotany for what it is - **the study of the relationship between people and plants**.

To illustrate the point, consider this: Some indigenous people don't even have a word for the forest or the environment, but regard them as an extension of themselves. To say, as we do, 'I am going into the forest' would be as absurd to them as to say, 'I am going into my skin', - they are already in it and a part of it. We have lost this ability to relate directly with the nature that surrounds us. We have become alienated and are suffering a mass-psychosis because of it. Perhaps Ethnobotany can help to heal the dichotomy between spirit and matter that is afflicting the 'civilized world' and provide a lifeline through which we can again begin to relate to nature and value the gifts of 'Mother Earth'.

Ethnobotany is the study of total natural relationship or traditional relationship between people and the plant wealth around them. In the recent past there has been an increasing interest in the scientific study of man-plant interaction in natural environment. The indigenous people are designated as aboriginals, natives, adivasi (original settlers), vanyajati (forest casteman), Adimjati (primitive caste), janjati (folk communities) and tribals. More than 300 million indigenous peoples live in more than 70 countries in habitats ranging from the Arctic to the rainforests of Asia and South America. In India alone there are 427 tribal communities. According to 1981 census, schedule tribe population in the country was 53.8 million, constituting about 7.5 percent of the total population. It is estimated that the predominant tribal areas comprise of about 15 percent of the total geographical area of the country.

Many tribal societies in India live in complete harmony with natural surroundings. They have their own notions, beliefs and way of conserving the plants. They harvest the natural resources of their surroundings in a moderate manner to maintain a sustained supply for their utilization. They have their own tradition of cultivation. The so called reluctance of the present day tribals in divulging the uses of their plants to field workers is mainly due to their conservational attitude towards their plant resources. It is observed that many tribals of West Bengal and Orissa have some traditional timetable for cutting the wood even for fuel. They have some dates when there will be a festival, which generally relates to the ripening of fruits of a particular species. It is after this festival that they can cut the branches of that species for fuel.

Conservation had been an essential part of forestry operations in ancient Indian tradition. It is believed that its root goes back to the pre-agricultural period. There are references in ancient literature of worship of herbs as well as tree species and also of certain spots in the forests, which are referred to as sacred trees or sacred groves. Many interesting observations have been published recently on these aspects and the knowledge of ethnobotany in environmental planning has been appreciated at the international level. It is worth mentioning that in "Manusmriti" cutting of trees has been treated as an offence and the penalty for the same has also been recommended. In 'Arthashastra' some system of management of forests is also mentioned.

Over the years indigenous people have developed innumerable arts, crafts and technologies. Their cultures, economies and identities are inextricably tied to their traditional land and resources. Forests have sustained the cultural, spiritual and economic needs of many of the indigenous people. Deforestation, semi-modernization, transmigrations and colonization have threatened the survival of traditional cultures. Every year, cattle ranching, logging, road building, hydroelectric development, mining, dams and farming destroy some 20 million hectares of tropical forest. Saving the indigenous people has become part of the campaigns to save the forests. The current devastation of indigenous people and the ecological systems that they have conserved, managed and intimately known for millennia, require new and drastic steps to reorient world priorities. The Convention on Biological Diversity signed by 167 States and the European Community at Rio on June 5 1992 accepts the idea that the first beneficiaries of the conservation and sustainable use of wild plant species should be the indigenous people whose traditional knowledge and respect for those resources has preserved them for centuries.

Ethnobotany offers a very effective approach to tropical forest conservation, since it may provide a wealth of data on nontimber products, which can be collected in a nondestructive manner. Commodities might thus be extracted from the forest with minimal ecological and/or environmental damage, yet provide some incentive for the conservation and rational utilization of the forest.

REFERENCES

Ashton, P. and Plotkin, M. 1982 *Technologies to Increase Production from Primary Tropical Forests and Woodlands*. Report to Office of Technology Assessment, U.S. Congress, Washington,DC.

Auchter, R.S. 1978. Summation of Conference, *Proceedings of a Conference on Improved Utilization of Tropical Forests. USDA, Madison,WI.*

Gilmour, D. 1971. Effects of logging on streamflow and sedimentation in North Queensland Rain Forest Catchment. *Commonwealth Forestry Review* SO(1)143:39-48.

Guppy, N. 1984 Tropical deforestation: A global view. Foreign Affairs. 928-965.

Schultes, R.E. 1977. The odyssey of the cultivated rubber tree. Endeavor, n.s., 1(3&4):133 – 138.

UNESCO, 1978. *Tropical forest Ecosystems: A state of knowledge Report Prepared by UNESCO/UNEP/FAO.* Paris: UNESCO.

Wilson, E.O. 1985. The biological diversity crisis. *Bioscience* 55(11): 700 – 706.

6

Mining Scenario in India-Environmental Problems and Prospects

*Prafulla Soni

Mining Scenario

India with its huge geographical area is fairly well endowed with mineral resources. It is the world's largest producer of Mica block sand splittings and is also a major producer of Chromite, Iron ore, Manganese ore, Barytes and Kyanite. The mineral sector with an outlay of about Rs. 98.2 billion in 1988-89 has grown at an average rate of 7.3% during the past decade. Of the 52 minerals produced, India is by and large self sufficient in 30 which include Coal, Bauxite, Iron and Manganese ore.

The mineral sector, which was hitherto reserved for the public sector, has now been opened for exploitation by the private sector. The minerals now open for investment by the private sector are iron ore, manganese, chrome, sulfur, gold, diamond, copper, lead, zinc, molybdenum, tungsten, nickel and platinum group of minerals. The revised policy has removed any restriction on equity holding by foreign nationals in a mining company. Mineral and metal processing units which wish to develop captive mines to secure assured supplies of raw materials will be allowed foreign equity participation in the manner and to the extent already permitted to such processing units.

The Mines & Minerals (Regulation & Development) bill recognises the need for providing greater stability of tenure to leaseholders. As such, for the first time, a minimum period of lease of 20 years has been made mandatory. The maximum period for grant of lease has been increased to 30 years. There is a provision of renewal of a mining lease for a period of 20 years. Further renewals are possible with approval of the Central Government. These amendments would be of particular benefit to the captive coal mines in the Power Sector, where a number of projects

* Ecology & Environment Division Forest Research Institute, Dehradun INDIA 248006

have been approved, and also to other mines attached to mineral processing industries like steel, cement etc. where an assured supply of raw material is essential. The period for which prospecting license can be granted has been increased to three years and these licenses can be renewed at the discretion of the State Governments so that the total period does not exceed five years. The bill has deleted 15 minerals hitherto listed in the First Schedule of the Act requiring prior approval of the Central Government before the State Government could grant a prospecting license or a mining lease. The minerals so deleted are Apatite and phosphatic ores, Barytes, Dolomite, Gypsum, Kyanite, Magnesite, Molybdenum, Nickel, Platinum and other precious metals. Sillimanite, Silver, Sulfur and its ores, Tin, Tungsten and Vanadium Ore. With this deletion (excepting the atomic minerals and fuel minerals) there would only be 11 minerals left for which the State Government would require prior approvals of the Central Government. These are Asbestos, Bauxite, Chrome ore, Copper Ore, Gold, Iron ore, Lead, Limestone (except where it is used in Kilns for the manufacture of lime as building material) Manganese ore, Precious stones and Zinc.

With increasing discovery of mineral deposits by application of modern exploration techniques, world grade mines are expected to be opened and constructed to produce minerals at costs, which are internationally competitive. This would help meet not only the needs of the domestic market but also provide sizable exports. The opening up of the Mining sector has evinced a keen response from countries like Australia, Canada, USA, South Africa, etc. It is expected that exploration and exploitation activities in respect of the 13 minerals now thrown open for exploitation by the private sector would be accelerated.

Nearly one million hectare of land in India is presently under the stress of mining activities of various kinds. Much larger area is disturbed by other activities associated with mining. There are approximately 9340 mining leases covering this area (Indian Mineral Year Book, 1999). However, the working mines in 2000 were only 8996 (table 1). This excludes coal and other minor minerals.

Table—1. Mineral wise distribution of mining leases (as on 31.3.2000)

Mineral code	*Mineral*	*No. of leases*	*Lease area (ha.)*
1	**2**	**3**	**4**
01	Agate	9	390.19
65	Amethyst	7	110.94
03	Apatite	1	1040.07
04	Asbestos	83	5306.71
05	Ball clay	37	1778.97
06	Barytes	150	4017.73
07	Bauxite	349	32743.21
08	Calcareous sand	15	269.64
9	Calcite	70	1687.56

(Contd...)

1	2	3	4
10	Chalk	150	638.68
67	China clay	551	49761.73
11	Chromite	36	9603.30
12	Clay (others)	106	2110.02
14	Copper ore	17	10756.51
16	Corundum	27	891.45
17	Diamond	2	576.51
18	Diaspore	23	444.53
19	Dolomite	367	10222.41
20	Emerald	1	129.50
69	Epidote	1	46.68
21	Felsite	7	195.69
22	Felspar	198	7499.52
23	Fireclay	435	23225.99
24	Fluorite	20	2566.18
26	Garnet	18	569.37
27	Gold	10	7441.99
28	Graphite	181	6510.90
29	Gypsum	187	28201.35
30	Iron ore	630	104817.89
31	Jasper	21	1212.86
32	Kaolin	4	679.43
33	Kyanite	25	3307.12
75	Laterite	100	1442.46
34	Lead & Zinc ore	12	8221.35
35	Lignite	9	7406.73
36	Lime kankar	34	1713.07
37	Limeshell	54	7762.21
38	Limestone	1861	146984.21
39	Magnesite	31	4301.10
40	Manganese ore	452	41885.65
41	Mica	325	16007.17
42	Moulding sand	87	2154.80
44	Ochre	158	4449.73

(Contd...)

1	2	3	4
72	Opal	1	200.00
81	Perilite	1	144.65
46	Phosphorite	10	3737.58
47	Pyrite	2	847.50
48	Pyrophyllite	85	2515.08
49	Quartz	681	14724.79
50	Quartzite	59	1424.88
	Rock phosphate	1	5.00
51	Rock salt	2	44.24
52	Sand (others)	87	7900.24
77	Shale	24	705.57
54	Silica sand	553	26158.78
55	Sillimanite	21	2681.53
56	Slate	1098.42	
58	Steatite	544	296721.21
60	Tin	2	57.12
78	Tourmaline	5	215.15
61	Tungsten	5	10.14
79	Vanadium	1	1346.80
62	Vermiculite	19	335.21
63	Wollastonite	1	101.80
	Total	**8996**	**655017.90**

Environmental Impacts and Impact Assessment

It is a paradox that mineral rich areas are also rich in forest resources. Therefore during the process of mining the first impact is on the forest vegetation. Although there may be minerals present in grasslands and pastures too, an overview of mineral resource distribution and forests in the country clearly shows that most of the minerals are present in areas where good forests happen to occur. Dhar (1990) reported that mining activities in India involve 1200 ha. of forest area every year. Prior to Forest (Conservation) Act, 1980 an area of 7896073.67 ha. of forest land was diverted to non forest use by various states/ UT governments during the period 1951 to 1980. This diversion works out to be 1.5 lakh ha. per annum. It is also felt that environmental losses on account of diversion of some forestlands were far more than the economic gains made from the projects for which diversions were made. To check this indiscriminate diversion of forestlands to non-forestry purposes, the Forest Conservation Act, 1980 was brought into force with effect from 25.10.1980. With the result, diversion rate went down to 15,500 ha per annum at present as against about 1,50,000 ha/annum between 1950 and 1980.Table 2 gives the Year wise details of forestland diverted for nonforestry purposes. During 1991-1995,about 9% of the diverted forest area was used for mining purposes (NFAP-India, 1999). A condition of compensatory afforestation was also stipulated in all the cases where Government of India has agreed to a proposal for diversion of forestland (Anon., 1987). Table 3 shows the status of various project proposals received, up to May'88, by Department of Environment and Forests for environmental approval. As per State of Forest Report there is no substantial decrease in forest area for non-forestry purposes including mining.

Table—2. Year wise details of forestland diverted for non-forestry purpose

Year	*Forest land diverted (in hectares)*
1980	Nil
1981	2672.64
1982	3246.54
1983	5702.01
1984	7837.59
1985	30608.07
1986	11963.11
1987	4822.30
Total	**7896073.67**

Table—3. Status of various project proposals received for clearance under Forest (Conservation) Act, 1980

Type of project	*Referred*	*Approved*	*Rejected*	*Pending*
River Valley Project	498	256	225	17
Thermal Power Project	191	129	38	24
Mining Projects	181	71	68	49
Industrial Projects	134	68	44	22
Atomic Power Projects	3	2	–	1
Ports, Harbors, Human Settlements,Tourism Development etc.	50	42	3	5
Total	**1064**	**568**	**378**	**118**

Once a promising mineral deposit has been identified, three detailed analyses are required: (i) a feasibility study, (ii)an environmental impact statement, and (iii) a proposed reclamation plan. The feasibility study determines if a proposed mine is technically possible and commercially viable. This study presents all of the technical information related to the development and operation of the mine. The feasibility study must be extremely thorough because it is used to assess the commercial viability of the mine. A decision is then made to proceed, postpone or cancel mine development.

The environmental impact statement is an analysis of the environmental and related socio-economic impacts of a projected mine. Depending on the results of these studies, further work can proceed only after all the required permits, which determine the terms under which an environmentally responsible mine may be developed, have been obtained.

The reclamation plan presents an action plan, to ensure that the site is reclaimed after mine operations cease. Restoration usually includes: demolition of structures, sealing of mine openings, regrading to approximate pre-development profile, revegetation for stabilization, water treatment, etc. Costs for the restoration are projected and a financial plan, designed to be implemented throughout the life of the project, is prepared. The proper management of environmental impacts is essential to modern mining operations. In order to reach that goal, activities such as environmental education, waste reduction and recycling, and monitoring and reporting, are essentially required to be promoted by the mining industry.

- *Education:* Companies actively promote environmental awareness. In addition, they train managerial and technical personnel in the environmental aspects of mineral development.

- *Waste Reduction and Recycling:* The minimization and treatment of wastes is the major environmental challenge facing mining operations. This challenge addresses the protection of the land, air and water.

- *Land:* In reclaiming open pits, overburden is piled adjacent to the pit, contoured, covered with topsoil and seeded with native species plants. The pit is allowed to fill up with water. In reclaiming strip mines, overburden is returned to the open cut after the mineral has been removed, and the land restored to pre-development conditions. In-situ mining is conducted from the surface. Some ore bodies containing metals (more commonly those containing copper and uranium) can be mined in situ by treating them with a solution that dissolves the desired metals. The solution is then pumped and processed above ground. With this method, low-grade deep deposits can be exploited economically with minimal environmental impact. In addition, personnel are isolated from broken ore and other radiation hazards.

- *Air:* All major mining operations in Canada have succeeded, by modifying operations, in considerably reducing air emissions. The Mining Association of Canada reports that 13 industry leaders have pledged to reduce discharges of 12 toxic air pollutants by the turn of the century. These companies, members of the ARET (Accelerated Reduction/Elimination of Toxics) program, report that they have reduced toxic air emissions by 43% since 1988 and plan to reduce them by a total of 71% by the year 2000. The coal industry is also committed to reducing air emissions. Three new coal-fired electrical-generating stations: Stony Plain, Alberta; Estevan, Saskatchewan, Point Aconi and Nova Scotia, are among the cleanest in the world.

- *Water:* Water from the tailing ponds is directed to the mill instead of continually drawing fresh water from nearby lakes and rivers. Used water is treated and neutralized before being returned to natural drainage basins.

- *Monitoring and Reporting:* Monitoring systems are designed to assess environmental performance at mining operations and to take corrective action when deviations are noted. Most mining companies include a discussion of environmental issues in their annual reports to keep the public informed about their environmental performance. The impact of extraction of minerals is reflected not only on the natural resources but the human population also gets affected through disruption of biogeochemical cycles at large and carbon cycle in particular. Ecological intervention to restore these mined wastelands on sustainable basis is therefore, urgently required. The conventional practice to restore these areas has been through planting of trees without giving a thought to the structure and functioning of the ecosystem. Understanding the site-specific ecology, including plant succession and soil development is crucial for expedient restoration on a sustainable basis. Revegetation of the spoils to restore the pre-mining ecological equilibrium or to return such areas to ecologically acceptable post-mining land use on a sustainable basis is a challenging problem.

During the first stage of mining activities that is the exploration stage, activities such as prospecting, drilling, and trenching, are undertaken in order to locate mineral deposits of economic interest. With proper planning, these activities can be carried out with minimal disturbance to land, water, vegetation and wildlife habitat.

Wildlife and Vegetation: Ecosystems are studied over a period of time to provide information on conditions before development;

Water: Both water use in the area and any potential water supply problems, are studied and analyzed. Streams, rivers and underground water sources are carefully documented;

Air Quality and Climate: Meteorological and air-quality conditions are documented;

Soils: The type of soil is determined by sampling, analyzing and mapping undisturbed soils. This information is essential to reclaimation planning;

Special Sites: Historical and archaeological sites, areas of cultural significance to aboriginal communities, and the presence of plant or animal fossils are documented.

These studies provide background information against which future conditions will be compared and evaluated. Now the question that arises is what will be the appropriate way to revegetate and utilize the mined areas? For arriving at a conclusion one has to evaluate the specific site for a number of factors like:

(*i*) **Land Use Determination:** Mining is a temporary change of landuse. Therefore, before taking up any restoration plan it is essential to survey and do detailed mapping of the area, so as to establish the land use class to which the mined land belonged prior to disturbance and the post mining land use class viz., agriculture, landscaping, fisheries or forestry. Infact, current regulations in many countries require land disturbed by mining to be reclaimed to pre-mine land use and to productivity equal to or greater than pre-mine land. In India, although no such regulations exist to effectively restore the self-sustainable productive ecosystem on disturbed mined lands, land use capability assessment is of utmost importance.

(ii) **Geological and Ecological Survey** of mining sites is required for an effective implementation of restoration. Geological survey may include the contour plans of the site, ground water aquifers and surface drainage patterns. Once potential toxic materials are known, their seepage should be immediately checked. Further analysis of parent rock and overburden helps in evaluating the soil characteristics and hence have a better and more effective selection of species for revegetating the site.

Ecological survey may include the major vegetative associations, chemical and biological characteristics of streams etc. The benchmark information will prove an asset in re-establishing the productive and sustainable use of the site. Wherever possible, seeds rhizomes, bulbils etc. of the pioneering species should be collected and preserved so that immediately after the mining operations, these can be used for ameliorating the site by broadcast seeding or propagation to provide immediate visual and ecological improvement of the mined sites.

(iii) **Socioeconomic impact of mining** include impact of mining on the social and economic status of local people through road construction, school facilities, hospital facilities, employment generation, depletion/availability of forest and agricultural resources, general health conditions etc. Since mining operations disturb the local human population, in most of the cases either local population is shifted by the company or if not they are deprived of their day to day requirements of fuel, fodder, fiber, cottage industry, medicines and many other natural products. Any reclamation programme should therefore be taken up after discussion and involvement of local populace. The modern challenge for the mining companies is to find, extract and process mineral and metal resources with the least possible disruption to our ecosystem and socio-economic structure of population.

Mining of raw materials and the industrial processes required to turn these ores into usable materials and products are the underlying basis of our modern industrialized civilization. These activities represent the first stages of minerals and metals' life cycles. Successive refining, use, and recycling complete this life cycle. The mining industry is actively developing cleaner technologies and new processes to ensure that mining activities are conducted in an environmentally responsible and sustainable way.

Technological Developments Towards Mine Restoration

In the past, little attention was given to the protection of the environment. Today, however, environmental protection is paramount in the modern mining activities of exploration, planning, operation, and restoration. It also underlines research activities oriented towards the safe and responsible use of our mineral resources.

Acid Rain

Almost all of our base metals are extracted from sulfide ores. During the smelting process, sulfide ores are oxidized, producing sulfur dioxide. This chemical compound is also produced when the coal is burned. Once in the atmosphere, sulfur dioxide reacts with other gases and moisture to form acidic precipitation (acid rain). When this precipitation falls to earth as rain or snow, it can affect surface and groundwater in areas where the nature of the bedrock cannot counteract the acidity of the rain. An obvious solution to this problem is to reduce the amount of gaseous sulfur dioxide produced at the source. Smelters can now be equipped with high-performance scrubbers that can reduce by 90 per cent the amount of sulfur dioxide released into the atmosphere, thus greatly reducing the pollution linked with acid rain. Coal-fired power generating stations using new technologies can also achieve a 90 per cent reduction in sulfur dioxide emission. Research toward the development of clean-air technology has been successful. Acid rain control technologies are now being incorporated into new and older units. The attention is now turning to technologies that will also reduce carbon dioxide emissions.

Acid Mine Drainage

Waste material from mining activities, discharged as tailings, can generate acid mine drainage. In the past, tailing areas were permeable, allowing seepage into surrounding ground waters. Today, dams built to retain tailings are constructed so as to be impermeable through the use of bentonite clays, compacted tills or synthetic fabrics (geomembranes). Research toward the development of new geomembranes and new technologies to safely contain waste material from mining activities is ongoing. Large quantities of water and various chemicals are used in the milling processes. Like mine water, these tailing waters must be treated and neutralized before they can be discharged to natural drainage basins.

Treating acidic drainage from tailings ponds is a costly and long term liability. Treatment systems developed to do so can be defined as passive or active. Active systems use lime to neutralize acid water in conventional water treatment plants. Passive systems use the natural ability of the ecosystems to heal themselves. Research emphasis is now on the understanding and development of passive systems. This research focuses on the field of biotechnology and explores the possibilities offered by bacteria, plants, various organic materials, and engineered wetland.

Bacteria

Biological processes for cleaning contaminated soils and ground waters use bacterial biodegradation to decompose organic contaminants into harmless compounds. Under some conditions bacteria can also be used to recover metals. For example, a group of bacteria eat and thrive on arsenic. The microorganism acquires energy for growth by reducing arsenate compounds into arsenite compounds. These compounds could then be flushed out with water and treated.

Plants for Metal Removal

Plants can be used to harvest metals from the soil, either as an economic resource or to remove contamination. In the U.K., trials have demonstrated that some plants from the cress genera have the ability to extract zinc, cadmium and nickel from the soils. When these plants are harvested and incinerated, their ash can contain up to 20% metal oxides which can be removed by conventional methods. This mechanism is so effective that, once reclaimed, the land should be suitable for agriculture. In California, the same approach is being tested, aiming at recovering nickel and cobalt from the soil with the aid of the wild mustard plant. Scientists expect the ash of the plants, sown on nickel-bearing soils, to contain between 15-20 percent of nickel, which they will then recover using standard leaching procedures.

Organic Materials

Another research avenue is to investigate the effectiveness of various organic materials (peat, compost from municipal solid wastes and lime stabilized sewage sludge) as covers to control acid generation in tailings. This new approach would see the application of one waste to help solve a problem with other wastes.

Engineered Wetland

An effective mechanism for treating runoff water from mine sites is to filter it through a biological polishing pond before release. The water flows through a swamp or marsh. The plants absorb any heavy metals in the water. With the unwanted material thus removed, clean water can be released into the environment. This method is virtually self-sustaining and can operate unattended for many years. However, the efficiency of engineered wetland is limited in northern countries such as Canada because harsh winters prevent the use of engineered wetland during winter months.

Research into wetland chemistry has demonstrated that several physical, chemical and biological processes contribute to the improvement in the water quality as it flows through the system. The simplest physical process is the filtering mechanism of the dense plant root system. The suspended solids and flocculated particles are thus trapped as they pass through the wetland. The main process of toxic metal removal however is bacterial biodegradation. For example, a specific bacterium is responsible for the oxidation of 90 % of available iron in the form of iron hydroxides on the bed of the wetland. Another bacteria is responsible for the reduction of metallic sulfates in solution to sulfides which form an insoluble precipitate that remains buried in the organic substrate of the wetland. Acid rain and acid mine drainage are acute environmental problems facing the mining industry. Active research is aiming at the reduction of acid rain and at the prediction, prevention and treatment of acidic drainage from mine sites.

Restoration of Mined Lands

Mine reclamation techniques consist of removing, relocating or demolishing buildings and physical infrastructure; closing pits and shafts; stabilizing underground workings, soils and slopes; treating tailings and waste water; and revegetating land. Till the last two decades, reclamation was not getting much importance and only partial or no back filling was being done after surface mining. However, now it is required that different horizons be segregated and replaced in the same order as removed during mining operation. A Mining company has to restore land approximately to original contour levels wherever restoration plans require them to do so. Till now, no such regulation in the country exists, which may force the miners to undertake this type of regrading. However, with the enactment of recent environmental laws it is hoped, that in the near future miners will take initiatives to reclaim the site in the best possible way. Current legislation and regulations of many countries require that productivity of the land be returned to the premining conditions. For example, US laws required that a land primarily designated as farmland is reclaimed as pasture or row crops according to existing agricultural practices. These regulations further make it obligatory on the part of miners, that the natural vegetative communities that existed on the site prior to mining must be re-established on the site. This helps in the complete restoration of natural ecosystem. The establishment of vegetation on abandoned and active mine site as well as refuse heaps is rather difficult because of altered pH, lack of organic matter, coarse rock fragments, poor drainage and many other adverse biological factors.

Restoration of mined land depends on a number of factors e.g. ore mined, composition of overburden/waste dumped, local climate, topography etc. Hence it is very site specific, technology evolved for one site can not be exactly used in the same way on another one. Prior to taking up rehabilitation, a number of factors are taken into account and the rehabilitation techniques are then applied. Application of organic matter, hydraulic seeding, chemical stabilization and vegetative restoration by planting are some of the ways of rehabilitation.

It is very essential to restore the productivity of land for future generations. If we do not adopt any ecological restoration measures right now, we may reach a point of no return (Saxena & Chatterjii, 1988). Establishment of vegetation cover on such degraded lands is the only way to restore the viability of the area (Dean et al 1973). Vegetative rehabilitation has definite advantage over physical and chemical methods due to the long term nature of the solution combined with low maintenance cost and improved aesthetics (Douheret, 1973).

A package of practices integrating both biological and mechanical measures according to the local needs is required. A variety of these, measures have been tried on sites with varying degrees of success. Although in most of the reclamation projects local species are given priority, some exotic species have been found successful. Depending on the specific mined site, most suitable revegetation practices may be tried.

Organic Mulches

In areas where topsoil is not available as surface layer, physical structure of the site is poor for rehabilitation. The surface layer is poor in organic matter and in most of the cases surfaces tend to get encrusted after rains and dry out rapidly, raising the temperatures. Mulching of surface may help to moderate temperature and conserve soil moisture. Mulching also helps to decrease erosion, add to fertility and detoxify toxic metals and moderate temperature near the ground.

A variety of materials are used for mulching depending on site conditions, availability, cost etc. The use of pulp, fiber, straw, sawdust, wood chips, cured hay, chemicals and *in situ* mulch are commonly used. Use of local plants, etc. is also suited to check soil erosion and is more suitable for seedling establishment. With the increased awareness about environment, and concern for its protection, the use of natural material for prevention of soil erosion is an area which is in vogue.

Coir Geotextiles

Geotextiles have the mechanical strength necessary to hold soil in place and prevent erosion. The netting breaks up run-off during heavy rains and dissipates the energy of flowing water.

Coir also promotes the growth of new vegetation by absorbing water and preventing the topsoil from drying out. Coir geotextile resembles natural soil in its capacity to absorb solar radiation. This means that there is no risk of excessive heating, which is sometimes a problem with synthetic geotextiles. When the coir geotextile eventually disintegrates, it leaves only humus. Thus there is no need for post-installation work.

Number of products for erosion control in degraded sloppy areas are available. These include mat, straw mat, coconut mat, polyester blanket, liquid polymer, jute mesh, paper mulch, polyethylene net, biofabric mat, vinyl mat, and wood fiber mulch.

Hydroseeding & Hydromulching

Recently hydraulic seeding is being successfully practiced to stabilize barren steep slopes and fragile unstable surfaces. The system requires large volumes of water, seed fertilizer and mulch, which are applied on the surface in the form of a fine spray. Non-toxic dyes can be used by the operator to confirm the area covered. Some studies suggest the use of hydraulic seeding device to provide a plug of vegetative material for transplanted trees on stony ground.

Any organic material which is capable of being shredded can be used in hydroseeding. Shredded bark wastes are mixed with fines to avoid clogging. Studies in U.S. have shown that wood mulch at 2.5 t/ha prevents soil crusting. In U.K. 2.5 t/ha of previously moistened sphagnum peat have been used on China clay wastes. Use of *Medicago sativa* hay or use of mixture of grass seed, fertilizer and mulch of composted garbage material has also been recommended by some workers.

Hydroseeding

Has been found to be most useful for establishment of grass cover, however, a few tree and shrub species also germinate and survive well.

A number of hydraulic mulching materials have been tried all over the world these include chemicals biopolymers and natural fibers.

Chemical stabilizers

A number of chemical stabilizers have been used as substitutes for mulch. Most common and useful among these are bitumen or tar materials used as sprays to hold the sand as an alternative to costly brush material, mulch, hydrophobic bituminous emulsion and hydrophillic polyacrylamide solution as soil conditioner. A number of stabilizers like polymers, lignins, and resins are used. However, chemical stabilizers have been found to be less effective than peat mulch when hydomulching combated the steep surfaces. Economically, also chemical stabilizers appear to be more costly if they are to be used on large scale as required in mine rehabilitation programmes.

Revegetation practices

The drastic changes in site conditions and micro environment after mining make it necessary to use pioneering species, which may give immediate/earliest possible restoration of the site. Natural succession on such sites is often slow reflecting the extreme edaphic conditions which mining creates. Hence attempts to rehabilitate the mine area should be based on the understanding of natural processes and must be modified accordingly. Plants, which appear on abandoned sites, provide a source, which is especially useful for unusual or intractable sites.

Establishment and cultural practices must be practical and economical. For acidic mine spoils *Pinus teada, P. echinata* and *P. virginiana* have been found suitable after treating with lime equivalent to 10 tons of lime/acre. Terracing technique has been used on the steep slopes, direct afforestation has been recommended with *Pinus sylvestris* and *P. nigra* as principal species.

Some attempts have been made to use algae, lichens and mosses in tailing stabilization. On pit heaps the rate of pioneer establishment is accelerated by presence of a moss lichen surface.

Leguminous species are important colonizers on abandoned mine spoils. A number of species have been listed for reclamation. The predominance of legumes of revegetated mine sites suggests that these have an advantage in colonizing spoils which are low in available nitrogen.

Restoration Technologies Adopted in India

In India, reclamation of mined areas has gained importance in the last two decades. As mentioned earlier till very recently environmental legislations were also not very stringent and hence in most part of the country large areas of abandoned mined lands are available. Out of the total degraded land area of the country i.e. 175 m ha the area under mining leases amounts of 9,25,000 hectares. However, even this relatively microscopic proportion of mined wasteland is significant due to its devastating affect on the ecology and environment.

Iron ore mines in Kudremukh which fall in tropical evergreen zone had drawn a master plan for forestry and ecology way back in 1977. Trial of *Eucalyptus* sp. *Acacia auriculiformis, Erythrina*, Bamboos and Agaves was undertaken. Visvesvaraya Iron and Steel Ltd. Bhadrawati have planned to stabilise old dumps to prevent soil erosion, water and air pollution. *Albizia lebbeck, Eucalyptus, Dalbergia sissoo* and *Acacia auriculiformis* are commonly used. At Bharat Aluminum company's bauxite mines in Madhya Pradesh reforestation work started in 1979. About 35 tree species have been tried in which *Eucalyptus camaldulensis* exhibit the best survival and growth followed by *Grevilliea pteridifolia, Acacia auriculiformis, Pinus caribaea* and *Grevilliea robusta*. In zinc mines of Zawar Rajasthan grasses like *Dicanthium annulatum* and *Cynodon dactylon* have been used to reclaim the tailings.

Studies on eastern sector iron ore mines in Jharkhand and Orissa have recommended the use of species of local plants like *Pithecelobium dulce, Lawsonia indica* and *Moringa oeifera* besides the shrub-grass mix of *Mimosa sp.,Wendlandia exerta,Trema politoria,Plectarenthus sp.,Pennisetum sp.,Saccharum spotaneum* , *Malva verticillata* and *Alysicarpus sp*. Preparing the surface according to appropriate soil and moisture conservation measures is essential (Soni et al,1999).

Reclamation approaches throughout the country stress mainly on tree plantation and not on ecological restoration which, should be the ultimate goal. Whereas, different technologies have been adopted all over the world to tackle the ecological problems of mined lands. It is significant to note that in most of the projects, stress has been given to tree planting. However, at Forest Research Institute, Dehradun technology for reclamation of opencast mined areas has been developed which is based on the ecosystem approach. The four important steps required for effective ecorestoration programme are (i) assessment of the problem (ii) ecological maintenance of essential processes, (iii) restoration of biodiversity and (iv) sustainable utilization of species and ecosystems.

REFERENCES

Anon, *Forest (Conservation) Act, 1980*

Anon 1999, *Indian Minerals YearBook 1998 and 1999,* Indian Bureau of Mines, Ministry of Environment & Forests, Govt. of India.

Dean, K.C., Havens, R., Harper, K.T. & Rosenbaum, H.B. (1973). Vegetative stabilization of mill mineral waste. Pages 119-136 *in Hutnik, R.J. & Davis, G. (eds), Ecology and reclamation of devastated land. Vol.* 2. Gordon and Breach, New York.

Dhar, B.B. (1990). Land reclamation planning- A note. 280-299. *In: Environmental Management of Mining Operations*. B.B. Dhar (ed.) Ashish Publishing House, New Delhi.

Douheret, K. (1973). Planting and grassing mining tips. *Revue Forestiere Francaise 25*(2): 153-162.

Saxena, S.K. & Chatterji, P.C. (1988). Ecological imbalances caused by mining in Rajasthan and their restoration. *Mining and environment in India*. Himalayan Research Group Nainital (UP).

Soni, Prafulla, S.D. Sharma & Veena Chandra (1999). Ecorestoration Studies in Bolani Ore Mines (Report of SAIL-ICFRE project).

7

Prospects in Reclamation of Iron ore Mine Waste Lands: Role of Arbuscular Mycorrhizal (AM) Fungi and Inoculation Procedures

***Rodrigues B.F., Khade Sharda W., Bukhari Mehtab, Jaiswal Varsha & Uday Gaonkar**

INTRODUCTION

Mining is one of the most common activities of ancient and modern world. Mining is regarded as the second largest industry after agriculture and has played a vital role in the development of civilization from ancient days. Most valuable materials for man such as metals, chemicals, fuels for energy, rocks and stones for building comes from mining (Trivedy, 1990). Land surfaces are inevitably disturbed in seeking to win ores from the earth. Mechanization and improved technology has brought increasingly large tracts of land into state of disturbance. With increasing demands, land has been constantly exploited for raw materials from the natural environment. Land is not a resource, which automatically renews itself like rainfall and sunlight. It is a finite resource, being diminished by the spread of industry and urbanization (Coleman, 1979).

The State of Goa with an area of approximately 3702 sq. km lies on the West Coast of India between 15°48′00″N and 14° 53′54″ N Latitude and 74°20′13″E and 73°40′33″E Longitude. In Goa, mining commenced in 1910. However, commercial exploitation began in 1947 when the erstwhile Portuguese Government granted over 700 concessions all over Goa. During those days the concept of pollution and conservation were not given due weightage and the people welcomed widespread mining activities as an additional source of income and employment. In a way, therefore the growth of mining has proceeded simultaneously with the growth of agriculture

* Department of Botany, Goa University, Taleigao Plateau, Goa- 403206.

and population. Goa has been a prime exporter since 1950, as much as 300 million tons of iron ore has been exported. Present production of iron ore is of the order of 15 million tons per year which constitutes 40% of the total iron ore production in the country and 50% of its export. The estimated reserves of iron ore as on today is around 400 million tons and is expected to last for another 25-30 years at the present rate of mining.

The mining operations are such that, two classes of wastes are produced *viz.*, (i) piles of surface overburden waste rock and lean ore, which constitutes the reject dumps, and (ii) a fine grained waste resulting from the ore beneficiation process and deposited in large man made basins called tailing ponds. The latter kind of waste materials are termed as tailings.

It is true that in the process of mining mineral resources from the earth, disturbance to environment and ecosystem is unavoidable. In other words, mining is to some extent an unavoidable destructive process. Though, there are problems of mine wastes in terms of erosion, environmental pollution, damage to adjoining agricultural fields, forests, *etc.*, many a times they are exaggerated. These hazards are within measurable limits and can be easily ameliorated to a significant extent by extensive research and planning to control the impacts.

MINING AND ITS IMPACT ON ENVIRONMENT

In terms of employment and foreign exchange earnings, mining industry plays an important role in Goa's economy. On the other hand, however, the wastes produced by mining activities are likely to pose a serious threat to the environment if proper measures are not taken to re-establish vegetation at the mining sites.

The tailings occupy large segments of the landscape in the vicinity of the mine and diminish the aesthetic quality of the natural landscape. The tailing basins may occupy upto 40% of a mine site land area (Shetron and Duffek, 1970). Essentially open caste mining involves excavation and movements of large volumes of earth's crust. A ton of iron ore mined for instance produces 2 to 3 tons of waste. Dean and Havens (1971) estimated that the total tonnage of such wastes in United States covers about 200 million acres. The annual accumulation exceeds one billion tons, which are distributed over an area of approximately 2 million acres. In the Western States, nearly one half million tons are being produced daily (Neilson and Peterson, 1972). The excavation of iron ore exposes large chunks of earth's crust to the atmosphere that intrude upon the landscape.

Mining accounts for a substantial proportion of the loss of land of primary production. In India, 7,85,000 hectares of land is reported to be under mining operations (Baliga, 1985). Indiscriminate mining since 1961 has destroyed 50,000 hectares of forest in Goa and it is estimated that during all these years as much as 900 to 1000 million tons of waste rock, low grade ores and tailings have been accumulated near mining areas. The waste materials consist mainly of laterites, phyllites, quartzites, manganiferous and other types of clays, slimes, *etc*.

With the present annual production of 15 million tons of iron ore, it is expected that 40-50 million tons of wastes have to be stored per year, and approximately 150 million cubic meters of water is to be discharged from pits to the drainage system. Mine waste dumps are biggest man-made hillocks, volume and height of such dumps increases every year. Most of the waste dumps rise up to 50-60 meters high with 50° -55° angle of repose. These being unconsolidated are prone to slumps and slides due to heavy monsoon rains.

Damage to environment my mining activity has been caused largely by reject dumps, pumping out of muddy waters from the working pits including those where the excavations have gone below the water level, and slimes from the beneficiation plant. The damage is more conspicuous during monsoon, when the rainwater carries out the washed out materials from the mine waste dumps to the adjoining agricultural fields and water streams. The slimes and silts, which enter the agricultural fields get hardened on drying, thus making aeration and root penetration difficult. Indiscriminate dumping of rejects has rendered over 10,000 hectares of agricultural land infertile, pollute the springs and wells, and also cause silting of waterways especially during monsoon. Such silting of waterways over the years have caused flooding of adjacent fields and inhabited areas during the dry seasons. The noise created due to the blasting operations, movement of heavy vehicles, operation of heavy machineries and dumping of iron ore poses a constant problem in the surrounding areas. Dust from blasting, crushing and transportation is the major cause of air pollution around the neighbouring villages of mining areas, sometimes reaching miles away with increasing wind velocities. Diseases such as silicosis, tuberculosis and allergic diseases like asthma are frequently common in inhabitants of the area and the mineworkers. The dust has its effects on nearby communities, industrial machinery and demanding effects on vegetation by blocking plant pores and hampers photosynthesis.

The most common hazards of open cast mining of iron ore have been the defacing of landform by development of depression and elevation or sloppy terrain. It also leads to large-scale deforestation, destruction of wild life and natural resources resulting in a fragile ecosystem lacking in flora and fauna. In the process of mining the topsoil is removed, leaving bare rock, thereby making it hard for vegetation to become re-established. Normally, natural processes would gradually recolonise the mine sites and spoils heaps building up the soil and reclothing the landscape with vegetation. However this can take a long time. Meanwhile, the unprotected surface is subjected to erosion leading to the clogging rivers and lakes with silt. Thus the common approach towards stabilization i.e. establishment of a permanent cover of vegetation involves not merely growing plants. But it necessitates bringing into a plant community that will maintain itself indefinitely without further attention or artificial aid such as irrigation. Such a performance could be achieved most advantageously, by selecting species adapted to growth, spread and reproduction under the inimical conditions. Most of the plants, which are desirable for the revegetation of these lands, are dependent on " Mycorrhizal Fungi".

ECOSYSTEM, PLANT SUCCESSION AND MINING

The vegetation together with the soil in which it has its roots, the associated fauna, and the environment that surrounds them form a closely interrelated and interdependent to the ecosystem. Although ecosystems are sensitive to the outside influences, they are self-sustaining. Once properly established, they need no further support. This is because of natural cycling of accumulated materials, which maintain the vegetation and the other organisms with it. After a major disturbance, vegetation slowly and gradually develops over a period of time: a process termed plant succession.

If the above two properties (self-sustaining and capacity to develop) of the ecosystem are considered, then one may presume that after mining disturbance, there is no need of any revegetational efforts *i.e.*, a self-sustaining vegetation cover will develop naturally. But the process of natural succession will take many years.

The aim of revegetation of mining sites is to achieve vegetation cover within a few years, so that the subsequent succession may take place at a rapid pace. Hence, it is obvious that one must look for the appropriate treatments and management strategy so that useful vegetation can be established quickly economically leading to a self-sustaining ecosystem.

CHARACTERISTICS OF MINE WASTES

For reclamation of any degraded area, knowledge of physico-chemical parameters of degraded and undegraded area in the locality is essential. However, the exact assessment of these parameters over the entire area is not easy, as the constitution of the soil varies even at the close proximity of the sampling sites due to the random dumping of the top soil overburden, rock waste and due to interaction of various factors.

Soil texture is used extensively as a guide to evaluate soil water storage, water availability, surface erosion, land stability and chemical properties (Shetron and Trettin, 1984). Natural soil consists of an inorganic framework of sand, silt and clay particles, intimately mixed with organic material. The physical analysis of iron ore wastes reveals that the tailings and rejects have high bulk and particle density which is normally the characteristic feature of metalliferous mine waste (Rodrigues and Bukhari, 1996 & 1997) (Table 1 & 2). The bulk density of natural soils fall within the range of 1.0 – 1.5 g cm^{-3} (Williamson *et al.*, 1982) and particle density 2.63 g cm^{-3} (Waddington, 1969). Bulk density is a useful measure of compaction to root penetration. Surface accumulation of fines in slim dams may give a bulk density as high as 7.5 g cm^{-3} with low infiltration (Ruschena *et al.*, 1974).

Cation exchange capacity (CEC) is important as it is a measure of total exchangeable cations (calcium, magnesium, potassium and sodium) in soil materials (Black, 1968). Low water holding capacity of the rejects and tailings can be attributed to the poor soil texture, structure and organic matter content which are known to be responsible for improving water holding capacity.

Maclean and Dekker, (1976) studied the pH of different wastes and reported large variations in acidity among different sites ranging from pH 1.5 to above 10. Varying soil pH changes the concentration of many nutrients and toxic ions in soil solutions as well as the concentrations of hydrogen ions (Russell, 1973). In solutions of acid soils, there are often higher concentrations of aluminium and manganese, and lower concentrations of calcium, magnesium and molybdenum as compared to that in alkaline soils (Porter *et al.*, 1987).

Nutrient deficiencies are widely reported as a major limitation, particularly in terms of a low or a complete lack of organic matter and nitrogen in mining wastes. Smith and Bradshaw, (1970) stated that micro-nutrient deficiencies are frequently encountered in the mine wastes. Wong *et al.*, (1983) showed that the tailings were alkaline, lacking in organic matter and nitrogen, but were rich in metals such a Fe, Zn, Cu, Mn, Mg and Ca.

Table—1 Some properties of iron ore mine rejects in Goa. (Rodrigues and Bukhari, 1997).

Properties	*Mean (S.D.)*
pH	6.02 (0.18)
EC (mS /cm)	0.051 (0.012)
Total N	93.2* (N.A)
Available N	3.8* (N.A.)
P	1.5 (N.A.)
SO_4^{-s}	<0.1 (N.A.)
Ca	1.76 (0.80)
Mg	0.92 (0.55)
K	0.76 (0.26)
Na	2.60 (0.54)
Cu	<0.05 (N.A.)
Fe	<0.1 (N.A.)

Concentrations in $\mu g.g^{-1}$ oven dry spoil.
N.A. = Not applicable.
S.D. = Standard deviation.
EC = Electrical conductivity.
* = Mean of two replicates taken from bulked samples.

Shetron (1983) reported that in iron ore tailings the organic matter and nitrogen are essentially non existent, phosphorus levels are low; Ca, Mg, K and metal range in availability; have alkaline pH and low cation exchange capacity.

MYCORRHIZAL FUNGI

In 1842, Vittadini proposed that tree rootlets are nourished by certain fungal mycelia, which mantle them, as observed by him more than a decade earlier. This hypothesis was elaborated to a theory of mutalistic symbiosis by Bernhard Frank (1885) who coined the term "mycorrhiza" to denote the symbiotic association formed by fungal mycelia with plant roots (Gr. myces = fungus; rhizo = roots). The concept of fungus-root symbiosis has since been a subject of extensive research. Though the word was introduced in 1885, mycorrhizae itself, of course are millions of years older. It is generally believed that Arbuscular mycorrhizal (AM) fungi evolved early in the history of vascular plants (Trappe, 1987, Morton, 1990). Despite of their geological age (Birch, 1986; Pirozynsky and Dalpe, 1989) and their crucial role in origin of plants very little is known about their phylogenetic origin (Sancholle and Dalpe, 1993). The members of Glomales are believed to have been present as early as the Cambrian period (Pirozynsky and Dalpe, 1989). They played a pivotal role in the origin of the terrestrial flora. Early Devonian plants showed the presence of the endosymbionts represented by non-septate mycelia, coiled hyphae, irregularly shaped thin walled spherical structures resembling the vesicles. Arbuscule like structures are recently been reported from plants preserved in Rhynie Chart (Remy *et al.*, 1995). In their symbiotic habit, the mycorrhizal fungi constitute a special group among root inhabiting fungi.

Table—2. Some properties of iron ore mine tailings in Goa. (Rodrigues and Bukhari, 1996).

Properties	*Mean (S.D.)*
pH	6.48 (0.07)
EC (mS /cm)	0.065 (0.018)
Total N	60.3* (N.A)
Available N	1.7* (N.A.)
P	1.9 (N.A.)
SO_4^{-s}	<0.1 (N.A.)
Ca	2.34 (0.57)
Mg	0.75 (0.17)
K	0.71 (0.30)
Na	4.85 (2.94)
Cu	<0.05 (N.A.)
Fe	<0.1 (N.A.)

Concentrations in $\mu g.g^{-1}$ oven dry spoil.
N.A. = Not applicable.
S.D. = Standard deviation.
EC = Electrical conductivity.
* = Mean of two replicates taken from bulked samples.

ADVANTAGES OF MYCORRHIZAL FUNGI

Mycorrhizal association is a universal phenomenon throughout the plant kingdom and is beneficial and even indispensable for the life and healthy growth of the host plants. Because almost all plant species of natural vegetation and the agricultural crop plants of the tropics live in association with fungi, it should be possible to increase productivity through manipulation of mycorrhizal systems.

Absence of suitable mycorrhizal fungi is one of the main reasons for the failure of afforestation programmes. Tree species, especially exotics, fail to establish on afforestation sites in the absence of their fungal symbionts. Repeated attempts for the last twenty years to raise pine plantation in Puerto Rico failed due to lack of ectomycorrhizae on their roots. The pine seedlings would grow to 5-30 cm in height, then become chlorotic, and die. Transfer of ectomycorrhizal soil from successful plantations in the mainland to Puerto Rico nurseries helped in the successful establishment of pine plantations on this island. The inoculated plants of slash pine grew healthy and reached a height of 2.0- 2.5 m in three years. Similarly, success in raising pine plantations in South Africa was achieved only when ectomycorrhizal inoculum from Netherlands was introduced in South African nurseries. Soil from South African pine nurseries was then used as soil inoculum in Kenya and this resulted in large scale successful pine plantations in that country.

Normal growth occurs only when mycorrhizal fungi present in the site colonize and establish mycorrhiza on the roots. It may be possible to grow mycotrophs without its fungal symbiont if nursery plants are given a high dose of fertilizers. However, such seedlings when transplanted in the field where mycorrhiza is absent and nutrient is discontinued for economic reasons, the growth of plants is retarded appreciably in many cases.

Mycorrhizal plants are more efficient to draw nutrient from soils, particularly soils poor in available phosphorus as compared to non-mycorrhizal plants. Mycorrhizae benefit their hosts in a variety of ways. They increase the absorptive surface of colonized roots through inducing profuse branching. The extrametrical hyphal growth of these fungi can explore large soil areas for acquisition of nutrients from the soil. The hyphae of mycorrhizal fungi are able to mobilize nutrients from the substrates where such nutrients are in unavailable form for absorption by the plant roots. The fungal hyphae assimilate nutrients from such substrates through their active metabolic process and these assimilated nutrients are then transported to the plant roots.

Besides, direct nutritional advantages, mycorrhizae have also been accredited with other benefits to the host plants such as ability of arbuscular mycorrhizal roots to overcome water stress by stomatal regulation in *Citrus* (Levy and Krukum, 1980). Mycorrhizal inoculation also stimulates rooting (Barrow and Roncadri, 1977) growth and transplant survival (Bryan and Kormanik, 1977) of cutting and seedlings raised in sterilized nursery media. It also increases disease resistance by depressing root penetration and larval development of nematodes (Sikora, 1978). In addition to this, mycorrhizal plants have shown to have greater tolerance to toxic heavy metals, to drought, to high soil temperature, to saline soil, to adverse soil pH than the non mycorrhizal plants (Schenck, 1984). Arbuscular mycorrhizal fungi also bind soil into semi stable aggregates, thus improving the structure of the soil. Because of these attributes, mycorrhizae are now considered important in the establishment of plants in inhospitable sites like mine wastelands.

Apart from nutrient benefits to plants derived from mycorrhizal symbiosis, it is known to impart disease resistance in plants. In *Pinus echinatus,* mycorrhizal roots resist infection due to *Phytopthora cinnamomi,* where as uninfected roots are highly infected. This is based on the hypothesis that the rhizosphere and the sheath surface of mycorrhizal roots possess microflora different from this in non-mycorrhizal roots. In order to derive maximum benefits from mycorrhizae, it is necessary that nursery raised seedlings have optimum level of mycorrhization on their roots for which introduction of suitable mycorrhizal species at nursery stage should form an integral part of nursery management practices.

BROAD CLASSIFICATION OF MYCORRHIZAE

Based on colonization anatomy, two major groups of mycorrhizae have been recognized. *Viz.,* Ectomycorrhizae and Endomycorrhizae.

Characteristics of Ectomycorrhizae

1. Prevalent in the temperate regions, in forest and ornamental tree species.
2. Characterized by the presence of a mantle of fungal tissue around the host rootlet, and intercellular penetration of the rootlet cortex (Hartig net), and not intracellular.

3. Fungi belong to Basidiomycetes, which form sporophores and release air-borne spores. The inoculum thus is easily spread by wind from one location to another.
4. Host plant roots with ectomycorrhizal colonization are short, swollen, dichotomously branched with distinctive colours *viz.*, white, black, orange, yellow or olive green.
5. It is fairly easy to isolate a number of ectomycorrhizal fungi in pure culture by using routine microbiological techniques and grow them saprophytically. Thus, it is possible to produce inoculum on a large scale.

Characteristics of Endomycorrhizae

1. These fungi are prevalent in the tropics.
2. The colonization is both intercellular and intracellular. In this case, there is no 'Hartig net' and no 'mentle'.
3. Endomycorrhizal fungi cannot be cultured in vitro conditions. The production of endomycorrhizal inoculum requires the growth of a susceptible host plant. The inoculum has to be produced under sterile glasshouse conditions.

Classification of Endomycorrhizae

The Endomycorrhizae are again classified into 4 major groups viz., i) Arbuscular mycorrhizae (AM); ii) Arbutoid mycorrhizae; iii) Ericoid mycorrhizae, and iv) Orchid mycorrhizae.

Arbuscular Mycorrhizae (AM) Fungi:

The arbuscular mycorrhizal fungi are non-septate and ubiquitous. Despite their near omnipresence, the AM fungi have, until recently, received very little attention, because the AM fungi can neither be cultured in the absence of a living root nor isolated on agar plates by standard microbiological techniques. It is now well established that many plants cannot grow adequately without AM fungi, especially in phosphate-deficient soils. Arbuscular mycorrhizal fungi are characterized by the presence of arbuscules and vesicles.

Arbuscules-are similar to haustoria, developed by repeated dichotomous branching of hyphae that enter in the cortical cells. Each arbuscular tip sometimes appears to be surrounded by a cloud of granular material. They remain viable or active only for a short period *i.e.* 4-15 days. Their main function is nutrient transfer between symbionts. The cause of their destruction is the digestion of the fungal cell wall by host chitinase activity.

Vesicles-develop as terminal or intercalary swellings of the inter- or intra-cellular hyphae. They may be spherical or oval or lobed. Vesicle size and shape usually depends on the host nutrient conditions. They are known to contain oil droplets. When young, they have thin walls and contain homogenous protoplasm. They remain thin walled and function as storage organs for food or develop into thick walled chlamydospores functioning as reproductive structures.

Arbutoid Mycorrhizae

This type of colonization is suggested to be a transition between ectomycorrhizae and endomycorrhizae. It is hence, sometimes called as ectendomycorrhiza. This type of colonization is characterized by intracellular penetration. There is also a Hartig net and occasionally a fungal sheath. It is structurally intermediate between ectomycorrhizae and endomycorrhizae. Due to this type of colonization, root dimorphism might occur, with colonized roots remaining shorter. Only a few fungal species are known to form arbutoid colonization *viz., Amanita, Cortinarius* and *Boletus.*

Ericoid Mycorrhizae

This type of colonization is prevalent in the members of the family Ericaceae and hence the term Ericoid mycorrhizae. In this type of colonization, the fungal hyphae penetrate the epidermal and cortical cells and the fungus ramifies within each cell to form a coil or knot of filaments occupying much of the volume within the cells. The stele is not invaded. Hyphal knots can be readily detected at 200 to 400x magnification when observed under a microscope after staining with 0.05% trypan blue stain. Presently only one fungal species *viz., Pezizella ericae* is known to cause this type of colonization.

Orchid Mycorrhizae

This type of mycorrhiza has been proposed as one of the most complex of the symbiotic interaction. Members of the genera *Neottia, Limodoron, Epipogon* and *Vanilla* are dependent on mycorrhizae for growth. The entry of the fungal hyphae is a must for further growth of the seedlings. The fungal members which exhibit this type of colonization include *Fomes, Corticum* and *Rhizoctonia.*

STAGES OF DEVELOPMENT OF AM FUNGI

Arbuscular mycorrhizal fungal spores occur in physiologically inactive stages in soil. The spore germinate, grows and multiplies in the presence of actively growing roots of plants. The development of AM fungi in roots can be divided into four stages (Tommerup and Briggs, 1988):

- Spore germination and hyphal growth from infective propagules of AM fungi.
- Growth of hyphae from soil to host roots. The mycelial systems surrounding the roots are dimorphic (Mosse, 1959; Nicolson, 1967).
- Penetration and successful initiation of colonization in roots. Hyphae penetrates mechanically and enzymatically into cortical cells (Kinden and Brown, 1975). At the point, penetrating hyphae may or may not form appresoria (Abbott, 1982).
- Spread of colonization and development of internal hyphal system, arbuscules, which bifurcate inside a cell and bring about nutritional transfer between two symbionts and vesicles which, develop as terminal or intercalary swellings in inter- or intra-cellular hyphae. They are responsible for storage and vegetative reproduction.

CURRENT STATUS OF AM FUNGI

Endomycorrhizae produced by the nonseptate fungi are commonly called as "Arbuscular Mycorrhizal (AM) Fungi". These fungi belong to the Family Endogonaceae of the Order Endogonales and Class Zygomycetes (Trappe and Schenck, 1982). It includes six genera viz., *Acaulospora, Gigaspora Entrophospora, Glomus, Sclerocystis* and *Scutellospora.*

Benjamin (1979) placed eight genera of endogonaceous fungi under Endogonaceae in a single family Endogonaceae. Morton and Benny (1990) divided this order in two independent Orders *viz.*, Endogonales and Glomales on the basis of their spore structure and mycorrhiza development (Table 1).

Table—3: Classification of AM fungi.

Old Classification (Gerdemann & Trappe, 1974; Benjamin, 1979; Warcup, 1990)		*Present Classification (Morton & Benny, 1990)*	
Order:	Endogonales	1. Order:	Endogonales
Family:	Endogonaceae	Family:	Endogonaceae
Genera:	*Endogone*	Genera:	*Endogone*
	Sclerogone		Sclerogone
	Glomus		
	Sclerocystis	2. Order:	Glomales
	Acaulospora	Sub-order:	Glomineae
	Entrophospora	(*i*) Family:	Glomaceae
	Gigaspora	Genera:	*Glomus*
	Scutellospora		***Sclerocystis***
		(*ii*) Family:	Acaulosporaceae
		Genera:	***Acaulospora***
			Entrophospora
		Sub-order:	Gigasporineae
		Family:	Gigasporaceae
		Genera:	***Gigaspora***
			Scutellospora

Taxonomy of Glomales is based on the structure of their spores/sporocarps (Morton, 1990a). Some authors lay more emphasis on the wall structure, which also is a valid criterion (Walker, 1983, 1992). Glomales are divided into sub-orders, Glomineae and Gigasporineae (Table 3). The six genera included in this order are placed under three families (Morton and Benny, 1990). Till present about 150 species of fungi are genuine and validly published. Schenck and Perez (1990) have scientifically done the accumulation of the data on identification of AM fungi.

TYPES OF ECTO- AND ENDO-MYCORRHIZAL INOCULA FOR NURSERIES

The following types of inocula being employed for introduction of ecto- and endo-mycorrhizal fungi in nurseries are described below:

I. Soil inoculum

Soil inoculum consists of mycorrhizal roots, spores, chlamydospores, hyphae, hyphae strands, rhizomorphs and other propagules of mycorrhizal fungi. This method has the disadvantage of transferring inadvertently pathogenic fungi to the nursery and the planting site. This is particularly objectionable if the transport is made between countries and it is not permissible under quarantine regulations. Also the soil inoculum is bulky, messy, and it is difficult to transport to the nursery and involves cost of transport. Finally, the same soil inoculum introduced into the same host species may induce different growth responses due to different fungal symbionts.

The endomycorrhizal inoculum can be chopped finely using hand tools. This inoculum is usually air dried to about 5-20% moisture and resembles granular fertilizer in its consistency.

II. Inoculation with spores

The spores of ectomycorrhizal fungi are easy to collect and store, cheap to transport and can be taken to long distances. Spore inoculum is commonly used in case of ectomycorrhizal fungi belonging to class Gasteromycetes, the members of which produce fruiting bodies in the form of puff balls and truffles. At maturity, these fruiting bodies are full of powdery mass of spores. These fruiting bodies can be stored at 4-6 ^{0}C for 6-12 months without losing their viability. The following methods are being used for spore inoculation of nursery soils.

(a) Spore Soil Mix

In this, approximately 2 μg of spore powder (containing approximately $1x10^{12}$ spores) is thoroughly mixed with one kg of sterilized or fumigated soil and this spore mix is used as inoculum. For inoculation of nursery beds or potting mixture, 1 kg of soil spore mix is evenly spread over one square meter of nursery bed soil and the soil is then thoroughly racked up to uniformly distribute the spore inoculum in the soil.

(b) Seed Encapsulation with Spores

Slurry is made by suspending spores in water and the sticker is added to the slurry. The seeds are then dipped in the slurry and kept overnight before sowing in nursery beds.

(c) Spore Pellets

In Philippines, inoculum of ectomycorrhizal fungi in the form of tablets have been prepared under the name "MYCOGROE". These tablets consist of basidiospores of ectomycorrhizal fungi combined with soil as carrier and palletized in a tabletting machine. Pellets containing arbuscular mycorrhizal inoculum have been successfully used to inoculate plants. These pellets can be prepared by mixing 20 parts mycorrhizal inoculum (finely ground roots, soil and spores, pot culture), one part sedimentary-loess clay (mean particle size 16 μ) and one part tertiary sedimantary clay (mean particle size 2-6 μ). Add water until malleable and then roll into pellets.

(d) Mycorrhizal Beads

In Philippines, scientists have produced ecto-mycorrhizal beads, under the trade name "MYCORRHIZAL BEADS" by entrapping mycelia of *Pisolithus tinctorius* grown in liquid fermenters in calcium alginate beads. The entrapped mycelial inoculum has many advantages over solid medium inoculum. With this technique, the mycelium is produced in 1-2 weeks while in solid technique, several months are required to produce the inoculum.

(e) Pure Culture Inoculum

It is easier to produce ectomycorrhizal inoculum on a large scale using routine microbiological techniques. However, for arbuscular mycorrhizal (AM) fungi, large scale production of pure inoculum has been a major drawback as these fungi are obligate symbionts and hence requires the presence of a living host plant root and cannot be cultured using routine microbiological techniques. This process is time consuming and labour expensive. An extensive research is required to make pure cultures at reasonable time and cost.

INOCULATION PROCEDURES FOR NURSERIES

I. Broadcast Inoculations

This involves spreading a known quantity of mycorrhizal inoculum over a given area of soil surface and then mixing the inoculum into the soil to a depth of 10-20 cm before seeding. Several inocula *viz.* duff (consists of mycorrhizal tree roots and fungal propagules), sporocarps and spores, and pure culture vegetative mycelium have been applied in this manner to obtain mycorrhizal seedling in nurseries.

II. Banding of Inoculum Below Seeds

This involves placing the inoculum below the seed in a layer or band. This facilitates concentration of inoculum near developing roots. The major advantage of this method is that it requires only one third as much inoculum as the broadcast method. In addition, it saves time and labour. The only disadvantage being the need of an additional machine.

III. Slurry Dips

Slurries of mycorrhizal inoculum can be prepared by mixing the mycorrhizal inoculum with water, and a carrier such as clay or soil. The seedlings are inoculated by dipping them into the slurry prior to planting.

MYCORRHIZAL INOCULUM TECHNOLOGY

Successful production of mycorrhizal seedlings is dependent upon type and age of inoculum used, time of inoculation, inoculum density, inoculum placement and a number of host and fungus interactions. The mycorrhizal inocula consist of soil inoculum, mycorrhizal seedlings and roots, sporocarps and spores, and pure cultures of mycorrhizal fungi.

Seedlings can be inoculated at three different stages *viz.*, i. before seeds are sown, ii. When seeds are sown and, iii. After seedlings emerge. The most efficient stage would be to inoculate the seedlings before or when the seeds are sown. An efficient time to inoculate cuttings is at the time of propagation. It also depends upon the economic considerations and on the ecology of mycorrhizal fungi. As regards to economics, the stage when the seeds are sown or cuttings are propagated requires least amount of inoculum per volume of growing medium. Again, the newly developed rootlets of seedlings or cuttings are receptive to mycorrhizal colonizations, as they are non-lignified. The mycorrhizal fungi are also known to increase the rooting in cuttings and increase the root developments during propagation. It is also cost efficient to develop a mechanical inoculation system for use when seeds are sown or cuttings are propagated than any other later stages. Inoculation at the time of transplantation is time consuming, requires more inoculum, and the introduced fungi must be compatible with the native microorganisms and climatic conditions of the planting site.

ROLE OF AM FUNGI IN RECOVERY OF MINE WASTELANDS

Nicolson (1967) suggested that plant growth in industrial waste could be improved by incorporating AM fungi. Khan (1978) reported similar results for Australian coal spoils, noting that some members of Proteaceae were successful non-mycorrhizal invaders. However, species vary in their degree of dependency on mycorrhizal endophytes. Janos (1980) has explained that during succession, three main types characterize a range of ecological dependency : non-mycotrophs, facultative mycotrophs and obligate mycotrophs. In this case, obligate mycotrophs could fail to become established in sites of vary low inoculum density and may only become established after endophytes have colonized the area. If this is so, then these organisms are determinants of community composition during early succession and they may in part, control the progress of succession (Reeves *et al.*, 1979).

It is seen that iron ore mine rejects are poor in nutrients as indicated in Table 1. In a survey conducted at Sanquelim iron ore mines belonging to M/s Sesa Goa Limited, all the herbaceous plants growing on a 12 year old reject dump showed AM fungal colonization (Rodrigues & Bukhari, 1997) (Table 4). In all, a total of 27 species of AM fungi belonging to five genera were recorded from the iron ore mines (Rodrigues, 2000) (Table 5) (Plate I). Glasshouse studies conducted to evaluate effect of two AM fungal species (*Glomus mosseae* Nicolson and Gerdemann and *Glomus fasciculatum* (Thaxter *sensu* Gerdemann) Gerdemann & Trappe) on biomass of nine tree species grown on mine

Table—4: Degree of root colonisation (%) in some naturally occurring herbaceous plant species of iron ore mine wastelands of Goa. (Rodrigues and Bukhari,1997a)

Sr. No.	*Plant species*	*Family*	*Degree of root colonization %)*	*Type of colonization*
1.	*Lygodium flexuosum* (L.)Swartz.	Schizaeaceae	72	H A
2.	*Polygala elongata* Klein ex Willd.	Polygalaceae	10	H A
3.	*Impatiens Kleinii* W.& A.	Balsaminaceae	81	H V
4.	*Hydrocotyle asiatica* L.	Apiaceae	42	H A V

(*Contd...*)

5.	*Neanotis foetida* Benth. & Hook.	Rubiaceae	79	H	V	
6.	*Spermacoce hispida* L.	Rubiaceae	37	H	V	
7.	*Blumea mollis* (D. Don) Merr.	Asteraceae	87	H	A	
8.	*Parthenium hysterophorus* L.	Asteraceae	81	H	V	
9.	*Vernonia cinerea* (L.) Less.	Asteraceae	11	H	V	
10.	*Canscora diffusa* (Vahl.) R.Br.	Gentianaceae	40	H	V	
11.	*Merremia tridentata* (L.) Hallier f .	Convolvulaceae	96	H	A	
12.	*Lindernia crustacea* (L.). F.Muell	Scrophulariaceae	90	H	A	
13.	*Lindernia parviflora* (Roxb.)Haines	Scrophulariaceae	60	H	A	
14.	*Ramphicarpa longiflora* Benth.	Scrophulariaceae	36	H	V	
15.	*Striga asiatica* (L.) Kuntze	Scrophulariaceae	93	H	V	
16.	*Centranthera hispida* R.Br.	Scrophulariaceae	56	H	V	
17.	*Justicia procumbens* L.	Acanthaceae	23	H	A	V
18.	*Gomphrena celosioides* C.Martius	Amaranthaceae	62	H	A	
19.	*Amorphophallus commutatus* Engler	Araceae	74	H	V	
20.	*Eriocaulon cinereum* R. Br.	Eriocaulaceae		29	H	V
21.	*Eragrostis amabilis* W & A.	Poaceae	29	H	V	
22.	*Heteropogon contortus* (L.) P. Beauv. Ex Roeme & Schultes	Poaceae	30	H	A	
23.	*Ischaemum semisagittatum* Roxb.	Poaceae	67	H	V	

Legend: H = Hyphae; A = Arbuscules; V = Vesicles.

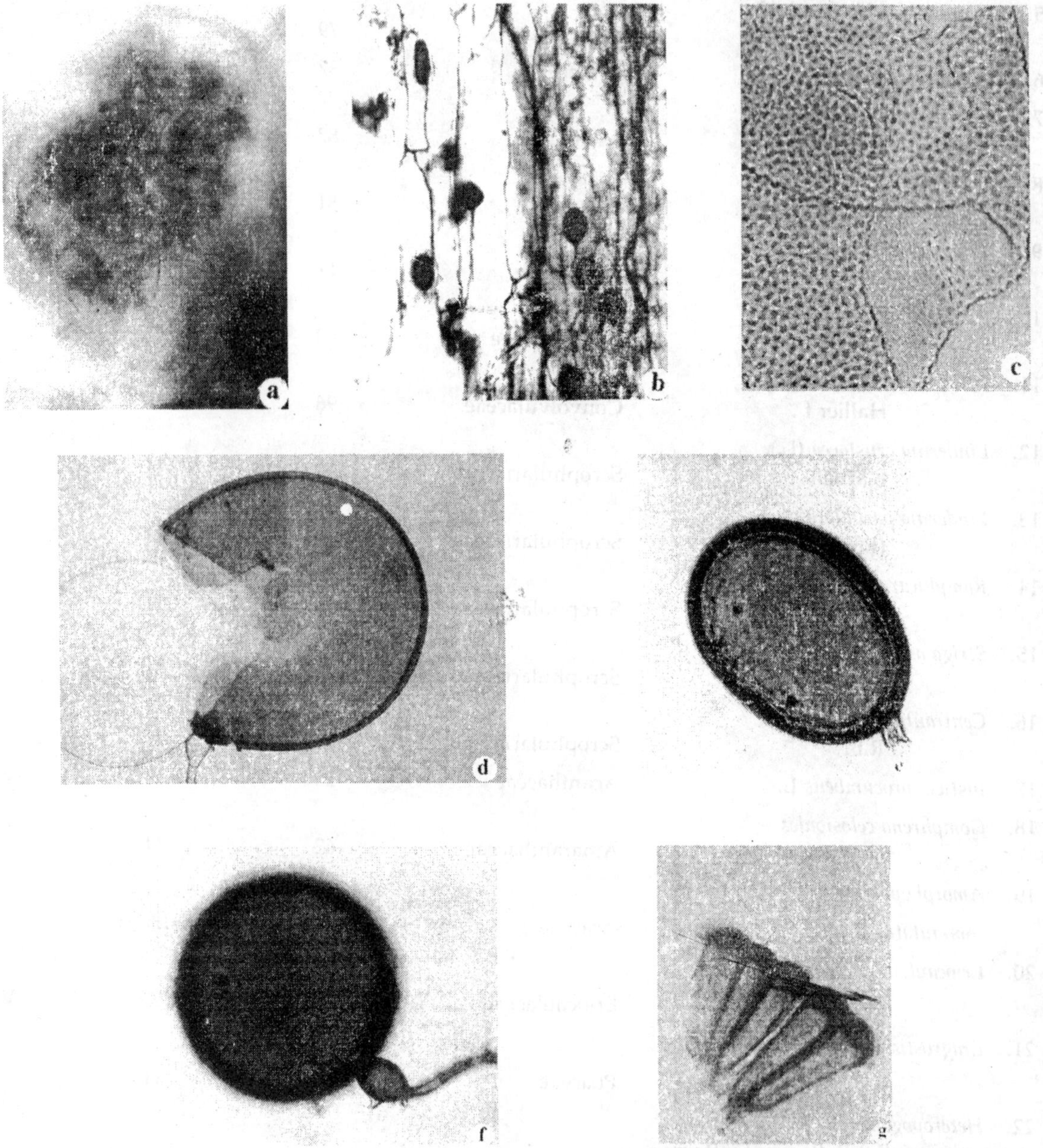

Plate I: a. Arbuscular Colonization (400x); b. Vesicular and hyphal Colonization (100x)
c. Surface ornamentation in *Acaulospora spinosa* spore (400x);
d. Spore of *Gigaspora margarita* (100x);
e. Spore of Glomus macrocarpum (400x);
f. Spore of *Scutellospora gregaria* (100x), and
g. Chlamydospores of *Sclerocystis taiwanensis* (400x)

rejects, revealed that the inoculated plants performed better than the uninoculated controls (Rodrigues, 1997). It was also noted that out of the nine tree species, *Glomus mosseae* Nicolson and Gerdemann appeared to be the best mycorrhizal inoculum for eight tree species (Table 6).

Table—5: Diversity of AM species found on iron ore mine reject dumps of Goa. (Rodrigues, 2000).

Sr. No.	*AM species*	*Sr. No.*	*AM species*
1.	*Glomus geosporum*	16.	*Glomus deserticola*
2.	*Glomus mosseae*	17.	*Gigaspora albida*
3.	*Glomus fasciculatum*	18.	*Gigaspora margarita*
4.	*Glomus hoi*	19.	*Gigaspora candida*
5.	*Glomus aggregatum*	20.	*Acaulospora nicolsonii*
6.	*Glomus australe*	21.	*Acaulospora spinosa*
7.	*Glomus reticulatum*	22.	*Acaulospora bireticulata*
8.	*Glomus clarum*	23.	*Acaulospora laevis*
9.	*Glomus constrictum*	24.	*Acaulospora morrawae*
10.	*Glomus caledonium*	25.	*Acaulospora foveata*
11.	*Glomus radiatum*	26.	*Acaulospora mellea*
12.	*Glomus etinucatum*	27.	*Scutellospora gilmorei*
13.	*Glomus albidum*		
14.	*Glomus monosporum*		
15.	*Glomus globiferum*		

Table—6: Effect of AM fungal species on dry weight (g) on various plant species grown on iron ore mine rejects. (Rodrigues, 1997).

Plant species	*Total dry wt. (g)*		
	Control	G. F.	G. M.
Delonix regia	1.48 (0.10)	2.28(0.14)	4.05(0.31)
Tamarindus indica	1.68 (0.06)	1.93(0.43)	2.54(0.25)
Acacia farnesiana	0.22 (0.03)	0.64(0.06)	1.00(0.13)
Acacia mangium	0.12 (0.00)	0.35(0.03)	0.38(0.03)
Adenanthera pavonina	0.82 (0.09)	1.37(0.10)	1.56(0.06)
Albizzia lebbeck	0.24 (0.02)	0.84(0.17)	0.86(0.12)
Leucaena leucocephala	0.31 (0.03)	0.74(0.05)	0.75(0.09)
Samanea saman	0.44 (0.07)	0.89(0.10)	2.36(0.26)
Ziziphus jujuba	0.35 (0.04)	0.78(0.07)	0.43(0.07)

Legend:

Control=Pure reject; **G.F.** = *Glomus fasciculatum* & **G.M.** = *Glomus mosseae* Mean of 5 replicates. Figures in the brackets denote Standard deviation values.

CONCLUSION

"Mine land is a fascinating challenge because the pre-existing ecosystems are extinguished". It is challenge to the biologists and engineers to replace them as they were. It is also a challenge to the soil scientists, ecologists and to the agriculturists to reconstruct an ecosystem from nothing at minimal cost.

Mine rejects are not true soils but are derived mostly from crushed bedrock and /or glacial deposits hence they are low in nutrients. In this relation, the role of microorganism in rehabilitation has received little attention than correction of nutritional deficiencies and imbalances, toxicity, moisture deficits and wind erosion.

Future revegetational research has to be oriented towards:

- Developing methods of maintaining inoculum level in soil.
- Developing techniques for introducing the endophytes in the soil.
- Naturally colonizing plant species should be given preference while considering revegetation strategies. Seedlings of such plant species should be inoculated with AM fungi in nursery stages and than transplanted to the target site. This would enhance plant growth and survival in the inhospitable sites.

In addition to this, alternative strategies such as reducing the angle of slope of reject dumps through terracing in order to improve water holding capacity, addition of organic materials like sewage sludge, sea weeds, green manure *etc.*, would help to elevate the soil status and enhance mycorrhization in spoils. Removal and storage of topsoil for reuse would make reestablishment of vegetation relatively easier as topsoil contains organic matter, plant nutrients, seed propagules and useful microbes. This would also lead to increase the inoculum potential of AM fungi thereby helping in plant growth and survival.

Thus mining industry need not lead to degradation of environment if those who are involved in such programmes apply a combination of imagination, care and scientific skill.

REFERENCES

Abbott, L. K. 1982 Comparative anatomy of vesicular arbuscular mycorrhiza formed on subterranean clover . **Australian Journal of Botany** 30: 485-449.

Baliga, B.P. 1985 Mining and mine areas as a conservation hazard and programme for rehabilitation. National Seminar, **Soil Conservation and Wasteland Management**. pp 191-200.

Benjamin, R.K. 1979 Zygomycetes and their spores. In: B. Kendrick Ed. The Whole Fungus. National Museums of Canada, Ottawa, Canada, pp. 573-622.

Black, C.A. 1968 **Soil-plant relationships.** John Wiley and Sons, Inc., New York pp 792.

Barrow, J. E and W. R. Roncadri 1977 Endomycorrhizal symbiosis of *Gigaspora margarita* Poisettia. **Mycologia** 69: 1173-1184.

Birch, S.N. 1986 Endogonaceae: Taxonomy, Specificity, Fossil record, Phylogeny. *In*: Mukerji, K.G., Pathak, N.C. and V.P. Singh Eds. **Frontiers in Applied Microbiology** Vol.II, Print House, Lucknow, India pp. 161-188.

Bryan, W.C. and P.P. Kormanik 1977 Mycorrhiza benefit survival and growth of sweet gram seedling in the nursery. **Southern Journal of Applied Forestry** 1: 21-23.

Colman, A. M. 1979 Land Use Planing. Success or Failure. **Architect Journal** 165: 91-134.

Dean, K.C. and R. Havens 1971 Stabilization of mineral wastes from processing plants. **Proceedings of Second Annual Mine Waste Utility Symposium.** pp. 205-213.

Frank, A.B. (1885). Uerber die auf Wurzelsymbiose beruhende Ernahrung gewisser Baume durch unterirdische Pilze. **Berichte der Deutschen Botanischen Gesellschaft** 3: 128-145.

Gerdemann, J.W. and J.M. Trappe 1974 Endogonaceae in the Pacific Northwest. Mycologia Memoire 5: 1-76.

Janos, D.P. 1980 Mycorrhizae influence tropical succession. **Biotropica 12: 56-64.**

Khan, A. G. 1981 Growth of endomycorrhizal onions in unsterilized coal waste. **The New Phytologist** 87: 363-370.

Kinden, D. A and F. M. Brown 1975 Electron microscopy of yellow poplar II. Intercellular hyphae and vesicles. **Canadian Journal of Microbiology** 21: 1968-1980.

Levy, Y. and J. Krukum 1980 Effect of VAM on *Citrus jambiri* water relations. **The New Phytologist** 85: 25-31.

Maclean, A.L. and A.J. Dekker 1976 Lime requirement availability of nutrient and toxic metals to plants grown in acid mine tailings. **Canadian Journal of Soil Science** 56: 27-36.

Morton, J.B. 1990 Evolutionary relationships among arbuscular mycorrhizal fungi in the Endogonaceae. **Mycologia** 82: 192-207.

Morton, J.B. 1990a Species and clones of arbuscular mycorrhizal fungi (Glomales, Zygomycetes), their role in macro- and micro-evolutionary processes. **Mycotaxon** 37: 493-515.

Morton, J.B. and G.L.Benny 1990 Revised classification of arbuscular mycorrhizal fungi (Zygomycetes): a new order Glomales, two sub-orders, Glomineae and Gigasporineae, and two new families, Acaulosporaceae and Gigasporaceae, with an amendation of Glomaceae. **Mycotaxon** 37: 471-491.

Mosse, B. 1959 The regular germination of resting spores and some observations on the growth requirement of *Endogone sp.* causing vesicular arbuscular mycorrhiza. **Transactions of British Mycological Society** 42: 273-286.

Neilson, R.F. and H.B. Peterson 1972 Treatment of mine tailings to promote vegetative stabilization. **Agricultural Experiment Station Utah State University Bulletin** 485: 1-2.

Nicolson, T. H. 1967 Vesicular Arbuscular Mycorrhizae – A Universal Symbiosis. **Science** 55: 561-581.

Reeves, F. V., Wagner, D. Moormen, T. and J. Kial 1979 The role of endomycorrhizae in revegetation practice in Semi-arid West.I Comparison of incidence of mycorrhizae in severely disturbed vs. natural environment. **American Journal of Botany** 66: 6-13.

Pirozynsky, K.A. and Y.Dalpe. 1989 Geological history of Glomaceae with particular reference to mycorrhizal symbiosis. **Symbiosis** 7: 1-36.

Porter, W.M., Robson, A.D. and L.K. Abbott 1987 Factors controlling the distribution of vesicular-arbuscular mycorrhizal fungi in relation to soil pH. **Journal of Applied Ecology** 24: 663-672.

Remy. W., Tailor, T.N. Hass H. and H. Kerp. 1995 Four hundred million years old vesicular-arbuscular mycorrhizae. **Proceedings of National Academy of Sciences, USA** 91: 11841-11843.

Rodrigues, B.F. and M.S. Bukhari 1996 Preliminary investigation into VAM colonization of plant species found growing in tailing pond. *In*: Villinayagam *et al.* Eds. **Micro-organisms in Sustainable Agriculture.** Madurai. pp. 25-28.

Rodrigues, B.F. and M.S. Bukhari 1997 Occurrence of VAMF colonization in herbaceous plant species growing on iron ore mine wastelands in Goa. *In*: S.M. Reddy *et al.* Eds. **Microbial Biotechnology** Scientific Publishers, Jodhpur pp.83-86.

Rodrigues, B.F. 1997 Comparative effect of two VA-mycorrhizal species on biomass of nine plant species grown on iron ore mine rejects. *In*: T.V Ramana Rao and I.L.Khotari Eds **Plant structure and Morphogenesis** pp. 163-167.

Rodrigues, B.F. 2000 Diversity of Arbuscular mycorrhizal (AM) fungal species from iron ore mine wastelands in Goa. **The Indian Forester** 126(11): 1211-1216.

Ruschena, L.J., Stacy, G.S., Hunter, G.D. and P.C. Whiteman 1974 Research into the revegetation of concentrator tailing dams at Mount Isa, Parkville, Victoria. **Australian Institute of Mining and Metallurgy.** pp. 16.

Russell, E.W. 1973 **Soil Conditions and Plant Growth**. 10th Edition. Longman, London.

Sancholle, N. and Y. Dalpe. 1993 Taxonomic relevance of fatty acids of arbuscular mycorrhizal fungi and related species. **Mycotaxon** 49: 187-193.

Schenck, N. C. 1984 **Methods and Principle of Mycorrhizal Research.** The American Phytopathological Society. St. Paula Minnesota Second edition.

Schenck, N.C. and Y. Perez. 1990 **Manual for the identification of VA Mycorrhizal Fungi.** 3rd eds. Synergistic Publications, University of Florida, Gainesville, Florida, USA, pp. 1-286.

Shetron, S.G. and R. Duffek 1970 Establishing vegetation on iron mine tailings. **Journal of Soil and Water Conservation** 25:227-230.

Shetron, S.G. 1983 Alfalfa, *Medicago sativa* L. Establishment in mine mill tailings. 1. Plant analysis of alfalfa grown on iron and copper tailings. **Plant and Soil** 73: 227-237.

Shetron, S.G. and C.C. Trettin 1984 Influence of mine tailing particle density on pipette procedures. **Soil Science Society of America Journal** 48(2): 418-420.

Sikora, R. A. 1978 Influence of the endotrophic mycorrhiza *Glomus mossene* on the host parasitic relationship of *Meloidogone incognita* on tomato. **Zeit-Sch.Fur. Zon. Pfln.** 85: 197-200.

Smith, R.A.H. and A.D. Bradshaw 1970 Reclamation of toxic metalliferous wastes using tolerant populations of grass. **Nature** 227:376-377.

Trappe, J.M. 1987 Phylogenetic and ecological aspects of mycotrophy in the angiosperms from an evolutionary stand point. *In*: Ecophysiology of VAM plants. C.R.C. Press Inc., Florida USA. pp. 5-26.

Trappe, J. M. and N. C. Schenck 1982 Taxonomy of the fungi forming endomycorrhizae. A. Vesicular arbuscular mycorrhizal fungi (Endogonales.) *In*: N. C. Schenck Ed. **Methods and Principles of**

Mycorrhizal Research. pp: 1-9 The American Phytopathological Society. St. Paula Minnesota Second edition

Trivedy, K. 1990 Mining and Environment Overview. *In*: Trivedy, K. & M. P. Sinha Eds. **Impact of mining on environment.** pp. 1-8. Ashish Publishing House, New Delhi.

Tommerup. J.C. and G.G. Briggs 1988 Influence of agricultural chemicals on germination of vesicular arbuscular endophyte spores. **Transactions of British Mycological Society** 76: 326-328.

Waddington, D.V. 1969 Soil and soil related problems. *In*: Hanson, A.A. and P.V. Juska Eds. Turf Grass Science. Agronomy Series No. 14. **American Society of Agronomy, Madison.** pp. 80-129.

Walker, C. 1983 Taxonomic concepts in the Endogonaceae: spore wall characteristics and species descriptions. **Mycotaxon** 18: 443-455.

Walker, C. 1992 Systematics and taxonomy of the arbuscular endomycorrhizal fungi (Glomales) – a possible way forward. **Agronomie** 12: 887-897.

Warcup, J.H. 1990 Taxonomy, culture and Mycorrhizal associations of some zygosporic Endogonaceae. Mycological Research 94: 173-178.

Williamson, N.A., Johnson, M.S. and A.D. Bradshaw 1982 Mine waste reclamation. **Mining Journal Books Ltd.,** London, England.

Wong, M.H., Lau, W.M., Li , S.W and C.K. Tang 1983 Root growth of two grass species on iron ore tailings at elevated levels of manganese, iron and copper. **Environmental Research** 30: 26-33.

8

A Holistic Bio-technological Approach in Farming Environments for Quality Agriproducts in the 21st Century

*Basavarajappa, S.

INTRODUCTION

Agriculture is the way of life for more than a million families in India. After the green revolution, the country continues to look only at short term benefits of higher food grain production, the profit of this has neglected the hazardous consequences of disturbing the delicate ecological balance (Chakraborty and Basu, 1997) in terrestrial environment. The current farming practices with indiscriminate use of chemical fertilizers and pesticides have had an adverse impact on the natural resources like air, soil and water in our surroundings (Basavarajappa, 1992).

Though India needs a second green revolution, it should be based on eco-friendly technologies. The increase in crop production and land productivity are essential steps to meet the food demand of the growing human population in India (Arya, 2000). The increased agricultural production has to come through the adoption of better management technology. The long term sustainability in agriculture is possible with diversified cultivable land with varied irrigation facilities only through: (i) The use of low cost farm grown inputs which work in hormony with the nature. (ii) The inputs which are available near the farm which help for better soil and water management and iii. The inputs easy to use and adopt for improved agricultural practices and in turn production of quality agriproducts sustainably.

India being a tropical country, possess more than 47 million hectares of dry land, spreading in various districts of Western (Rajasthan, Gujarath), northern (Punjab, Haryana and Uttar Pradesh), Central (Madhya Pradesh, Maharashtra) and Southern (Karnataka, Andhra Pradesh

* Dept. of Applied Zoology Kuvempu University, B.R.Project-577 115, Karnatak. India.

and Tamil Nadu) states (Teotia *et al.*, 1992). In India, the dry lands exhibit complex seasonal pattern of land use due to their varied climate, ecology, floristics, vegetational and agronomic conditions. Moreover, the rate of evaporation and transpiration are greater than precipitation in these regions. The soils of dry lands constitutes red, black, laterite and lateritic soils, which are deficient in Nitrogen, Phosphorus, Potash and Organic matter (Teotia *et al.*, 1992). Further, the rainfall is highly erratic, which develops further pressure on dry soils due to harsh climate and poor irrigation facilities. It results in high incidence of crop failure on account of frequent occurrence of drought and floods. Thus, the crop productivity is below 50% than that of fertile soils. Agriculture, therefore, faces a challenge, not simply to produce dry land crops (viz., Cotton, Rice, Groundnut, Sugarcane, Maize, Jowar, Wheat, Millets, Pulses and Oil seeds), but to assure undisturbed soil fertility and pollution free environment. The aspect is signified in the term 'Sustainability' (Naik, 1993).

Sustainable agriculture can be defined as the "Successful management of resources for agriculture to satisfy changing human needs while maintaining and enhancing the quality of environment and conserving natural resources" (Naik, 1993). Hence, we have to look for alternatives viz., 1. Dry land, barren land and waste lands for growing crops such as cereals, pulses, vegetables, sugarcane, oil-seed crops and horticultural crops (Teotia et al., 1992), and 2. Biological means of fertilization and pest control (viz., Green manure, Vermicompost, Organic manure, Biofertilizer, Biopesticides use) in agriculture. Such a holistic bio-technological approach enable us to achieve a better balance between quality agriproducts, their productivity and sustainability. Keeping this in view, biological means of nutrient supply to agricultural farming ecosystems has been prescribed to replace the chemical fertilizers and pesticides, for better terrestrial ecosystem to future generation. In this regard, use of green manure, organic manure, vermicompost, biofertilizers and biopesticides in the farming environment has been suggested as an effective substitutes for chemical fertilizers and pesticides so as to safegaurd the delicate ecological balance in the multifaced terrestrial environment.

CULTIVABLE SOIL, FERTILIZERS AND PLANT NUTRIENTS

Cultivable Soil System

Agricultural productivity is controlled by climate (consists of rainfall, air, sunlight) and soil (Bongale, 1991). However, continuous farming activity requires regular fertilizer input for optimising the yield and productivity per unit area. Soil constitutes the natural medium, which supports the growth of plants on earth's surface. The soil contains large number of elements which are very essential for the normal growth and development of food crops. Certain elements viz., Carbon, Hydrogen, Oxygen are supplied by the air and water. The elements like Iron, Manganese, Zinc, Copper, Molybdenum, Boron, Chlorine etc., are obtained by plants from the soil. Nitrogen, Phosphorus and Potassium are called major nutrients. Whereas Calcium, Magnesium and Sulphur are known as secondary elements. Soils must contain all these elements required by plants in available form and in balanced proportions (Sinha *et al.*, 1992). The absence or deficiency of any one of these elements decreases the fertility of the soil (Krishna and Bongale, 2001) and affects growth of crop plants.

However, soils of tropical regions vary with respect to their physical, chemical and biological properties. The content of available forms of nutrients differ not only from one farm to another, but also from field to field in the farm (Sinha *et al.*, 1992). The poor soil fertility, hot climate and dry weather are characteristic of the tropics, which present a physiological stress on the plant system (Bongale, 1991.) This results in poor production from plant system.

Importance of Fertilizers

Fertilizers constitute the kingpin of agricultural production under varied terrestrial conditions. The effect of fertilizers on soil physical properties in the long run requires careful consideration. It is known that unbalanced use of fertilizers and manure affect soil aeration and crop growth by modifying physical properties such as bulk density, porosity, aggregation, and water retention capacity of soil.

Fertilizer practices require many intricate details such as soil type, crops, fertilizers and prevailing climate. The availability of nutrients is very much dependent upon the pH of the soil. In acidic soil, availability of Fe^{+3}, Al^{+3} and Mg^{+2}, is high but nitrogen, phosphorus and potassium are poorly availability. Nitrogen and potassium are available in moderate acid range (pH 6.00) and later decrease and at pH 4.00 they become unavailable because they get fixed. Phosphorus is absorbed by the plants only in the form of HPO^{4-} which remains in the pH range of 6.5 to 7.5 It changes into another ionic form which cannot be utilized by the plant (Khare *et al.*, 1992). If we ignore the prevailing soil condition and apply the fertilizers at random, the desired output is not achieved.

Plant Nutrients

Plants require sixteen elements for the growth and development (Khare et.al. 1992). Nitrogen, phosphorus and potassium are most important elements. Nitrogen is a constituent of amino acids, nucleotides and co-enzymes. Its deficiency interferes with protein synthesis and consequently affect biochemical activities and growth. Phosphorus plays a key role in energy metabolism and occurs in phospholids, in sugar phosphates, various nucleotides and co-enzymes. The potassium plays the principal role as activator of numerous enzymes. Moreoften, nitrogen - the much required mineral nutrient in soil gets lost in the natural process of volatalization, deep percolation, leaching, etc., and necessitates extra input to ensure proper plant growth resulting in escalation of cost of cultivation (Sridhara *et al.*, 1995).

The essential micro-nutrients for plant growth are Iron, Copper, Manganese, Zinc, Molybdenum, Boron and Chlorine. All these micronutrients play a significant role in several essential physiological processes of plants like photosynthesis, respiration, carbohydrate, protein and nucleic acid metabolism, disease resistance, enzymatic activities etc. In micronutrients deficient soils, the plants suffer from various physiological disorders and fail to complete their life cycle.

GREEN MANURE

In recent years, importance is given to replace the chemical fertilizers with organic ones, for reasons of better soil management. One of the effective ways is to use green manure (Sinha et.al. 2000). A large number of leguminous crops viz., Sunhemp *(Crotalaria juncea)*, Dhaincha *(Sesbania aculeata)*, Indigo *(Indigofera tenctoria)*, Pillipesara *(Phaseoalus trilobus)*, Cow pea *(Vigna sinensis) and Wild Indigo (Tephrosea purpurea)*, leaves of Karanja *(Pongamia glabra, P. pinnata)* and Water

Hyacinth *(Eichhornia crassipes)* are used for green manures. Green manures become: **(a)** best substitute for compost and are cost effective. **(b)** their fast growth supresses the growth of weeds which are vigorously growing competing with main food crops. **(c)** these plants also enrich the soil by fixing atmospheric nitrogen.for example: Sunhemp fix 80 kg. of nitrogen per hectare (Sinha *et al.*, 2000). Rainfed areas where farmers can't afford to apply recommended doses of compost/ FYM to their agriculture fields green manuring can be a better and suitable substitute.

The commonly occurring weed 'Water hyacinth' is suggested for the use in agriculture as "Green manure". The water hyacinth *Eichhornia crassipes* is a ' German weed' which grow in a vast areas of water in the tropics and sub-tropics. It is indigenous to south America and was introduced into India from Brazil in 1896 in the form of an ornamental free-floating pond plant (Sharma and Habib, 1994). It is a fast growing aquatic weed, found in the rivers and other fresh water bodies. Its growth has reached alarming proportions in some parts of India. It has posed serious problem by choking the irrigation and drainage channels and irrigation pumps. Scientists have initiated research on various aspects of water hyacinth to identify its beneficial nature. Water hyacinth can be converted into agriculturally useful products. Field experiments using sun dried water hyacinth and whole or chopped water hyacinth for soil amendment and for source of plant nutrient revealed encouraging results in Pearl millet crop fields (Sharma and Habib, 1994). The yield increased considerably under weed treatment (2700 kg/ha of dried water hyacianth) than chemical fertilizer treatment. The concentration of C, N, K, Ca, Mg, Mn and Cl and cation exchange also increased when it is applied to agriculture fields. Pressed, chopped and dehydrated material contains 12-18% crude protein, 25-35% ash, free crude contents and 11% ash. Nitrates, oxalates and other minerals were also identified in water hyacinth, which are useful to commercial crops (Sharma and Habib, 1994). Utilization of water hyacinth as useful product in agriculture fields is important step in combating this evil (Sharma and Habib, 1994). The efficacy of water hyacinth was also tested in mulberry garden soils by several workers. Water hyacinth will not only improve the soil fertility but also increases the mulberry leaf yield (Sinha *et al.*, 2000). Thus, use of green manuring crops not only improve the soil fertility but also increases the quality crop production.

ORGANIC MANURE

Organic sources of manures are a gift of nature to the agriculturists (Sinha *et al.*, 1992). Organic manure includes farmyard Manure (viz., Livestock, dung, urine, bedding material, straw and etc.), compost (all organic farm wastes, straw, dry weeds etc.), press mud (bye-product of sugar industry), poultry manure and others supply balanced macro nutrients (viz., Nitrogen, Phosphorus, Potassium, Calcium and etc.) and micro nutrients (Viz., Boron, Manganese, Copper, Zinc, Molybdenum and etc.) (Table -1) to plants. The water holding capacity, physical properties of soil and availability of nutrients present in the soil improve (Sinha *et al.*, 1992). In addition, chemical and biological activities, nutrients retention in the soil also increase. Loss of nutrients by volatilisation and leaching is minimised. Organic manures also have the residual effects in the next crop. Furthermore, depletion of organic manure is quick in tropical soils due to the high sunshine and humidity. Thus, use of organic manure is necessary to make essential nutrients available to plant for a longer period (Sinha *et al.*, 1992). In this regard, farmers should be educated to make optimum use of energy and plant nutrients from the organic materials available with them.

Table—1. Nutrient contents of Manure from different source

Sl. No.	Source of Manure	Percentage content		
		Nitrogen (N)	Phosphoric acids (P2O5)	Potash (K2O)
1.	Poultry Manure fresh	1.0 – 1.8	1.4 – 1.8	0.8 – 0.9
2.	Sheep dung fresh	0.5 – 0.7	0.4 – 0.6	0.3 – 1.0
3.	Horse dung fresh	0.4 – 0.5	0.3 – 0.4	0.3 – 0.4
4.	Cattle dung fresh	0.3 – 0.4	0.1 – 0.2	0.1 – 0.3
5.	Sewage sludge dry	2.0 – 3.5	1.0 – 5.0	0.2 – 0.5
6.	Sewage sludge activated dry	4.0 – 7.0	2.3 – 4.2	0.5 – 0.7
7.	Farm Yard Manure (FYM)	1.0	0.6	1.0
8.	Compost	1.2	0.7	1.2
9.	Press Mud	3.2	8.4	0.9
10.	Poultry Manure Dry	4.0	3.2	1.2

Source: Krishna and Bongale, 2001

VERMICULTURE

The new concept of sustainable agriculture is fast replacing the practices that depended on inorganic fertilizers and more emphasis is being laid on application of farmyard manure, compost, green manure, green leaf manure (Basavarajappa, 2001 a & b). Of late, vermiculture has gained considerable popularity and become an economical substitute. In 'vermiculture' the earthworms are cultivated for rapid conversion of organic wastes into value added manure (Das *et al.*, 1993). It holds out a good prospect for transforming the colossal sericultural wastes (Das *et al.*, 1997) agricultural, residential and other wastes of organic origin into vermicompost. The products of vermiculture are rich in Nitrogen, Phosphorus, Potash, micronutrients, enzymes and vitamins suitable for crop growth. U.S.A., Canada, U. K., Mexico, Germany, Japan, Taiwan and East Asian countries have already commercialized vermiculture to produce vermicasts on large scale basis. In India, there is no dearth of organic wastes and man-power to do vermiculture. It is being practised widely in southern states, Maharashtra and West Bengal. Further, the agricultural universities, national and state level research institute scientists are doing research on the vermiculture technology for efficient use of agricultural wastes.

Vermiculture can be practiced with the involvement of educated unemployed youths, rural masses without much financial involvement. In this technology, earthworms are used as versatile natural 'bioreactor' for rapid conversion of organic wastes and to obtain organic fertilizer in short duration. Several species of earthworms (viz., *Eudrillus euginae, Isenia foetida, Perionyx excavatus, Kisenia fetida*) are mass multiplied and used in a vermary. These voracious eaters convert any decomposable, and bio-degradable waste such as city refuses, cattle and horse dung, sheep and poultry droppings, agriculture waste such as sugarcane thrash, banana stem, decomposed vegetables, fruits, silkworm litter, mulberry waste, waste paper, waste cardboards, saw dust, semi-rotten kitchen refuse, shredded coconut coir into valuable manure within a span of about six weeks. The cement tanks, wooden boxes, stone lined pits or plastic tubes of convienent size can be used (Kale, 1995) for vermiculture process to produce vermicompost. The bedding and feeding materials collected from various source should be mixed thoroughly and allowed to decompose for about 2-3 weeks. To assist uniform decomposition, the decomposing material should be turned 3 or 4 times and when the material becomes quite soft, the matured earth worms are introduced into the tanks. Temperature is maintained at 20 to 30° C. and 40-50% moisture and pH of 6.8 to 7.5. The contents are mixed once in every 10-15 days for aeration. Depending upon the softness of the material, the worms may require 15-20 days to convert the entire materials into vermicompost (Kale, 1995).

Vermicompost contains good amount of nutrients (Table - 2) which play a vital role in the development of plants. It acts as an excellent base for the establishment and multiplication of symbiotic microbes (Kale, 1995). The vermicompost provide balanced nutrition to plants, increase organic matter, improve soil structure, aeration, water drainage and water holding capacity. This helps ground water recharge. Soil erosion and salinization can be minimised by using vermicompost in farming practices (Kale, 1995). The vermicompost benefits are thus multidimensional.

Table—2. Useful nutrients for plants and microbes in vermicompost contents

Sl. No.	*Contents*	*Composition (%)*		
	I. Chemical nutrients :			
1.	Organic carbon	9.20	-	18
2.	Total Nitrogen	1.5	-	1.9
3.	Phosphorus	0.3	-	0.6
4.	Potassium	0.6	-	0.1
5.	Sodium	0.06	-	0.3
6.	* Calcium	22.7	-	46.7 (MEC/100gm)
7.	* Magnesium	22.7	-	46.7 (MEC/100gm)
8.	* Copper	2.0	-	9.5 (PPM)
9.	*Iron	2.0	-	9.3 (PPM)
10.	* Zinc	5.7	-	11. (PPM)
11.	* Sulphur	128	-	548 (PPM)
	II. Microbes: - Considerable amounts			
1.	Symbiotic microbes			
2.	Vesicular Arbuscular Mycorrhizal (VAM) fungi			
3.	Plant Growth Promoting Substances (PGPS)			
4.	Humus forming microbes and others			

Note:* - Not in percentage; Source : Kale, 1995 and Das *et al.*, 1997

BIOFERTILIZERS

Biofertilizers are the microbial inoculants, which contain living cells of species of various micro-organisms viz., Bacteria, Fungi etc. These organisms have the ability to mobilise nutritionally important elements from non-usable to usable form through biological process (Basavarajappa, 2001 a & b).

The advent of biofertilizers is as old as a century, the need of its commercial exploitation was not felt in traditional agriculture. However, the development of microbial inoculants for improved supply of nitrogen, phosphorus and vital trace elements for the plants is a significant development in past few decades (Arya, 2000). In recent years, biofertilizers have emerged as an important component of Integrated Nutrient Supply System (INSS) and hold a promise to improve the crop yields and nutrient supplies. In the recent years, the growing nutrient imbalance poses a major threat to sustain soil health and crop productivity. The recent fall in fertilizer consumption due to unprecedented hike in the prices of phosphorus and potassic fertilizers has further aggravated the need of adoption of Integrated Nutrient Supply System (INSS) which involves the combined use of different nutrient sources such as chemical fertilizers, organic manures, farmyard nanure (FYM), green manure, vermicompost, biofertilizers and other eco-friendly management packages in agriculture.

The biofertilizers act as perpetually renewable inputs helping in better crop nutrient management and maintenance of soil health (Arya, 2000). Further, microbes influence life process of all the living systems. Their dynamic inter-relationships with plants have been identified mainly in three ways i.e., saprophytic, pathogenic and symbiotic means. Application of symbiotic microbes as 'biofertilizer' become a major alternative, which tends to minimise the ecological disturbance (Arya, 2000). They play a key role in maintaining the biological equilibrium on earth (Lal Sing, 1999).

Bacteria

Free living nitrogen fixing bacteria are classified into aerobic, anaerobic and facultative aerobes depending on the growth and survival of the organisms in the presence or absence of oxygen. *Azotobacter* and *Azospirillum* are aerobic bacteria. *Azospirillum* is a micro-aerophilic nitrogen fixer (Table-3) in the rhizosphere of soils and different plant parts such as roots, stems.

The combined application of *Azospirillum*, Vesicular Arbuscular Mycorrhizae (VAM) and Phosphate Solubilizing Bacteria (PSB) has contributed to increased plant height, and plant spread. Increased productivity due to enhanced leaf yield in mulberry has been observed by several workers. The evolution of tripartite system (host VAM – diazotroph) is beneficial for nitrogen and phosphorus economy in natural environments. The VAM explores the soil volume more thoroughly and finds more sources of phosphorus The ability of different microorganism especially PSB to stabilize bound phosphates has been well established.

Diazotrophs

Diazotrophs (Table-3) are free living as well as associated forms which fix nitrogen in the rhizosphere of a variety of crop species viz., stem and leaves of sugarcane, rice, cereals, grasses, pearl millet (Okon, 1968). Species of Cyanobacteria (Table 3) also fix nitrogen in crops like rice which grow in plenty of water (Arya, 2000). The cyanobacterial flakes are dried and mixed at the rate of 10 15 kg ha-1 one week after rice is planted.

Table—3. Agents of Biofertilizers in Agriculture and other farming activities

Sl. No.	*Group*	*Species*	*Used in*
1.	Nitrogen fixing bacteria	*Azorhizobium* spp.	Cereals
		Azospirillum lipoferum	Legume crops
		Azotobacter chroococcum	Pigeon pea & others
2.	Rhizobial bacteria	*Rhizobium* spp.	Wheat, Rice, Maize
		Bradyrhizobium spp.	Oil seed rape
		Azorhizobium spp.	- do -
3.	Cyanobacteria	*Anabaena* spp.	Most of the crops which grow in plenty of water
		Nostoc spp.	
		Scytonema spp.	- do -
		Aulosira spp.	- do -
		Nodularia spp.	- do -
		Lyngbya spp.	- do -
		Mastigocladus spp.	- do -
4.	PSB (Phosphate. Solubilizing Bacteria)	*Bacillus* spp., *Acrobacter acrogenes*	Wheat & other Leguminous Crops
		Pseudomonas spp., *Serretia* spp.	
		Brevibacterium spp., *Nitrobacter* spp.	
		Bacillus megaterium var. *phosphaticum*	- do -
		Pseudomonas striata	- do -
		Bacillus polymyxa	- do -
5.	PGPR (Plant Growth Promoting Rhizobacteria)	*Pseudomonas fluorescens*	Used as microbial pesticide against *Pythium* & *Rhizoctonia*
6.	Ectomycorrhizal fungi	*Amanita muscaria*	Agriculrural and forestry farms
		Laccaria laccata	Mulberry and other commercial crops
		Pisolithus tinctorius	
		Russula aeruginea	
7.	Endomycorrhizal fungi	*Glomous fasiculatum*	Different soils of Baroda during cropping
		Endogone vesicular	In agriculture field
		Arbuscular mycorrhizal	- do -
		fungi (VAM)	- do -
8.	Mycorrhizal Fungi	*Amanita muscaria*	- do -
		A. pantherina	- do -
		Suillus granulatus	- do -
		Lactarius deliciosus	- do -
9.	Actinomycetes (Ray fungi)	*Frankia* spp	Higher plants
10.	PSF(Phosphate Solubilizing Fungi)	*Aspergillus awamori*	Coconut and Cocoa

(Contd...)

11.	VAM (Vescicular Arbuscular Mycorrhizal Fungi)	*Glomus* spp. *Gigaspora* spp *Acaulospora* spp. *Enterophospora* spp *Sclerocystis* spp. *Acaulospora* laevis *A. dicolsoni, A. lubis*	Commercial crops like Rice, Jowar, Pearl millet, Soybean, Groundnut, Barely, Sunflower, Zingiber and etc.
		Glomus aggregatum, G. albidum *G. mosseae,* *Glomus caledonium*	Oil seed crop – Safflower, Mug, Green gram, Urd bean
12.	Basidiomycetes	(Ectomycorrhizae Fungi)	*Boletus* spp. *Hebeloma* spp. Forestry plants
		Tricholoma spp. *Cortinarius* spp. *Suillus* spp. *Russula* spp. *Laccaria* spp. *Lactarius* spp.	
13.	Gasteromycetes	*Rhizopogon* spp. *Scleroderma* spp. *Alpova* spp. *Pisolithus* spp.	- do -
14.	Blue Green algae	*Anabaena azolla*	Rice
15.	Diazotrophs	*Thiobacillus* spp.	Pearl millet,
		Desulphovibrio spp	Variety of crop species
		Herbaspirillum spp.	Sugarcane and Rice
		Azospirillum brasilense	-do-

Source: Monoharachary and Rani, 1992; Lal Singh, 1999 and Arya, 2000

A large number of heterotrophic and autotrophic soil microorganisms are now known to have the capacity to solubilize inorganic phosphates through their metabolic activities directly or indirectly. *Bacillus* spp., *Psdeudomonas* sp., *Brevibacterium* sp., *Acrobacter* sp., *Nitrobacter* sp., are important phosphate solublizing and phosphate mobilizing microbes (Table-3). The product of *Bacillus megaterium* var. *phosphaticum* used as '*Phosphobacterin*' (Arya, 2000).

Actinomycetes

Actinomycete members also solubilize phosphates. They are is used in the rhizosphere of coconut and cocoa (Table-3). The Plant Growth Promoting Rhizobacteria (PGPR) are non-colonizers, belonging to Pseudomonads which play a major role in plant productivity.

Fungi

In modern agriculture system, the term "Mycorrhizae" (Fungus root) is gaining importance due to their role in plant growth and health. The term 'Mycorrhizae' was coined by a German Botanist Albert Bernad Frank in 1885. It referes to a symbiotic association between plant roots, and fungi

Ectomycorrhizea: About 5000 species of fungi have been identified as ectomycorrhizae, occuring on 2000 species of the woody plants (belongs to families – Pinaceae, Fagaceae, Salicaceae, Juglandaceae, Myrtaceae, Ericaceae.(Lal Singh, 1999 and Marks and Cordel, 1989). In the developed countries the ectomycorrhizal fungi with greater potential have been identified as mycorrhizal biofertilizers. The pure cultural mycelial inoculum is introduced to deficient soils and for the nursery practices. Forest Research Institute, Dehradun, India has initiated the work on mycorrhizal association in late fifties (Bakshi, 1974). Later, several workers have studied different aspects of ectomycorrhizal association with Pinus species (Sharma and Lakhanapal, 1981; Kumar and Lakhanpal, 1983), mulberry (Padma *et al.*, 2000) and other plants.

The ectomycorrhizal association enhance the water absorption, uptake nutrients, increase the tolerance to drought, high soil temperatures, organic and inorganic soil toxins, and extremes of soil acidity caused by high levels of sulphur or aluminium (Lal Singh, 1999). Mycorrhizal technology, has brought waste land and marginal lands under green cover in many developing countries. In India, there is a wide scope for artificial mycorrhization, where selection, propogation, manipulation and management of superior strains of ectomycorrhizal fungi are paramount.

Endomycorrhiza

Endomycorrhiza are the most widespread types and comprise three groups Ericaceous, Orchidaceous and Vesicular Arbuscular Mycorrhizae (VAM), found in agricultural crops (Table-3). About 90% of the 30,000 species of vascular plants form VAM. Whereas Ectoendomycorrhizae are intermediate types with features of both Ecto and Endomycorrhizae. In recent years, a considerable work has been carried out on the beneficial values and the influence of VAM on commercial crop plants. The fungus is present in the soil as sporocarps or chlamydospores. During favourable conditions, the spores germinate and hyphae enter the roots through the root hairs and produce finger shaped structures with the cortical cells called arbuscules or bladder shaped which are either inter or intra cellular and perform absorptive and storage function (Padma *et al.*, 2000). The root system of majority of angiospermic plants are colonized by network of mycorrhizal fungi. The efficacy of VAM and its influence on mulberry cultivation has also been done (Padma *et al.*, 2000).

The plants in the top soils of the cultivated fields harbour endomycorrhiza (VAM) in their root system. Mycorrhizal plants develop extensive network of hyphae which ramifies in the soil as an accessory root system of plants helping in efficient absorption of water, mobilization of nutrients (viz., phosphorus, nitrogen, potash, zinc, magnesium, calcium and sulphur). The VAM also plays an important role in the biological control of root pathogens and has a crucial role in reducing the nematode infection at the plant roots due to the production of chemicals, physiological and pathological changes in host plants. The VAM also secretes antibiotics like Diatretynenitrite, growth hormones (viz., Auxins, Cytokinins, Gibberellins) and growth regulating B vitamins. Therefore, VAM promotes good root growth, overall development of plants and inturn enhances the productivity. Furthermore, it has been suggested that fungus increases the rate of CO_2 fixation and helps in maintaining the CO_2 level of atmosphere which is most important from the environment point of view (Koch and Johnson, 1984). Thus, to protect our environment, a multibeneficial organism 'VAM' can be exploited and put into use as 'Biofertilizers'.

Advantages of Biofertilizers

1. Microbial processes are quick.
2. Biofertilizers are less energy consuming.
3. Biofertilizers are formed by myriad of organisms.

4. Soil microorganisms help the plants to obtain nutrients from the soil by means of their intimate relationships with the root system.

5. The micro-organisms present in the soil, water and air when brought into contact with plants inturn influence plants growth and functions which are mostly beneficial.

6. They degrade plant residues.

7. The biofertilizers can be used in different field crops.

8. Biofertilizers act as perpetually renewable inputs for crop nutrient management and maintenance of soil health.

9. Biofertilizers sustain soil fertility

10. They play a key role in maintaining the biological equilibrium on earth.

11. They posess astonishingly rich physiological qualities.

12. They are ubiquitous.

13. Since biofertilizers comprise microbes, they influences life process of all other living systems.

14. Microbial organisms help in increasing the quality and crop yield through:

    (i) Biological nitrogen fixation,

    (ii) Mobilising plant nutrients, and

    (iii) Degrade ligno-cellulosic wastes

    (iv) Accelerate the recycling of organic resources.

PLANT GENETIC ENGINEERING TECHNIQUE

Plant Genetic Engineering Technique (PGET) is a biotechnological approach, where Genetically Modified Crops (GMC) grow can be produced which can grow under stress conditions (Ravikumar and Sreekantah, 2001). The Plant Genetic Engineering (PGE) has varied applications in the field of agriculture. It can improve the nutritional quality, resistance to pests, diseases and climatic stresses. The technique involves the insertion of a desired gene or group of genes into suitable vector system chosen for its ability to which will eventually transfer Deoxyribonucelic Acid (DNA) into plant cell. In this regard, a considerable work has been done on agriculture crops such as corn, soyabean, potatoes, tomatoes, coffee, rapeseed oil and rice. Thirteen countries from six continents are using genetically improved crops in 44.2 million hectares of land. Due to gradual increase of total global hectarage of the transgenic plants by molecular farming, the genetically engineered seed market has reached 500 million US$ (Ravikumar and Sreekanth, 2001). This new technique can circmvent the limitations of conventional farming activities and inturn may solve the problem of environmental pollution.

BIOPESTICIDES

Crop protection is a vital component of farming (Basavaraju, 2001) under diversified terrestrial environment. Over the last few decades, the use of chemical pesticides have been used for pest control in agriculture fields. Various chemicals such as insecticides, fungicides, herbicides, weedicides, rodenticides, nematicides, molluscicides, acaricides are being applied to control the pests. The pesticides belong to different classes of chemical compounds, which affect not only the pests but also the non-target organisms like economically important insects (Silkworm *Bombyx mori*, Honey bee *Apis* spp. Lac insects, etc.). and wild life including mankind. Although chemical pesticides are essential for increasing the food grains production, their excess use affects the ecological balance due to their high toxicity. Appearance of pesticide resistant strains of pests have compounded the problem along with residual effect of systematic insecticides, chemosterilants, metabolic pesticides and contact poisions.Biological magnification of pesticides has further added to the problem.

Biopesticides with low toxicity and less harmful to biological systems have to be developed. The development and production cost of biopesticides is much less as compared to chemical pesticides (Chakraborty and Basu, 1997). It is imperative now to discuss the eco-friendly biological products for pest control which are produced from natural or synthetic means to increase quality food grains production and prevent environmental pollution and health hazards.

Biological control of insect pests has a long history. Farmers prepared the biopesticides from seeds of resistant plants. Natural predators were used for the control of pests as early as 12th century. In 1762 the Indian Myna bird was imported by Mauritius for the control of red locusts. In 1888 predatory Vedalia beetle (*Rodolia cardinalis*) was exported from Australia to California to control cottony cushion scale insect in Citrus orchards. Popularisation of biopesticides is going on in a big way in India for controlling various pests of vegetables, cotton and other commercial crops. India imported parasitoid (*Aphelins mali*) from England to control apple wooly aphid (*Rodolia cardinalis*) cottony cushion scale. Serious efforts have been made in U.S.A. and other developed countries to replace chemical pesticides with environmentally safer pest management methods. Use of parasitoid (*Prospeltella perniciosi*) against San Jose scale infesting apple in USA (1958) parasitoid (*Telenomus* spp.) against castor semilooper in New Guinea (1964) and predator (*Platymeris*) against coconut rhinoceros beetle from Zanz. Bar (Subhash Chander, 1999) are some examples. Directorate of Plant Protection, Quarantine and Storage, Faridabad, India has involved in mass rearing and utilization of natural enemies of commercial crop pests through its Central Integrated Pest Management Centres (CIPMCs) located all over India (Subhash Chander, 1999). Similarly Indian Council of Agricultural research (ICAR) Institutes and state agricultural Universities also took up mass rearing of natural enemies. All India Co-ordinated Research Project on Biological Control of Crop Pests and Weeds (AICRPBC) is in operation since 1977. Many of the Non-Governmental Organisations (NGOs) such as pest control (India) Ltd., Sonepat co-operatve Sugar Mills Ltd., Simbhaoli Sugar Mills Ltd., Sriram Research Farm, Tamil Nadu Co-operative Sugar Federation, Chigleput and Nawanshaher Co-operative Sugar Mills are carrying out research and mass rearing of natural enemies (Subhash Chander, 1999).

Table—4. Modern insecticides with minimum hazards.

Sl. No.	*Source*		*Name of insecticide*	*Target pest/ organisms*
1.	Plant origin (Neem product)	*(i)*	Azadirachtin	Nematodes Root borers
		(ii)	Meliacin	Root rot of
		(iii)	Nimbin	Brinjal, Groundnut,
		(iv)	Salannin	Onion, Stem rot of
		(v)	Meliantriol	Jute, leaf curl of Tomato
		(vi)	Neem kernel	and Chilli
2.	Synthetic			
	(a) Herbicides:	*(i)*	Flufenacet	weeds, grasses
		(ii)	Azafenidin	-do-
	(b) Fungicides:	*(i)*	Kresoximmethyl	Leaf spot, Leaf rust causing pests.
		(ii)	Famoxadine	-do-
		(iii)	CGA 279.202	-do-
	(c) Insecticides	*(i)*	Etooxazole	-do-
		(i)	CGA 293.343	-do-
		(iii)	Carbosulfan,	-do-
		(iv)	Imidacloprid 17.8SL	-do-

Source:Chakraborty and Basu, 1997

Microbes for Harmful Insects Control

Many microbial organisms (Table-6) are used as potent biocontrol agents. They are called as antagonists (Chakraborty and Basu, 1997). They are toxic to an array of commercial crop insect pests. They are specific to a particular insect and thus are harmless to man and domestic animals. Baculoviruses are an important group of biocontrol agents which are highly invertebrate specific and safe for poultry, fish and honey bees. There are many Baculoviruses which possess large crystals inside their protective coat. The Nuclear Polyhedrosis Virus (NPVs) are virulent, highly specific, and have been found to be very effective against caterpillars like gram pod borer and hairy caterpillars on several crops (Table-5). The first viral insecticide was marketed in 1975 by Sandoz under trade name Elcar. In recent years a number of insect specific viruses are used. The most important among them are NPV and Granulosis virus (GV). Some of the commercially known viral insecticides are given in Table-5. However, NPVs and GVs can cause damage to silkworms. Thus, the safety tests must be conducted so as to assess their effect on non target organisms before their release.

Table—5. Commercially known Bioinsecticides available in the market

Sl. No.	*Source*	*Trade name*
1.	Viral based formulations	(*i*) Elcar (*ii*) Binab T (*iii*) P. g. Suspension
2.	Bacterial based formulation	(*i*) Thuricide HP (*ii*) Biotrol BTB (*iii*) 25W (*iv*) Thuricide dust (*v*) Bt
3.	Fungal based formulations	(*i*) Boverin (*ii*) Metaquino (*iii*) An-27

Source: Chakraborthy and Basu, 1997)

The entomopathogenic bacterial strains were first recognised by the Japanese biologist S. Ishiwata in 1901 in silkworm *Bombyx mori* (Lambert and Peferoen, 1992). In 1911, the German biologist E. Berliner has described the pathogenic bacteria found in pupae of Mediterranean flour moth and other insect larvae living in stored grains in the city of Thuringen (East germany). After realising its pathogenic nature of this bacteria, it was named as *Bacillus thuringiensis* by Berliner in 1911. Upto 1976, it was known that, the *B. thuringiensis* strains are pathogenic to lepidopterans alone. In 1977, Goldberg and Margalit have shown that *B. thuringiensis* could also be pathogenic to Dipterans. This discovery has gained immediate practical importance world wide (Sur and Bihari, 1997) and opened a new horizons to produce bio-insecticidal preparation on large scale basis. The bacterial formulations of *B. thuringiensis* and *B. Sphaericus* are now being widely used as bio-insecticides. These bio-insecticides are toxic to many agents in virtually every country where such pests are a severe problem (Sur and Bihari, 1997).

The bacterium *B. thuringiensis* (Bt) produce delta endotoxin which is in a crystalline form called parasporal body and spores are highly toxic to more than 182 species of insects (Jyothi and Patil, 1997). It is used as high potential bio-insecticide because of its ability to produce toxin which can be used against an array of pests such as bollworms of cotton, tobacco hornworm and pests of vegetables like cabbage, cauliflower, borers of peas etc.

The spore and the parasporal body together constitute the bioinsecticide and these have been commercialized in different trade names. Recently a number of foreign and Indian Research Institutes have developed different non-sporulating strains of Bt, which are toxic to specific insects. The commercially available bacterial formulations are used in many countries to control insect pests. However, *B. thuringiensis* (Bt) has banned in India in sericulture areas, where it causes 'sotto' disease in silkworms(Jyothi and Patil, 1997)

Table—6: Microbial agents for insect pests control

	Type of Microbes		*Microorganism*	*Target pests*
1.	Viruses	*(i)*	NPV (*Nuclear Polyhedrosis Virus*)	Caterpillars like Gram pod borer, Hairy caterpillars on several crops
		(ii)	GV (*Granulosis virus*)	- do -
2.	Bacteria	*(i)*	*Bacillus thuringiensis* (bt)	Lepidopterous larvae Viz., Diamond black moth, Army worm,Butterfly, Fruit borer, Leaf miner, Shoot & fruit borer in Brinjal and ladies finger
		(ii)	*B. sphaericus*	- do -
		(iii)	*B. thuringiensis* var. *israelensis*	Black flies
		(iv)	*B. thuringiensis* var. *kurstaki*	Lepidopteran larvae
		(v)	*Clostridium bifermentans*	Mosquitoes, plants juice sucking insects
3.	Fungi	*(i)*	*Beauvera* spp.	Hairy caterpillar, Banana Rhizome borer
		(ii)	*Metarrhizium* spp.	Sugarcane Pyrilla
		(iii)	*Fusarium* spp.	Mealy bug, Hadda beetle on Brinjal
		(iv)	*Entomophthora* spp.	Aphids, Rice brown plant hoppers
		(v)	*Verticillium* spp.	Coffee green bug
		(vi)	*Asperqillus* spp.	Pentatomid bugs
		(vii)	*Verticillium lecanii*	Several crop pests
		(viii)	*Hirsutela thompsonii*	-do-
		(ix)	*Trichoderma harianum*	-do-
		(x)	*Gliocladium vivens*	-do-
4.	Protozoans	*(i)*	*Nosema* spp.	Grass hoppers, Boll weevils, Citrus mites, Mosquito larvae.
		(ii)	*N. infesta*	Sod webworm
		(iii)	*N. locustae*	Grasshoppers, Crickets
		(iv)	*N. fumiferanae*	Forest pest Spruce budworm
		(v)	*N. necatrix*	Tobacco budworm
		(vi)	*Glugea gasti*	Boll weevil
		(vii)	*Pleistophora culicis*	*Culex pipiens fatigans*
		(viii)	*Thelohania hyphantriae*	Fall webworm
		(ix)	*Vairimorpha necatrix*	*Spilosoma* spp and others

Source:Bhagat and Bala, 1993; Chakraborty and Basu, 1997 and Subhash Chander, 1999)

Several species of Fungi (Table-6) have been found to be effective control agents of several crop pests. Only few fungi are commercially used as bioinsecticides in a number of countries. In India, *Trichoderma harianum* and *Gliocladium vivens* are under process of commercialization (Chakraborty and Basu, 1997). These fungi are not harmful to man. Their application against commercial crop pests is expected to be cheaper than synthetic pesticides.

Plant Products as Biopesticides

More than 600 plant species have been identified in the world to control harmful insects (Chakraborty and Basu 1997). Some of the plant flowers act as an insect repellent, (*Tagetus spp.*). The seed kernel of Neem (*Azadiradicta indica*) has been found to be quite effective against more than 200 pests which cause root rot of groundnut, brinjal, onion, stem rot of jute, leaf curl of tomato and chilli. The Neem cake is more effective against sheath blight in rice, red gram, green

Table—7. Plant products used for the control of pests and plant diseases

Sl. No.	*Plant*	*Scientific Name*	*Products*	*Target pest*	*Plant disease*
1	**2**	**3**	**4**	**5**	**6**
1.	Black pepper	*Piper nigrum*	Seed extract	Agricultural pests	-
2.	Brown mustard	*Brassica juncea*	Leaf extract	-	Burn spot in Rice
3.	Castor	*Ricinus communis*	Seed extract	Weevil, Aphids, Caterpillars	-
4.	Chrysanthemum	*Chrysanthemum cinerarietolium*	Dried flowers	House hold pests	
5.	Citronella grass	*Cymbopogon nardus*	Leaf & Bark extract	-	Stem & root rot in Jute, Groundnut & Ornamental crops
6.	Custard apple	*Annona squamosa*	Leaf & Seed	Diamond black moth pod borer in Bringal, Rice, Wheat	-
7.	Datura	*Datura metal*	Leaf extract	-	Fungal infection, brown rust in Rice/Wheat
8.	Garlic	*Allium sativum*	Bulb, Leaf extract	Fruit flies	Powdery mildew in pea grains and fruit trees
9.	Grape fruit	*Citrus paradise*	Seed extract	Colarado Potato beetle, gram pod borer ant etc., in Potato, Tobacco	-

(Contd...)

Table—7 (Contd...)

1	2	3	4	5	6
10.	Guava	*Psidium guayava*	Leaf extract	-	Brown rust in Rice/Wheat
11.	Ipomoea	*Ipomoea* spp.	Leaf powder Leaf extract	Earhead bug in Cotton, Rice	White fly sheath rot in Rice & Cotton
12.	Jowar	*Andropogon sorghum*	Leaf extract	-	Twig rot in Groundnut
13.	Karabi	*Nerium indicum*	Leaf dust & extract	-	Brown rust in Rice
14.	Lalpata	*Euphorbia pulcherrima*	Leaf extract	-	Burn spot in Rice
15.	Kurchi	*Holarrhena antidysenteria*	Bark exract	-	Root knot in vegetables
16.	Lantana	*Lantana camera*	Leaf extract	Leaf miners in Potato,Brinjala, Chillies, Onion, Tomato and others.	-
17.	Naochi	*Vitex negundo*	Leaf powder and Extract	Earhed bug, pests of stored grains	-
18.	Papaya	Carica papaya	Leaf dust & extract	Caterpillar, Aphid, Weevil, & Garden bug in Rice, Wheat, Maize & Ornamental crops	-

(Contd...)

Table—7 (Contd...)

1	2	3	4	5	6
19.	Turmeric	Curcuma domestica	Seed dust	–	Root rot in Rice, Wheat & vegetables
20.	Tulsi	Ocimum basilicum	Leaf extract	Leaf miners in vegetables	
21.	Periwinkle	Catharanthus roseus	Leaf, flower root extract	White fly in Cotton	
22.	Neem	Azadiracta indica	*(i)* Neem cake	Leaf folder, Earhead bug, pod borer in Rice, Red gram, Bengal gram White fly, leaf hopper in Cotton & Castor	Sheath blight, Sheath rot, Powdery mildew in Rice, Red gram, Bengal gram.
			(ii) Seed Kernel	Rhinocerous beetle in Coconut	
			(iii Leaf extract	Leaf curl in Tomato Chilli, Groundnut, Jute	Root & Stem rot in Groundnut Jute and Vegetables

Source: Chakraborty and Basu, 1997

gram, bengal gram and others. Some of the plant products which can be used as substitutes are recommended in Table 4.

Table—8. Animals used as biocontrol agents

Sl. No.	*Type of Biological agent*	*Scientific name*	*To control*
1.	Egg Parasitoid	*Trichocramma* spp.	Stem borer of Paddy, Maize, Sorghum & Sugarcane
2.	Egg Parasitoid	*Trichocramma japonicum* *T. chilonis*	Rice stem borer Cotton bollworms
3.	Larval parasitoids	*Apanteles* spp. *Bracon* spp.	Cotton bollworms & defoliators
4.	Parasitoid	*Aphelinus mali*	Apple wooly aphid
5.	Parasitoid	*Prospeltella perniciosi*	San Jose scale infesting apple
6.	Parasitoid	*Telenomus* spp.	Castor semi-looper
7.	Predatory beetle	*Rodolia cardinalis*	Cotton cushion scale insect in Citrus orchards
8.	Predator	*Platymeris*	Coconut Rhinoceros beetle
9.	Predator	Lady bird beetle, Syrphids & Lacewings	Aphids on several crops
10.	Predator	Lizard	Grasshoppers, Locust
11.	Predator	Spiders, Rove beetle, Ground nut beetle, Mirid bug, Dragon fly, Fish, Frog, Toad and Duck	Devour Paddy pests
12.	Predatory mite	*Phytoseiulus persimilus* *Typhlodromus occidentalis*	Phytophagous tetranychid mites. -do-
13.	Nematodes	*Neoaplectana*	Insect pests
14.	Wasp	*(i)* *Trichogamma* spp. *(ii)* *T. Chilonis* *(iii)* *Neopletana carpocapsae*	Caterpillar in cotton, corn, apple, spruce and avocado Internode borer of sugarcane Carpenter worm insect
15.	Spider	Wolf spider	Brown plant hoppers (Asia's most destructive vice pest)
16.	Frog	*Rana* spp	to attract and kill harmful insects
17.	Nematodes	*(i)* *Steinernema* spp *(ii)* *Heterorhabdits* spp	Insect pests Insects pests

Animals Used as Biocontrol Agents

An insect can also control insect pests. About 11,000 insects have been identified as biocontrol agents against harmful insects (Chakraborty and Basu, 1997). Some of the predators, and parasitoids are enlisted in Table-8.

CONCLUSION

The promotion of biological control would thus greatly reduce dependence on chemical pesticides, and reduce the problem of chemical residues in food and help alleviate environmental pollution. This helps to manage pests in an eco-friendly and sustainable manner (Subash Chander, 1999). India being an agriculture country, has proclaimed her self sufficiency in food grain production. Her food grain production is expected to cross 240 million tons by 2002. In the past few decades, however the increase in food production has declined from 30 million tons to about 12 million tons annually. A certain percentage of people are facing problem of toxic compounds in agriproducts and food adultration. A good nutrition is therefore an indispensible component of healthy life, it comes through use of quality agriproducts. Therefore, production of quality agriproducts is a prerequisite in the 21st century, where sustainable agriculture can be made through modern biological/bio-technological methods which are effective substitutes for chemical fertilizers and pesticides. Conventional and modern bio-technological methods in agricultural activities could help to conserve the environment and produce toxic free agriproducts for our future generation in this century.

ACKNOWLEDGEMENTS

I am thankful to the Chairman, Department of Applied Zoology, Kuvempu University, B.R. Project for encouragement.

REFERENCE

1. Arya A. 2000. Biofertilizer for sustainable plant development. Everyman's Science. 35(1): 19-25.
2. Bakshi B. K. 1974. Mycorrhiza and its role in forest. FRI. Dehradun.
3. Basavaraju S. N. 2001. Pesticides, mulberry plants and silkworm. Indian silk. 39(11) 9-10.
4. Basavarajappa S. 1992. Atmospheric toxicity. J. NIE. Lucknow. 2(11): 17-20.
5. Basavarajappa, S 2001a. Krushi Khsethradhalli Jaivikagobaragala Balake. Prasaranga, Kuvempu University, Shankaraghatta. Pp. 1-50.
6. Basavarajappa S. 2001b. Sukshmaanu aekakoshajeevi Bacteriagalu. Vijnana Sangathi. 8(6):28-29.
7. Bhagat R. C. and P. Bala 1993. Protozoans as microbial insecticides. Everyman's Science. 28(1): 14-16.
8. Bongale U. D. 1991. Mulberry sericulture zones and agro-climate in Karnataka. Indian Silk. 30(5): 7-12.
9. Bongale U. D. and S.B. Dandin 1993. Nitrogen fixing Bacterial Biofertilizer. Indian Silk. 32(2): 28-33.
10. Chakraborty I. and S. Basu 1997. Biopesticides: for our healthier Tomorrow. Everyman's Science. 32(1): 16-26.
11. Das, P. K. K. Bhogesha, P. Sundareswaran, Y.R. Madhava Rao and D.D. Sharma 1997. Vermiculture: Scope and potentiality in Sericulture. Indian Silk. 36(2): 23-26.
14. N.B. Jyothi and C.S. Patil 1997. *Bacillus thuringiensis* as a Bio-insecticide - A Threat?. Indian Silk. 35(12): 11-12.
15. Kale R. D. 1995. Vermicomposting has a bright scope. Indian Silk. 34(5): 6-8.
16. Khare, R.; Md. Isa and K. Thangavelu 1992. The soil of tassar growing belt of Chotanagpur. Indian Silk. 30 (9): 37-38.

17. Koch K.and C.R. Johnson 1984. Plant Physiol. 75:26.

18. Krishna M.and U.D. Bongale 2001. Role of organic manure on growth and quality of mulberry leaf and cocoons. Indian silk. 40 (2) : 11-12.

19. Kumar S. and T.N. Lakhanpal 1983. Tree Science. 2:38.

20. Lal Sing 1999. Ectomycorrhizal symbiosis: Introduction and use in forestry. Everyman's Science. 34(2): 62-65.

21. Lambert B. and Peferoen 1992. Bioscience. 42:112.

22. Marks D.H. and Cordell C.E 1989. Biotechnology of fungi for improving plant growth. Cambridge Uni. Press. Cambridge. pp. 1-25.

23. Monoharachary C. and J. Rani 1992. Management of Microbes . Natl. Symp. Allahabad.

25. Naik M.K. 1993. Sustainability and the environment with special reference to food production in the Semi-Arid Tropics. Everyman's Science. 28(1): 11-13.

26. Okon Y 1968. Biological nitrogen Fixation: Recent developments. Oxf. And IBH. Publ. Co. Delhi.

27. Padma, S.D.; J. Sukumar and T.D. Nagendra Kumar 2000. Vesicular Arbuscular Mycorrhiza & in mulberry cultivation. Indian Silk. 39(1): 11-12.

28. Ravikumar G. and B. Sreekanth 2001. Geneticaly modified crops – Emerging issues and concerns. E. News. 24(29):1-3.

29. Sharma B.K. and Iqbal Habib 1994. Water hyacinth - use, abuse and control. Everyman's Science. 29(2): 38-40.

30. Sharma J.R. and T.N. Lakhanpal 1981. Proceeding of Symposium on Improvement of Forrest Biomass. Soc. Tree Scientists, Solan. pp. 455.

31. Sinha, M. A.; K. R. Subbaswamy, M. Munirathanam Reddy and N.R. Singhvi 1992. Soil fertility management in mulberry gardens. Indian Silk. 30(9): 13-17.

32. Sinha, K.K.; S.N. Sinhadeo, N. Chakraborty, B.D. Dash, K.V.S.N. Rao, D.P. Dasmohapatra and K.C. Brahma 2000. Response of azotobacter to mulberry in cool and dry regions of Orissa under rainfed conditions. Indian Silk. 39(3): 5-8.

34. Sridhara, S. H.R. Shivakumar and C. Doreswamy 1995. Mulberry leaf yield and Nitrogen application. India Silk. 34(2): 23-24.

35. Subhash Chander 1999. Biological control of insect pests in Agriculture. Everyman's Science. 34(3): 127-131.

36. Sur B.and V. Bihari 1997. Microbial control of Mosquitoes. Everyman's science. 32(3): 105-107.

37. Teotia, R.S. K.P. Singh and S.K. Sen 1992. Sericulture can meet challenges of drought in Dry lands. Indian Silk. 31(6):40-43.

9

Ecofriendly Pesticides-An Emerging Trend

Shalini Khanna, M.M. Srivastava & Shalini Srivastava

Dr. **Norman E. Borlaug,** the Noble Prize winner for peace (1970), who was designated as Father of "**Green Revolution**", stated in an address to **United Nation's Food and Agriculture Organization** that use of pesticides is indispensable for rise in agricultural production. It is truly reflected in almost all the countries of the world. It is a well-acknowledged fact that pesticides are one of the most important components of modern agricultural technology and have proved important chemical tools for world agricultural production. The major use of pesticides is to combat the pests that are competitors of our crops. The activities of harmful organisms result in one-third of World's agricultural losses. About 13.8% of all losses fall to the share of insects and mites. Nearly 70,000 species of insects and mites attack all the parts of agricultural plants during their entire period of vegetation and storage. About 10,000 species of them causes a substantial economical harm.

We, therefore, had no choice but rely heavily upon the use of pesticide chemicals. The enormous economical loss resulting from the diversity of the species of harmful insects, mites and the feature of their structural, biological and ecological parameters have usurped mankind to develop and use various types of insecticides, pesticides, acaricides etc. as a means of protection against them. The world is now peppered with various types of insecticides and pesticides. The use of these pesticidal chemicals in agriculture makes it possible to save approximately more than one third of our crop.

What are Pests and Pesticides?

In general, the term pest means any insect, rodent, nematode, fungus, weed or any other form of terrestrial, aquatic plant, animal life, virus, bacteria or other micro organism that harms

* Department of Chemistry, Faculty of Science, Dayalbagh Educational Institute, Dayalbagh, Agra-282005-India.

garden plants, trees, food stuffs, house hold articles and vector of several diseases. However, for farmers, pests include insect and mites that feed on crops, weeds in fields, aquatic plant that clog irrigation and damage ditches, plant diseases caused by fungi, bacteria, viruses, nematodes, snails, slugs and rodents that consume unbelievable quantities of plant seedlings and grains. **George Ware,** a well-known environmentalist coins the term pesticides as man's intentional additions to his environment quality for himself, his animals and his plants. According to **United State Federal Environment** the term pesticide is any substance used for preventing, destroying, repelling any insect, rodents, nematodes, fungus or any form of terrestrial and aquatic plant, animal life, micro organism which agency declares to be a pest. However, in general, they may be defined as substance used to control organism, which adversely effect public health, food and material or substance, which exert toxic action to the pests.

Owing to wide diversity, it is really a difficult task to provide a general scheme for the classification of various pesticides. However, pesticides are classified on the basis of their evolution process, mode of action, chemical nature and target species. Fig 1 presents a schematic view of categorization of pesticides in different classes.

National and International Status of Consumption of Pesticides

The era of chemical plant protection came into being after **World War II.** This infact gave a great impetus to the development of synthetic pesticides. DDT was the first synthetic pesticide used and it turned out to be the most powerful and successful chemical. DDT was first synthesized by a German student in 1873 while its insecticidal activity was explored by a Swiss entomologist **Dr. Paul Muller** in 1939. Simultaneously, organophosphate compounds were discovered in Germany during World War II, while studying the materials, which are related with the nerve gases, sarin, soman and tabun. Since then pesticidal research has attracted the attention of scientific community. Today, a wide range of pesticides has been synthesized, formulated and tested for their bioefficacy against various target species. According to **United State Environmental Protection Agency**, there are more than 1,180 pesticides of which 435 are herbicides, 335 are insecticides and 410 are fungicides. These are sold in more than 32,800 formulations. Worldwide pesticide consumption pattern in 1965 shows that its consumption was 172 MT of which USA consumes about 24%, Europe 46% and remaining 30% by developing countries. Pesticide consumption was increased to about two fold in 1980 and three fold in 1990 compared to 1965. This increase has given a similar rise in agricultural production and farm practices causing revolutionary change leading to incredible possibility that hunger can be vanished from the world.

India is the third largest consumer of pesticides in the world and highest among the South Asian countries. Our country is the second largest manufacturer of pesticide chemicals in Asia, next to China, and number twelve globally. About 155 pesticides have been registered under Insecticide Act 1968, till March 2001, which includes 57 insecticides, 44 fungicides, 33 herbicides, & rodenticides, 4 plant growth regulator, 4 fumigants, 3 acaricides, 1 soil sterillents, 1 molluscicide, and 1 nematicide. During the last four decades, the consumption of pesticides in India has increased several hundred folds from 154 MT in 1953-54 to 800.00 MT in 1994-95.

Table—1 Consumption of Pesticides in India

Pesticides	Years		(Consumption in tonnes)
	1986-87	1989-90	1994-95
BHC	44750	47000	47000
Aldrin	250	300	350
Lindane	35	50	80
Diazinon	50	100	150
Carbaryl	200	1000	1500
Aldicarb	30	50	75
Malathion	6050	6550	7550
Phosalone	250	300	400
Chlordane	75	100	150
Heptachlor	75	100	150
Carbofuran	375	500	650

Source: Indian Chemical Statics, Department of Chemicals and Petrochemicals, Ministry of industry, Govt. of India, New Delhi.

The average per hectare consumption of pesticides in agriculture (calculated on the basis of total consumption of technical grade pesticides divided by the gross cultivated area) was 1.2 gm ha^{-1} in 1953–54. It has grown from 3.2 gm ha^{-1} in 1954-55 to 336 gm ha^{-1} in 1980 and increased to 431 gm ha^{-1} in 1992– 93. The leading chemical used in India during 1995–96 was BHC, which accounted for more than 40% of the total pesticides consumed followed by malathion, methyl parathion, endosulfan, carbaryl and dimethoate. The states of Uttar Pradesh, Punjab, Andhra Pradesh, Haryana, Gujrat, West Bengal and Maharashtra account for 79.4 % of the pesticide consumption covering only 53.4 % of the total cultivated area. The total pesticide consumption remained highest in Andhra Pradesh (12,775 MT) which alone accounted for 15.7% of the total consumption, followed by Uttar Pradesh (11,500 MT), Punjab (7,600 MT), Haryana (5,390 MT), West Bengal (5,338 MT) and Maharashtra (4,898 MT).

A study conducted by Planning Commission, Govt. of India estimated the pesticide consumption by 2000 AD as 1,18,000 tonnes out of which 97,000 tonnes for agriculture and 21,000 tonnes for public health. Therefore, in order to fulfill the current requirement, Indian pesticide industry has installed capacity of 1,16,000 tonnes per annum, of which about 70,000 tonnes is in the organized sector, whereas the rest is in some 500 odd units belonging to the small sector. Imports are currently about 2,000 tonnes only.

Deleterious Effects of Pesticides

The increased consumption of pesticides was considered to be a good sign of progress in agricultural production. Later man realized the ecological principle that every poison we put

The demand pattern of major pesticides used in India during 1999–2000 is given below:

Table—2 Demand pattern of major pesticides in India (1999-2000)

Pesticides	Technical grade (MT)	Pesticides	Technical grade (MT)
Insecticides		**Herbicides**	
Phorate	2378	Isoproturon	2544
Carbaryl	854	Butachlor	2464
Malathion	3089	2,4-D	688
Quinalphos	2210	Anilophos	457
Dimethoate	1794		
Endosulfan	3794	**Fumigants**	
Chlorpyriphos	1762	Aluminium phosphide	272
Phosphamidon	2501		
Methyl parathion	2962	Ethylene dibromide	114
Monocrotophos	4149		
Fungicides			
Thiram	420		
Mancozeb	2828		
Sulphur(dust, wp)	2671		
Carbendazim	812		

Source: Directorate of Plant Protection and Quarantine and Storage, Faridabad.

out into the environment comes right back to him. The general population can be exposed to pesticides in three ways:

- Through vector control
- Through residues in environment
- Through residues in food and water

The segment of population tends to have only a low- dose, chronic exposure, but larger doses can be transmitted if the exposure is persistent and bioaccumulative. Persistent pesticides move through air, soil and water, finding their way into living tissues, where they undergo biological magnification. The deterioration of ecosystem by continuous use of fertilizers and pesticides has been seriously observed. **How disgusting is to hear that, almost all the surface water in our country except at mountains is unfit for human consumption.** The quality of river water worsens almost as soon as it enters the plains. At every step, water is abstracted from the river and large loads of polluted water is returned to it. As the river progresses downstream, its quality degrades constantly. The discharges change the physico-chemical properties of water and adversely effect aquatic life. Chemical fertilizers and pesticides constitute one of the major pollutants of the river. Although the water is treated before supply, the treatment does nothing to remove pesticide traces and industrial pollutants present in the water. It is really very difficult to get rid off the chemicals and biological contamination of water. Burden of water borne disease is about 30.5 million of **DALY (Disabled life years)**. Leaching from agricultural fields has been the most important non-point source of pollution to the aquatic environment. Traces of HCH and DDT have been found in rivers of the US and Europe, where they have been banned for more than two decades. Residues of the persistent organochlorines, long banned in the West, have been found to be present in alarming levels in the rivers of our country. The harmful beta isomer of these pesticides accumulates in river water due to its stable chemical structure, low vapour pressure and high resistance to microbial degradation.

India consumes around 86,311 tonnes of technical grade insecticides each year to cover 182.5 million hectares (mha) of cultivated lands. The pesticides are highly persistent and found in the aquatic systems for decades. The continued use of synthetic broad-spectrum insecticides on crops resulted in the resurgence of insect species. They are hazardous due to their potential toxicity and environmental contamination. In India, high residual levels of benzene hexachloride, lindane, heptachlor, endosulphan and dieldrin etc. have been found in just about everything necessary for life from food to water. The list includes tea in which DDT residues are so high that its export has been refused. Significantly high levels of food contamination with HCH, DDT, aldrin and dieldrin were evident throughout India. Delhites have one of the highest levels of DDT bioaccumulated in their body fat. The average daily intake of HCH and DDT were higher than those observed in most developed nations. In our country 1 to 2 Kg of pesticides are used each year for every man, woman, child. For a country like India, where human exposure to pesticides is reportedly one of the highest in the world, this is a serious concern.

Table—3 Acceptable daily intake of Pesticides for human beings

Pesticides	Acceptable daily intake in ppm
DDT	0.0050
Captan	0.1000
Thiram	0.0250
Systox	0.0025
Ovotran	0.0100
Lindane	0.0125
Carbaryl	0.0100
Malathion	0.0040
Parathion	0.0200
Dimethoate	0.0025
Methoxychlor	0.1000
Chlorobenside	0.0200
Methyl Parathion	0.0050

Source: Tolerance limit, P.F.A Rules 1955-56 Govt. of India, Ministry of Health

The use of organochlorine pesticides is completely banned in many western countries but in India some of these are still being used on farms for the control of insect pests in forage and other crops.

Table—4 Pesticides restricted in West but used in India (1994-95)

Pesticides	Use in agriculture (tonnes)	Use in public health (tonnes)
BHC	24,000	6,305
Carbafuran	280	
DDT		8,181
2,4-D (Dichlorophenoxy-acetic acid)		1,200
Dichlorovos	1,500	
Dimethane	1,900	
Endosulphan	4,600	
Lindane	50	
Methyl Parathion	2,600	
Monocrotophos	6,296	
Mancozeb	4,000	
Paraquat	400	

Source: All India Coordinated Research Project on Pesticides Residues, Agnihotri, IARI, New Delhi.

Recently, pesticides and related compounds have been correlated with cancer causing agents. These pesticides slowly enter our bodies and after years cause cancer, immune system disorders, hormonal or reproductive system disorders and affecting even foetus. Cancers of lymphatic, haematopoietic system and brain are also associated with the exposure to pesticides. A link between environmental pollution by pesticides and cancer has led to stringent regulations against air, water and soil pollution **[National Cancer Research Institute, US]**. Pesticides enhance the risks of cancer in two ways:

- Direct way—by acting as carcinogens themselves
- Indirect way—by suppressing the immune system, which has ability to destroy the process of tumour creation in the body.

Studies have linked the rising incidence of **Non-Hodgkin's Lymphoma, (NHL)** a form of cancer, due to the increased use of organophosphate pesticides and phenoxy herbicides and the cumulative effects of these pollutants on the human system.

Frequent use of herbicides, particularly 2,4–dichlorophenoxyacetic acid (2,4-D) has been associated with a 200-800 percent increased risk of NHL. In addition to people working in agriculture, the general population is also at a heightened risk of NHL because of the use of these pesticides in homes. The use of pesticide particularly 2,4-D, 2,4,5- trichlorophenoxy acetic acid, 2-methyl-4-chlorophenoxy acetic acid and organophosphate pesticides have increased over the last 40 years. This increase has played a significant role in contributing to the rising incidence of NHL. Pesticides are capable of altering the functions of the immune system, which may otherwise partially or completely abrogate the processes of tumorigenesis. "Pesticides are reported to cause a variety of cancers through an immunological mechanism".

Table—5 Pesticides causing lymphatic cancer

DDT	Dieldrin
DDE	Heptachlor
TCDD	Picloram
Perthane	Trichlorophenol
Strobane	Hexachlorocyclane
Toxaphene	

Source: World Health Organization. Washington.

Adding to the concern about carcinogenic effects of pesticides, the latest finding is a new discipline of science called immunotoxicity, which studies substances with a negative impact on the immune system. Reduced immunity influences incidence of cancer. A weak or devasted immune system allows cancerous cells to escape and form a tumour. Farmers have significantly higher risks for **Hodgkin's disease, melanoma, multiple myeloma, leukaemia** and cancers of lip, stomach and prostrate. One can only imagine the kind of havoc pesticides can play in a country where a large percentage of population is malnourished and, hence, suffers from immunodeficiency. It becomes mandatory to screen the agricultural products for such contaminants. To allay the public concern and to protect environment, regulatory restrictions are imposed on these groups of pesticides. Consumption and sale of some pesticides have been banned, withdrawn or severely restricted (United Nations, 1985).

Changing pattern of Conventional Pesticides Consumption

No doubt the extensive use of chemical biocides has helped us to increase the agricultural production and destroy variety of pathogens but at the same time their consumption is raising the level of chemicals in the environment and thus resulting into the Chemical pollution of Environment, which is now becoming threat to the survival of non-target species including human beings. The adverse effects of the chemicals in the environment are not only limited to ourselves but also passed on to future generation by way of genetic mutation, birth defects, inherited diseases and so on. The pests have developed resistance against conventional pesticides and hence they need higher and higher doses.

Table—6 Journey of Pesticides [1995-2000]

Organochlorine	40 - 14.5%
Carbamates	15 - 4.5%
Synthetic Pyrethroids	10 - 5%
Organophosphate	30 - 74%
Natural Pesticides	0 - 2%

Source: Government of India, Gazette

Considering the demerits of conventional pesticides and related disease management, our country has shifted steadily towards environmentally safe products. In India the average per hectare consumption of pesticides in agriculture was 1.2 g/ha in 1953-54 which increased to 431 g /ha in 1992-93. However, thereafter consumption gradually declined to 288 g/ha in 1999-2000. There is a definite trend of decline in the consumption of pesticide in India in all states particularly the ones which were earlier using high amounts of pesticides like Andhra Pradesh, Tamil Nadu, Karnataka and Maharashtra. The typical example is Andhra Pradesh where pesticide consumption declined from 12,775 to 7,000 MT. In general the decline is more conspicuous in the case of pesticides. Consumption of many organochlorine and organophosphorous pesticides has either been banned or restricted because of general consciousness about the environment and international pressure on export of food commodities having undesirable residues. Tendency is to discard the use of persistent and toxic pesticides and replace them with Natural insecticide. Natural compounds have increasingly become the focus of those interested in discovery of insecticide. As a result the scientists, the world over are turning towards ecologically sound pest control technologies based on emerging trends on the use of non conventional, ecofriendly, non pollutant and biodegradable pesticides.

Various new strategies for plant protection and insect control are as follows:

- Insect Repellents
- Insect Attractants
- Juvenile Hormones
- Pheromones
- Synergists
- Pesticides of plant origin

Insect Repellents

Control of insects can be achieved by means other than causing rapid deaths. Plants produce many compounds that are insect repellents or act to alter insect feeding behaviour, growth, development (molting) and behaviour during mating and oviposition. Insect repellents are the groups of chemical agents, which prevent damage to plants or animals by making the conditions unattractive or offensive to the pests. The first chemical repellent was discovered soon after man became acquainted with fire. Plant derived insect repellents have also been explored. Most insect repellents are volatile terpenoids – 4 ol. In some cases, the same terpenoids can repel certain undesirable insects while attracting more beneficial insects. Insect repellents include dimethyl phthalate or pyrethrum (ingredient of mosquito repellent cream/ coils), naphthalene or p-dichlorobenzene (ingredient of mothballs or flakes), mercury (ingredient of tablets for protecting stored grain) etc. Extracts of citronella plant *(Andropogon nardus)* contains gerniol, cintronellol, borneol, and terpenes which are considered to be the principal mosquito repellents. Repellents are used in the form of ointments, pastes, solutions, emulsions and arosols.

Some important insect repellents commonly used are as follows:

Benzyl Benzoate

It is used as repellent for clothing treatment against chiggers and mites.

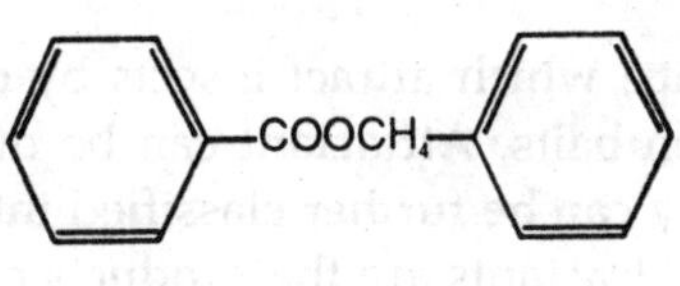

Butoxypolyproplene Glycol

It is used as a repellent for flies against cattle.

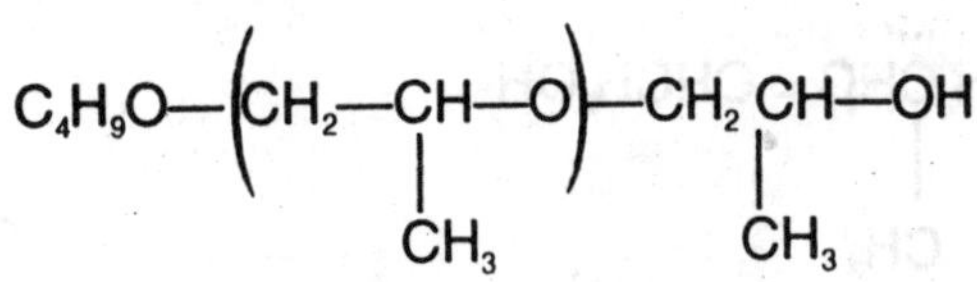

Dimethyl Phthalate

This repellent is used as effective mosquito repellent. It causes irritation to eyes and mucous membrane.

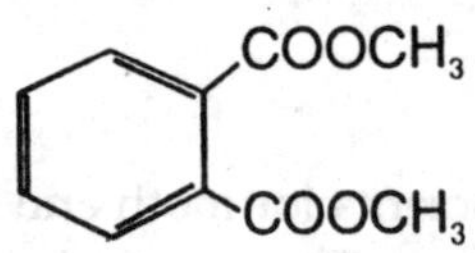

Indalone

Butyl 3,4-dihydro-2,2-dimethyl-4-oxo-2H-pyran-6-carboxylate is used as a repellent against mosquito and bitting insects.

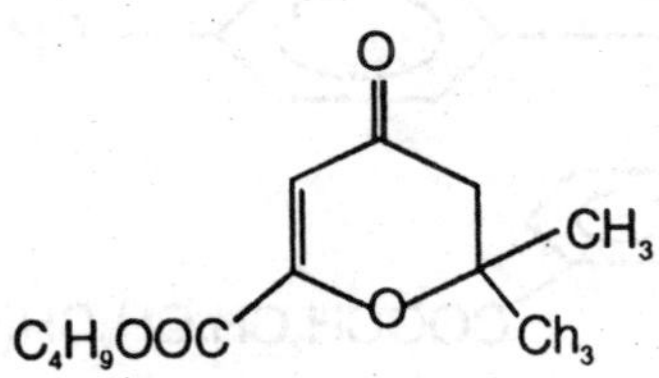

Rutgers 612

2-Ethyl-1, 3-hexanediol is used as a repellent against flies and mosquitoes.

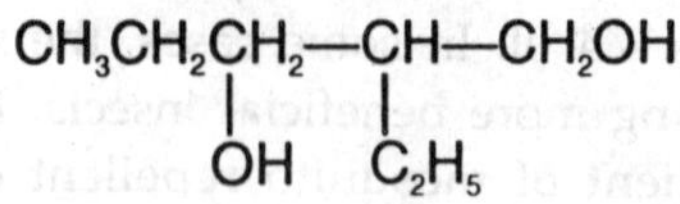

n-propyl-N-N-diethylsuccinamate

It is used as a repellent for mosquitoes.

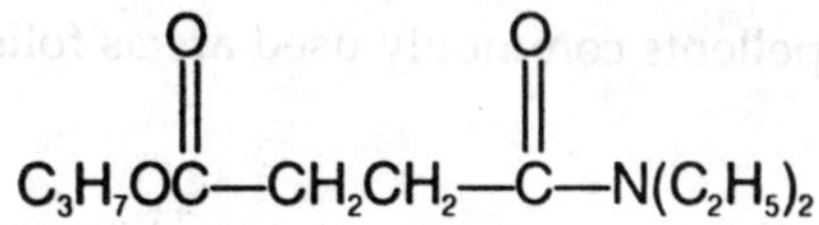

Insect Attractants

These are the chemical agents, which attract insects by olfactory stimulation. They attract the insects into traps or to poison baits. Attractant can be defined as the chemical substances whose vapors attract insects. They can be further classified into two groups **food attractant** and **sexual attractants**. Usually food attractants are the products of decomposition of food products, proteins, enzymes and molasses. The important synthetic attractant is Eugenol, which attracts fly. The 9:1 mixture of geraniol and eugenol forms an attractant for Japanese's beetle adults.

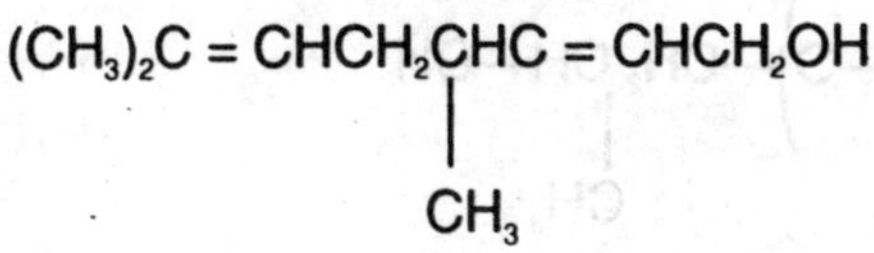

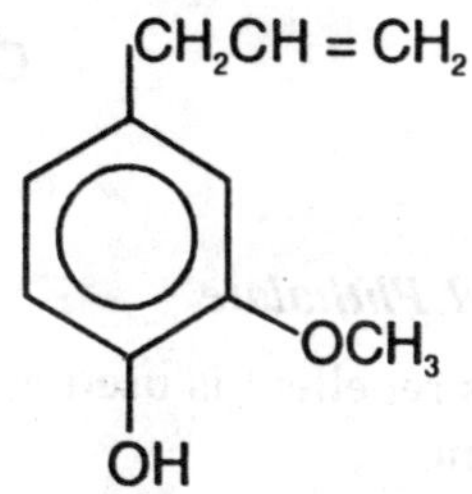

Geraniol Eugenol

Fermenting sugars serve as attractants for moth and butterflies. They may be supplemented by essential oils for specific requirements. Thus anethole is added for cooling moth while isoamyl salicylate for tomato and tobacco hornware moths.

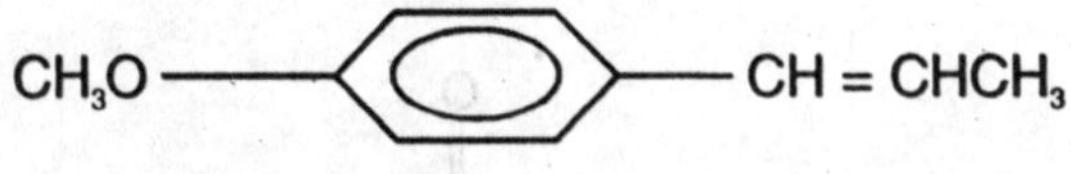

Metaldehyde, $[CH_3CHO]_4$, is used as an attractant in poison baits for snails and slugs. Trimedlure is an attractant for melon fly.

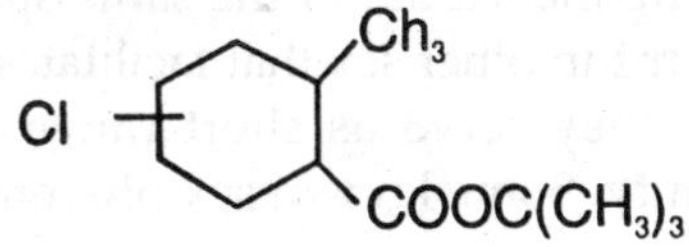

Trimedlure

Juvenile Hormones (JH):

Insect hormones are a group of physiologically active group of organic compounds, which regulate the specific development of insects from egg to the adult insect. Juvenile hormones are one of the important members of insect hormones. These are secreted by corpura allata (a part of brain) and control complicated postembryonic developments. Juvenile hormones cause developmental disturbances during ontogenetic stage, which prevent insect reproduction. Such an arrest of reproduction is a better and timely approach. They do not have direct toxicity to insects. Normally they do not kill cause but check them from reproduction. Moreover, they do not cause any danger to man and warm-blooded animals. JH like compounds are occurring in human organs and in our daily edibles like milk, cream and bread.

JH like compounds form a new class of compounds characterized as insect sterillents sp. All kind of developmental disturbances like female sterility, ovicidal effects, adult emergence failure and finally inability of mating. JH along with mounting hormone (MH) maintain larval character of growing insects. They also regulate intensity of motor activity by affecting central nervous system. JH like compounds have also been reported to cause reversal of adult in larval stage in some insect species and are capable of preventing the metamorphosis of insects. In addition to their JH activity, their chemical stability and retention of activity under field conditions make them suitable as **Fourth generation of insecticides.**

Various synthetic analogue of aliphatic and aromatic nature of JH have been synthesized. A good amount of attention has been paid on structure-relativity relationship of these compounds. Both acyclic and cyclic juvabione derivatives having several structural modifications have been synthesized and explored for their bio-efficacy. JH like compounds are commercially available in the market. Juvenoid has almost double the activity of pyrethrum.

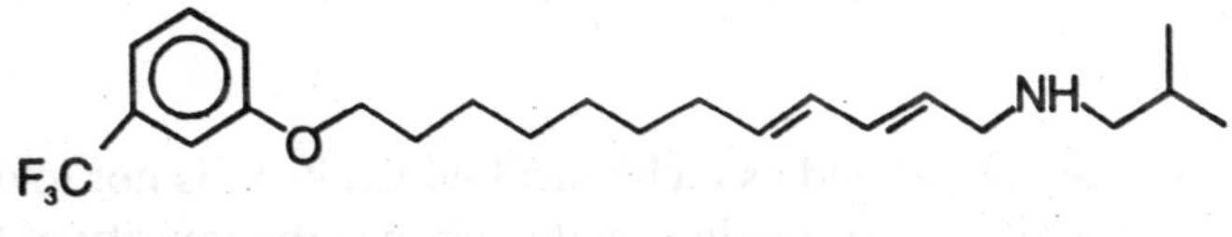

Phenoxy Amino Juvenoid

Pheromones

These are the chemicals that are secreted into the external environment by an animal and that elicit a specific reaction in a receiving individual of the same species. These are released by one sex only and trigger behaviour patterns in other sex that facilitates mating. They are produced by certain male insect species. Mostly they serve as short range mating stimulant after the individuals have come into close proximity. Some known sex pheromones are as follows:

Pink Bollworm:

$$(CH_3(CH_2)_2)_2Cl = CH(CH_2)CH = CHO(CH_2)_3(CH_2)_6O\overset{\overset{O}{\|}}{C}—CH_3$$

10-propyl-trans-5, 9-tridicadienyl acetate

Silkworm:

$$CH_3(CH_2)_2CH = CHCHCH = (CH_2)_8CH_2OH$$

trans-10-cis-12-hexadecenol

Gypsymoth:

$$CH_3(C_2H_5)_2CH—CH_2CH = CH(CH_2)_5CH_2OH$$

10-acetoxy-cis-7-hexadecenol

Cabbage looper:

$$(CH_3(CH_2)_3CH= CH(CH_2)_5CH_2O\underset{\underset{O}{\|}}{C}—CH_3$$

Cis-77-dodecenyl acetate

Pheromone enables pests to be controlled specifically, their small quantities avoid contamination of food and fodder as well as environment. They can be used to attract insects to a site where they are destroyed by treatment with insecticides.

Synergists

The term synergist may be defined as a chemical which itself is not toxic to insects at doses used but when combined with an insecticide greatly enhance the toxicity of the insecticide. They reduce the amount of pesticides required for killing insects and simultaneously it reduces the risk of environmental pollution. It has lead to use of pesticides. Of the several different groups of chemicals able to synergise insecticides, the most widely studied are methylenedioxyphenyl derivatives. Some common examples of synergists are piperonyl butoxide, safrole, sulfoxide, tropitol and thanite.

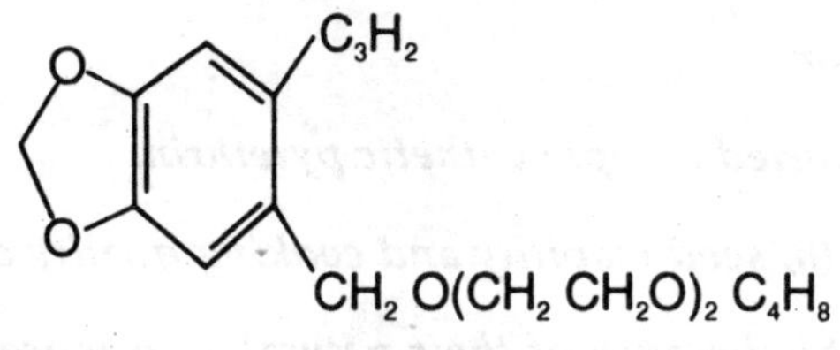

Piperonyl butoxide

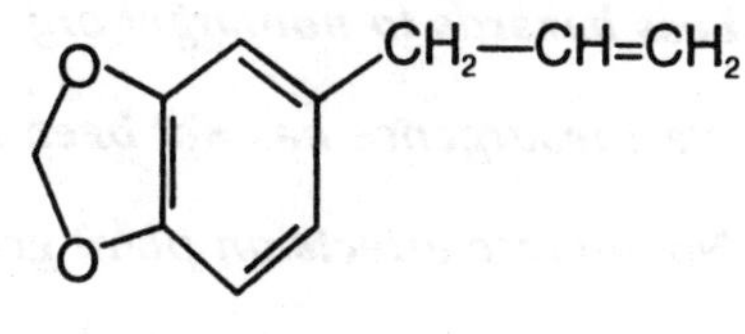

Safrole

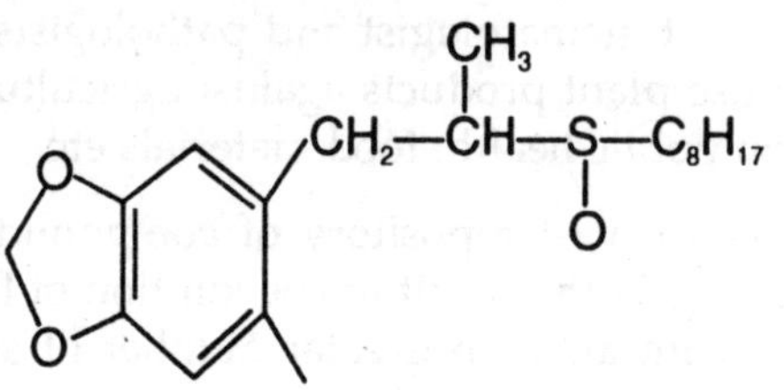

Sulfoxide

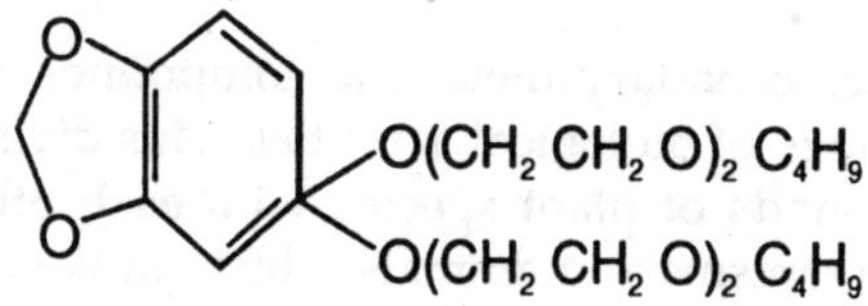

Tropitol

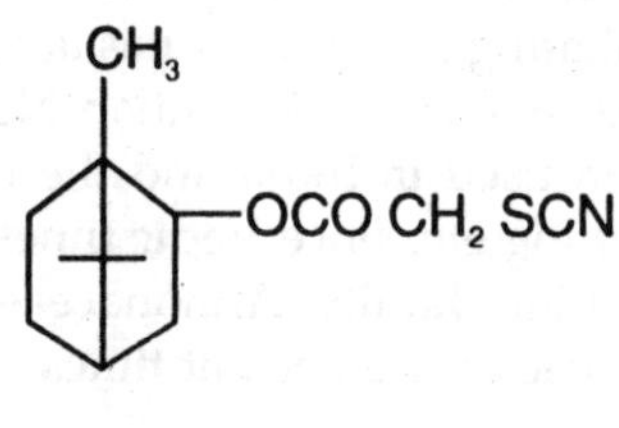

Thanite

Pesticides of Plant Origin

Use of natural plant products is now emerging as one of the prime means to protect crop and the environment from pesticidal pollution. Plant pesticides possess an array of properties including insecticidal activity, repellency to pests, antifedency, and insect growth regulation, toxicity to nematodes, mites and other agricultural pests. Some of these indigenous resources have been in use for over a century to minimize losses due to pests and disease in agricultural production. Herbal pesticides have many advantages over synthetic pesticides. Some of them are as follows:

- ***Plant pesticides, in general, possess low mammalian toxicity and have least or no health hazards and environmental friendly.***
- ***There is practically no risk of developing pest resistance to these products, when used in natural forms.***
- ***Less hazards to nontarget organism.***
- ***Pest resurgence has not been reported except synthetic pyrethrin.***
- ***No adverse effects on plant growth, seed viability and cooking quality of grains***
- ***Less expensive and easily available because of their natural occurrence especially in oriental countries.***

Due to prohibitive cost of synthetic pesticides and the problems of environmental pollution caused by continuous use of these chemicals, there is a renewed interest in the use of plants for crop protection. A number of agricultural entomologist, nematologist and pathologists all over the world are now actively engaged in research to use plant products against agricultural pests and disease to minimize losses caused by them to on public health, food materials etc.

The secondary metabolic compounds of plants are vast repository of compounds with a wide range of biological activities. This diversity is largely the result of coevolution of hundreds of thousands of plant species with each other and with an even greater number of species of microorganisms and animals. Thus, unlike compounds synthesized in the laboratory, secondary metabolites from plants are virtually guaranteed to have biological activity and that activity is highly likely to function in protecting the plants from a pathogen, herbivores, and competitors. The knowledge of the pests to which the producing plant is resistant may provide useful leads in predicting what pests may be controlled by compounds from a particular species. Empirical knowledge of plants useful for combating insect pests has accumulated over millennia in different cultures in different parts of world. For example, Indian Neem tree (*Azadirachta indica A.* Juss; family: Meliaceae) leaves have been used in India and the neighboring countries for control of insect infestation in food grain, clothing etc, since Vedic times (1500 B.C). The dried seed powder of custard apple (*Annona squamosa* Linn; family: Annonaceae) has been used as an insecticide in India and many other tropical countries since ancient times.

Nicotine from tobacco plant (*Nicotiana tabacum* Linn; family: Solanaceae) was first isolated in 1828, and its aqueous extracts for plant protection have been used in Europe since 16th century. Nicotine and nicotinephosphate have been marketed since1901. It is effective against a wide range of insects and kills them through feeding, contact and fumigation. In the early 19th century 2200 tonnes of the alkaloid was used as a crop protection agent.

Another important insecticide from plant is Pyrethrum (*Chrysanthemum cinerarifolium* Vis., *C.coccineum* Wild; family: Compositae). It is the most important and commercially cultivated in a large scale in several countries including Kenya, Tanzania, Ecuador, Brazil, erstwhile USSR, Japan and India. The insecticidal properties of the several *Chrysanthemum* species were known for centuries in Asia. Even today, powders of the dried flowers of these plants are sold as insecticides. Products based on Pyrethrum are essentially used in indoor applications and are especially valued for their rapid knock-down (of flying insects) effect, photoability and low mammalian toxicity. Formulated preparation contains synergists such as piperonyl butoxide that enhance their insecticidal activity.

Various plants of genus *Derris, Lonchocarpus, Milletia, Mundulea, Tephrosia* (family: Leguminosae) possess rotenone, a potent insecticide. Rotenone was first isolated in 1902 from *Derris chinensis* Benth. In 1930s preparations of roots from the genera *Derris, Lonchocarpus,* and *Tephrosia,* containing rotenone were commercial insecticides.

Ryania (*Ryania speciosa* Vahl; family: Flacourtiaceae) consists of the powdered roots and stems of the South American plant. It has been used as a contact and stomach poison to insects and used as a commercial insecticide against European corn borer. Sabadilla consists of the powder barley like seeds of the South and Central American plant (*Schoenocaulony officinale* Gray; family: Liliaceae), which the local people have been using for a long time to control insect-infestation.

There are many other plant species identified which contain insecticidal principles and used at one time or the other by different people in different regions of the world. With the introduction of modern scientific methods of research, our knowledge of insecticidal plants has expanded vastly. Natural compounds have increasingly become the focus of those interested in discovery of pesticides. Over 2000 plant species belonging to several families are now known to possess insecticidal properties.

Table—7 Plant Families with Insecticidal Activity

1. Acanthaceae	19. Cucurbitaceae	37. Papaveraceae
2. Agavaceae	20. Dioscoreaceae	38. Piperaceae
3. Annonaceae	21. Ebenaceae	39. Poaceae
4. Apocynaceae	22. Euphorbiaceae	40. Polygonaceae
5. Araceae	23. Flacourtiaceae	41. Polypodiaceae
6. Aristolochiaceae	24. Guttiferae	42. Ranunculaceae
7. Asclepiadaceae	25. Helleboraceae	43. Rosaceae
8. Balanitaceae	26. Hippocastanaceae	44. Rubiaceae
9. Berberidaceae	27. Hypericaceae	45. Rutaceae
10. Boraginaceae	28. Labialae	46. Sapindaceae
11. Brassicaceae	29. Lamiaceae	47. Sapotaceae
12. Burseraceae	30. Lauraceae	48. Simaroubaceae
13. Capparaceae	31. Leguminosae	49. Solanaceae
14. Capparidaceae	32. Liliaceae	50. Stemonaceae
15. Celastraceae	33. Loganiaceae	51. Taxaceae
16. Chenopodiaceae	34. Lycopodiaceae	52. Theaceae
17. Compositae	35. Magnoliaceae	53. Umbelliferae
18. Convolvulaceae	36. Malvaceae	54. Verbenaceae

Source: Naturally Occuring Insecticides. Jacobson, M. 1971.

Tens of thousands of secondary products of plants have been identified and there are estimates that hundreds of thousands of these compounds exist. Plants are known to produce a wide range of secondary metabolites such as alkaloids, terpenoids, polyacetylenes, flavonoids, quinones, phenylpropanoids and amino acids. There is growing evidence that most of these compounds are involved in the interaction of plants with other species-primarily the defense of the plant from plant pests. Thus, these secondary compounds represent a large reservoir of chemical structures with biological activity. This resource is largely untapped for use as pesticides.

Most commercially successful pesticides have been discovered by screening compounds synthesized in the laboratory for pesticidal properties. The average number of compounds that must be screened to discover a commercially viable pesticide has increased dramatically.

Being environmentally benign, phytochemicals are ideally suited for incorporation in integrated pest management programmes, particularly in developing countries, which are rich in plant biodiversity and many phytochemical biopesticides can be developed from indigenous plants. This approach has led to the discovery of several commercial pesticides derived from plants.

Combinatorial Approach – A Bright Future

However problems are also associated with plant produced phytotoxins as potential insecticides. Number of bioactive principles have been isolated from various plant species, but only a few have reached to the level of commercialization. The most common problem is that the amount of purification initially conducted is a variable for which there is no general rule. The isolation of bioactive principle depends on the particular growth stage of a plant. The methods used for isolation of an active molecule are lengthy, time consuming and depending on the source. A small amount of compound is generally isolated and is generally weak in comparison to the synthetic chemicals. Collection of natural sources, storage, production, formulation and application (spray) is a tedious piece of work. Plant based pesticides show slow rate of efficacy compared to the conventional chemical pesticides. The biopesticides are to be highly specific because certain stages of the life cycle of the pest may be almost immune. Most of the pesticides require damp climate, initial spray distribution and movement of their hosts for dissemination. In case of few natural products, structural identification might be proved as a difficult task. The major draw back of biopesticides is limited phytochemical stability i.e. temperature, UV or sunlight, rainfall, humidity and other environmental factors readily affect them. They need various stabilizers like anti oxidants, ascorbic acid, retionic acid and flavonoids etc to obtain desired persistence. Molecular complexity, limited environmental stability, low activity of plant based pesticides, compared to synthetic pesticides are discouraging. It appears that none of the approaches is suitable in larger sense. The best choice is the combination of two i.e. the use of **Semisynthetics**. Plant derived chemicals are recently used as model compounds for development of semisynthetic insecticides. Plants contain a virtually untapped reservoir of pesticides that can be template for synthetic pesticides. Structural manipulations in the bioactive principles derived from plant-based pesticides are supposed to improve the activity, toxicological properties with no or less deleterious environmental effects. Considering the demerits of synthetic and plant base pesticides it might be better to prepare synthetic analogue of the biologically active principle derived from the plant. Semisynthetic pesticides can prove to be an excellent alternative approach to synthetic pesticides having favorable toxicological and environmental expectations.

Physostigmine, a carbamate alkaloid, isolated from calabar beans (*Physostigma venenosum* Balf) was recognized for its insecticidal (anticholinesterase) activity. Structural modifications were made on the bio-active principle of this plant derived carbamates which resulted into synthetic analogue isolan. Isolan was found to be non-persistent, non-toxic to human beings. At present twenty-five different products are in the market and their worldwide production is about 35,000 tonnes per year.

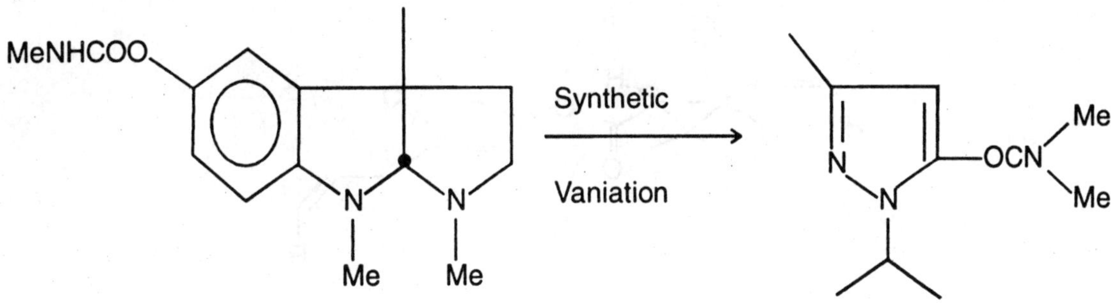

Physostigmine Isolan

Nereistoxin, a biocide metabolite of the annelid (shell fish) *Lumbriconereis heteropoda* has also served as the starting skeleton for developing synthetic insecticide. Bis-thiocarbamate (cartap) is synthetically tailored molecule of nereistoxin. It is found to be highly effective insecticidal agent against rice leaf beetle and non-toxic to human beings.

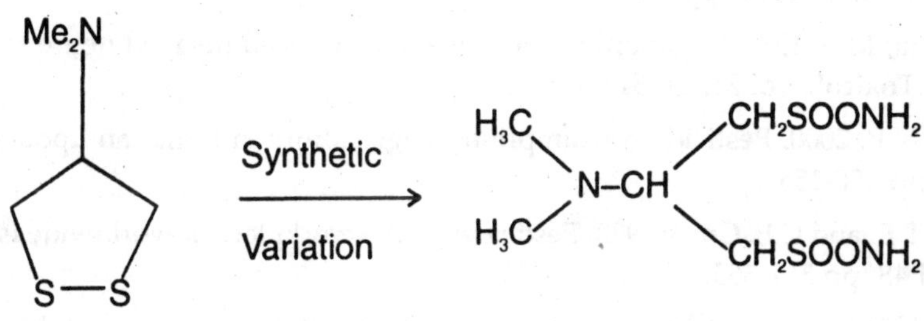

Nereistoxin Cartap

Pyrethrin, a plant derived insecticide is found to be photolabile. Its use was prohibited due to its undesirable half-life in the open environments. In order to make it photostable, safe and environment non-persistent, synthetic variations have been carried out and two classes of pyrethrin analogues have been synthesised. One class includes bioallethrin, bioresmethrin and biophenothrin. In this class of compounds the acid component i.e. chrysanthemic acid of pyrethrin retained and only the alcohol moiety has been changed. These are recommended for household, public heath, industrial and certain agricultural uses. While in second class of compounds both the acid and alcohol moieties differ from those present in natural compounds. In deltamethrin and cypermethrin halogen atoms have replaced two vinyl methyls. In fenvalerate even the cyclopropane ring has been dispensed with. They are registered for agricultural applications and about eighteen compounds of this class are in the market. Most of the products are mixtures of isomers, all of them do not have activity. Cypermethrin is a mixture of eight isomers and active component constitutes only 18-25%. Deltamethrin is a single chemical entity and is the most important pyrethroid to date.

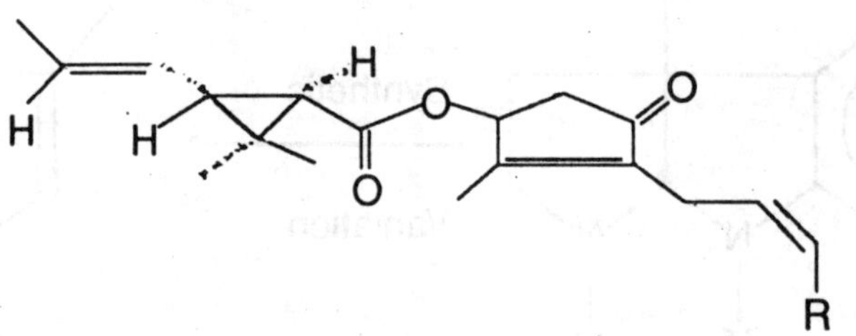

This combinatorial approach seems to be a time tested strategy for the development of biologically active synthetic analogues from a naturally occurring biologically active molecule. In this endeavor natural products having biological activities have served well. Sincere efforts towards the research on the above lines are required.

REFERENCES

1. Abott, D.C. and R. Goulding 1968. Organochlorine pesticide residues in human fat in Great Britain. Br. Med. J., vol 3, pp 146-149.

2. Abou- Donia, M.B. 1981. Organophosphorous ester- induced delayed neuro toxicity. Annu. Rev. Pharmacol Toxicol., vol 21, pp 511-548.

3. Agnihotri, N.P. 2000. Pesticide consumption in agriculture in India- an update. Pesticide Res. J., vol 12 (1), pp 150-155.

4. Albertson, T.E. and C.E. Cross 1993. Pesticides in the workplace: a worldwide issue. Arch Environ Health, vol 48, pp 364-365.

5. Ando, K. 1983. Pesticide chemistry: Human welfare and the environment. Miyamoto J. and P.C. Kearney (Eds), vol 2, pp 253. Pergamon, Oxford.

6. Anon 1995. Funding safer farming: Taxing pesticides and fertilizers. Center for Science in public interest, Mimeo.

7. Arnason, T., G.H.N. Towers, B.J.R. Philogene and J.D.H Lambert 1983. The role of natural photosensitizers in plant resistance to insects. Amer. Chem. Soc. Symp. Ser., vol 208, pp 139-151.

8. Balandrin, M. F., J. A. Klocke, S. Wurtele and W. H. Bollinger 1985. Natural plant chemicals: Sources of industrial and medicinal materials. Science 228, pp.1154-1160.

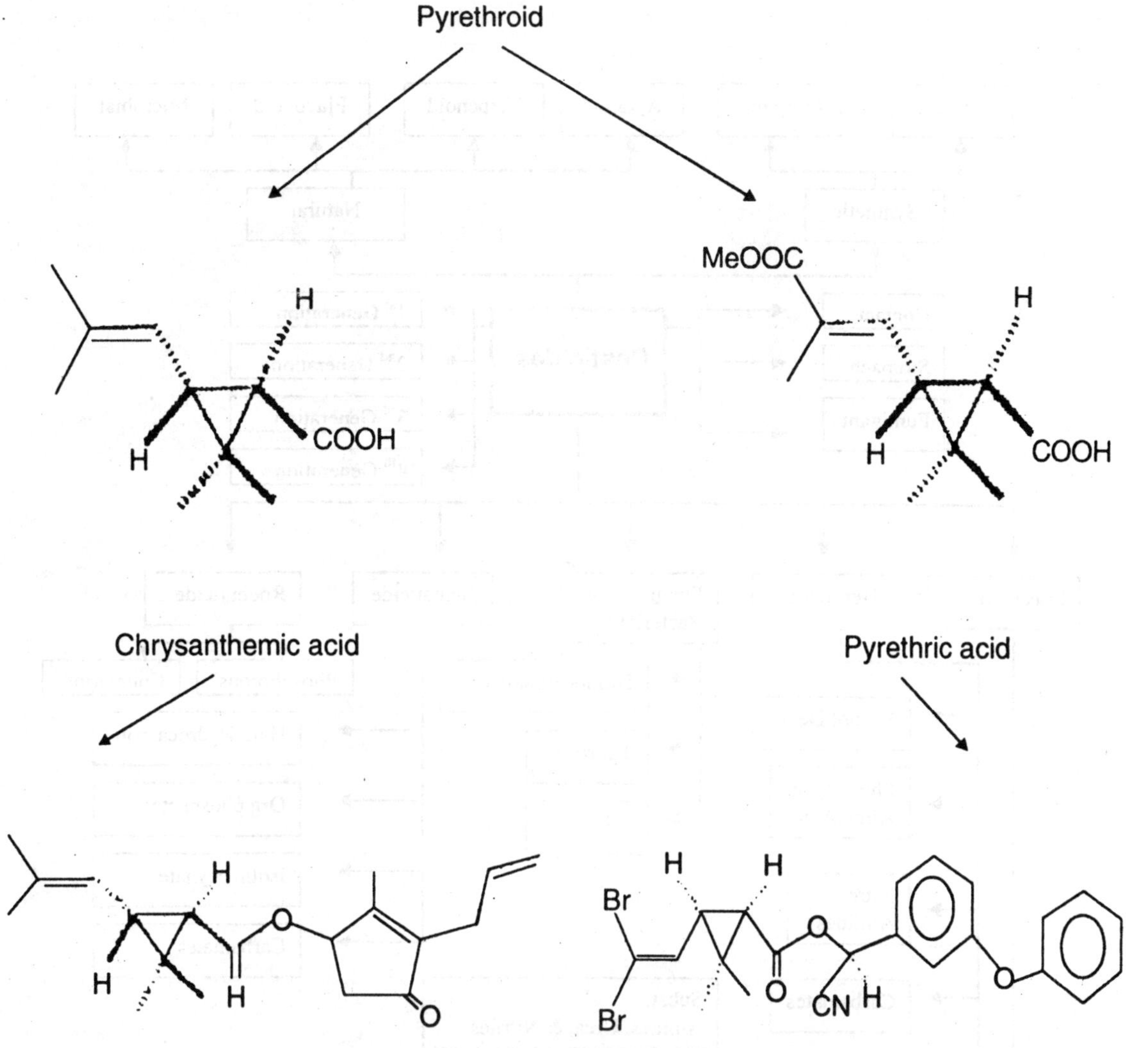

9. Berenbaum, M. R. 1989. North American ethnobotanicals as sources of novel plant based insecticides. Insecticides of plant origin. Arnason J. T., B. J. R. Philogene and P. Morand (ACS Symposium series No. 387), pp. 11-24. American Chemical Society, Washington.

10. Bhat, B. K. and P. N. Pandita 1982. Variation in components of yield of pyrethrins in pyrethrum. Cultivation and utilization of medicinal plants. Atal, C. K. and B. M. Kapur (Eds), pp. 704-707. Regional Research Laboratory, C.S.I.R., Jammu.

11. Bidstrup, P.L., J.A. Bonner and A.G. Beckett 1953. Paralysis following poisoning by a new organophosphorous insecticide (Mipafox). Br. Med. J., vol 1, pp 1068-1072.

12. Brandson, C. and K. Hommann 1995. The cost of invocation: Valuing the economy – wide cost of environmental degradation in India, Asia Environment division, World Bank, October 17, Mimeo.

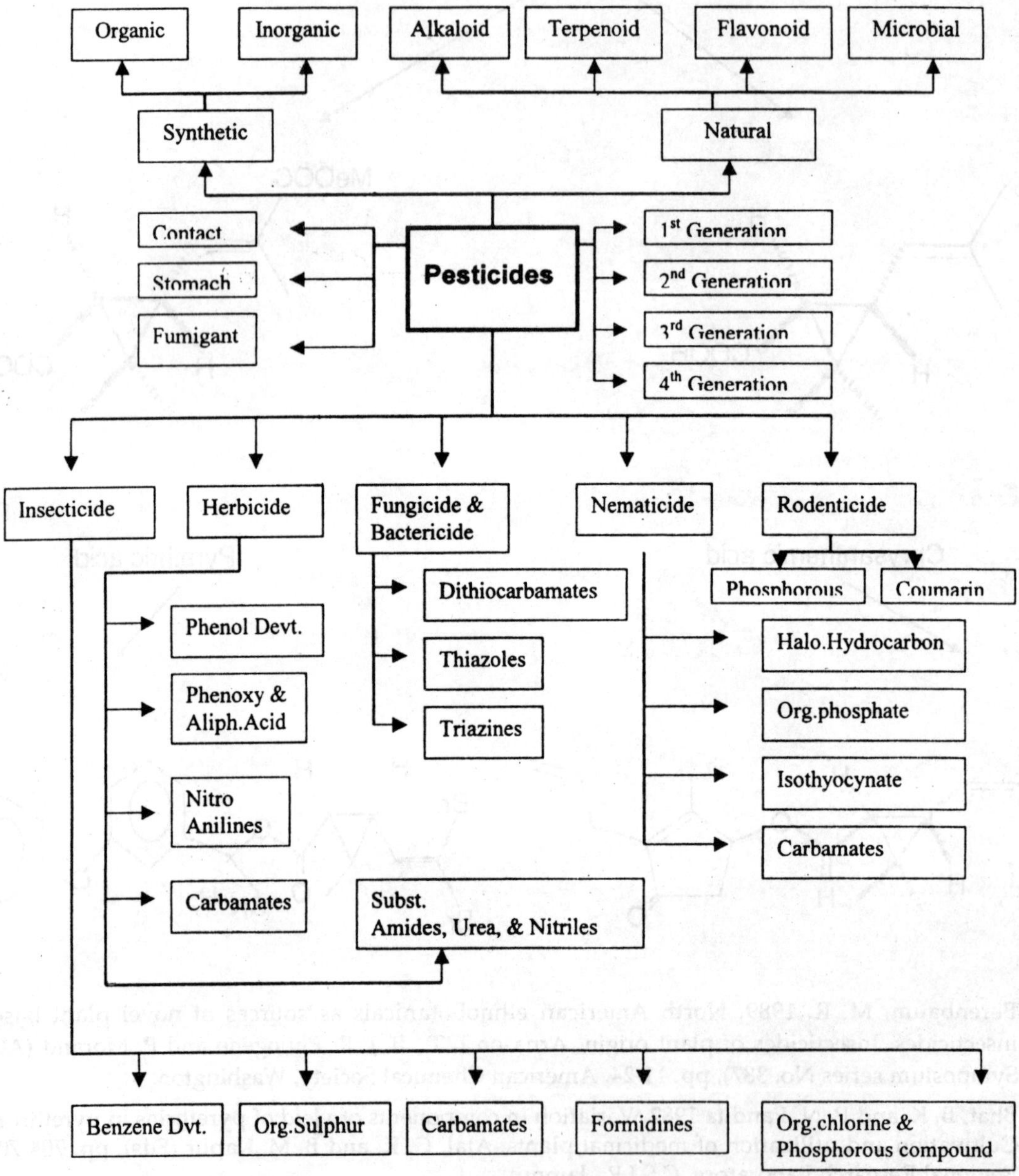

Fig. 1 CLASSIFICATION OF PESTICIDES

13. Brignal, W.J., A. Rachwal and V. Aldridge 1992. Pesticide removal from potable water, Symposium on pesticides and water industry, Loughborough University of Technology, UK, April 1992, Mimeo.

14. Caisda, J. E. 1973. Pyrethrum, the natural insecticide. Academic press, New York.

15. Caisda, J.E. 1976. Prospects for new types of insecticides. The future for insecticides: Needs and prospects. Metcalf, R.L. and J.J. Mckelvey (Eds), pp. 349-366. Wiley- Interscience, New York.

16. Caisda, J.E. 1979. Pesticide research to maintain and improve plant protection. advances in pesticide science. H. Griessbuhler (Eds), vol 1, pp 45-53. Pergamon, Oxford.

17. Chang, L.W., K.R. Reuhl and G.W. Lee 1977. Degenerative changes in the developing nervous system as a result of utero exposure to methyl mercury. Environ Res., vol 14, pp 414-423.

18. Dev S., A. P. S. Narula and J. S. Yadav 1982. Handbook of Terpenoids: Monoterpenoids. vol 2, pp. 34-35. CRC Press, Boca Raton.

19. Dev, S. and O. Koul 1997. Insecticides of natural origin. Harwood academic publisher,New York.

20. Duke, S.O. 1985. Biosynthesis of phenolic compounds—Chemical manipulation in higher plants. Amer. Chem. Soc. Symp. Ser., vol 268, pp 113-131.

21. Duke, S.O. 1986. Naturally occurring chemical compounds as herbicides. Rev. Weed Science, vol2, pp 15-44.

22. Elsohly, H.N., S.O. Duke, K.C. Vaughn and E.M. Croom, Jr. 1987. Artemisinin, a constituent of annual wormwood (*Artemisia annua*), is a selective phytotoxin. Weed Science, vol 35, pp 499-505.

23. Paul, R.N., S.O. Duke and S.M. Lee 1988. Terpenoids from the genus *Artemisia* as potential pesticides. Amer. Chem. Soc. Symp. Ser., vol 380, pp 318-334.

24. Elliot, M. 1977. Synthetic pyrethroids. American Chemical Society, Washington.

25. Friebele, E. 1996. The attack of asthma, in environmental health perspectives. Down the Earth, vol 104 (1) pp 22-25.

26. Grainage, M. and S. Ahmed. 1988. Handbook of plants with pest-control properties. Wiley-Interscience, New York.

27. Green, M.B., G.S. Hartley and T.F. West 1987. Chemicals for crop improvement and pest management. Pergamon, New York.

28. Handa, .K. 1999. Principles of pesticide chemistry. Agrobios, India.

29. Hardell, I. 1981. Malignant lymphoma and exposure to chemicals, especially organic solvents, chlorophenols and phenoxy acids. British journal of cancer, vol 43, pp 169-176.

30. Hartge, P. and S.S. Devesa 1992. Quantification of the lipact of known risk factors on time trends in Non-Hodgkins Lymphoma incidence.cancer research, World Resources Institute, Washington DC.

31. Hoar, S.K. 1986. Agricultural herbicides use and risk of Lymphoma and soft tissue Sarcoma. Journal of the American medical association, vol 54, pp 214-231.

32. Jacobson, M. 1986. The neem tree: natural resistance par excellence. Amer. Chem Soc. Symp. Ser., vol 296, pp 220-232.

33. Jacobson, M. and D.G. Crosby 1971. Naturally occurring insecticides. Marcel Dekker, New York.

34. Jedrychowski, W. 1995. Prospective epidemiological study on respiratory diseases in children in Cracow- pilot study. Prezegl Epidemiol, vol 49(3) pp 331-339.

35. Karrer, W. (1976) Konstitution and Vorkommen der organischen Pflanzenstoffe, pp. 573 – 574. Basel: Birkhasuer Verlag.

36. Klocke, J.A. 1987. Natural plant compounds useful in insect control. Amer. Chem Soc. Ser., vol 330, pp 396-415.

37. Kudesia, V.P. and M.U. Charya 1993. Pesticide pollution. Pragati prakashan, India.

38. Lydon, J. and S.O. Duke 1987. Herbicides from natural compounds. Weed Technol., vol 1, pp 122-28.

39. Lydon, J. and S.O. Duke. 1988. Porphyrin synthesis is required for photobleaching activity of the *p*-nitrosubstitued diphenyl ether herbicides. Pestic. Biochem. Physiol., vol 31, pp 74-83.

40. Lydon, J. and S.O. Duke. 1989. Pesticide effects on secondary metabolism of higher plants. Pestic. Sci., vol 25, pp 361-373.

41. Matringe, M. and R. Scalla. 1988. Studies on the mode of action of acifluorfen-methyl in non-chlorophyllous soybean cells: Accumulation of tetrapyroles. Plant Physiol., vol 86, pp 619-622.

42. McLaren, J.S. 1986. Biologically active substances from higher plants: status and future potential. Pestic. Sci., vol 17, pp 559-578.

43. Miyakado, M., I. Nakayama and N. Ohno 1989. Insecticidal unsaturated isobutylamide from natural products to agricultural leads. Insecticides of plant origin. Arnason, J.T., B.J.R. Philogene and P. Morand (ACS Symposium Series No 387), pp 173-187. American Chemical Society, Washington.

44. Mohapatra, S.P., V.T. Gabhiye, N.P. Agnihotri and M. Raina 1995. Insecticide pollution of Indian rivers. Environmentalist, vol 15(14), pp 41-44.

45. Morgan D.P. and C.C. Roan 1970. Chlorinated hydrocarbon pesticide residue in human tissues. Arch Environ Health, vol 20, pp 452-457.

46. Naumann, K. 1990. Synthetic pyrethroid insecticides. Structure and properties. Springer-Verlag, Berlin.

47. Nexo, B. A. 1995. Risk assessment methodologies for carcinogenic compounds in indoor air, in scan journal of work. Environment and health, vol 21(5), pp 376-381.

48. Pearce, F. 1996. A heavy responsibility. New Scientist, vol 151(2040), pp 12-13.

49. Rao, P.V. and M. K. Chakraborty 1982. Tobacco cultivation improvement and utilization of tobacco waste for production of nicotine. Cultivation and utilization of medicinal plants. Atal, C.K. and B.M. Kapur (Eds), pp. 669-679. Regional Research Laboratory, C.S.I.R., Jammu.

50. Rebeiz, C.A., A. Montazer-Zouhoor and H.J. Hopen 1984. Photodynamic herbicides: 1: Concept and phenomenology. Enzyme Microb. Technol., vol 6, pp 390-401.

51. Rebeiz, C.A., J.A. Juvik and C.C. Rebeiz. 1988. Porphyric insecticides: Concept and phenomenology. Pestic. Biochem. Physiol., vol 30, pp 11-27.

52. Repetto, R. and S. Baliga 1996. Pesticides and the immune system: The public health risks, World Resources Institute, Washington DC.

53. Rice, E.L 1984. Allelopathy. Academic Press, New York.

54. Rosenstock L., M. Keifer and W.E. Daniell 1991. Chronic central nervous system effects of acute organophosphate pesticide intoxication. Lancet, vol 338, pp 223-227.

55. Russell, G.G. 1986. Phytochemical resources for crop protection. J. Technol., vol 2, pp 127-134.

56. Saxena, R.C. and D. Dale 1989. Plant defense mechanisms against insects. Plants and society. Swaminathan M.S. and S.L. Kochhar (Eds), pp. 350-366. Macmillan, London.

57. Wecker, L., T. Kiauta and W-D. Dettbarn 1978. Relationship between acetylcholinesterase inhibition and the development of a myopathy. J. Pharmacol Exp Ther, vol 206, pp 97-104.

58. Yang, R.Z. and C.S. Tang. 1988. Plants used for pest control in China: A literature review on Econ. Bot., vol 42, pp 376-406.

59. Zahm, S.H. and A. Blair 1992. Pesticides and Non Hodgikn's Lymphoma. Cancer Research, World Resources Institute, Washington DC.

10

Oxygen Supplementation Through Soil: A Boon to Agriculture in New Millennium

*Neelam A. Patil

Plants, like other living beings, require oxygen for respiration and the energy released during the process is utilised for growth. The aerial plant parts get plentiful supply of oxygen through atmosphere (21%) but the underground plant parts are dependent on the oxygen available in soil, which is governed by soil texture, soil porosity and water content. Gas-filled pores in well-drained, well-structured soil readily permit the diffusion of gaseous O_2 to depth of several meters. The O_2 concentration deep in the soil is similar to that in humid air. When drainage is poor or irrigation or rain is in excess, soil becomes water logged. Water then fills the pores and blocks the diffusion of O_2 in gaseous phase. Oxygen diffuses so slowly in water that only a few centimetres of soil near the surface remain oxygenated. At temperature more than 20° C under water logged condition, oxygen consumption by plant roots, soil fauna and soil micro-organisms can totally deplete the oxygen from the bulk of soil water within 24 hours. Crop yields can be severely reduced under such condition as observed by Taiz & Zeiger, (1984) in garden pea with 50% loss in yield.

Oxygen supply determines growth rate at each stage of plant life cycle. When dormant, the seeds require very low oxygen supply. When germination starts, more oxygen is needed to meet the demands of increased respiratory rate. This is a very critical stage in the life cycle and ample supply of oxygen at this stage determines sturdiness of the plant during rest of the life.

When the seedling establishes itself in soil, it produces leaves and begin to photosynthesize. About 30% of the total photosynthetic products are excreted in soil via roots in the form of sugars, amino acids, organic acids, vitamins etc. The content of exudate is in accordance with the type of soil microbe required by the plant in vicinity of its root zone for the specific interaction with

* Botany Department, Tuljaram Chaturchand College, Baramati (Maharashtra) 413102, India.

soil environment e.g. sulphate solublisation, nitrogen fixation, conversion of urea into nitrates, killing of soil pathogen etc. The soil microbes like bacteria actinomycetes, algae, fungi etc. grow in vicinity of root system and get nourishment from the root exudates. They show manifold increase in population density within a few days as a result of this symbiotic relationship. In turn, they carry out various processes essential for proper plant growth such as nitrogen fixation, phosphate solubilization, synthesis of vitamins and antibiotics, degradation of organic matter to release macro- and microelements. These products are supplied to the plant. Antibiotics kill the soil pathogen causing plant diseases and thus create an atmosphere favourable for the growth of host plants. Oxygen demand of soil microbes is four times greater than that of the root cells. This means that the soil should maintain five times more oxygen content than that required by root cells for proper growth of soil microbes and roots. This oxygen demand increases tremendously when a chemical fertiliser like urea is used in the form of soil application because one molecule of urea requires five molecules of oxygen for its conversion into soluble nitrate. Thus most of the oxygen available in soil is used by urea leading to mass killing of microbes. By this, further conversion of urea into soluble nitrates ceases in absence of both, oxygen (required for process) and soil microbes (carrying out this conversion). Moreover, intermediate toxic compounds are formed and are assimilated by the soil. Under such conditions, though urea is applied to soil as a source of nitrogen, it is not available to the plant instead toxic residual substances accumulate in the soil. Thus under soil hypoxia, the application of urea becomes not only a wasteful activity but it also kills the useful soil microbes and in addition, hinder metabolic activities of root cells. This reduces soil fertility and crop yield. Realising this fact, the developed countries have banned the use of urea. The developing countries, on the other hand, are compelled to increase the crop productivity using hybrid seeds which essentially require high dosage of nitrogen for higher yields and N_2 urea is cheapest source of N_2 with highest nitrogen percentage (46%). So urea application is unavoidable inspite of its known disadvantages to the soil properties. Thus the developing countries are caught in a vicious circle. The consequences are very clear in agricultural scenario of India. Green revolution is successful in increasing the productivity from 50 million tonnes to 200 million tonnes during 1960^s to 1990^s by using hybrid seeds, chemical fertilisers and ample irrigation, but unfortunately soil management is totally ignored. Air to water ratio (3:1) is not maintained properly because of small field areas, less mechanised cultural practices, improper land cultivation, poor drainage and high labour cost accompanied by non-judicial use of water and chemical fertilisers. This led to soil infertility. Romheld (1999) stated that salt accumulation and poor water management can result in the dramatic increase in unproductive agricultural land. According to National Commission on Agriculture, in India out of 328 million hectares of agricultural land, 150 million hectares of land has lost its fertility. 59 million hectares of land is eroded while the remaining land has lost its fertility due to mismanagement. Faulty cultivation practices led to destabilisation of physical structure of soil, resulting in inadequate aeration, low oxygen supply and accumulation of CO_2 and other toxic substances. So the crop has to spend more energy just to survive resulting in lower yields. Thus, it is the need of the hour to give emphasis on protection of soil environment which will be automatically followed by increased yields.

Deficiency in oxygen content of soil environment leads to reduction in metabolic activities of both soil microbes and plants. Deterioration of ecological system and loss in soil microbe population lead to decrease in both availability of nutrients in soil and the absorption capacity of roots. As a result, the fertilisers applied to soil are not fully utilised and they are unable to exhibit expected results. The increasing difficulties in soil management and greater environmental awareness are compelling us to find a suitable solution to this problem. Under these conditions, oxygen supplementation through soil is the most reliable method for the stability of soil : microbe : plant ecosystem and for the efficient use of fertilisers. It not only protects the soil environment but also maximises the crop yield due to maximum utilization of inherent genetic potential of the crop species.

A commercial preparation of oxygenated peptone for soil application (released in 1996 under brand name Chaitanya, meaning vitality) releases oxygen slowly and steadily for 40 - 50 days when injected in soil at the depth of 10 cm near the root zone under moist conditions. Each gram of the powder contains 50 mg of oxygen and 600 mg of peptone. Release of oxygen is slow due to peptone coating. Supply of oxygen to soil environment improves microbial activity, availability of nutrients for root uptake and extension of root system. The impact of oxygen on germination is well known but the effectiveness of the technique of oxygen supplementation to soil environment for crop management is not studied yet. To fill up the lacuna, an investigation was carried out to study the influence of oxygenated peptone in various crops at different stages of plant life cycle viz. germination, vegetative growth, and reproductive growth.

The experiments were conducted on variety of crops such as cereals, pulses, vegetables, floricultural crops and fruit crop at different stages of growth. The oxygenated peptone was employed in various ways to test its versatility in application methods.

Cereals, pulses and oil crops responded well to pre-soaking treatment . The increase in germination percentage was evident in gram, sorghum and mustard and marginal in cowpea. Wheat and groundnut, however, had no scope to show increase as they already had a high germination percentage of 96% under control condition.

There was a consistent increase in the length of radical, plumule, number of secondary roots and biomass in all crops studied. Pulses exhibited better response in terms of biomass as compared to oil crops and cereals. Cowpea, sorghum, groundnut and mustard showed better enhancement in number of secondary roots. This character is significant for absorption of water and minerals by roots. Gram, cowpea and groundnut showed a significant increase in length of radical and plumule. The root system exhibited increase in dry weight in the crops studied. Moreover, when the experiment was continued in the same variety of soybean (var.J.S.335) for next 12 days, a tremendous increase from 25.03% to 90.90% was observed in the dry weight of root system. Aster, jasmine and gerbera showed an increase in chlorophyll and carotenoid contents. This was reflected in the enhancement in photosynthetic capacity which resulted in higher carbohydrate content. Seed dressing of cotton enhanced number of leaves and shoots per plant, diameter of stem and length of root system. This led to increase in fruit setting. Soil application of oxygenated peptone was found to boost the vegetative and reproductive growth of chilli and brinjal . Soil application in these crops was equally useful as the seed dressing in cotton, both leading to higher yield.

In aster, oxygenated peptone in the form of soil application increased the quality of flowers in terms of increase in diameter and stalk length of flower . It also induced early flowering and better yield.

In onion *(Allium cepa)*, there was increase in perimeter and weight of bulb accompanied by better keeping quality . Chrysanthemum *(Chrysanthemum morifolium)* is very sensitive to root rot. Under high rainfall conditions, non-treated plants showed 50 % reduction in yield while the treated plants showed four-fold increase.

Citrus *(Citrus medica)* and coconut *(Cocos nucifera)* showed reduction in immature fruit drop. This response was accompanied by increase in fruit size. Decrease in the length and number of aerial roots, both desirable characters in grapevines are achieved using soil application of oxygenated peptone. This was reflected in about 50% increase in number of bunches. Lodging during heavy rainfall is a basic problem in papaya *(Carica papaya)* cultivation. This problem was solved by soil application of oxygenated peptone. In addition, there was considerable increase in fruit setting in first and second flush.

Germination is the first and most important stage in the life cycle of the plant. The environmental conditions, mainly moisture and oxygen status available during germination, decide the pattern of vegetative and reproductive growth of plant, which later reflect in the yield.

The experiments showed that seed dressing as well as soil application of oxygenated peptone was useful in increasing the germination percentage of seeds. This is predominantly due to the role of oxygen in breaking seed dormancy. The embryo synthesises ethylene required for breaking dormancy. Ethylene synthesis requires higher level of oxygen, which is available by seed dressing as well as soil application of oxygenated peptone. In addition, soil application of oxygenated peptone creates aerobic soil environment in which fungi and bacteria produce appreciable quantities of ethylene, helpful in breaking seed dormancy. Secondly, the germination inhibitors present within the seed coat are oxidised in oxygen rich atmosphere (but not in air) and dormancy is broken. Sometimes, secondary dormancy develops under low oxygen supply. This type of dormancy is avoided by using oxygenated peptone.

The rate of respiration increases during germination, showing high demand for oxygen. Consequently, the rate of germination process is found to increase as a result of application of oxygenated peptone. This is supported by the observation of Bewley and Black (1994) who reported sequential increase in oxygen consumption by maize embryo when incubated in water for 6 h from start of germination. The enhancement in oxygen requirement during initial stages of germination is due to increase in mitochondrial oxidation as a result of repair and activation of mitochondria already existing within mature dry seed as well as production of new mitochondria. Ample supply of oxygen during germination allows the operation of citric acid pathway, liberating more energy which becomes available for growth process. The pre-soaking of cereals, pulses and oil crops with oxygenated peptone solution gave the same advantage. This vigour is reflected in consistent increase in length of radical, plumule, number of secondary roots and biomass of seedling when germination studies were carried out under laboratory conditions in petriplates. Seed dressing and soil application of oxygenated peptone brought about the same effect in cotton, chilli and brinjal under field condition. Moreover, the same vigour is maintained in these plants throughout the life cycle, which is reflected in extension of root and shoot system.

Increased level of chlorophylls, carotenoids and carbohydrates throws light on the significance of the technique of soil application of oxygenated peptone. Chlorophyll level is a good index to meet an overall evaluation of any crop for its photosynthetic ability. Chlorophyll content, decides the ability of plant to harness solar energy. Carotenoids allow the utilisation of light over a wider spectral range (Siefermann - Harms, 1985) and protect chlorophylls from photooxidation. The photosynthetic capacity of the plant is reflected in carbohydrate synthesis. The sugars formed during photosynthesis are translocated to root to maintain the energy status of existing roots and for the formation of new roots. Root growth plays a decisive role in the spatial availability of most nutrients because of the restricted movement of nutrients by diffusion through soil (Marschner, 1995, 1998). In fact, according to Helal and Mengal (1981), and Meiri & Plaut (1985), improved photosynthesis increases root growth and nutrient acquisition.

Soils with high mechanical impedance, soil compaction strongly inhibits root growth and nutrient uptake. This can be partially compensated by enhanced root hair formation and root exudation (Sauerbeck and Helal, 1986). All agronomic measures to improve nutrient acquisition have to be considered to improve root growth, particularly at the early seedling stage and thus improve fertiliser use efficiency (Muehlig-Versen *et al.*, 1979). Moreover, optimum nutritional status of a plant ensures a better stress tolerance and also a better water use efficiency (Romheld, 1999).

The energy required for absorption by roots is liberated during respiration in root cells using oxygen available in soil pores. In poorly managed soils (like water logged, poorly drained or compact soils), pores in the soil particles are blocked. In water logged condition, at temperature above 20° C, plant roots, soil fauna and soil microorganisms totally deplete soil oxygen within 24 hours. The growth and survival of many plant species are greatly depressed under such conditions and crop yields are severely reduced. The same thing is observed in Chrysanthemum under high rainfall condition in open fields showing 50% of loss in yield. In fact, under anaerobic condition, citric acid cycle stops, yet carbohydrate utilisation rate doesn't decrease but increases leading to decline in energy status of cell (Taiz and Zeiger, 1984). The fermentation takes place converting pyruvate into lactate. Accumulation of lactate lowers cellular pH and favours ethanol synthesis with production of less energy (2 ATPs per hexose molecule utilised). Thus, injury to root metabolism by oxygen deficiency originates from lack of ATP to drive basic metabolic processes. Under conditions of extreme O_2 deficiency, protons usually leak from vacuole (pH 5.8) into cytoplasm (pH 7.4) making it acidic (Roberts *et. al.*, 1984). These changes are associated with the onset of cell death. The reverse process (transport of protons from cytoplasm to vacuole) requires more energy, which is not available under anaerobic conditions. Application of oxygenated peptone to soil not only protected *Chrysanthemum* crop from anaerobiosis but also led to maximum yield, providing the most favourable soil atmosphere for growth of both roots and aerobic soil microbes. When oxygen supply is ample, the processes like synthesis of nitrates and sulphates are favoured by aerobic soil microbes supplying nourishment to plant but under anaerobic conditions, the soil microbes favour the processes like denitrification and sulphate reduction and the products of these process are toxic to plants at higher concentration (Drew and Lynch, 1980).

Interference with root respiration and root metabolism occurs even before oxygen is completely depleted from root environment. When O_2 concentration is below critical oxygen pressure (cop) at which the respiration rate first slows by O_2 deficiency and the centre of root becomes anoxic. Anoxic or hypoxic roots lack sufficient energy to support physiological processes on which the shoot depends (Jackson and Drew, 1984).

The failure of roots to absorb nutrient ions and transport them to shoot leads quickly to shortage of ions within developing and expanding tissues. Soil application of oxygenated peptone relieves the anoxic and hypoxic conditions and enhances physiological processes within root leading to better shoot growth. This is clearly exhibited by cotton; brinjal and chilli in terms of better shoot growth leading to better yield. Auxin controls root initiation and differentiation of conducting tissue system. Auxin transport requires oxygen. With ample oxygen supply, there is more growth of both, roots and conducting tissue system as auxin transport is enhanced .

Oxygen deficiency in roots stimulates abscisic acid synthesis, (Glinka and Reinhold, 1971), and brings about stomatal closure (Zhang and Davies, 1987), which has adverse effect on uptake and transport of nutrients from roots to leaf. Slow release of oxygen from soil applied oxygenated peptone improves the situation. It checks abscisic acid synthesis, consequently stomata open and transpiration pull is established. More water enters the root system through increased ability of water absorption and extension of root system. This situation demands for increase in the girth of stems for transport of water and nutrients from roots to leaf which is well evident in the experiments with herbaceous plants like cotton, brinjal and chilli. In grapes, it is reflected in terms of decrease in number of aerial roots, a significant advantage to the grape growers. In onion, it leads to increase in weight and size of bulb.

Herbaceous species have a network of intercellular air spaces running from the leaves to roots to provide a continuous channel for movement of oxygen by diffusion. During the day, atmospheric oxygen enters through open stomata, diffuses through intercellular spaces of cortex and reaches the root region. During night, the stomata are closed and oxygen is not available to roots through stomata but the demand for oxygen is very high as growth processes mainly take place during night. Possibly, the roots absorb oxygen from soil during night and this oxygen diffuses through intercellular system in the pith to the shoot system. The anatomical structure of pith with thin walled parenchymatous cells enclosing bigger intercellular spaces facilitates the easy oxygen transfer by diffusion. The oxygen is then transported to different aerial parts. For smooth running of this system, the soil should be rich with oxygen. Use of oxygenated peptone releases O_2 which diffuses into root and is transported to shoot for growth. This is the reason for higher yields in all plants studied.

Formation of aerenchyma is induced by oxygen deficiency in non-wetland cereals like wheat and maize (Taize and Zeiger, 1984). Hypoxia induces ethylene synthesis in root cells, which promotes break down (lysis) of cells in root cortex. As the root extends into oxygen deficient soil, aerenchyma is formed continuously just behind the root tip (that is why papaya shows lodging under heavy rainfall condition). This aerenchyma allows oxygen movement from leaf to root upto apical zone. In absence of barriers like deposition of lignin or suberin, the root tips are permeable to oxygen and thus internal oxygen becomes insufficient for aerobic respiration in root cells. The oxygenated peptone controls ethylene synthesis and lysis of root cells (as evident by decrease in lodging in papaya by use of oxygenated peptone.) Some photographs showing the results are presented at the end of the paper.

Germination: Effect of pre-treatment of oxygenated peptone on germination of ground-nut

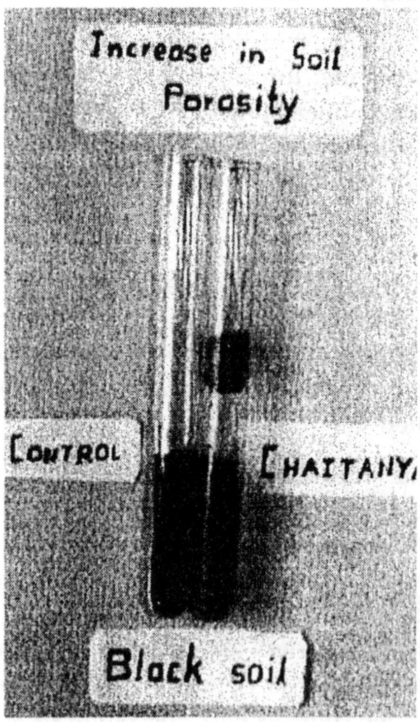

Increase in soil porosity in black soil using oxygenated peptone

Increase in vegetative growth and fruiting in chilli (var. Jwala) using oxygeneted peptone

Increase in rooting and fruiting in soybean (var. J. S. 335) using oxygeneted peptone

Enhancement in vegetative and reproductive growth in brinjal (var. Ankur 55) using oxygenated peptone

Increase in number of flowers in Chrysanthemum using oxygenated peptone

CONCLUSION

The oxygen supplementation through soil improves oxygen supply, microbial activity and nutrient availability, which encourages the growth of aerobic soil microorganisms useful to plants and at the same time discourages the growth of anaerobic microorganisms harmful to plants. In poorly managed soils, it enhances uptake of water, minerals and oxygen, exhibiting efficient use of fertilisers applied to soil. In this way, this method improves the soil properties using less cost (Rs. 1000/ha or $20/ha) and maintains ecological balance of soil, microorganisms and plant ecosystem. It induces better vegetative growth, early flowering and fruiting with enhancement in quality and quantity of flowers and fruits exhibiting full exploitation of genetic potential of the plant. This cheap technique is universal in application as it can be used in any soil, any crop at any growth stage. It can also be used for soil amendment. Certainly, it is a boon to agriculture in new millennium.

REFERENCES

Bewley, J. D. and Black M. (1994) Seeds Physiology of Development and

Germination 2d Edn. Plenum Press, New York. pp. 147- 229.

Drew, M. C. and lynch, J. M. (1980) Soil anaerobiosis, microorganisms and root function. *Annu. Rev. Phytopathol.* 18 : 37–66.

Helal, H. M. and Mengel, K. (1981) Interaction between light intensity and NaCl salinityand their effects on growth, CO2 assimilation and photosynthate conversion in young broad beans. *Plant Physiology*, 67 : 999–1002.

Glinka, Z. and Reinhold, I. (1971) Abscisic acid raises the permiability of plant cells to water. *Plant Physiology*, 48 : 103–105.

Jackson, M. B. and Drew, M.C. (1984) Effects of flooding on growth and mechanism of herbaceous plants. In : Flooding and Plant Growth.

Kozlowski, T. T. ed. pp. 47–128. Academic Press, New york.

Marschner, H. (1995) Mineral Nutrition of Higher Plants, 2d Ed. Academic Press, London. pp. 889.

Marschner, H. (1998) Soil–root interface: Biological and biochemical processes. In : Soil Chemistry and Ecosystem Health. Soil Science Society of America, Madison Special Publication 52.

Meiri, A. and Plaut, Z. (1985) Crop production and management under saline conditions. *Plant and Soil*, 89 : 235–271.

Muehlig–Versen, B., Buerkert, A. , Batino, A., and Marschner, H. (1979) Crop residue and phosphorus managementin millet based cropping systems on sandy soils of Sahel.

In : Renard *et. al.* (eds.) Soil Fertility Manegement in West African Land

Use Systems. Margraf Verlag, Weltershein, Germany, pp. 31–40.

Roberts, J. K. M., Callis, J., Jardetzky, O., Walbot, V. and Freeling, M. (1984)

Cytoplasmic acidosis as a determinant of flooding intolerance in plant. *Proc. Natl. Acad. Sci.* USA 81 : 6029–6033.

Romheld, V. 1999 Morphological and physiological aspects of nutrient acquisition under stress conditions. Proceedings of Dahlia Greidinger International symposium on Nutrient Management under Salinity and Water Stress. Haifa, Israel, 1–4 March 1999. pp. 55–66.

Sauerbeck, D. and Helal, H. M. (1986) Plant root development and photosynthate consumption depending on soil compaction. Trans XIII

Congress. International Soil Science Society, Hamburg. 3 : 948–949.

Siefermann _ Harms, D. (1985) Carotenoids in photosynthesis. I. Location in photosynthetic membrane and light harvesting function. *Biochim.*

Biophys Acta, 811 : 325–355.

Taiz, L. and Zeiger, E. (1984) Plant Physiology. The Benjamin/Cummings Publishing Co. Inc. New York. pp. 365–367.

Zhang, J. and Davis, W. J. (1987) ABA in roots and leaves of flooded pea plants. *J. Exp. Bot.* 38 : 649–659.

Table—1: Response of different crops to soil application of oxygenated peptone (2.5 kg/ha in onion and chrysanthemum and 5g/plant for fruit crops).

Sr.No.	*Crop*	*Responses*
1	2	3
1	Onion	*(i)* Increase in weight of bulb (from 110 g to 235 g) *(ii)* Increase in perimeter of bulb (from 14.0 cm to 19.5 cm)
2	Chrysanthemum (Double Korean)	*(i)* Control plot showed 50% decrease in yield under heavy rainfall condition. *(ii)* Treated plot showed bumper crop (1246 flowers sq.m)
3	Citrus	*(i)* Reduction in immature fruit drop. *(ii)* Increase in yield. *(iii)* Increase in perimeter of fruit (from 10.2 cm. to 13.85 cm.)
4	Coconut	*(i)* Increase in yield. *(ii)* Reduction in immature fruit drop.
5	Grapes	*(i)* Decrease in number of aerial roots (from 107 to 20) *(ii)* Decrease in length of aerial roots (from 1.6 cm. To 2.4 cm.) *(iii)* Increase in number of bunches per vine (from 22 to 32).
6	Papaya	*(i)* Decrease in lodging of tree under heavy rainfall condition (from 13% to 1%). *(ii)* Increase in number of fruits per plant in first flush (from 8 to 34). *(iii)* Increase in percentage of fruit setting during second flush (from 25 to 80%).

(Contd...)

1	2		3
7	Wheat	*(i)*	Control and experimental plants with 45 and 69 grains per ear respectively, exhibiting 45% increase.
8	Aster	*(i)*	Increase in total chlorophylls, carotenoids and carbohydrates of leaves.
		(ii)	76% increase in flowering.
9	Cotton (Var-Daptari-29)	*(i)*	Enhanced vegetative growth
		(ii)	40% increase in flowering
10	Chilli (Var-Jwala)	*(i)*	Enhanced vegetative growth
		(ii)	increase in flowering
11	Brinjal (Var-Ankur 555)	*(i)*	Enhanced vegetative growth
		(ii)	5 times increase in flowerings
12	Gerbera	*(i)*	Enhancement in total chlorophylls, carotenoids and carbohydrates in leaves
		(ii)	Increase in number of flowers
13	Jasminum	*(i)*	Enhancement in total chlorophylls, carotenoids and carbohydrates
		(ii)	Isncrease in quality and quantity of flowers.

11

Blue Green Algae from Soils of Karnataka

*Shankar P. Hosmani & Partha S.

INTRODUCTION

Since time immemorial blue green algae have been of great significance to man. They occupy an important position in agriculture and are of great interest scientifically. Remote islands contain blue green algae and stand first in succession of plants. Blue green algae are ubiquitous in nature. They live in environments ranging from fresh water to hypersaline aquatic ecosystems and temperate to tropical region of terrestrial ecosystem covering most diverse types of environmental conditions on earth. The importance of soil algae can be attributed to their mear presence in considerable numbers. The pioneering role of blue green algae in barren or divided areas is note worthy. One major agronomic contribution of certain blue green algae is their capacity of fixing atmospheric nitrogen which increase the fertility of soils. Yet little has been done to utilize the soil algae as nitrogen fixers in soils deficient of nitrogen. Today due to hazards of toxic chemicals, contaminated water supply and unwanted wastes, algal population is becoming endangered in crop fields. One of the main threat to blue green algae is the excessive use of pesticides and fungicides to destroy insects, pests and fungi. Blue green algae are very sensitive to even low concentrations of these chemicals and their continuous use is found to bring about undesirable changes in physiological and biochemical pathways of the organism, thus causing damage to both algae and the crop. In contrast to such a situation, growth of technology during the later half of the 20th century has witnessed research work which is diverted towards the study of the effect of pesticides, herbicides, antibiotics, extracellular products and the nature of cellular components of blue green algae in soils.

During the last five to six decades considerable work has been done on soil algae. The available data has been reviewed by Peterson (1935), Bharadwaja (1934), Lund (1962), Shields

* Department of Studies in Botany University of Mysore Manasagangotri, Mysore–570006 (India)

and Durrell (1964) and Subbaraju (1972). Schmide (1901) and Beck and Zahlbruchner (1897) made extensive studies on blue green algae collected by Hansgirg (1888). Other records of blue green algae from India were by Turner (1892). Other Indian workers that need mention are Gonzalves and Gangla (1949), Datta and Venkataraman (1958), Pandey (1965), Marathe (1967), Anantani and Marathe (1972), Bharathi and Pai (1972), Grover and Pandhal (1975), Dabral et al (1978), Angadi (1982) and Partha (2000).

The present review aims at exploring blue green algae from various districts of Karnataka in order to evaluate the composition of the blue green algal flora and their relation to certain physico-chemical parameters; the frequency of occurrence of blue greens both heterocystous and non heterocyotous and their relation to the soil types, major crops, fertilizers and the source of water.

The results obtained from extensive field trials on algalization experiments under different agronomic conditions have shown that it is economically useful. The beneficial effect of algal inoculation on the growth and yield of many crops is well documented. In order to ensure the introduction and establishment of algae in the fields, it is necessary to search for efficient local strains with better ecological adaptations. This objective can be achieved by surveying the algal flora of the various moist environments so that isolation of better nitrogen fixing forms is possible.

SAMPLING

A total of 120 soil samples from eight districts of Southern Karnataka were collected mainly during the summer months. The large lumps of soil were crushed to a fine powder and sieved in a 2 mm sieve and used for further investigation. About 25 mgs soil from each sample was inoculated into sterile petriplates. Since sterile distilled water did not facilitate sufficient growth of algae, De's modified Beneck's medium was used to raise the algal cultures. Periodical collections of algae grown on the soil surface were made after fifteen days and consequently after 2 days and were preserved in 4% formaldehyde solution for further observation. This process was repeated for about 60 days or until no further growth of algae were recorded. Camera lucida drawings of only blue green algae was made and were identified using standard monographs. Table 1 indicates the details of the eight districts studied. The soils were also analyzed for a few important physico-chemical parameters. As many as 151 blue green algae have been reported (Plate 1 to 18). Checklist of frequently accuring blue green algae is provided in Table 3.

DISCUSSIONS

A statistical correlation "correlation matrix" provides us with generalized relations between physico-chemical factors as well as the blue green algae. The results of the correlation matrix (Table 2) are indicative of the fact that 7 among the 11 parameters are not of great significance in controlling the growth of blue green algae in these soils. Nitrogen available within the cells and not external, has an impact on their growth. However Mg++ involved in nitrogen fixation as (Mg+ATP) may have a role in regulation of the growth of blue greens, pH of soils in association with phosphorous, potassium and magnesium plays a significant role in controlling the growth of blue green algae in them.

The frequency of occurrence of blue green algae in the soils is represented in Table 3. *Chroococcus indicus* among the unicellular blue green algae was found in 18 soils, whereas among

Table—1: General characteristics of the eight districts studied

District	Location	Rivers	Tributaries	Common crops	Rainfall (mm) (average annual)	RH%	Soil types
1. Mysore	11°36′ and 13°35′ North latitude 75o51′ and 77o20′ East longitude	Cauvery	Hemavathi, Lokapavani, Simsha, Kabini Suvarnavathi	Ragi, Tobacco Sugarcane, Aracanut, Coconut	40 to 42.3	74.6 to 64.1	BCS, RSL, RCL
2. Mandya	12°13′ and 13°04′ North latitude 76°19′ and 77°20′ East longitude	Cauvery	Hemavathi Kabini	Rigi, Tobacco Sugarcane, Aracanut, Coconut.	39.5 to 43	80.4 to 81.0	RSL
3. Hassan	12°31′ and 13°33′ North latitude 75°33′ and 76°38′ East longitude	Hemavathi		Ragi, Tobacco Sugarcane, Aracanut Coconut.	51.2 to 76.8	57.2 to 65.9	RSL BCS
4. Chikkmagalur	12°54′ and 13°55′ North latitude 74°38′ and 76°40′ East longitude	Tungabhadra	Yagachi, Veda, Avathi	Ragi, Tobacco Sugarcane, Aracanut, Coconut, Tea & Coffee	69.1 to 70.1	67.6 to 67.5	RSL BS
5. Shimoga	13°27 and 14°39′ North latitude 74°38 and 77°2′ East longitude	Tungabhadra Sharavathi		Ragi, Tobacco Sugarcane, Aracanut, Coconut, Rice, & Jowar.	73.1 to 73.2	52 to 53	SL CL
6. Chitradurga	13°35′ and 15°2′ North latitude 75°38′ and 77°2′ East longitude	Tungabhadra (Partly)	Vanivilas Sagar, a large reservoir	Ragi and Aracanut.	36.5 to 36.6	55 to 56	BCS
7. Tumkur	12°45′ and 14°6′ North latitude 76°21′ and 77°28′ East longitude	Talperges (perennial springs)		Coconut, Aracanut	35.5 to 36.6	60.2 to 60.3	RS
8. Bangalore	12°51′ and 13°30′ North latitude 77°4′ and 77°59′ East longitude	Cauvery	Penna Arkavathi	Coconut, Aracanut, Oil seeds.	43.2 to 43.5	65.6 to 65.7	RSL latentic

BCS–Black cotton soils, RSL–Red sandy loams RCL–Red clay loams, BS–Black soils (Medium).

Table—2: Correlation matrix (Zero is dependent variable)

	X1	*X2*	*X3*	*X4*	*X5*	*X6*	*X7*	*X8*	*X9*	*X10*	*X11*	*X12*
X1	1.000											
X2	-.647	**1.000**										
X3	0.090	0.053	1.000									
X4	0.011	0.176	0.060	1.000								
X5	-.047	0.153	-.040	0.191	1.000							
X6	**0.060**	**0.106**	**0.005**	**0.143**	0.416	**1.000**						
X7	0.147	-.003	0.206	0.084	-.140	0.157	1.000					
X8	-.165	0.059	0.052	0.075	-.003	-.026	0.031	1.000				
X9	0.289	**-.185**	**0.074**	**0.036**	**0.078**	**0.039**	**0.091**	**0.031**	**1.000**			
X10	0.293	**-.070**	**-.045**	**-.047**	**0.121**	**0.029**	**0.058**	**-.056**	**0.174**	**1.000**		
X11	0.163	-.153	0.045	0.143	0.019	0.054	-.030	0.053	-.000	-.081	1.000	
X12	0.335	**-.165**	**0.001**	**0.073**	**0.102**	**0.062**	**0.019**	**-.001**	**0.126**	0.666	0.690	**1.000**

X1 – pH, X2 – Phosphorous, X3 – Chloride, X4 – Organic Carbon, X6 – Potassium, X7 – Nitrogen, X8 – Calcium,

X9 – Magnesium, X10 – Heterocystous blue green algae, X11 – Non-heterocystous blue green algae, X12 – Total blue green algae (Bold numbers indicate significance at 5% level) (Data obtained after feeding results of 121 soil samples)

Table—3: Algae of frequent occurrence

	Algae recorded	*No. of soils in which represented*	*Other authors reported*
	Unicellular	(24%)	
1.	*Aphanocapsa biformis*	05	3
2.	*Chroococcus indicus*	18	3
3.	*Chroococcus miuitus*	07	1
4.	*Chroococcus schizodermaticus*	05	1
5.	*Chrooccus turgidus*	05	6
6.	*Gloeocapsa rupestris*	06	3
	Non-heterocystous	**(12%)**	
1.	*Oscillatoria chalybae var insularis*	06	3
2.	*Lyngbya dendrobia*	05	4
3.	*Lyngbya mesotrichia*	05	1
	Heterocystous	**(64%)**	
1.	*Anabaena ballygunglii*	05	2
2.	*Anabaena circinalis var crassa*	12	2
3.	*Anabaena oryzae*	05	6
4.	*Anabaena doliolum*	05	4
5.	*Anabaena variabilis*	05	4
6.	*Calothrix braunii*	05	1
7.	*Calothrix epiphytica*	05	3
8.	*Calothrix marchica*	05	5
9.	*Hapalosiphon welwetzsheii*	05	2
10.	*Microchaete aequalis*	05	9
11.	*Nostoc calisicola*	05	12
12.	*Nostoc commune*	05	14
13.	*Nostoc muscorum*	07	12
14.	*Nostoc paludosum*	05	1
15.	*Nostoc punctiforme*	05	3
16.	*Tolypothrix tenuis*	05	3
17.	*Tolypothrix nodasa*	05	3

Total algae recorded 15%, total soils observed 120

the heterocystous forms, *Anabaena circinalis* var *crassa* was high in number. *Chroococcus minutus* and *Nostoc muscorum* were the next in the frequency. If the number of species occurring frequently is considered, heterocystous forms were more, followed by unicellular blue greens and non

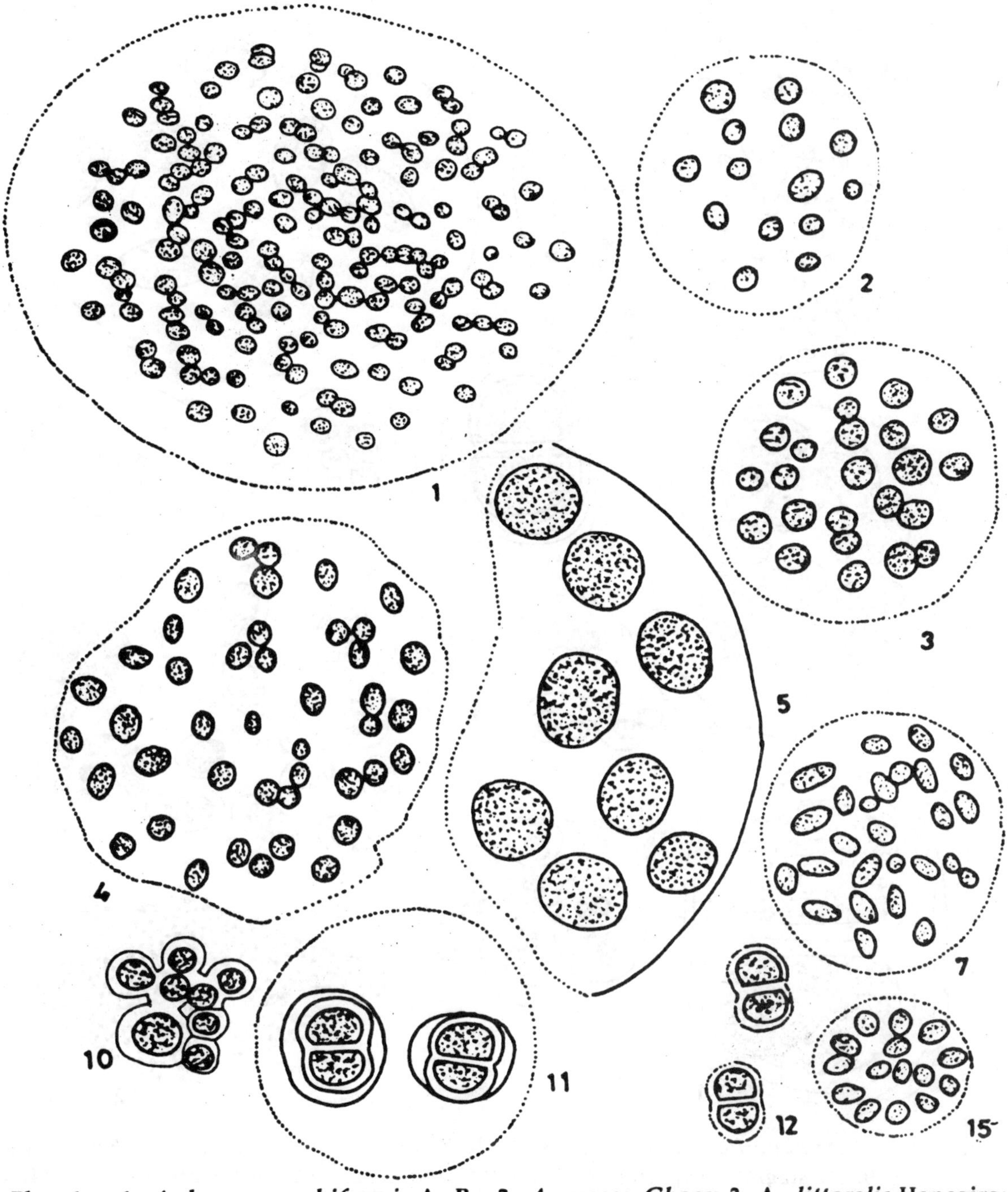

Plate 1: 1. *Aphanocapsa biformis* A. Br; 2. *A. crassa Ghose*; 3. A. *littoralis* Hansgirg; 4. *A. montana Cramer*; 5. *A. roseana* De. Barry; 7. *A. microscopica* Nag; 10. *Chrococcus indicus* Zellar; 11. *C. limneticus* Lemn; 12. *C. minutus* (kutz) Nag; 15. *Chroococcidiopsis indica*.

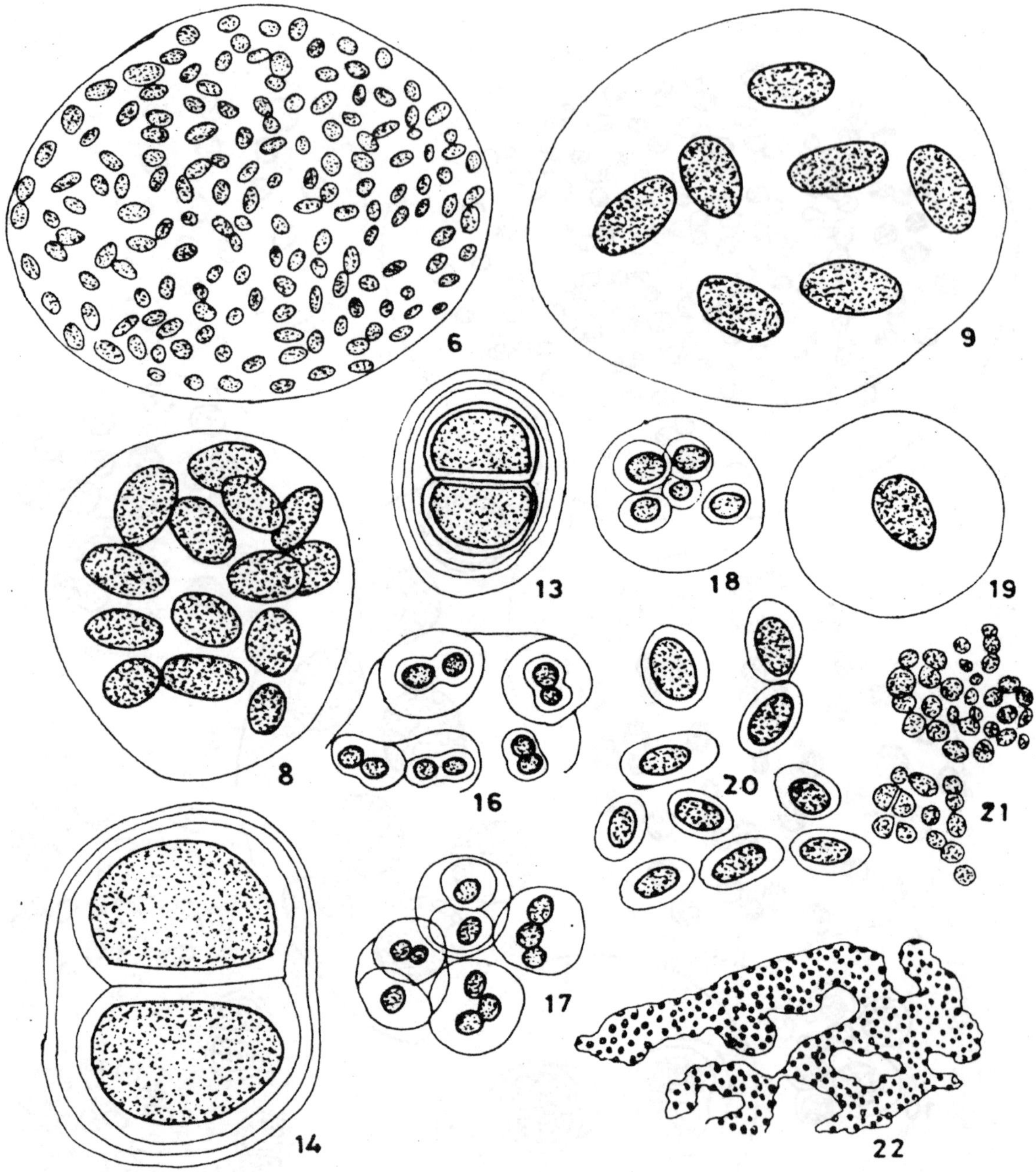

Plate 2: 6. *Aphanothece castagnei* (breb) Rab; 8. *Aphanocapsa pallida* (Kutz). Rab; 9. *A. stagnina* (Spreng) A. Br; 13. *Chroococcus schizodermaticus* West; 14. *C. turgidus* (Kutz) Nag; 16. *Gloeocapsa calcaria* Tilden; 17. *G. gilatinosa* Kutz; 18. *G. kuetzingiana* Nag; 19. *G. rupestris* (Kutz) Bornet; 20. *G somoensis* Wille; 21. *Hydrococcus rivilaris* Kutz; 22. *Microcystis aeruginosa* Kutz.

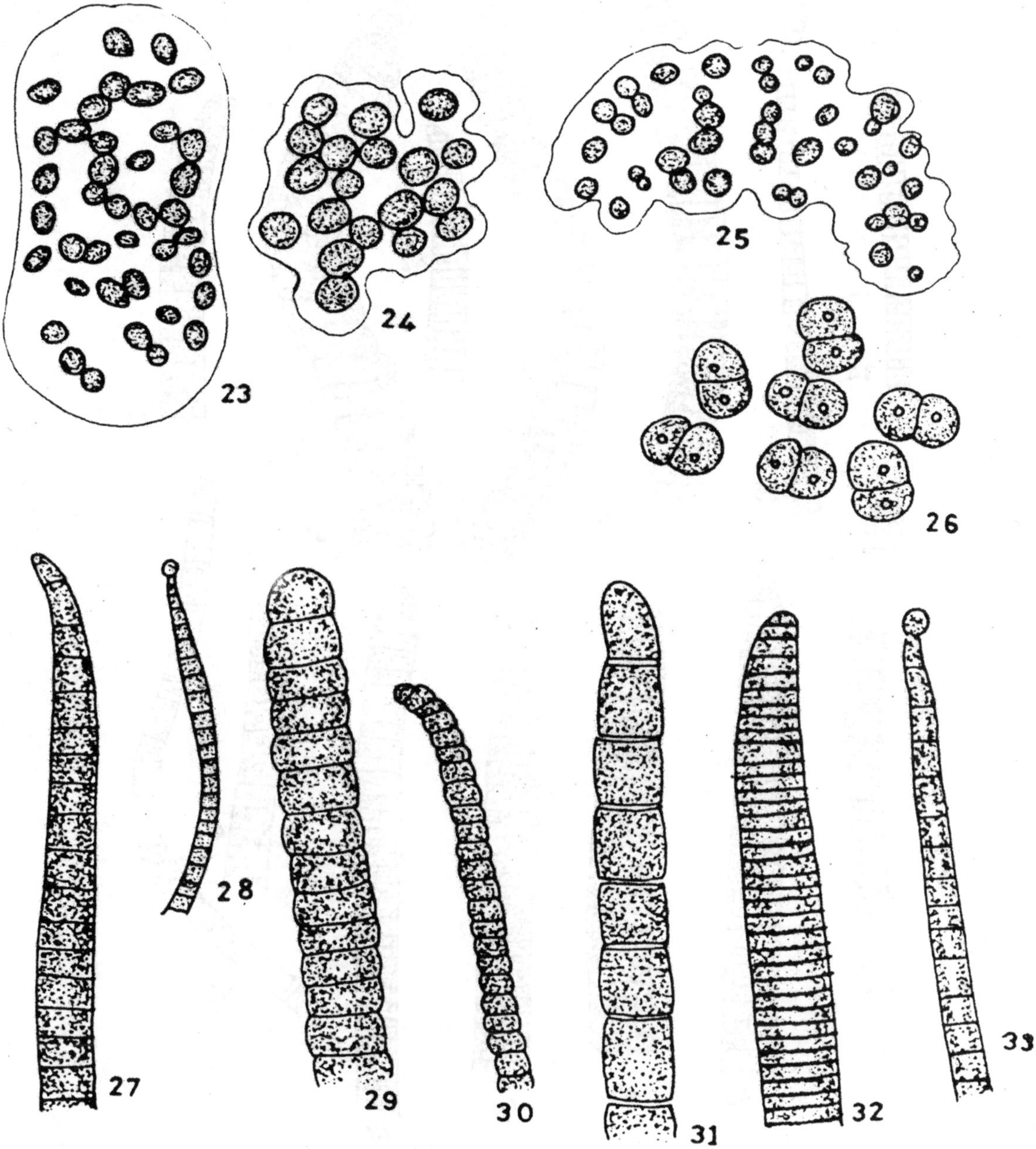

Plate 3: 23. *Microcystis elongata* Delik; 24. *M. flos-aquae* (Wittr.) Kir; 25. *M. viridis* (A. Br.) Lemn; 26. *Synechocystis aquatilis* Sauv; 27. *Oscillatoria acuminata* Gomont; 28. O. *amoena* (Kutz) Gomont; 29. *O. annae Van Goor*; 30. *O. chalybaea v. insularis* Gardner; 31. *O. chlorina* Mutz ex Gom; 32. *O. curviceps* Ag. ex Gomt; 33. *O. irrigua* Kutz.

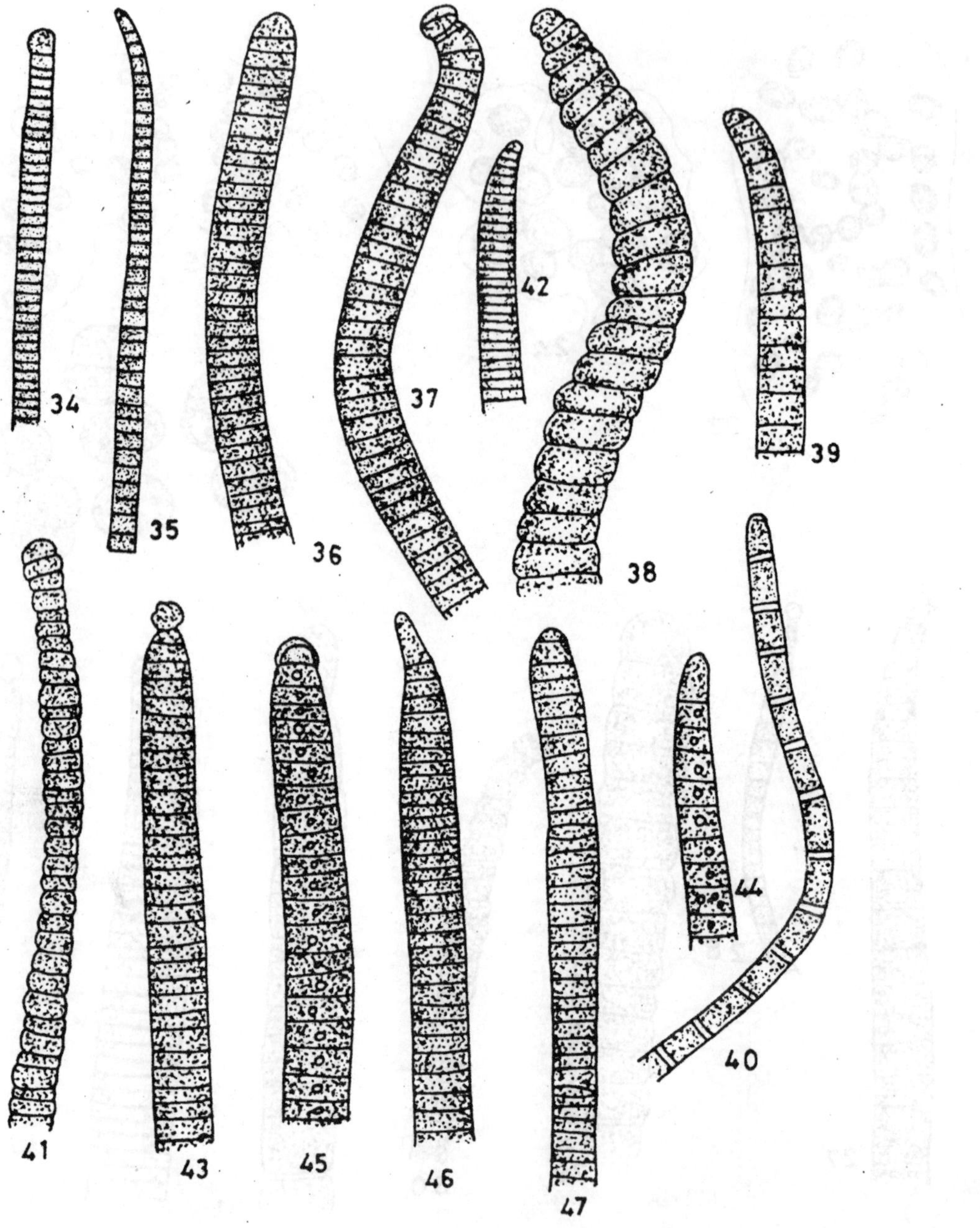

Plate 4: 34. *Oscillatoria limnetica Lemn*; 35. *O. laete-virens var, minimus* Biswas; 36. *O. limnosa* Ag-ex Gomont; 37. *O. martini* Fremy; 38. *O. ornato* Kutz ex Gomont; *O. obscura* Brul. Bis; 40. *O. pseudogerminosa* var. *unigranulata* Bis; 41. *O. proteus* Skuja; 42. *O. princeps* Vaucher Gom; 43. *O. proboscidea* Gomont; *O.* raciborskii Wol; 45. O. *rubescens* De ex Gom; 46. *O. salina* Bis; 47. *O. subbrevis* Schmidle.

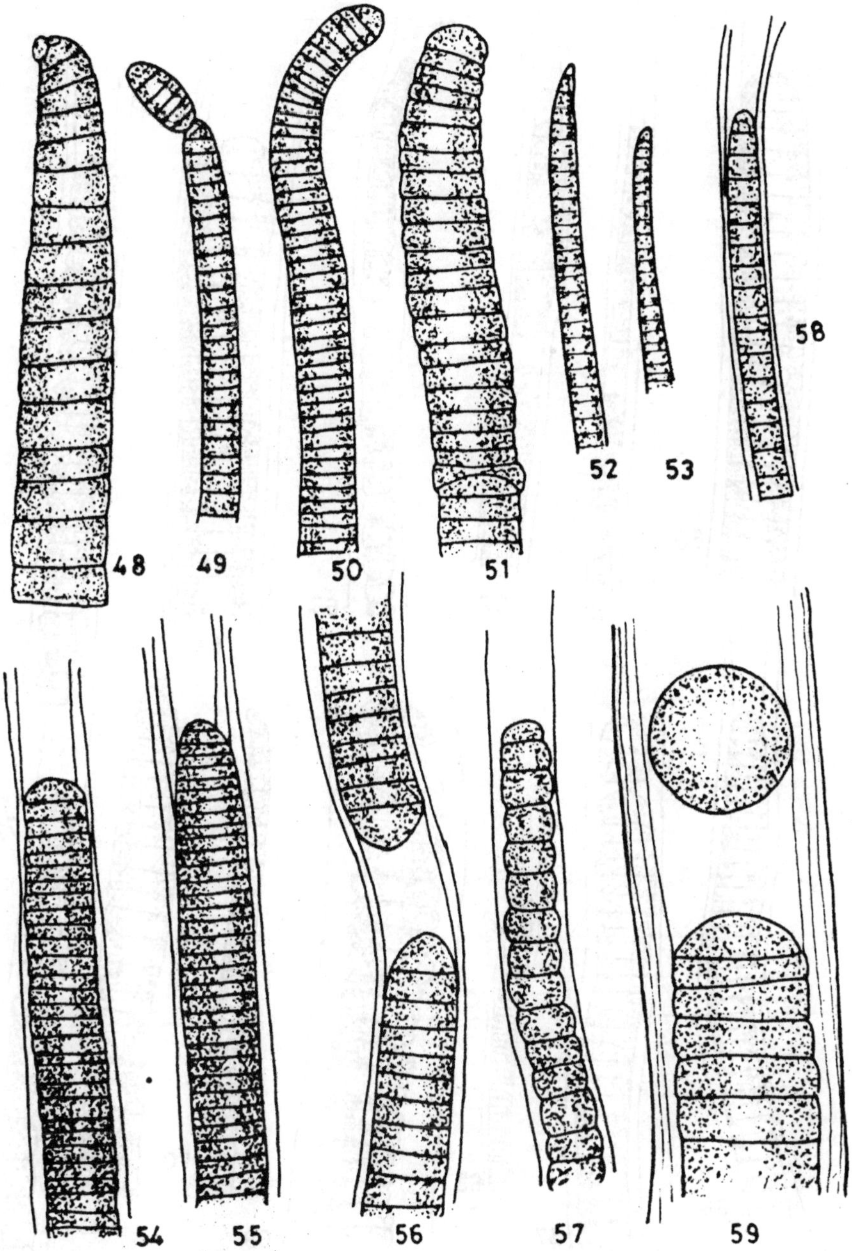

Plate 5: 48. *Oscillatoria subiliformis* Kutz. ex Gomont; 49. *O. tanganyikae* G.S. West, 50. *O. terebriformis* Ag. ex. Gom; 51. *O. vizagapatensis Rao;* 52. *O. acuta* Braul et. Bis; 53. *O. chilkensis* Bis; 54. *Lyngbya bergei* Smith; 55. *L. confervoides* C. Ag. ex Gom; 56. *L. dendrobia* (Ag. ex Gom; 57. *L. gravilis* (Mengh.) Rabenh; 58. *L. lagerhamii* (Mob) Gom; 59. *L. majuscula var. chakiaensis* J. De Toni.

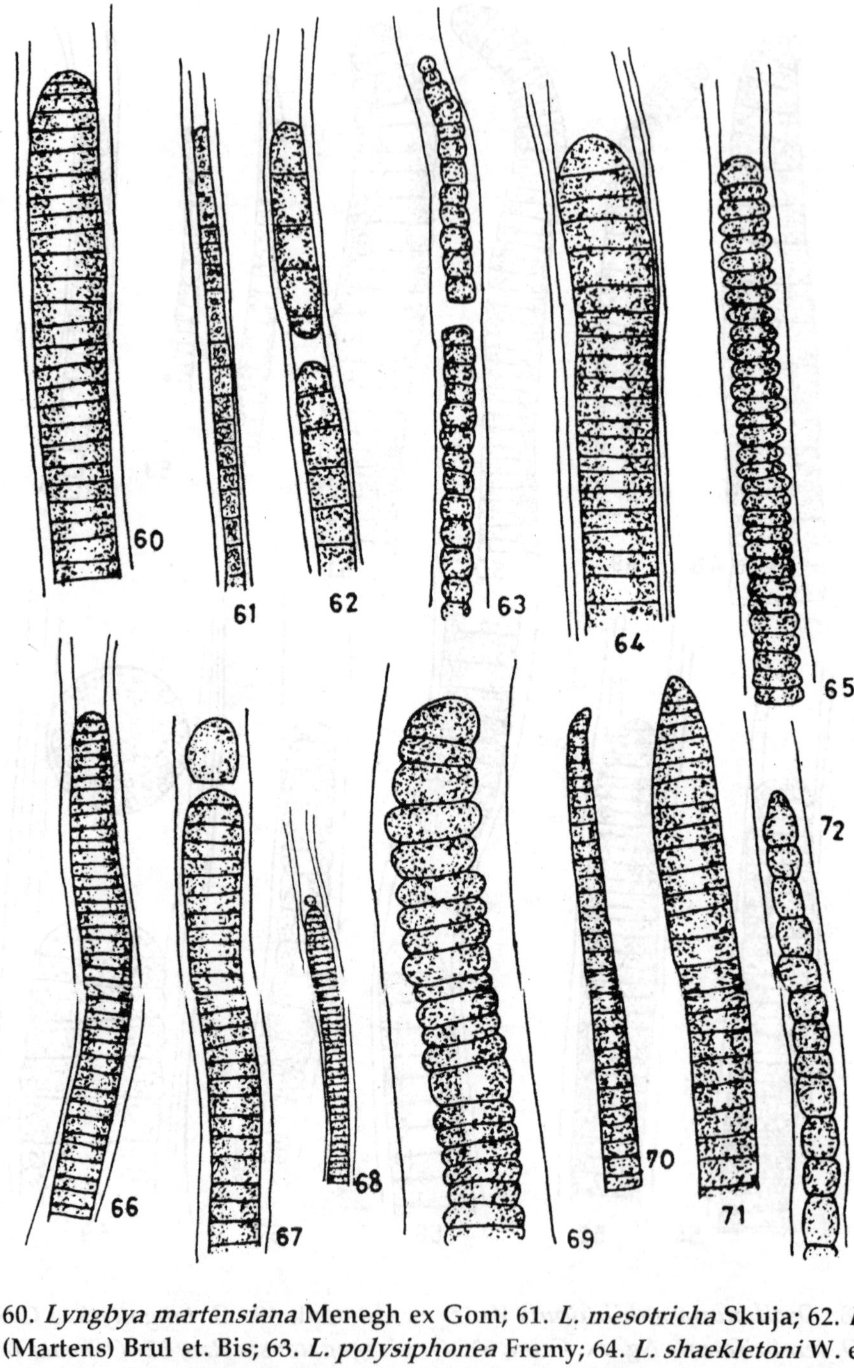

Plate 6: 60. *Lyngbya martensiana* Menegh ex Gom; 61. *L. mesotricha* Skuja; 62. *L. palmarum* (Martens) Brul et. Bis; 63. *L. polysiphonea* Fremy; 64. *L. shaekletoni* W. et G.S. West; 65. *L. sordida (*Zonard) Gom; 66. *L. spirulinoides* Gomont; 67. L. *spiralis* Geitler; 68. *L. aestuarii* Liebon; 69. *L. nordgardhii Wille;* 70. *Phormidium ambignum* Gomont; *P. inundatum* Kutz. Ex. Gomont; 72. *P. inundatum* Kutz. ex. Gomont.

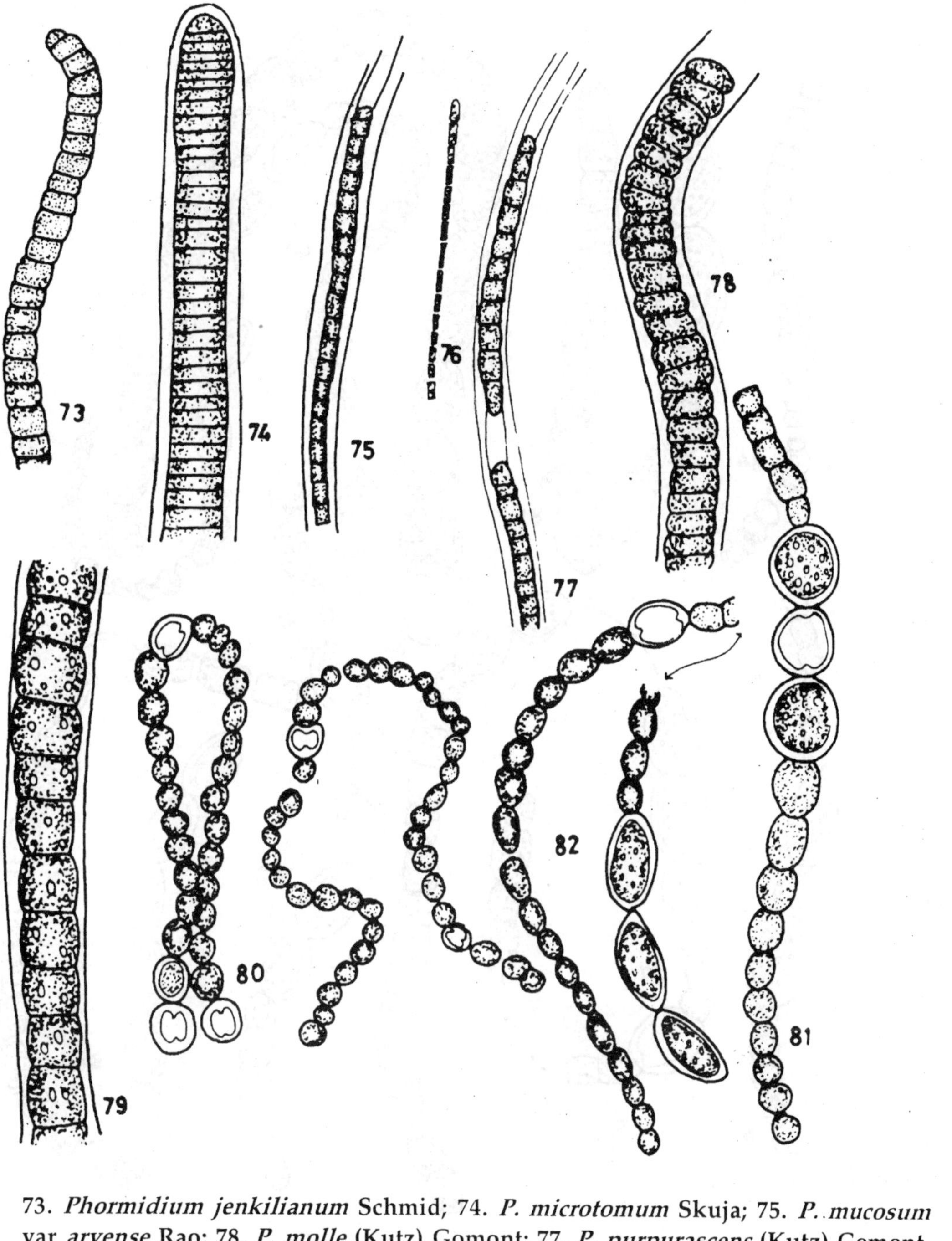

Plate 7: 73. *Phormidium jenkilianum* Schmid; 74. *P. microtomum* Skuja; 75. *P. mucosum* var *arvense* Rao; 78. *P. molle* (Kutz) Gomont; 77. *P. purpurascens* (Kutz) Gomont. 78. P. *uncinatum* (Ag) Gomont; 79. *Symploca muscorum* (Ag.) Gomont; 80. *Anabaena anomala* Fritsch; 81. *A. ballygunglii* Banerji; 82. *Aanabaena circinalis* var. *crassa* Ghose.

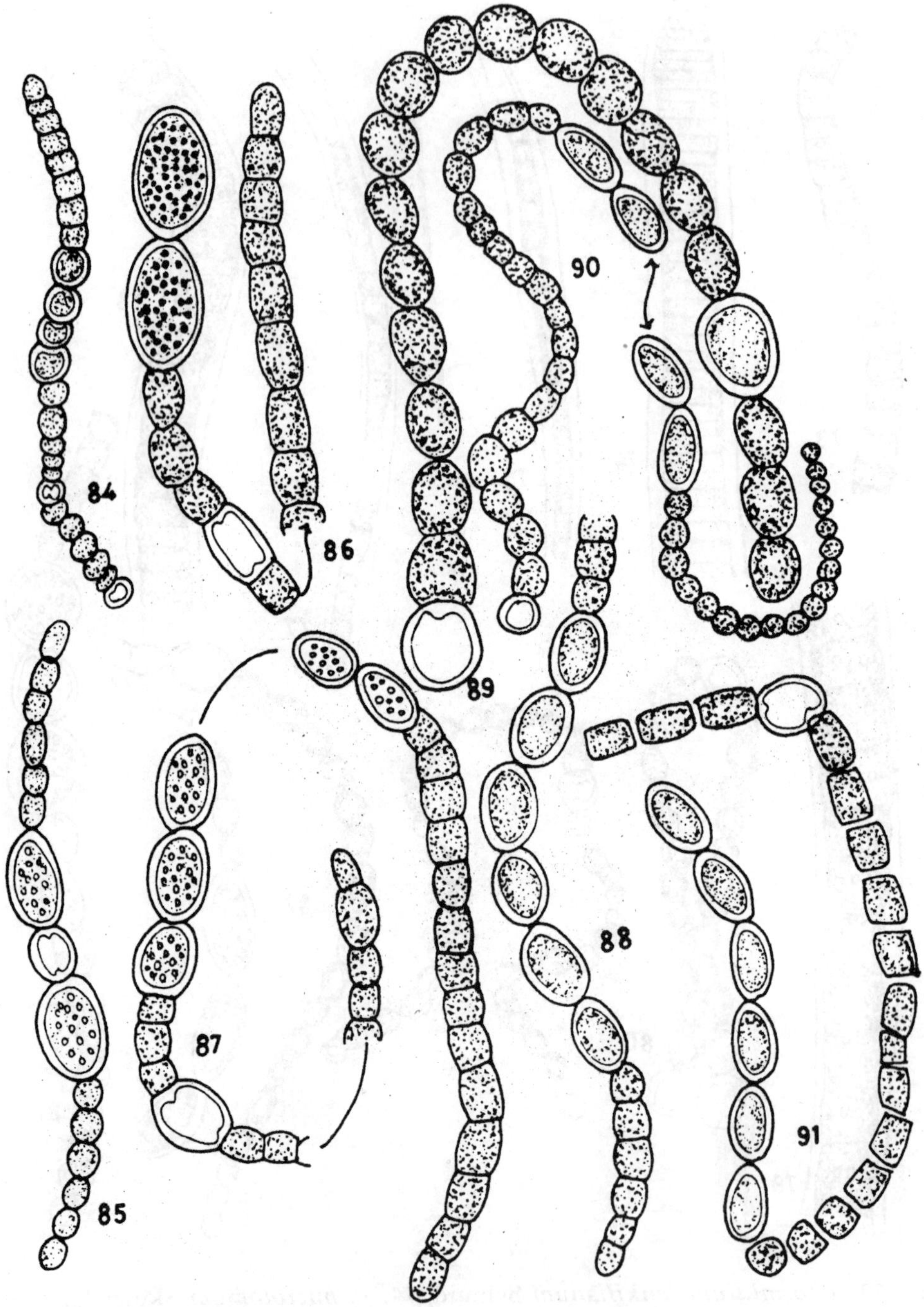

Plate 8: 84. *Anabaena fertilissima* Rao; 85. *A. jyengari* var. *tenius* Rao; 86. *A. navaculorides Fritsch;* 87. *A. iyengarii* Bharadwaj; 88. *A. oryzae* Fritsch; 89. *A. sphaerica* var. *attenuata* Bharadwaj; 90. *A. torulosa* (Carm); 91. *A. variabilis* var. *ellipospora* Fritsch.

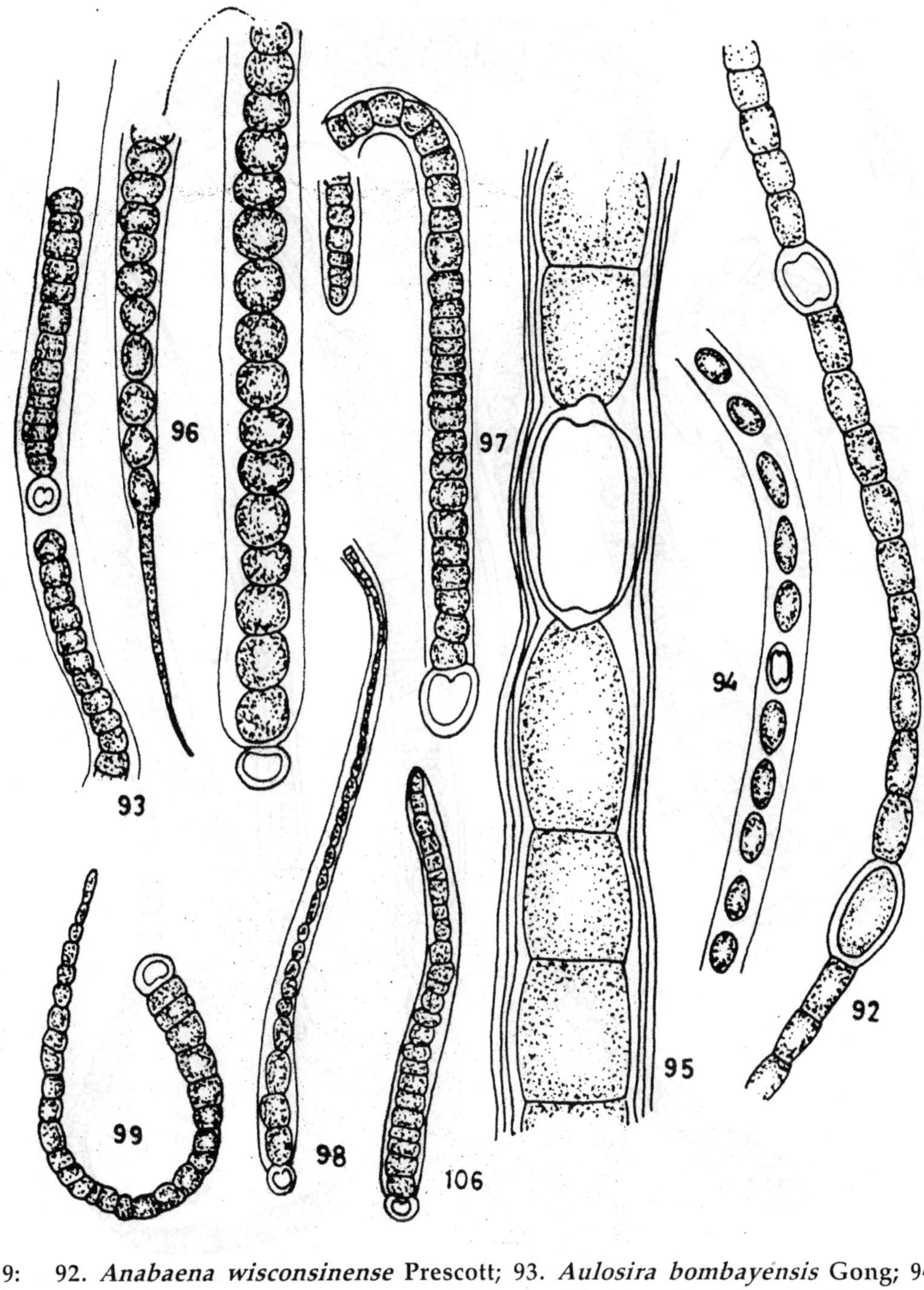

Plate 9: 92. *Anabaena wisconsinense* Prescott; 93. *Aulosira bombayensis* Gong; 94. *Au. fertilissima* Ghose; 95. *Au. implexa* Born et Flah; 96. *Calothrix braunii* Born et. Flah; 97. *C. brevissima* West; 98. *C. castellii* var. *somastipurense* Rao; 99. *C. clavata* West; 106. *C. marchica Lemn.*

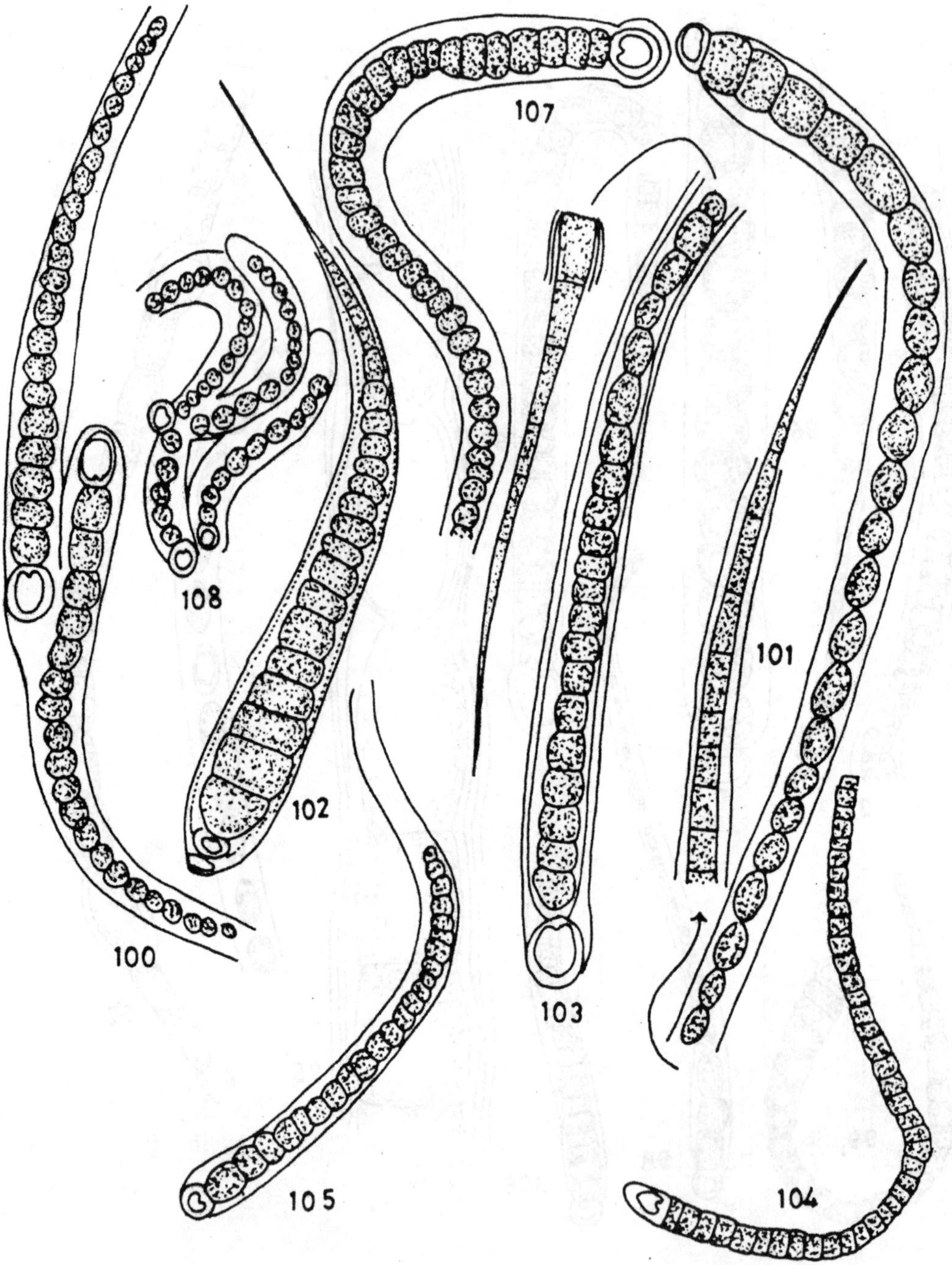

Plate 10: 100. *Calothrix elenkinii* Koss; 101. *C. epiphytica* W. et G.S. West; 102. *C. fusca* var *carassa* Rao; 103. *C. geitonos* Skuja; 104. *C. javanica* De Wilde; 105. *C. linearis* Gardner; 107. *C. marchica* var. *crassa* Rao; 108. *C. membranaceae* Schmidle.

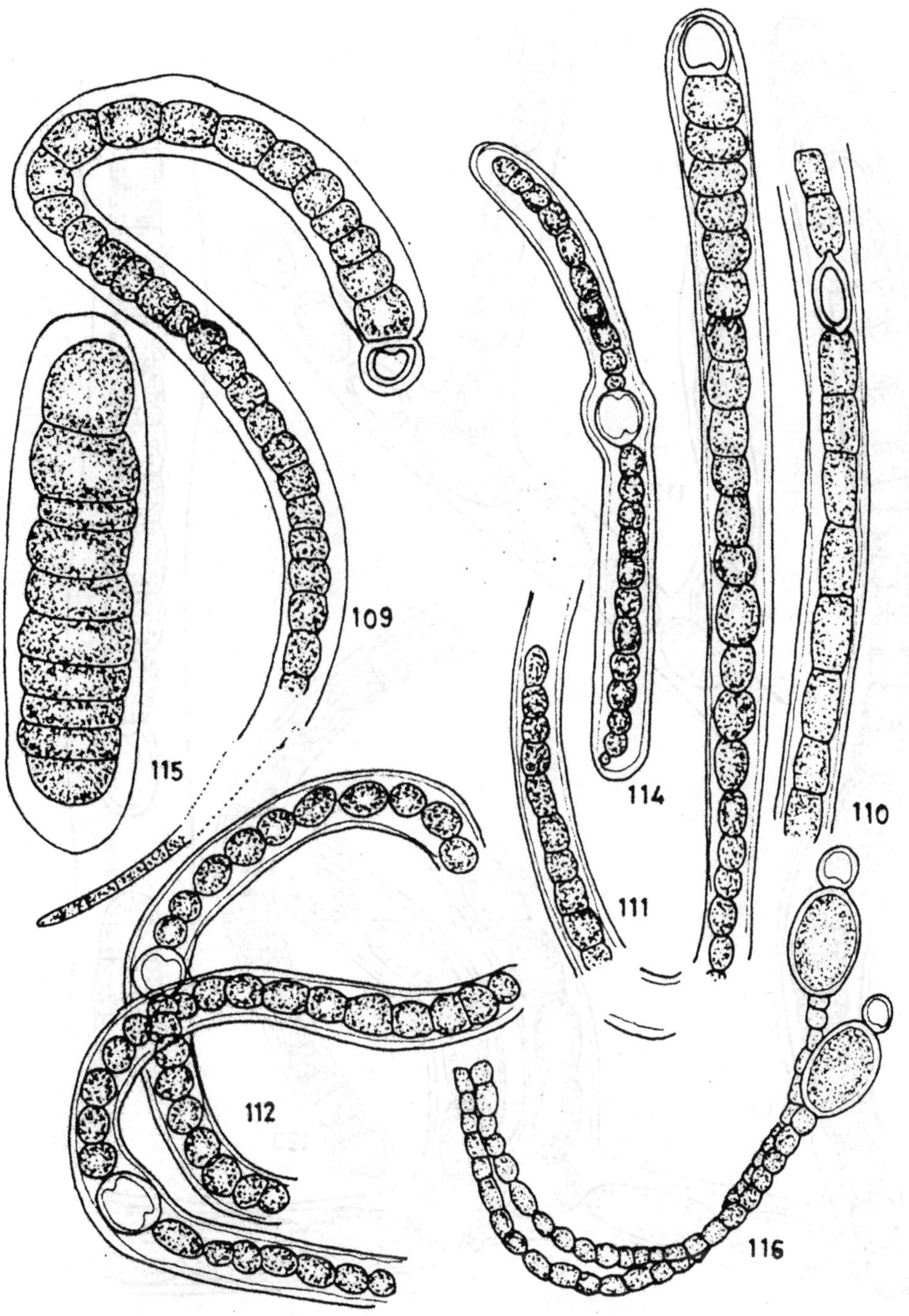

Plate 11: 109. *Calothrix scopulorum* Born et Flah; 110. *C. jhernalis;* 111. *C. weberi* Schmidle; 112. *Camptylonemopsis minor* Delik; 114. *Ca. indicum* Schmidle; 115. *Crinalium magnum* Fritsch and John; 116. *Cylindrospermum muscicola* Kutz. et. Flah.

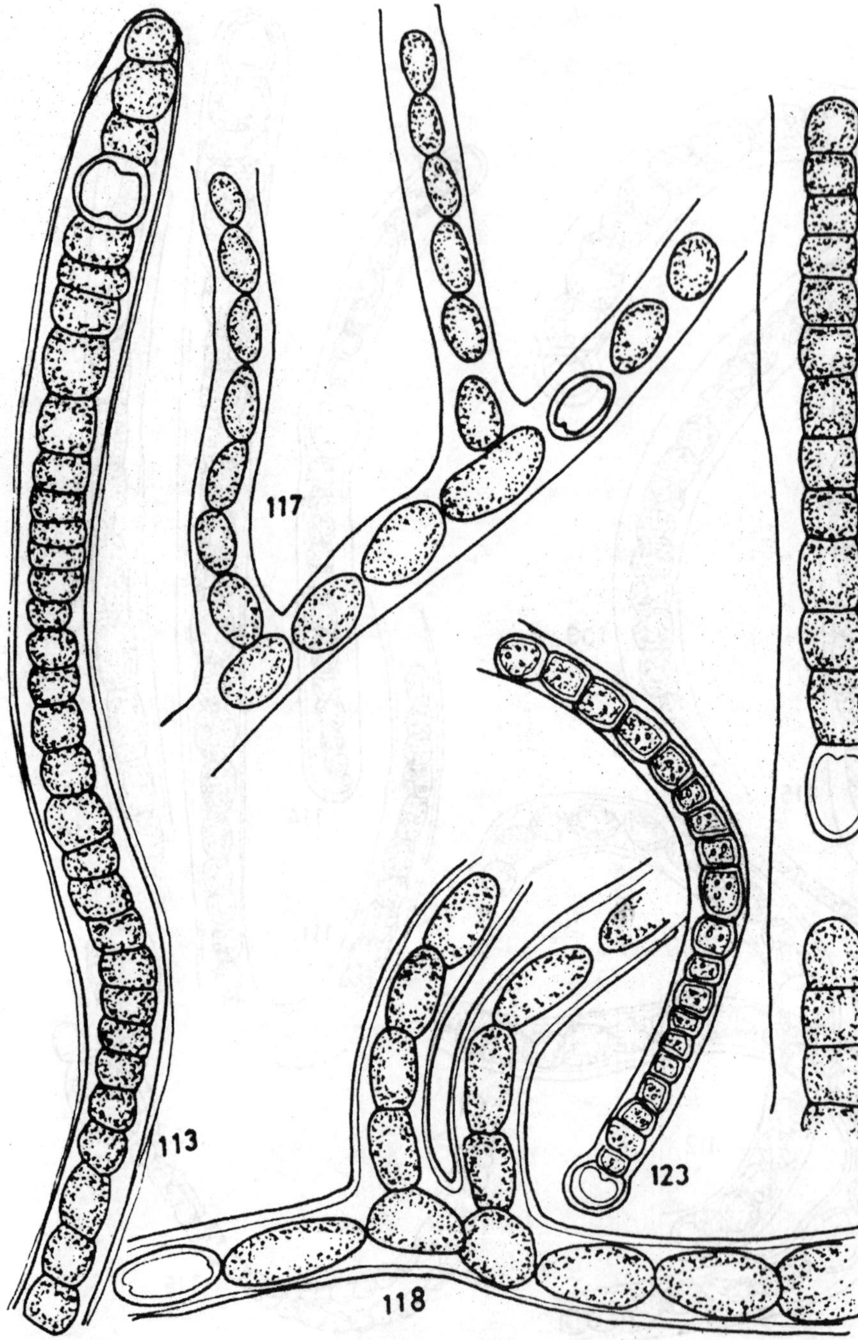

Plate 12: 113. *Camptylonemopsis pulneyensis* Desikachary; 117. *Hapalosiphon aurious* W. et. G.S. West; 118. *H. stuhmannii* Hieron; 121. *Microchaete aequalis* (Fremy) Comb. 123. *M. tenera* Thuret ex Bom et Flah.

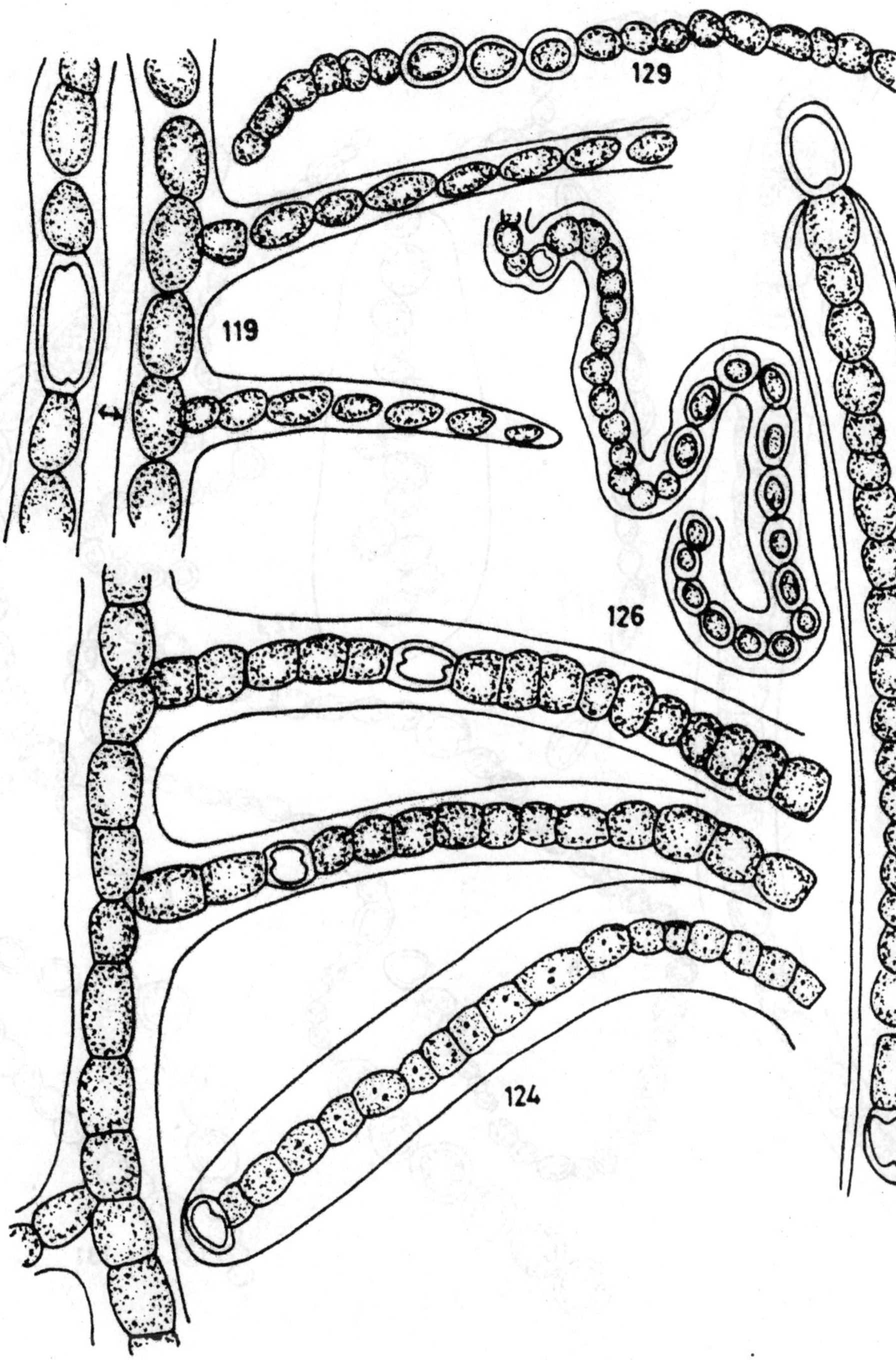

Plate 13: 119. *Hapalosiphon flagelliformis* (Schmidle) Forti; 122. *Microchate calothrichoides* Hansgirg; 124. *M. violacea* Fremy; 126. Nostoc *amplissimum* Sctchell; 129. *M. commune* Vaucher ex Born et Flah; 120. *Hapalosiphon welwetzschii* W. et. G.S. We

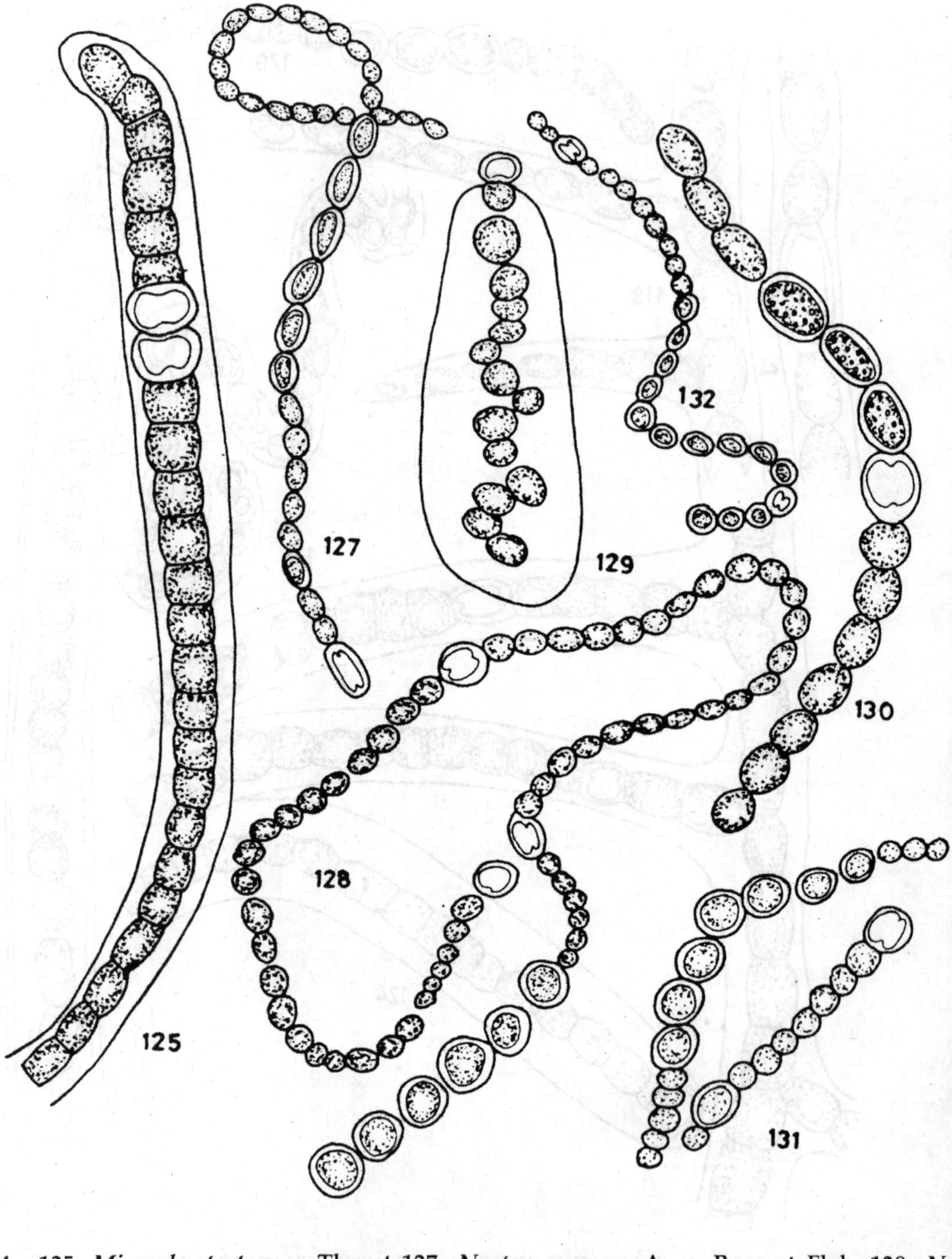

Plate 14: 125. *Microchaete tenera.* Thuret 127. *Nostoc comeum* A. ex Born et Flah; 128. *N. calcicola* Breb; 129. *N. commune* Vaucher ex. Born et. Flah; 130. *M. ellipsosporum* var. *violocea* Rao; 131. *N. entophytum* Born et. Flah; 132. *N. linkia* var. *orvenee* Rao.

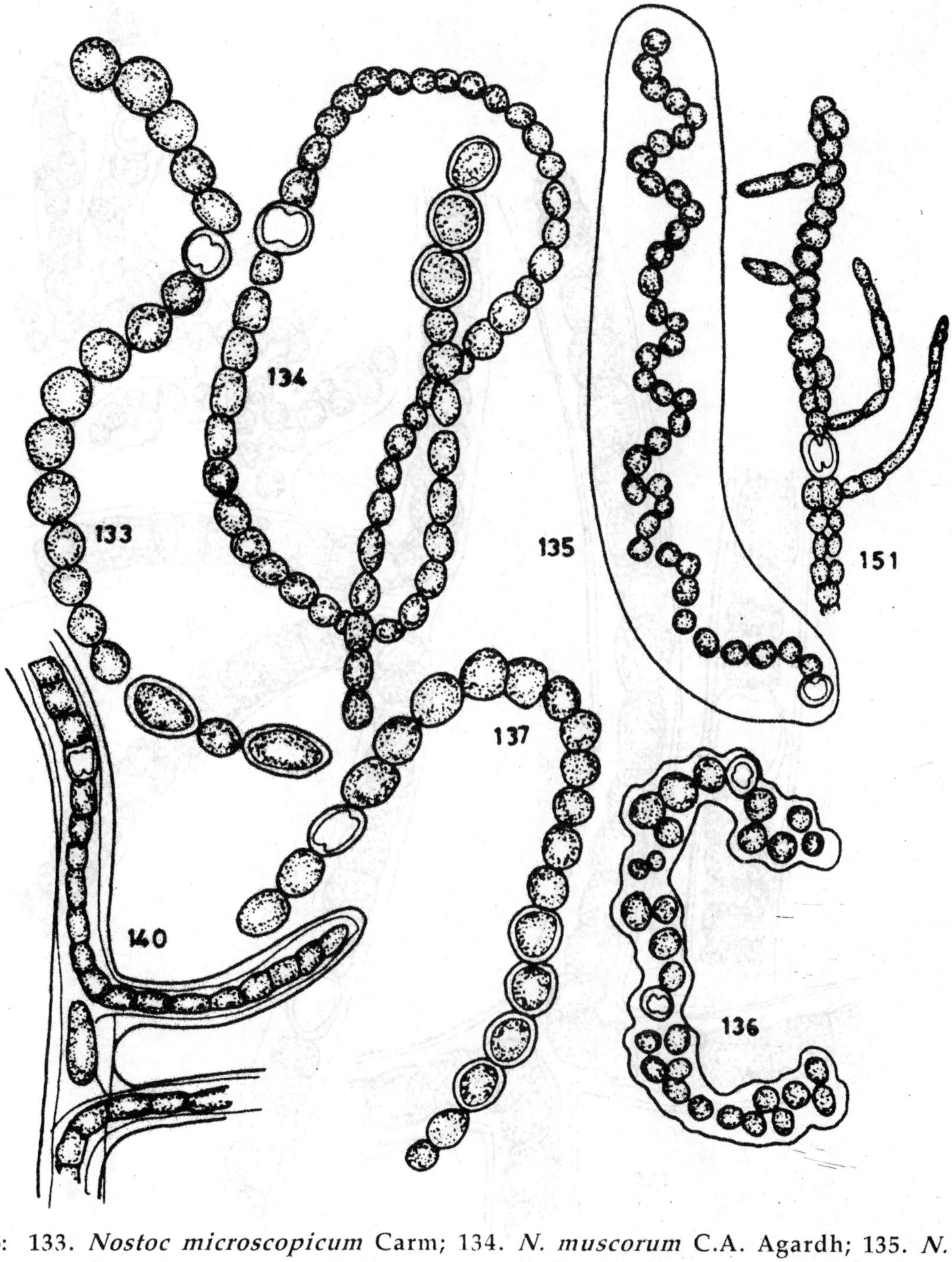

Plate 15: 133. *Nostoc microscopicum* Carm; 134. *N. muscorum* C.A. Agardh; 135. *N. paludosum* Kutz. Born et. Flah; 136. *N. punctiforme* (Kutz) Hariot; 137. *N. rivulare* Kutz ex. Born et. Flah. 140. *Scytonenna fritschii* Ghose; 151. *Westiellopsis prolifica* Janet.

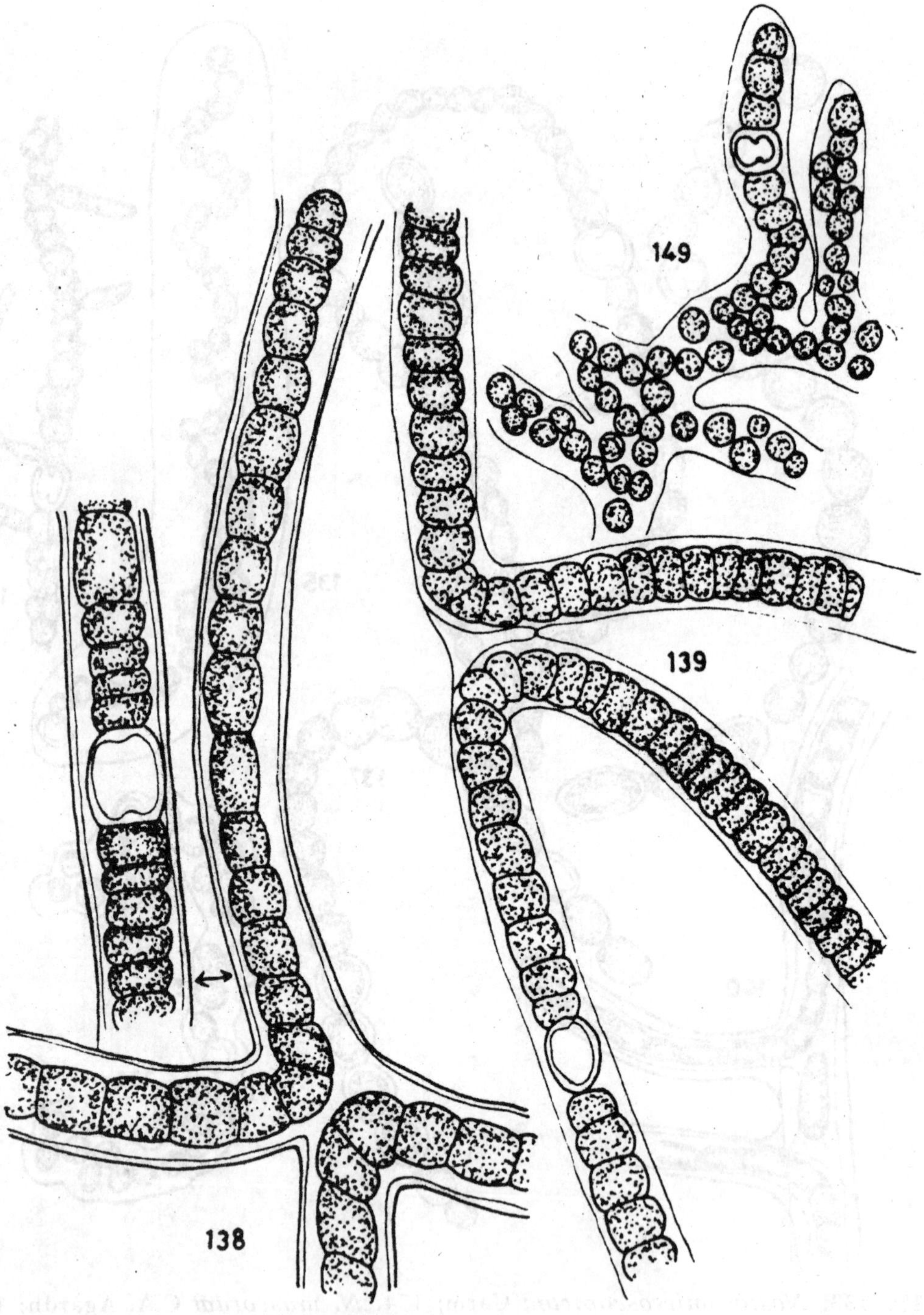

Plate 16: 138. *Scytonema burmonicum* Skuja; 139. *S. coatile* Montagene ex. Born et. Flah. 149. *Stigonema dendroideum Fremy.*

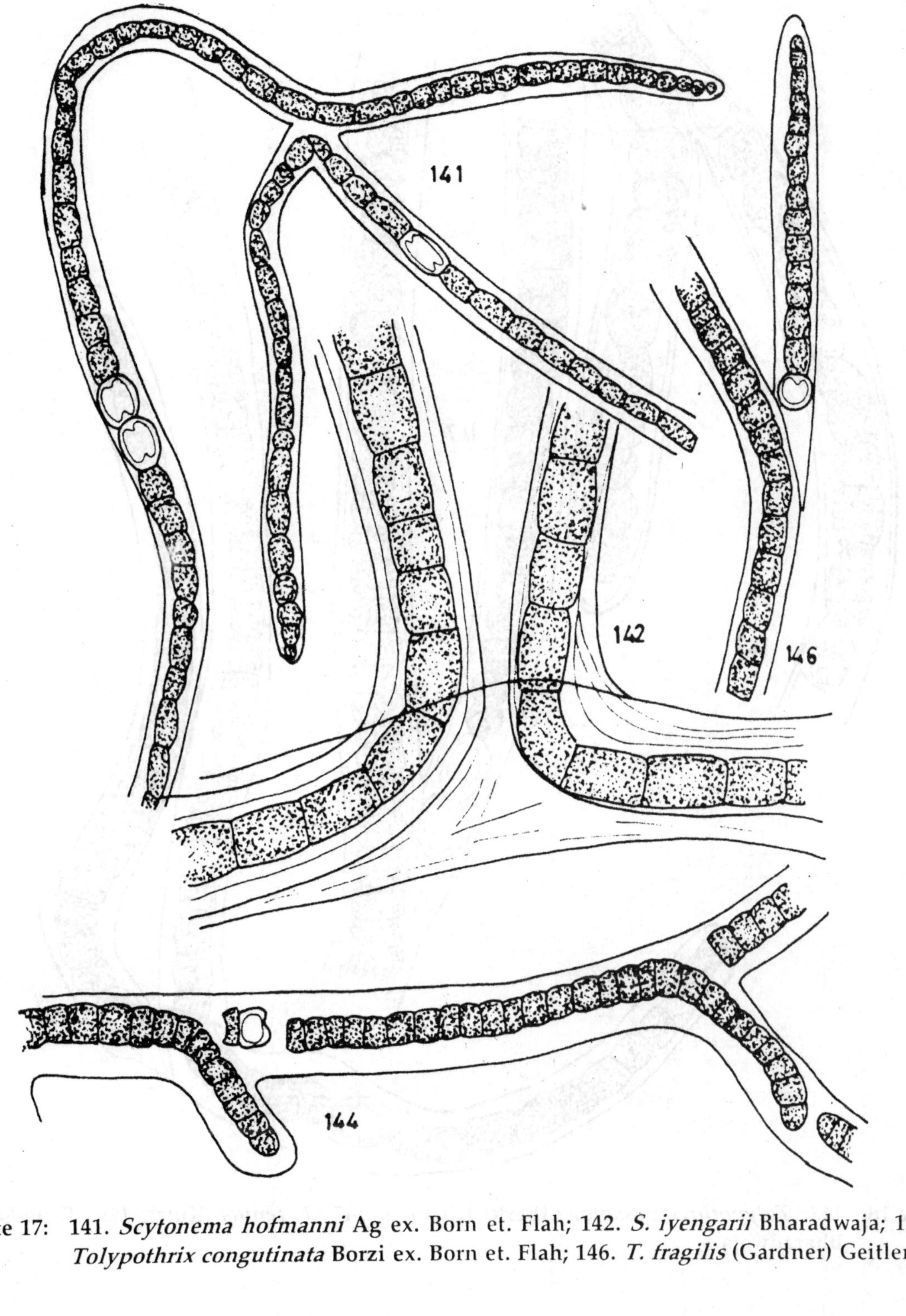

Plate 17: 141. *Scytonema hofmanni* Ag ex. Born et. Flah; 142. *S. iyengarii* Bharadwaja; 144. *Tolypothrix congutinata* Borzi ex. Born et. Flah; 146. *T. fragilis* (Gardner) Geitler.

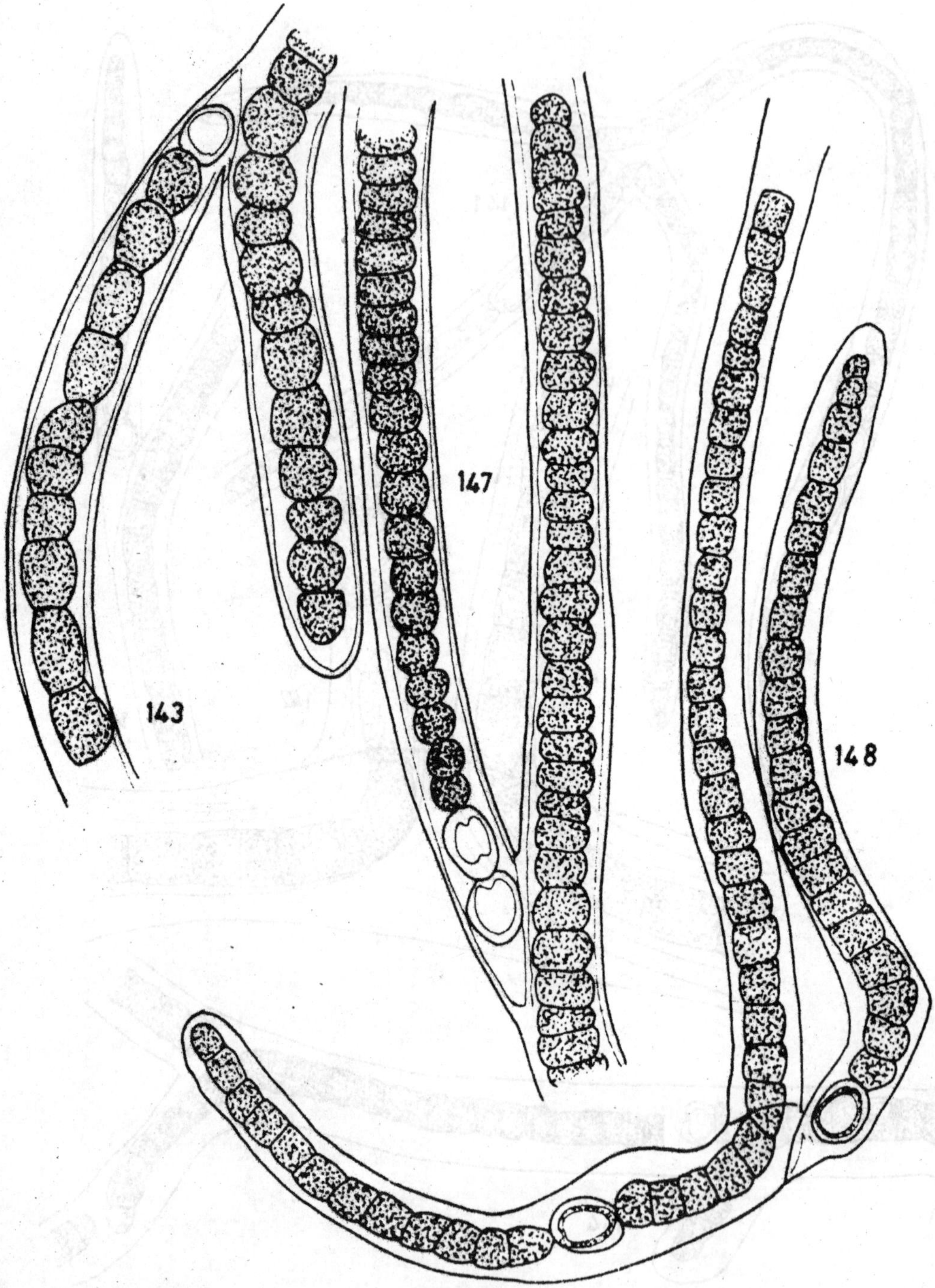

Plate 18: 143. *Tolypothrix byssoidea* (Berk) Kirchner, 147. *T. tenuis* (Kutz); 148. *T. nodosa* Bharadwaja.

heterocystous forms. Among the heterocystous forms, species of *Anabaena* were more along with a few species of *Nostoc, Hapalosiphon welwetzschii* and *Calothrix,* while the non-heterocystous forms were represented by *Oscillatoria chalybae* var. *insularis, Lyngbya dendrobia* and *Lyngbya mesotrichia.* The percentage of frequently occurring blue green algae in these soils indicates that heterocystous forms represent 84%, unicellular 24% and the remaining the nonheterocystous algae. Out of the 72 heterocystous forms recorded only 16 forms appear frequently and this representation is not significant as far as the nitrogen economy of soils is considered.

If the relationship of bluegreen algae to the soil types is considered, it is found that the highest number of heterocystous algae occur in black cotton soils followed by medium black soils. This may be attributed to their water retaining capacity (Table 4). Heterocystous forms are least represented in lateritic and red soils. On the contrary non-heterocystous forms are abundant in red soils and red clay loams.

Table—4: Blue green algae in relation to soil types

	Soil type	*Cyanophyceae*	
		Heterocystous	*Non-heterocystous*
1.	Lateritic soils (9)	16.9%	17.8%
2.	Red soils		
	(*a*) Red Clay loams (8)	16.2%	26.45%
	(*b*) Red sandy loams (75)	15.1%	24.8%
3.	Black soils		
	(*a*) Medium black (22)	20.2%	21.93%
	(*b*) Black cotton soils (6)	31.4%	17.8%

(Figures in parentheses indicate total number of soils collected in each groups)

The blue green algae collected in the present study are classified on the basis of their growth in different crops (Table 5). The greatest number of blue greens appear in paddy fields among them heterocystous forms are 31%, non-heterocystous 25% and unicellular 10%. Paddy and sugarcane fields harbour abundant blue green algae dominated by heterocystous forms. Species of *Calothrix marchica, Nostoc comenum, Nostoc muscorum, Aulosira bombayensis* and *Hapalosiphon stuhlmanni* were of common occurrence. *Chroococcus schizodermaticus, Gloeocapsa somoensis, Oscillatoria chalybae, Lyngbya bergii* among the non-heterocystous algae and *Anabaena* ballygunglii and *Anabaena circinalis* among the heterocystous forms were poorly represented.

Table 6 represents the occurrence of blue greens in relation to the various sources of water. It is found that heterocystous algae are predominant in all waters. In abundance however *Anabaena circinalis* and *Tolypothrix conglutinata* were significant. The non heterocystous bluegreens include many species of *Oscillatoria.* Although the heterocystous bluegreens account for 94% of the total soils collected, the number of representative species are very less. The assemblage consisted of *Anabaena ballygunglii, Calothrix brassii, Hapalosiphon welwitzschii, Microchaete aequalis, Nostoc*

Table—5: Occurrence of blue green algae in relation to the different crops

Crops	*Unicellular Blue green algae*	*Non-heterocystous blue green algae*	*Heterocystous blue green algae*
Paddy	16 (10.59)	39 (25.82)	47 (31.12)
Sugarcane	10 (6.62)	18 (11.92)	22 (14.56)
Ragi	04 (2.64)	06 (3.97)	06 (3.97)
Jowar	05 (3.31)	07 (4.63)	08 (5.20)
Banana	07 (4.63)	11 (7.28)	18 (11.92)
Coconut	09 (5.96)	07 (4.63)	12 (7.94)
Coffee	03 (1.98)	01 (0.66)	07 (4.63)
Vegetables	04 (2.64)	13 (8.60)	18 (11.92)
Total	**(38.37)**	**(67.51)**	**(91.35)**

(Figures in parentheses indicate percent of total occurrence of blue green algae)

Table—6: Occurrence of blue green algae in relation to the different sources of water used

Source of water	*Unicellular blue green algae*	*Non-heterocystous blue green algae*	*Heterocystous blue green algae*
Borewell	06 (3.97)	12 (7.94)	21 (13.90)
Lakes	06 (3.97)	08 (5.29)	12 (7.94)
Rivers	13 (8.60)	34 (22.51)	36 (23.84)
Ponds	09 (5.96)	08 (5.29)	16 (10.59)
Rainfall	01 (0.66)	07 (4.13)	13 (8.60)
Open wells	06 (3.97)	09 (5.96)	16 (10.59)
Channel	13 (8.60)	30 (19.86)	26 (17.21)
Total	**(35.73)**	**(70.98)**	**(94.67)**

(Figures in parentheses indicate percent of total occurrence of blue green algae)

commune and *Tolypothrix tenuis*. Among the habitats, river bed soils show the highest frequency followed by soils receiving channel water. These soils support about 8% unicellular, 35% non heterocystous and about 30% heterocystous algae.

Soils examined in the present study are fed by a wide variety of fertilizers based on the types of crops grown. Some of these are unmanured, while most of them are manured by synthetic and natural fertilizers. The present study observed that soils manured with only cowdung had about 53% heterocystous and an equal number of non heterocystous forms (Table 7). Soils manured with only chemical fertilizers had about 42% heterocystous and 37% non-heterocystous blue greens. Soils manured with both chemical and natural fertilizers may interact and produce an antagonistic effect on the growth of blue green algae and are therefore poorly represented.

Table—7: Blue green algae in relation to the sources of fertilizers used

Manure used	*Cyanophyceae*	
	Heterocystous	*Non-heterocystous*
1. Manured with chemical fertilizers (N:P:K, Phactomphous urea, Murate of Potash) (45)	41.2%	36.6%
2. Manured with chemical fertilizers (N:P:K, Phactomphous urea, Murate of Potash) along with cowdung (14)	6.4%	10.9%
3. Manured with cow dung only (61)	52.3%	52.4%

(Figures in parenthesis indicate number of soils collected)

CONCLUSION

Cyanophyceae dominate soils having an alkaline pH, low phosphate and sodium and higher concentration of chlorides. A direct correlation exists between percent organic carbon and average occurrence of cyanophyceae. They are capable of tolerating high nitrogen content. Very high values of the Pearsall's Basic Ration hinders the abundance of myxophyceae. The results of the correlation matrix are indicative of the fact that 7 out of 11 parameters studied are not of great significance in controlling the growth of blue green algae in soils. Nitrogen available within the cells, and not outside has an impact on the growth of these algae. pH of the soils in association with phosphorous, potassium and magnesium plays a significant role in controlling the growth of blue green algae in soils. Soils manured with cow dung or chemical fertilizers have dominance of blue green algae whereas soils manured with both, cow dung and synthetic fertilizers have an antagonistic effect on blue green algal growth. Heterocystous algae dominate black cotton soils. The percentage of frequently occurring blue green algae indicates that heterocystous forms represent 64%, unicellular about 24% and non heterocystous about 3%. Out of 72 heterocystous forms recorded only 16 forms appear frequently. Paddy fields and sugarcane fields have dominance of heterocystous forms. Soils receiving river and channel water harbour more blue green algae. Rainfed soils have the least unicellular bluegreen algae. The survey indicates that out of the 151 blue green algae only 47% were heterocystous, while the remaining 53% constituted unicellular and non heterocystous blue green algae and hence an enrichment of nitrogen fixing bluegreen algae in most soils is essential.

REFERENCES

Anantani Y.B. and Marathe K.V. 1972. Observation on algae of some arid and semiarid soils of Rajasthan. *J. Univ. Bombay Sci.* XLI (68): 88-93.

Angadi S.B. 1982. Studies on soils algae. Ph.D. thesis, Karnatak University, Dharwad.

Beck, G. and Zahlburcher, A. 1897. Shedue ad karypotoganes exiceate editae a musco palatino vindoboncnsis *Ann. De. KK. Naturhist* Hofnuseum 12: 75-98.

Bharadwaja, Y. 1934. The taxonomy of Scytonema and Tolypothrix and a report on new species from India and Ceylon. *Rev. Algal* 7: 149-178.

Bharathi S.G and K.M. Pai 1972. On the occurrence of some Oedogonaceae in certain soils of Mysore state. *Phykos* 17: 132-139.

Dabral, P.K. Bendre, A.M. and Lokendra Singh 1978. Cyanophyceae of Rishikesh and its adjacent foothills. *Bot. Prog.* 1(1-2): 41-44.

Datta, N. and Venkataraman, G.S. 1958. An explanatory study of the algae of some cultivated and uncultivated soils. *Ind. J. of Agronomy* III. 109-115.

Gonzalves, E.A. and Gangla K.S. 1949. Observation on the algae of paddy field soils. *J. Uni. Bom.* 18: 51.

Grover I. S and Pandhal R.K. 1975. Algal flora of paddy fields of Ludhiana and its adjacent areas. *Phykos* 14(1 & 2): 89-97.

Hansgirg, A. 1888. Synopsis of erenum subgerne rungque myxophycearum (cyanophycearum) hucusqune congniforum cup descritione genervies nov. *Dactyloccopis. Notarisia* 3: 548-570.

Lund, J.W.G. 1962. Soil algae, in Physiology and biochemistry of algae. Ed. R.A. Lewin Academic Press, 759-770.

Marathe, K.V. 1967. Studies on soil algae of India I. Soil algae from cultivated fields of Dhulia. *J. Univ of Bombay* 35: 100-105.

Pandey, D.C. 1965. A study on the algae of paddy field soils of Ballia and Ghazipur district of Uttar Pradesh, India. Part I. Cultural and Ecological considerations. *Nowa Hedwiga*, 10(1-4): 299-334.

Partha S. 2000. Studies on physiological and biochemical effects of fungicide and pesticide on some cyanobacteria species. Ph.D. thesis, University of Mysore, Mysore.

Peterson, J.B. 1935. Studies on the biology and taxonomy of soil algae. *Dansk. Bot. Arkiv.*, 8: 1-180.

Schmidle, W. 1901.Uber drei algaen genera. Ber. Atschi. *Bot. Gen* 19: 10-24.

Shields, L.M. and Durrell, L.W. 1964. Algae in relation to soil fertility. *The Bot. Rev.* 92-128.

Subbaraju, M. 1972. The blue green algae from the soils of Inida. Taxanomy and biology of the blue green algae. 1st International Symposium on Taxonomy and Biology of bluegreen algae. Madras Uni. Pp. 332-352.

Turner, W.B. 1892. The fresh water algae of East India *K. Vetensk Akerae Handl* 25: 1-187.

12

Plants and Air Pollution

Jitendra G. Gavali & Deepa (Saha) Gavali

INTRODUCTION

Among the terrestrial ecosystems, human settlements have virtually impoverished the environments and industrialization though has apparent evidence for providing human intelligence an edge over all living beings, it failed to maintain serenity among natural resources and environments. Soil, water and air pollution can be put in a series based on the dispersal of pollutants. where soil pollution could be at on very small scale or area basis depending on the mobility of pollutants while air being vector, atmospheric pollutants are known to travel quiet longer distances (i.e. transboundary) depending on wind velocity and direction.

Generation of energy by burning fossil fuels in power stations and engines, various types of industrial processes, the biodegradation of wastes, farming and forestry operations lead to the release of thousands of different air pollutants into the atmosphere. Current causes of concern are for the increasing levels of carbon dioxide (CO_2) and other compounds related to 'green house effect', phytotoxic air pollutants (O_3, NO_x, NH_3, SO_2) and other gases (CH_4, CFC_s). The Air (Prevention and Control) Act, [1981 C1 (b)] describe air pollution as "The presence of any solid, liquid or gaseous substance in the atmosphere as may be or tend to be injurious to human beings or other living creatures or vegetation or property or environment". This will not only cover the industrial air pollution rather it will include all kinds of operations like burning of rice and wheat straw in the fields etc. causing air pollution (Garg and Tiwana, 1995). Adverse effects of air pollutants on biota and ecosystems have been documented worldwide. In last three decades, environmental scientists have begun to appreciate the complex interactions of air pollutants with abiotic and biotic components of an ecosystem (Agarwal and Agarwal, 2000).

* Ecology Lab, Department of Botany, Faculty of Science, M S University of Baroda, Vadodara–390002. (India)

Air Pollution Status, Vegetation Exposure and Subsequent Consequences

There is an alarming rise in the air pollutants, especially in highly industrialized nations. In Poland the annual emissions of SO_2 are 3 to 4 million tons (Oleksyn and Reich, 1994). In Europe,highest levels of sulphur deposition is found in the "Black Triangle", which includes South Western Poland, Germany and Czech Republic (Hettenlingh et al, 1993). In Scots pine (*Pinus sylvestris* L.) forests of Dutch region, deposition of atmospheric ammonia is 100-200 Kg N ha^{-1} y^{-1}, while the critical load for nitrogen is 10-20 kg N ha^{-1} y^{-1} (Pervez-Soba, 1995). With the increasing levels of primary pollutants, the concentration of secondary pollutants particularly of ozone is increasing. The concentration of ozone in the troposphere continues to rise at a rate of 1-2 % per annum (Hough and Derwent, 1990). In developing countries, emissions and ambient air pollutant concentrations are decreasing due to national and international policies; however, sulphur dioxide concentrations in many cities in developing countries are still increasing. Ambient air concentrations of nitrogen oxides are increasing in cities in both developed and developing countries around the world (GEMS, 1988). The atmosphere of cities and areas near heavy industries usually contain mixture of sulphur and nitrogen dioxides (Murray et al., 1994[c]).

Forest decline and alteration in any natural ecosystem could be responsible for biodiversity loss particularly because of local level pollution and subsequent habitat degradation. This generally happens in areas around industrial set-ups. Yunus et al., (1996) gave revision to the consideration of larger scale air pollution problem form a narrow local context. They say "global" does not necessarily imply that pollutant encompasses the entire world, but similar air pollution problems occur at various locations around the globe. These small patches with similar air pollution problems have evolved in the form of industrial set-ups over the years. Their continuous operation leads to what is known as a "long-term exposure" of vegetation present in the vicinity of industrial area.

Due to high SO_2 emissions, as many as 2500 plant species are endangered ≅ 25 % of all species) and approximately 228 ≅ 20 % of the tree species) have been extirpated in Poland (Oleksyn and Reich, 1994). Ozone is implicated as the factor contributing to the widespread decline of high-elevated forests in Central Europe and Eastern North-America (Barnes and Davison, 1988). Pollution is not only responsible for widespread forest decline, but also playing a major role in altering the ecosystem. Barnes et al., (1996) have reported that excess of nitrogen deposition in the Netherlands has resulted in the transition of larger areas of *Calluna* dominated heath land into grassland dominated by nitrophilous species Atmospheric air pollutants under the influence of different environmental factors can pollute soil and water bodies on much more larger scale. Thus it is imperative to check air pollution by technological means by the modification at level of source itself. It is suggested that advanced technologies should imply optimum energy conversion techniques with minimum pollution release. This obviously demands an efficient fuel consumption and an appropriate technical advancement in the processing. Industrialized nations have set up air pollution acts and banned the production of hazardous chemicals. Results of such strict measures are already visible. In Britain and other parts of Europe, there has been a decrease in sulphur emissions over the past decade (Davison and Barnes, 1992). In India too, the Air Pollution Act was laid down in 1986. Under this act, several industries violating the norms have been asked to shut down. Gujarat High court passed an order to close down 14 small-scale industries at the Nandesari industrial estate near Vadodara and some of the major industrial giants like the fertilizer company has been asked to slash production by 30-50 % unless and until they curtail emissions (Saha, 1998). Such steps though have local level impact, these small scale pollution abatement acts are very important in its 'global ecological sense'.

IMPACT OF AIR POLLUTION ON PLANTS

Vegetation as a tool for air pollution monitoring

The impact of pollutants on vegetation is much apparent due to their stationary nature. There is ample literature on plants being used as indicators or studied for air pollution monitoring. They have been studied under field conditions, semi-field status and under fully controlled laboratory regimes. Investigation methods used in air pollution assessment on crops have their limitations with field experiments never offering proof of casual relationship, while controlled laboratory conditions are highly artificial (Houlden et al., 1990). Earlier prognoses were made pragmatically for one pollutant impact on one or more plant species. Later on, synergistic effect are experimented bounteously. Plant's morphological, phenological, physiological and anatomical studies have been conducted under laboratory conditions (Pearson and Mansfield 1993, Murray et al. 1994[a], Smeulders et al. 1994). Pure ecological studies under field conditions for atmospheric deposition have also been carried out (Steubing et al. 1989, Shaw et al. 1993, Edwards et al. 1994, Fredericksen et al. 1995). Many others have studied atmospheric depositions for broader base (Lindberg and Owens 1993). Erison (1992) has developed an atmospheric deposition model for the Netherlands by using satellite data and receptor monitoring on open fields and forests. Draaijers (1993) has estimated canopy interactionns for three types of forest (i.e. Oak, Pine and Spruce) stands by assessing atmospheric depositions onto the forests. After knowing the physiological alterations that are taking place in experimental plants under the laboratory conditions due to air pollutant treatment, one can compare damage due to field level air pollutants. Field level estimates however reveal how the air pollutants act on wider scale such as forest ecosystems. The 'field level' studies on ecosystems and the studies that include plants as 'air pollution-monitoring tool' should be given utmost importance.

Some of the common visible symptoms are chlorosis, necrosis, burnt tip of branches, defoliation etc. In fact the visual symptoms are the result of ongoing *subtle* changes within the plant system. Subtle changes start right from the entry of pollutants through the stomata and their direct interaction with the mesophyll cells.

Distortion of stomata, disintegration of chloroplasts, enlargement of plastoglobuli are some of subtle injuries seen in plants on exposure to pollutants (Krishnayya and Bedi, 1989; Saastomoinen and Holopainen, 1989; Ballach at al., 1992). Other associated changes are alterations in antioxidants and carbohydrate levels. Experiments on natural sites have shown that antioxidative defense system is a reliable indicator of oxidative stress (Schmieden et al., 1993). Puccinelli et al., (1996) have shown the use of peroxidase activity as a marker to edit maps of air quality. Hampp (1992), and Kainulainen et al., (1995) have suggested that carbohydrate levels indicate effects of pollution at an early stage and can be used to detect injury to the plants. The study of subtle injury indicates changes at the cellular level long before the appearance of visual symptoms. These changes are further explicated in the overall growth of trees. Trees having higher sensitivity have increased rate of leaf shedding with rise in ethylene concentration and temparature (Savada et al., 1989). Gunthardt-George et al., (1993) were of the opinion that ozone can induce xeromorphic changes in broad leaves. Silvennoinen et al., (1995) used pine reflectance spectra to classify pine growing in polluted and less polluted environments by using remote sensing technique. They found different spectral resolutions due to phenological shifts, which are studied for spatial and temporal environmental classifications. Thus plants can be very useful 'tool' for air pollution monitoring which might range from the analysis of cellular components such as chlorophyll in the laboratory to the observations for crown transparency in the forests.

Plants Responses to Mixture of Air Pollutants in the Field Conditions

In general plant response to a mixture of air pollutants is studied under laboratory conditions, but field level research always have been preferred because of its practicality and realistic nature. In natural state, air pollutants are presenting various mixtures that have synergistics effects. Given artificial laboratory conditions studies often land in unnatural hypothetical assumptions. Brown and Bell (1990) stressed that interaction with air pollutants is best served in realistic environment than in a closed chamber. Even the open top fumigation chambers used extensively in crop loss assessment programmes in USA and Europe have been criticized. Sanders et al., (1995) viewed that in open-top chamber there is alteration in the microclimate resulting in higher influx of pollutants, thereby showing greater damage as compared to ambient conditions. Thus they have documented altered microclimate in open-top chambers that was responsible for plant growth changes. Sellden and Pleijel (1995) have reported higher pollutant fluxes in chambers than in field conditions. Dawidson and Wu (1990) summarized dry deposition data from fields and chambers and opined that chamber experiments have limitation in understanding field status. Schmieden and Wild (1995) concluded that physiological and biochemical reactions of damaged and preconditioned trees in natural systems frequently differ from those of non-preconditioned trees.

Many others (i.e. Kroft et al., 1989; Olszyk et al., 1989; Cape, 1993; Krupa et al. 1994) have observed similar type of over simplicity and unmatched field conditions in artificial pollutant treatment experiments. The major reason for obscure concentrations of air pollutants in natural conditions is that there are many environmental factors that co-exist with the pollutants. Field studies are fraught with many complexities like changes in water, nutrient availability and meteorological conditions etc. that have impact on over all on plant growth. Van Den Driessche and Langebartales (1994) observed that drought stress reduces damages to trees when exposed to ozone. Pearson and Mansfield (1994) found reduction in growth when ozone was combined with water stress and predicted continuous longterm exposure to have negative impact. Thus these differences in responses to water stress and ozone itself explain the complexity in combine impact of pollutant and environmental factor over plant growth. Ashenden et al. (1995) have reported varied plant responses to the mixture of pollutants. Chappelka and Freer-Smith (1995) have reveiwed environmental stresses such as light, drought and cold. They observed the existence of air pollutant-cold tolerance interactions and concluded that trees with long life cycles experience many periods of drought and episodes of pollutant exposure of varying duration and severity. This type of long exposures of trees around industrial areas have resulted in reduced growth parameters that have further resulted in over all tree growth parameters such as canopy cover and canopy depth (Gavali, 1998).

Air Pollution, Soils and Vegetation

Soil can be a source or a sink for green house gases depending on land use and management (Lal, 1999). Air pollutants are known to change chemical and physical properties of soils. Enhanced soil electric conductivity as a consequence of higher ionization due to air pollutant depositions is reported by Misra et al. (1993). Sulphur dioxide is a major gaseous contributor among sulphur containing compounds that deposit sulphur to the soils. Thus soils of polluted area may have higher sulphur content (Rennenberg, 1984; Gavali, 1998). Zoettl and Huettl (1991) have recorded air pollutant effect on soil processes. Matzer and Murach (1995) suggested reductions in emission levels of SO_2, NH_3 and No_x to avoid changes in soil composition which may retard plant growth (Leith et al., 1995). At times temporal changes in pH of soils is expected due to the presence of acid forming gaseous pollutants. The well-known 'acid rain' cause leaf burnings and death of apical buds which results in severe damages to forest trees and may lead to forest decline.

Combined effect of pollutants depend upon the type and the concentration of air pollutant as well as on plant species and their variety (Kosta-Rick and Manning, 1993). No_x gases could be responsible for the formation of secondary pollutants such as ozone (Krupa and Manning, 1988). It is likely that the presence of other air pollutants (other than SO_2 and No_x) can also act synergistically. Wahid et al. (1995) reported the presence of other unmonitored air pollutants and held them responsible for excessive damage to vegetation. Toxic effects of SO_2 were found to be additively increased by the presence of NO and NO_2 (Wellburn, 1990). In UK, all soil survey maps are based on sulphur and nitrogen deposition (Ashenden et al., 1995). SO_2 and O_3 were responsible for leaf abscission (Bosac et al., 1994) and stem injury (Kargiolaki et al., 1991). Those soils, which are deficient in sulphite and nitrite have advantage by having sulphur and nitrogenous air pollutants. Both sulphur dioxide and nitrogen dioxide have been reported as potential nutrients after uptake and successive oxidation in the form of sulphite and nitrite respectively (Murray et al., 1994[c]). Wolfenden and Mansfield (1991) have reported accumulation of sulphite and nitrite in spruce trees even during dormancy. Therefore, common gaseous pollutants like SO_2 and NO_x though brought about some predictable damage pattern at times serve or convert as substitutes for plant nutrients. Besides sulphur and nitrogen, hydrocarbons and ammonia are major air pollutants around giant industries like oil refineries and fertilizer companies. Hydrocarbon such as methane is responsible for heavy leaf fall. Savada et al., (1989) have found an increased leaf abscission among plants growing around an oil refinery. Wahid et al., (1995) have predicted that ammonia released from the fertilizer industries was responsible for significant plant growth retardation.

Deposition and Entry of Air Pollutants and Subsequent Physiological State of Plants

Pollutants and their interactions with plants are process-mediated, form their deposition to entry. Deposition of the pollutants on the receptive surface (Vegetation) takes place by dry, wet (rain) or occult (fog/dew) deposition. Meteorological factors affecting the deposition of air pollutants are radiation, temperature, dew, wind speed, and wind direction (Erisman, 1992). The height of the chimney also plays a role in the dispersal of pollutants. For taller chimneys, the deposition nearer to the source is less and increases downwind from the source until it reaches a maximum and decreases again. For chimneys with lesser height, dry deposition with high concentrations takes place near the source and gradually decreases downwind. Once deposited on the foliar surface the entry of the pollutants into the leaves occur either via stomata or by permeating through the cuticular layer (Wellburn, 1988; Rieder, 1989). Selective uptake of ions through the cuticle has also been suggested (Evans et al., 1985). The ability of leaf surface to absorb substances varies between species, depending on epicuticular wax content, chemistry and other cuticular characteristics (Turunen and Huttunen, 1991). Embedded intracuticular waxes produce a non-permeable barrier for solutes (Lendzian and Kerstein, 1991). Rennenberg and Polle (1994) have suggested the difference in concentration gradient between the atmosphere and substomatal cavity as the factor for diffusive flux of water-soluble pollutants like SO_2. Sulphur dioxide may pass the plasmalemma and enter the symplasm either in the form of sulphate mainly by active, carrier-mediated transport (Cram, 1990) or by diffusion (Spedding et al., 1980).

EFFECTS OF AIR POLLUTANTS ON

(1) Photosynthesis and Related Activities

Darrell (1989) observed irregular stomatal functioning. Gerant et al., (1996) have reported modified carbon metabolism enzymes in plants on exposure due to change in gas exchange capacity i.e. net CO_2 assimilation and stomatal conductance. Edwards et al., (1994) have reported reduced gas exchange rates and photosynthesis in red Oak trees due to ozone. Sulphur dioxide acts directly on the chloroplast (Alscher et al., 1997). Chloroplast is the most sensitive organelle and the damage is due to the reactivity of SO_2 or anions formed during its hydration, and such reactions are irreversible (Veljovic-Jovanovic et al., 1993). Chlorophyll **'a'** is more sensetive to aqueous SO_2 than chlorophyll **'b'**. Any destruction to the chloroplast ultimately affects the pigment content and rate of photosynthesis. The contents of photosynthetic pigments decreased in chronic and acute treatments compared to control in Hedera helix L. exposed to O_3 (Della-torea et al., 1996).

Heimler et al., (1989) found that two years exposure of *Picea abies* to pollutants ($SO_2 + NO_2 + O_3$) the chlorophyll content decreased considerably without the appearance of visible injury. Khan and Khan (1994) reported reduction in pigments exposed to pollutants. Tiwari and Bansal (1993) have observed similar type of changes in *Mimusops elengi* exposed to SO_2. In a study by Pandey and Agarwal (1994), the chlorophyll concentration in tomato plants increased initially but declined after longer period of exposure ($SO_2 + NO_2$). The photosynthesis was progressively inhibited as the SO_2 concentration increased (Veljovic-Jovanovic et al., 1993). Ashenden et al., (1995) reported lower rate of net photosynthesis in white clover exposed to gaseous pollutants.

Contrary to the ill effects of pollutants on chlorophyll pigments, the presence of nitrogen in pollutant mixture often has a positive effect. Wingsle and Hallgren (1993) reported increase in chlorophyll content on NO_2 exposure and suggested that increment is due to the nutritional effect of NO_2. Van Hove et al., (1989) reported higher photosynthetic rates in poplar trees exposed to 100 mg NH_3 m^{-3}. Kelly et al., (1993) have reported increase in photosynthetic rates per unit leaf area with increasing O_3 exposure.

(2) Carbohydrate Metabolism and Starch Accumulation

Saha et al., (1999[a]) observed reduced chlorophyll content resulting in the reduced starch and carbohydrate. Photooxidants shift the activation of CO_2 fixation enzymes to oxidised state and thus decrease the fixation capacity for CO_2 (Hampp, 1992). Alscher and Amthor (1988) have proposed two possible mechanisms to explain the altered carbohydrate metabolism: (1) The ability of the photosynthetic mechanism to fix carbon is affected. (2) carbon fixed will be more rapidly catabolised to support the maintenance and repair processes. The diminished levels of assimilates (Starch and Sucrose) will result in diminished export to growing and storage tissues.

Alterations in the carbohydrate levels have been reported (Katzel and Loffter, 1996; Nerg et al., 1996). Changes in carbohydrates under the influence of SO_2 are very diverse and depend on the SO_2 concentration and fumigation duration, as well as on the sensitivity of different tree species and provenance of SO_2 (Kainulainen et al., 1995). Schmieden and Wild (1994) reported that trees growing nearest to the source contained significantly higher amounts of starch and sugar than control. Steubing et al., (1989) found accumulation of starch content in beech exposed to a mixture of O_3, SO_2, and NO_2 with a significant decrease in glucose content. Anttonen et al., (1996) have reported higher starch content. The alteration in leaf carbohydrates is dependant on leaf age with older leaves responding more strongly than younger leaves (Bucker and Ballach, 1992). Sulphur dioxide has been reported to affect the traslocation of carbohydrates by blocking the pholem loading, thus leading to starch accumulation in the chloroplast of mesophyll cells (Steubing et al., 1989). Likewise, Ozone also induces changes in the flux of photoassimilates, by the impairment of translocation within the assimilation tissue, impairment of phloem loading, increased allocation towards locations with increased repair activity (Schmieden and Wild, 1995). Inhibition of the traslocation of photosynthates by ozone has been described earlier (Adams et al., 1990). Forschner et al., (1989) reported impairment of the transport of photoassimilates in damaged spruce trees. Retention of translocates in the conifer needles under pollution stress is reported by many researchers (McLaughlin et al., 1982; Gorrisen et al., 1991; Friend and Tomlinson, 1992; Kelly et al., 1993). There was less phloem transport among the trees growing in polluted area, which was revealed by least translocation of soluble sugars and carbohydrates to stems via leaves in phloem exudates (Saha et al., 1999[a]). A depleted carbohydrate reserve has an impact on growth rates either by reducing the ability to withstand environmental extremes or by reducing the ability to resist disease and insects or both. This ultimately exacerbates the susceptibility of plants to multiple stresses (Kelly et al., 1993). Kainulainen et al., (1993) reported higher occurrence of conifer aphids in polluted environments. The increased pest attack is due to an increase in the concentrations of reducing sugars and free amino acids, which are important nutrients for aphids.

(3) Foliar Surface and Internal Structures

Foliar surfaces are directly exposed to the pollutants and are vulnerable to damage. Foliar surface comprises of cuticle made up of cutin and waxes and appendages such as trichomes. The functions of surface structures of higher plants are protection against desiccation, frost and pathogen infestation. The cuticular wax layer of leaf surface represents the primary point of attack for air borne pollutants. This leads to the deterioration of surface wax structure, disturbances in stomatal functions and visible injury (Raddi et al., 1992). Percy et al., (1990) have reported that acid deposition induces delayed cuticular development. It damages the tubular structures of epicuticular waxes in epistomatal chamber of needles. Delayed development of the epistomatal waxes and four types of deformed stomatal closure were found in Scots pine needles exposed to simulated acid rain treatment (Turunen and Huttunen, 1991). Erosion of epicuticular waxes have been reported by Schmieden and Wild (1995). The erosion of wax and structural changes to stomata can change needle wettability which alter cuticular and stomatal diffusion resistance of the needles leading to altered transpiration (Paoletti et al., 1996[a]). Significant changes in the chemistry of epicuticular waxes in plants exposed to nitric acid were found indicating that nitric acid could enter plants through cuticle in addition to stomatal uptake (Bytnerowicz et al., 1996). Saha et al., (1999[a]) found increased epicuticular wax quantity on foliar surfaces of trees growing in polluted environment. An increase in stomatal density with decline in their size and epidermal cell density was also reported (Saha et al., 2000). Trichome length and density increased in plant populations of *Cercis canadensis* inhabiting polluted habitats (Sharma 1996). The increased trichome density increases the receptive area (Bitterlich and Upadhyaya, 1990). Prasad and Inamdar (1990) reported decrease in stomatal frequency and increase in trichome frequency in *Pisum sativum* exposed to pollution. Sharma (1992) has described similar types of results. Paoletti (1996[b]) concluded that xeromorphic adaptations, such as thick cuticle and dense pubescence further promote self-protecting mechanism in Holm oak. Studies have shown that pollutants alter the foliar internal structures. Some of the changes include reductions in mesophyll cell size (Sutinen et al., 1990), decline in the intercellular spaces of spongy layer (Back, 1994) and increase in epidermis thickness (Antonelli et al., 1996; Laakso et al., 1996). Percy et al., (1990) have reported an increase in cuticular membrane thickness following exposure to O_3 in red spruce needles.

(4) Biomass Accumulation or Growth Measurements and Altered Phenology

The growth reductions are generally expressed in the perennial plants. Woodbury et al., (1994) recorded reduced leaf area in *Poplus* plants under the influence of ozone. The leaf number increased with decreased basal diameter, height and weight of the stem. This was reflected in an increase in shoot to root ratio. Higher concentrations of oxides of nitrogen treatments had reduced growth as compared to lower concentration (Pearson et al., 1994). Ayer and Bedi (1991) reported decreased growth performances in wheat cultivar at all the ages of plants. Kord et al., (1993) have observed decreased growth parameters after fumigation of radish by hydrogen sulfide (H_2S). When tomato plants were exposed to different combinations of O_3 and SO_2 both gases singly or synergistically have significantly depressed dry matter (Khan and Khan, 1994). Tiwari and Bansal (1993) have recorded decrease in root/shoot ratio, fresh and dry weight in *Mimusops elengi*. Rantanen et al., (1994) have studied combined exposure of SO_2 and NO_2 and found an uptake of sulphur and nitrogen in trees with reduced growth. Trees growing around industrial area have reduced leaf area that was reflected in reduced biomass accumulation by means of leaf and branch dry weight (Gavali et al., 1997). They also exhibited phenological alterations and the species that have fared better could over come biomass losses through sudden shooting ups during favourable conditions like leeward wind directions and rain fall (Gavali et al., 1998). Climatic fluctuations, atmospheric emissions originating from the industrial estates and summer concentrations of ozone were responsible for bad conditions of Holm oak tree crowns of Mediterranean forests (Bussotii et al., 1995). Ferretii et al., (1995_a) have reported phenological abnormality wherein all leaves were shed (Viz. Increased crown transparency) and replaced during sprouting period that had abundant male flowering. Ozone exposure accelerated autumn leaf fall in old leaves (Klumpp et al., 1994; MacLean, 1990; Mortensen and Skre, 1990; Wiltshire et al., 1993).

(5) Levels of Antioxidants

The entry of air pollutants lead to an increase in cellular oxidants that induce increased synthesis of non-enzymic antioxidants and antioxidant enzymes (Foyer et al., 1997) and cause early ageing of leaves(Gilliemaut et al., 1992) and finally to early senescence (Nie et al.). Substantial amount of work has been carried out on the changes in the antioxidants level of conifers exposed to pollutants (Madamanchi and Alscher, 1991; Van Hove et al., 1992; Polle et al., 1993; Fangmeier et al., 1994; Polle et al., 1994[b]). These studies revealed an increment in antioxidants level.Ballarin-Denti (1996) estimated ascorbic acid, peroxidase, superoxide dismutase and gluthathione reductase to study different degrees of forest decline in Italy. Rabotti and Ballarin-Denti (1996) concluded that leaves of beech, extend life span of their tissues by increasing the levels of different antioxidants. Nast et al., (1993) reported increased levels of peroxidase and gluthathione in Norway spruce exposed to single pollutant and combination treatments of SO_2 and NO_2. Studies on the tropical trees have shown an increase in perioxidase and ascorbate in plants growing in polluted environment (Krishnayya and Bedi, 1990; Rao and Dubey, 1990; Saha et al., 1999). Klumpp et al., (1994) reported reduction in ascorbic acid, increase in gluthathione and peroxidase content in broad leaved species near industrial area of Cubato, Brazil. Terry et al., (1995) interpreted that the increased activity of enzymes and subsequent contents of substrates are to scavenge the free radicals generated. Contrary to the findings that antioxidants level have increased due to the pollution stress, Wingsle and Hallgren (1993) reported no significant changes in membrane bound or water soluble antioxidants such as a - tocopherol, carotenoids and gluthathione on exposure of pine needles to SO_2 and NO_2. Polle et al., (1994[a]) failed to record significant differences in the activites of superoxide dismutase and total perioxidase in the needles of Scots pine from three SO_2 polluted sites. Antioxidants level increased only when the levels of pollutants were high.

Polle and Rennenberg (1993) suggested that to avoid damage, it is necessary that plants contain sufficient concentrations of antioxidants in those sub-cellular compartments in which reactive oxygen species are generated. Injury is observed in the absence of maintenance of antioxidant level. Rabotti and Ballarin-Denti (1996) reported higher injury levels in beech leaves with lower ascorbate peroxidase activity. Likewise May and Lever (1993) observed that in the absence of large reduced GSH pool, hydrogen peroxide accumulates in the cytosol and causes oxidative damage. Thus, when the influx of pollutants exceeds the detoxification capacity and repair mechanism by enzymatic and non-enzymatic systems in the cell, several types of injury are conceivable (Wingsle and Hallgren, 1993). The radicals attack the proteins and lipids leading to their peroxidation. Lipid peroxidation as measured by malondialdehde content serves as an indicator of damage to cellular membrane system (Weigel et al., 1989). Increase in lipid peroxidation due to exposure of plants to SO_2 & NO_2 has been reported (Sandmann and Gonzales, 1989; Price et al., 1990; Navari-Izzo et al., 1992). The higher rate of lipid peroxidation leads to destruction of membranes. Weckx and Clijsters (1996) suggested that lipid peroxidation leads to an increase in phase transition temparature of the membranes, resulting in enhanced membrane leakiness.

PLANT PHYSIOLOGY DICTATED BY AIR POLLUTANTS

(1) Free Radicals and its Generation

The entry of pollutants escalates the formation of **free radicals** in plant cell. Oxidation of sulphite into sulphate in chloroplast can give rise to the formation of superoxide anion (O_2) (Asada, 1980). Superoxide being a powerful nucleophile can react with protons to form the more toxic hydroperoxyl radical. In the hydrophobic environment of membranes, O_2 can react non-enzymatically with H_2O_2, thereby forming hydroxyl radical (Sgherri et al., 1996). Hydroxyl radicals lead to lipid peroxidation. The toxic superoxide anion is removed catalytically in the presence of enzymes and results in the formation of H_2O_2. Further, H_2O_2 can react with superoxide to form dangerous radicals (Cakmak et al., 1993). Thus the generation of radical continues. The active oxygen species are part of an alarm signaling process in plants (Foyer et al., 1994).

(3) The Detoxification Mechanism

To counteract the damaging effects of free radicals, plants possess natural scavengers that remove the free radicals at various steps to minimize the damage. Tolerance mechanisms and chemical constituents have been identified which scavenge free radicals. These comprise of low molecular weight antioxidants (gluthathione, ascorbate, a - tocopherol) and protective enzymes (ascorbate peroxidase, superoxide dismutase). This system operates in the cytoplasm and chloroplast of plant cells in a series of reactions. Foyer and Halliwell (1976) have successfully devised a mechanism to show how the entire enzyme system functions. The enzyme system involves detoxification of H_2O_2 to H_2O by ascorbate peroxidase, regeneration of ascorbic acid catalysed either by monodehydroascorbate or dehydroascorbate reductase at the expense of NADH or reduced gluthathione respectively. For the regeneration of reduced glutathione, glutathione reductase functions by using **NADPH**.

(4) Free Radical Scavengers

Ascorbate, reduced glutathione and peroxidase are important superoxide scavengers in the chloroplast. The significance of each of these scavengers is dependent on their concentration and rate constant for the conversion of superoxide radicals (Rennenberg and Polle, 1994). **Glutathione** (GSH) is a tripeptide (g - L - glutamyl L-cysteinyl glycine). It is widely distributed in plant cells. Through its particular structure glutathione renders its-SH group less susceptible to any attack by oxidants than other - SH group containing macromolecules (Smith et al., 1989). It functions in concert with ascorbate. It acts to protect labile macromolecules from the attack of free radicals and hydrogen peroxide. One of the main functions of GSH in a cell is to protect - SH groups in enzymes against oxidation to sulphide and structural proteins either by acting as a scavenger for the oxidising agents or by repairing oxidised - SH group via the GSH disulphide exchange reaction (Alscher and Amthor, 1988).

Ascorbic acid (AA) is a major metabolite in plants. It is an antioxidant and in association with other components of the antioxidant system, protect plants against oxidative damage resulting from aerobic metabolism, photosynthesis and a range of pollutants (Smirnoff, 1996). It occurs in the cytosol, chloroplasts, mitochondria and apoplast (Anderson et al., 1993 and

Rautenkranz et al., 1994). In addition to its role in photoscavenging cycle, AA can act directly as a scavenger of hydroxyl radicals. Ascorbate inhibits chain initiation by scavenging phenoxy radicals, sulphite radicals and superoxide radicals, which can be formed by reactions of sulphite radical with oxygen (Takahama et al., 1992). It also functions in the regeneration of a - tocopherol, a compound of the thylakoid membrane, which acts to scavenge toxic molecular species formed as a result of oxidative damage (Kunert and Ederer, 1995).

Peroxidase (POD) are ubiquitous enzymes which can oxidise donor molecules with the consumption of $H_2 O_2$. Peroxidases have been implicated in a broad range of metabolic functions such as cell elongation, lignification, pathogen defense, wound healing etc. Ascorbate peroxidase exists both in chloroplast and cytosol. Peroxidase uses ascorbic acid as an electron donor in the detoxification of H_2O_2. Perioxidase is labile in the absence of ascorbic acid (Polle et al., 1990)

The Fate of Sulphur Dioxide

Sulphur dioxide due to its higher solubility enters rapidly into the cell. Inside the cell hydrated SO_2 is dissolved into H^{+}, HSO_3 and SO_3^{-2} . Chloroplasts of the mesophyll cells are capable of oxidising (Asada and Kiko, 1972) and reducing (Schwenn et al., 1976) sulphur dioxide during light phase. Oxidation of SO_3^{2-} leads to the formation of sulphuric acid and finally into sulphate anions. Accumulation of sulphate occurs in the vacuoles (Polle and Rennenberg, 1993). Reduction of SO_2 leads to the formation of sulphur-containing amino acid, which is finally incorporated into proteins and glutathione (Diltirch et al., 1992). Reduction also results in the release of H_2S gas, which accounts for 10 % of the influx of SO_2 (Kindermann et al., 1995). Oxidation of SO_2 into sulphate is a fast process compared to the reduction, and is preferred (Takahama et al., 1992). Kaiser et al., (1993) also stressed that 68 % of SO_2 taken up through the leaves was oxidised to sulphate. The remaining 38 % is detoxified by reduction either to H_2S or organic sulphur compounds.

The presence of sulphate within the plants would signify the presence of higher ambient SO_2 and its entry. Ratio et al., (1995) found that total sulphur content in Scots pine needles showed a high correlation with the estimated mean SO_2 concentration in the air. The sulphur content of dried leaves and branches were as high as 3-4 times in the samples of polluted localities when compared with that of the least polluted one (Gavali, 1998). The amount of sulphate accumulated in the shoot is dependent on SO_2 concentration applied, the duration of exposure, the species analyzed and the sulphur nutritional status (De Kok, 1990). Thus plants detoxify SO_2 either into oxidised or reduced form.

Tolerance Among Plants and Pollution Accumulation

The antioxidants level has a direct role in the **tolerance** of a species (Weigel et al., 1989). Pfanz et al., (1996) reported that the sensitivity of the trees to the pollutants differed greatly between species, mainly by their detoxification capacity. Rao and Dubey (1993) studied response of two soybean cultivars (CV-Punjab 1 and CV-JS 7244) to SO_2 exposures. Their results indicate a relation between the ability of a plant to maintain reduced glutathione and ascorbate contents, and SO_2 tolerance. The species with elevated antioxidants could tolerate the stress of SO_2 induced free radicals. Madamanchi and Alscher (1991) on a similar work on Nugget concluded **that** resistant variety had higher glutathione content then the sensitive variety. Thus the resistant species can prevent the oxidative stress by increasing the antioxidants level.

Pollution accumulation by plants is studied widely (Wolfenden et al., 1991; Szarek, 1995). Pollutant uptake often proved harmful (Friend and Tomilinson, 1992; Ashenden et al., 1995) or beneficial (Kosta-Rick and Manning, 1993; Murray et al., 1994[b]). SO_2 is a toxic gaseous air pollutant (Wellburn, 1990) and plants can detoxify it and store it in the form of sulphate (Rennenberg, 1984). Liblik et al, (1995) have recorded foliar sulphur uptake that was comparable with SO_2 concentration zonation depending on wind direction from the source. Similarly, foliar sulphur content was correlated with mean SO_2 concentrations in Russia and northern Europe (Raitio et al., 1995) Even in case of Holm oak, there was significant increase in crown density as the distance from the industrial zone increased and leaf sulphur concentrations decreased (Feretti et al., 1995[b]). Polle et al., (1994[a]) reported two-three fold higher sulphate from pine needles collected from polluted sites. Sturaro et al., (1993) have found detoxification of SO_2 into sulphate that remained in barley leaves. Higher plants emit H_2S into atmosphere in the presence of SO_2 or when supplied with excess of sulphate or L-Cysteine (Rennenberg, 1984). SO_2 may be detoxified by oxidation, which yields sulphate or by reduction that finally yields organic compounds. Gymnosperms in an atmosphere polluted by SO_2 have shown the accumulation of sulphate (Kindermann et al., 1995). Bell et al., (1995) treated plants with radiolabelled SO_4^{-2} for studying partitioning and redistribution of sulphur during sulphur stress. They found that sulphur was transported to roots from the abraded surfaces of mature leaves while that applied to the young leaves was mostly retained and was not transported. Sheppard (1994) has also worked on age-dependent sulphur uptake in leaves and recorded similar observations. Solvik et al., (1995) found stomatal sulphur uptake is dependent on SO_2 emission, stomatal conductance and canopy features that could vary with the time (i.e Seaon and subsequent Phenology). Sergeichik and Sergeichik (1993) have noted assimilation of emitted nitrogen (Via nitrogen metabolism) and metal accumulation in leaves of trees and bushes those are planted in the vicinity of factories emitting NO_x, NH_3 and SO_2. Accelerated sorption of nitrogen from polluted soils by plants and detoxification of heavy metal ions is well documented (Evdokimova, 1993). In general, all these studies indicate that tolerant species have assimilated the emitted nitrogen. Okano et al., (1989) have studied absorption of nitrogen for various broad leaved species and found difference in nitrogen sorption. They also describe purification of the atmosphere by planting NO_2 absorptive species and pointed out the synergistic effect of pollutants.

Dust capturing is one of the most important roles of trees in pollution abatement. The areas with heavy dust pollution, thick canopy cover can effectively capture the dust and keep the environment clean. There are studies (Kennedy, 1980; Brabec et al., 1981; Yunus et al., 1985) describing quantification of dust cover on vegetation. Even chemically inert dust can affect photosynthesis and transpiration by covering and plugging stomata (Hirano et al., 1995). Shukla et al., (1990) have noticed stunted growth without foliar damage in *Brassica* plants when trated with cement dust. Gavali (1998) has observed increased dust deposition on leaves of trees growing around industrial areas; coinciding with an increased suspended particulate matter of ambient air during summer. He also recorded that the species with simple, hairy and broad-leaves had maximum dust deposition.

Reduction of emission levels itself is the best measure for preventing air pollution impact and it has to be cost effective (Kageson, 1995). Zundel et al., (1995) discussed cost effective measures to reduce CO_2 emission, which according to them will also result in significant reduction of SO_2 and NO_x. They stressed that restructuring of energy system is a major emission reduction option that includes abatement technology and high efficient energy conversion technology. The cost effectiveness of these is always on the higher side. Vegetation can act as "natural sink" for most of the atmospheric pollutants while acting as a chief living component of biogeochemical cycling in terrestrial ecosystems. Dry and wet deposition of air pollutants on forest trees is widely studied (Draaijers et al., 1993; Lindberg and Lovett, 1992). Canopy buffering has been discussed in detail by Pederson et al., 1995. They have seen buffering pH was affected by nitrogen. The fluxes measured for Cl^-, SO_4^{-2}, Na^+ and Mg^{+2} were highest during dormant season and less during growing season due to elevated levels of organic compounds. Transport of pollutant from source to the receptor (forest) begins with turbulence diffusion into the atmosphere at the source during which substantial transformation of chemical form can take place (Murphy jr and Sigmon, 1990). Atmosphere deposition onto vegetation occur by dry and wet deposition, of which, dry deposition accounts for a major fraction (Dasch, 1989; Schaefer and Reiners, 1990. For high level (tall) sources, the deposition near the sources is less and increases with the down-wind until it reaches the maximum and decreases again. In case of ground level sources, the dry deposition takes place near the source where concentrations are highest and decreases down-wind (Erisman, 1992). Dry deposition in the forest edge was higher and reduced down-wind at the interior (Draaijers, 1993). These studies indicate the importance of forests in the cleaning the atmosphere. Vegetation not only purifies the air by taking up air pollutants but also acts as a barrier that check their free dispersal. Schmieden and Wild (1995) considered trees are complex organisms that possess numerous mechanisms such as buffering, filtering, detoxification and repair by which they are able to avoid or impede permanent damage. Tree plantations or green belts are suggested as one of the strategy for air pollution amelioration (Clarke and Murray, 1990; Ahmed et al., 1991; Bhattacharya, 1994). Green belt plantation around air polluting industries can never be a claim for the removal of air pollutants in that region. However, it can potentially remove considerable amounts of air pollution. A great deal of work has been done. A holistic approach is required to control the impact of air pollution on ecosystem.

REFERENCES

Adams M.B, N T Edwards, G E Taylor jr. and B K Skaggs 1990. Whole plant 14C-photosynthate allocation

in *Pinus taeda*. Seasonal patterns at ambient and elevated ozone levels. Can. J. For. Res.,**20**, 152-158.

Agarwal M and S B Agarwal 2000. Effects of air pollution on plant diversity. Environmental Pollution and Plant Responses (Eds. S B Agarwal and M Agarwal).Lewis Publishers,Boca Raton, Florida.pp. 137-152.

Alscher R. G and J S Amthor 1988. The physiology of free radical scavenging maintenance and repair processes. Air pollution and Plant Metabolism (Eds. S Schulte - Hostede, N M Darrall, L W Blank and A R Wellburn). Elservier Applied Science, London, pp, 94-114.

Alscher R.G, J L Donahue and C L Cramer 1997. Reactive oxygen species and antioxidants: Relationship in green cells. Physiol. Plant., **100**, 224-233.

Anderson J W, C H Foyer and D A Walker 1993. Light dependent reduction of dehydrascorbate and uptake of exogenous ascorbate by spinach chloroplasts.*Planta*, **158**, 442-450.

Antonelli F, F Bussotti, D Grifoni, P Grossoni, B Mori, G Zipoli and C Tani 1996. Oak ***Quercus robur*** L.) seedling responses to realistic increase in UV-B radiation under open space conditions. In 17th International Meeting for specialists in Air pollution effects on forests ecosystems. Firenze,Sept.14-19, Italy.

Anttonen S, M Kittila and L Karenlampi 1996. Impacts of ozone on Aleppo Pine needles: Visible symptoms, stomatal resistance and starch concentrations. In 17th International meeting for specialists in Air pollution effects on forests ecosystems, Firenze, Sept 14-19, Italy.

Asada K and K Kiko 1972. Inhibition of aerobic oxidation of sulfite by illuminated spinach chloroplasts.

Eur. J Biochem; **33**, 253-257.

Asada K, 1980. Formation and scavenging of superoxide in chloroplasts, with relation to injury by sulphur dioxide in studies on the effects of air pollutants on plants and mechanisms of phytotoxicity. Res. Rep. Nat. Inst. Environ. Stud., **11**, 165-179.

Ashenden T W, S A Bell and C R Rafarel 1995. Responses of white clover to gaseous pollutants and acid mist: implications for setting critical levels and loads. *New Phytol*., **130**, 89-96.

Ayer S K and S J Bedi 1991. Effect of industrial air pollution on *Triticum aestivum* L. Cultivator J -24 (Wheat). Proc. Natl. Acad. Sci., India. Sect(B), **61**, 223-229.

Back J, 1994. Effects of acidic precipitation on scots pine and Norway spuce needles and their low temparature tolerance. Academic Dissertation, Univ. Of Oulu, Finland.

Ballach H J, J Mooi and R Witting 1992. Premature aging in *Populus nigra* L after exposure to air pollutants Angew Bot., **66**, 14-20.

Ballarin-Denti a, 1996. Tree physiology parameteres related to pollution agents in a forest system subjected environmental stress. In 17th International meeting for specialists in Air pollution effects on forests ecosystems, Firenze, sept 14-19, Italy.

Barnes J D and A W Davison, 1988. The influence of ozone on the winter hardiness of Norway spruce [*Picea abies* (L) Karst.] New phytol., **108**, 159-166.

Barnes J D, MR Hull and AW Davison 1996. Impacts of air pollutants and elevated carbon dioxide on plants in winter time. Plant response to Air pollution, (Eds. M Yunus and M Iqbal). John Wiley & Sons ltd,

Chichester, pp, 135-166.

Bell C I, D T Clarkson and W J Cram 1995. Partitioning and redistribution of sulphur during S-stress in *Macroptilium astropurpum* cv. Siratro. J. Expt. Bot., **46**, 73-81.

BhattacharyaAK 1994.Efficacy of tree species towards gaseous pollutants and its significance in air pollution control by plantations of pollution resistant trees. Indian Forester, **120**, 658-667.

Bitterlich I and M K Upadhyaya 1990. Leaf surface ultrastructure and susceptibility to ammonium nitrate injury.Can.J. Bot. **68**, 1911-1915.

Bosac C, J A Roberts, V J Black and C R Black 1994,. Impact of O_3 and SO_2 on reproductive development in oil rapeseed (*Brassica napus* L.) II. Reproductive site losses. New Phytol., **126**, 71-79.

Brabec E, P Kover and A Daeb Kova 1981. Particle deposition in three vegetation stands: a seasonal change. Atmos. Environ., **15**, 583-587.

Brown V C and J N Bell, 1990. The value of closed chamber experiments for studying biotic interactions with plants and air pollution. Environmental Research with closed chambers(Eds H D Page,T pfimann and P Mothy) CEC, Air Pollution Research report, **26**, 322-329.

Bucker J and H Ballach, 1992. Alterations in carbohydrate levels in leaves of *Populus* due to ambient air pollution. Physiol. Plant. **86**, 512-517.

Bussotti F, A Bottacci, P Grossoni and C Tani 1995. Morpho-anatomical alterations in leaves collected from beech trees growing in conditions of natural water stress. Environ. Expt. Bot, **35**, 201-213.

Bytnerowricz A, M Krywult, P Padgett, K Percy and G Riechers 1996. Nitric acid vapor effects on forest trees deposition, cuticular and metabolic changes. In 17th International meeting for specialists in Air pollution effects on forests ecosystems, Firenze, Sept 14-19, Italy.

Cakmak I, D Strbac and H Marschner 1993. Activities of hydrogen peroxide-scavenging enzymes in germinating wheat seeds. J. Expt. Bot, **44**, 127-132.

Cape J N 1993. Direct damage to vegetation caused by acid rain and polluted cloud:Defination of critical levels for forest trees. Environ. Pollut., **82**, 1167-80.

Chappelka A H and P H Freer-Smith 1995. Predisposition of trees by air pollution to low temperature and moisture stress. Envirn. Pollut., **87**, 105-117.

Cram W J 1990. Uptake and transport of sulfate. Sulfur nutrition and sulfur assimilation in higher plants. (Eds. H Rennenberg, C Brunold, De Kok L J, I Stulen), SPB Acad. Publ., The Hague, pp. 3-11.

Dann MS and EJ Pell 1989.Decline of activity and quantity of Ribulose bisphosphate carboxylase/ oxygenase and net photosynthesis in ozone-treated potato foliage. Plant Physiol., **91**, 427-432.

Darrall N M 1989. The effect of air pollutants on physiological processes in plant. Plant, Cell and Environ.,

Dasch J M 1989. Dry deposition of sulphur dioxide or nitric acid to oak. Elm and pine leaves. Environ. Pollut.,**59**, 1-16.

Davison A W and J D Barnes 1992. Patterns of air pollution: Critical loads and abatement strategies. In: Managing the Human Impact on the Natural Environment: Pattterns and processes (Ed. M D Newsom) Bellhaven press, London. pp, 109-129.

Dawidson C I and Yee-LinWu 1990. Dry deposition of particles and vapors. Acidic Deposition: Vol 3-sources, Deposition and Canopy Interactions. (Eds. S E Lindberg, A L Page and S A Norton),Springer Verlag, New York, pp, 103-215.

De Kok L J 1990. Sulphur metabolism in plants exposed to atmospheric sulphur. In: Sulfur nutrition and assimilation in higher plants:Fundamental, environmental and agricultural aspects(Eds. H Rennenberg C Brunold, De Kok, L J, I Stulen). The Hague, SBP Academic Publishing, pp,111-130.

Della Torea G, F Ferranti, M Lipattelli, N Pocceschi, A Figoli, C Nali and G Lorenzini 1996. The effects of Ozone treatments on morpho-anatomy, biochemistry and physiology in *Hedera helix* L. In 17th International meeting for specialists in Air pollution effects on forests ecosystems, Firenze, Sept. 14–19, Italy.

Diltrich A P M,H Pfanz and U Heber 1992. Oxidation and reduction of sulfite by chloroplasts and formation of sulfite addition compounds. Plant Physiol., **98**, 738-744.

Draaijers G P J 1993. The variability of atmospheric desposition to forests. The effects of canopy structure and forset edges. Ph D Thesis, University of Ultrecht, Ultrecht, The Netherlands.

Edwards G S,S D Wullschleger and J M Kelly 1994. Growth and physiology of northern red oak preliminary comparisons of mature tree and seedling responses to ozone. Environ. Pollut., **83**, 215-221.

Erisman J W 1992. Atmospheric deposition of acidifying compunds in the Netherlands. Ph D thesis. Univ. of Utrecht, Utrecht, the Netherlands.

Evans L S, K A Santucci and M J Patti 1985. Interactions of stimulated rain solution on leaves of *Phaseolus vulgaris* L. Environ. Expt. Bot, **25**, 31-40.

Evdokimova G A 1993. Nitrate pollution of plants on soils with elevated heavy metal concentrations. Pochvovedenie, 104-107.

Fangmeier A, S Bruschon and H J Jager 1994. Time course of oxidant stress biomarkers in flag leaves of wheat exposed to ozone and drought stress. New phytol., **126**, 63-69.

Ferretti M, F Bussotti, A Cozzi and E Cenni 1995. Forest decline and environmental pollution in Italy - A critical reassement. *Response of Plant to air pollution*. (Eds. G Lorenzini and G F Soldatini). Procee-dings of conference held in pisa, 7-8 April.

Forschner W, V Schmitt and A Wild 1989. Investigations on the starch content and ultrastructure of spruce needles relative to the occurrence of novel forest decline. Bot. Acta, **102**, 208-221.

Foyer C H and B Halliwell 1976. The presence of glutathione and glutathione reductase in chloroplasts: A proposed role in ascorbic acid metabolism. Planta, **133**, 21-25.

Foyer C H, P Descourvieres and K J Kunert 1994. Protection against oxygen radicals: an important defense mechanism studied in transgenic plants. Plant Cell Environ., **17**, 507-523.

Foyer C H, H Lopez-Delgado, J F Dat and I M Scott, 1997. Hydrogen peroxide- and glutathione-associated mechanisms of acclimatory stress tolerance and signalling. Physiol. Plant., **100**, 241-254.

Fredericksen T S, B J Joyce, J M Skelly, K C Steiner, T E Kolb, K B Kouterick, J E Savage and K R Snyder1995. Physiology, Morphology and ozone uptake of leaves of black cherry seedlings, saplings and Canopy trees. Environ. Pollut., **89**, 273-283.

Friend A L and P T Tomlinson 1992. Mild ozone exposure alters 14C dynamics in foliage of *Pinus taeda* L. Tree physiol., **11**, 214 -227.

Garg M R ans N S Tiwana 1995. The Air (Prevention and Control) Act, 1981 – Lacunaes and Remedies. In Environmental pollution and protection (Eds M R Garg, V K Bansal and N S Tiwana), Deep & Deep Publs., New Delhi, pp 3-20.

Gavali J G, 1998. Quantification of growth pattern of tree species growing in and around Baroda industrial area and their suitability for green belts. Ph D Thesis submitted to The M S University of Baroda, Baroda, India.

Gavali J G, D Saha, and N S R Krishnayya 1998. Changes in growth pattern of *Mangifera indica* L. growing in polluted environment. Ecol. Environ. & Cons., **4** (1-2), 39-43.

Gavali J G, D Saha, and N S R Krishnayya and S J Bedi 1997. Phenological alterations and biomass reductions in two deciduous tree species growing in polluted environ.Ecol. Environ. & Cons., 3(1)19-GEMS 1988. *Assessment of urban air quality*, UNEP, Nairobi.

Gerant D, M Podor, M Dixon, D Le Thicec, F Antoni, V Fontaine, DS Martin, G Corbel, P Grieu, J P Garrec and P Dizengremel 1996. Interaction between ozone and mild drought stress on coniferous gas exchanges and carbon metabolism. Comparison between spruce (*Picea abies* L. Karst.) and Aleppo pine (*Pinus halepensis* Mill.) In 17th International meeting for specialists in Air Pollution effects on forests ecosysytems, Firenze, Sept 14-19, Italy.

Gillimaut S, F W Lotfi, D Blache, M prost, B Rether and a Dietrich 1992.Conifer decline in the north-east of France: Characteristic changes in chloroplast protein pattern and absence of anti-oxidative defense Capability point to an involvement of ozone. Physiol. Plant., **85**, 215-222.

Gorissen A, N N Joosten and A E Jansen 1991. Effects of ozone and ammonium sulphate on carbon partitioning of mycorhyzal roots of juvenile Douglas fir. New Phytol., **119**, 243-250.

Gunthardt-Goerg M S, R Matyssek, C Scheidegger and T Keller 1993. Differentiation and structural decline in the leaves and bark of birch (*Betula pendula*) under low ozone concentration. Trees, 7, 104-114.

Hampp R 1992. Comparitive evaculation of the effects of gaseous pollutants, acid deposition and mineral deficiencies on the carbohydrate metabolism of trees. Agric. Ecosys. & Environ., **42**, 333-364.

Heimler D, M Michalozzi and V Boddi 1989. Quantitative TLC determination of chlorophylls in spruce needles under mild pollution condition. Chromatographia, **28**, 148-156.

Hettenlingh J P, R J Downing and P A M de Smet 1993. Maps of critical loads, critical sulphur deposition and exceedances. RIVM (Eds. R J Downing, J P Hettelingh and P A M de Smet),National Institute of Public Health and Environment, Bilthovan, The Netherlands. pp, 6-19.

Hirano T, M Kiyota and I Aiga 1995. Physical effects of dust on leaf physiology of cucumber and kidney bean plants. Environ. Pollut. **89**, 255-261 .

Hough A M and R G Derwent 1990. Changes in the global concentration of tropospheric ozone due to human activities. Nature, **344**, 645-648.

Houlden G, S McNeill, m Aminu-Kano and J N B Bell 1990. Air pollution and agricultural aphid pests. I: Fumigation experiments with SO2 and NO2. Environ. Pollut., **67**, 305-314.

Kageson P 1995. Control techniques and strategies for regional air pollution from the transport sector - The European case. Water, Air and Soil Pollu., **85**, 225-236.

Kainulainen P, H Salka, A Mustaniemi, J K Holopainen and J Oksanen 1993. Conifer aphids in an air-polluted environment. I. Host plant quality. Environ. Pollut., **80**, 193-200.

Kainulianen P, J K Holopainen and J Oksanen 1995. Effects of SO_2 on the concentrations of carbohydrates and secondary compounds in scots pine (*Pinus sylverstris* L.) and Norway spruce [*Picea abies* (L.) Karst] seedlings. New phytol., **130**, 231-238.

Kaiser W M, M Hofler and W Heber 1993. Can plants exposed to SO_2 excrete sulphuric acid through the roots. Physiol. Plant., **87**, 61-67.

Kargiolaki R, D J Osborne and F B Thomson 1991. Leaf abscission and stem lesions (intumescences) on Poplar clones after SO_2 and O_3 fugimation: A link with ethylene release?J. Expt.Bot., 42, 1189-1198

Katzel R and S Loffler 1996. The influence of NH_3 - emission on the physiology of pines (*Pinus sylvestris* L.) in field studies and fumigation experiments. In 17th International meeting for specialists in Air Pollution effects on forests ecosystems, Firenze, Sept 14-19, Italy.

Kely J M, G E Taylor, Jr N T Edwards, M.B.Adams and A L Friend 1993. Growth, Physiology and Nutrition of Loblolly pine seedlings stressed by O_3 and acidic precipitation: A summary of the ROPIS - South project. Water, Air and Soil pollut., **69**, 363-391.

Kennedy R A 1980. Ash from Mt. St. Helens. Nature, **287**, 581.

Khan M R and M W Khan 1994. Single and interactive effects of O_3 and SO_2 on tomato. Environ. Expt. Bot **34**, 461-469.

Kindermann G, K Huve, S Slovik, H Lux and H Rennenberg 1995. Emission of hydrogen sulfide by twigs of conifers - a comparison of Norway spruce [*Picea abies* (L.) Karst] and Blue spruce (*Picea pungens* Engelm.). Plant and Soil., **168-169**, 421-423.

Klumpp A, G Klumpp and M Domingos 1994. Plants as Bioindicators of air pollution at the Serra Do Mar near the industrial complex of Cubatao, Brazil. Environ. Pollut., **85**, 109-116.

Kord M, F Fahmy and H Abbas 1993. Effect of hydrogen sulfide on growth and some biochemical changes in rocket and radish plants. Egypt. J Physiol. Sci., **17**, 117-137.

Kosta-Rick R and W J Manning 1993. A model for studying plant responses to air pollutants and other environmental stresses. Environ. Pollut., **82**, 107-138.

Krishnayya N SR, and S J Bedi 1989.Effect of sulphur dioxide and ascorbic acid on the plastid ultrastructure of *Azadirachta indica* leaves. Annals of Bot., **64**, 311-313.

Krishnayya N S R, and S J Bedi 1990. Differential response of three tree species growing at low levels of ambient air pollutants. Indian j. Ecol., **17**, 1-5.

Kroff M J, J Mooi, J Goudriaan, W Smeets, A Leemans and C Kliffen 1989. The effects of long-term open air fumogation with SO_2 on a field crop of broad beans(*Viciafaba L.*)II:Effects of growth components leaf area development and elemental composition. New phytol., **113**, 345-351.

Krupa S V nad WJ Manning 1988. Atmospheric ozone formation and effects on vegetation.Environ. Pollut., **50**, 101-137.

Krupa S V, M Nosal and H Legge 1994. Ambient ozone and crop loss: Establishing a cause-effect relationship. Environ.pollut., **83**, 269-276.

Kurnert K J and M Ederer 1995. Leaf ageing and lipid peroxidation: the role of the antioxidants vitamin C and E. Physiol. Plant., **65**, 85-88.

Lal R, 1999. Soil managemnet and restoration for C sequestration to mitigate the accelerated greenhouse efffect. Progress in Environmental Sciences.**1 (4)**: 307-326.

Laakso K, H Kinnunen and S Huttnen 1996. Effects of ultraviolet radiation on the growth of Scots pine and Norway spruce.In 17th International meeting for specialists in Air pollution effects on forests ecosystem Firenze, sept 14-19, Italy.

Leith I D, L J Sheppard and J N Cape 1995. Effects of acid mist on needles from mature sitka spruce grafts. Part II. Inflence of development stage, age and needle morphology on visible damage. Environ. Pollut., **90**, 363-370.

Lendzian K and G Kersteins 1991.Sorption and transport of gases and vapour in plant cuticles.Rev. Environ Contamination Tox., **121**, 65-128.

Liblik V, A Ratsep and H Kundel 1995. Pollution sources and spreading of sulphur dioxide in the North-Eastern Estonia. Water, Air and Soil pollut., **85**, 1903-1908.

Lindberg S E and G M Lovett 1992. Deposition and forest canopy interactions of air borne sulfur: Results from the Integrated forest study. Atmos. Environ., **26**, 1477-1492.

Lindberg S E and J G Owens 1993. Throughfall studies of deposition to forest edges and gaps in montane ecosystems. Biogeochemistry, **19**, 173-194.

MacLean D C 1990. Joint action of ozone and hydrogen floride on foliar senescene in maize. Environ. Pollut., **63**, 283-292.

Madamanchi N R and R G Alscher 1991. Metabolic bases for differences in sensitivity of two pea cultivars to sulphur dioxide. Plant Physiol., **97**, 88-93.

Matzer E and D Munach 1995. Soil changes induced by air pollution deposition and their implication for forest in Central Europe. Water, Air and Soil pollut., **85**, 63-76.

May M J and C J Lever, 1993. Oxidative stimulation of glutathione synthesis in *Arabidopsi thaliana* suspension cultures. Plant Physiol., **103**, 621-627.

McLaughlin S B, R K McConathy, D Duvick and L K Mann 1982. Effects of chronic air pollution stress on photosynthesis, carbon allocation and growth of white pine trees. For. Sci., **28**, 60-70.

Mishra J, V Pandey, S N Singh, M Yunus and K J Ahmad 1993. Growth responses of *Lycopersicon esculentum* to cement dust treatment. J. Environ.Sci. Health, **28**, 1771-1780.

Monk R J and F Murray, 1995. The relative tolerance of some Eucalyptus species to ozone exposure. Water Air and Soil Pollut., **85**, 1405-1411.

Mortensen L M and O Skre 1990. Effects of low ozone concentrations on growth of *Betula pubescens* Ehrh. Betula Verrucosa Ehrh. And *Alnus incana* (L.) Moench. New phytol., **115**, 165-170.

Murphy Jr. C E and J T Sigmon 1990. Dry deposition of sulphur and nitrogen oxide gases to forest vegetation In:Acidic precipitation Vol.3:Sources,Deposition and canopy interactions(Eds.S E Lindberg,A L Page And S A Norton), Springer Verlag, New York, pp, 83-95.

Murray F, R Monk and C D Walker 1994[a]. The responses of shoot growth of Eucalyptus species to concentration and frequency of exposure to nitrogen oxides. For. Eco. And Manage., **64**, 83-95.

Murray F, S Wilson and S Samaraweera 1994[b]. NO_2 increases wheat grain yield even in the presence of SO_2 Agric. Ecos. And Environ., **48**, 115-123.

Murray F, R Monk, K Clarke and M Qifu 1994[c] . Growth responses of N and S deficient white clover and burr medic to SO_2, NO and NO_2 . Agric. Ecos. And Environ., **50**, 113-121.

Nast W,L Mortensen,K Fischer and I Fitting, 1993. Effects of air pollutants on the growth and antioxidative system of Norway spruce exposed in open-top chambers. Environ. Pollut., **80**, 85-90.

Navari-Izzo N, M F Quartacci, R Izzo and C Pinino 1992. Degradation of membrane lipid components and antioxidant levels in *Hordeum vulgare* exposed to long-term fumigation with SO_2. Physiol Plant., **84** 73-79.

Nerg A, T Holopainen and J K Holopainen 1996. Effects of ozone exposure and mycorrhizal infection level on herbivores and chemical quality of scots pine seedlings.In 17th International meeting for specialists in Air pollution effects on forests ecosystems, Firenze, sept. 14-19, Italy.

Nile G V, M Tomasevic and N R Baker 1993. Effects of ozone on the photosynthetic apparatus and leaf proteins during leaf development in wheat. Plant, Cell and Environ., **16**, 643-651.

Okano K, T Machida and T Totsuka 1989. Differences in ability of NO_2 absorption in various braod-leaved Tree species. Environ. Pollut., **58**, 1-17.

Oleksyn jr and P B Reich 1994. Pollution, habitat destruction and biodiversity in Poland. Conservation Biology, **18**, 943-960.

Olszyk D M, A Bytnerowicz and B K Takemoto 1989. Photochemical oxidant pollution and vegetation: Effects of mixtures of gases, fog and patrticles. Environ. Pollut., **61**, 11-29.

Pandey J and M Agarwal 1994. Growth responses of tomato plants to low concentrations of SO_2 and NO_2. Sci. Hortic. (Amsterdam), **58**, 67-76.

Paoletti E, P Raddi and La Scala 1996[a]. Relationships between leaf wettability, stomatal damage and transpiration in declining beech trees. 17th International meeting for specialists in Air pollution effects on forest ecosystems. Sept. 14-19, Firenze, Italy.

Paoletti E, P Raddi and La Scala 1996[b] . UV-B and acid rain effects on beech (*Fagus sylvatica* L.) and Holm oak (*Quercus celix* L.) leaves. 17th international meeting for specialists in Air pollution effects on forest ecosystems. Sept. 14-19, Firenze, Italy.

Pearson M and T A Mansfield 1993. Interacting effects of ozone and water stress on stomatal resistance of Beech (*Fagus sylvatica* L.) New Phytol., **123**, 351-358.

Pearson M and T A Mansfield 1994. Effects of exposure to ozone and water stress on the following season's growth of beech (*Fagus sylvatica* L.), New Phytol., **126**, 511-515.

Pearson M, R T Besford D W Hand 1994.The effects of oxides of nitrogen and carbon dioxide enrichment on growth and content of Ribulose-1, 5-Biphosphate carboxylase-oxygenase and nitrite reductase in glass house lettuce. J. Hort. Sci., **69**, 257-266.

Pedersen L B, K Hansen, J Bille-Hansen, M Lqber and M F Hovemand 1995. Throughfall and canopy buffering in three Sitka spruce stands in Denmark. Water, Air, and Soil Pollut., **85**, 1593-1598.

Percy K E, C R Krause and K F Jensen 1990.Effects of ozone and acid fog on red spruce needle epicuticular wax ultrastructure. Can. J. For. Res., **20**, 17-120.

Perez-Soba M, 1995. Physiological modulation of the vitality of scots pine trees by atmospheric ammonia depositon. Ph D Thesis, university of Groningen, The Netherlands.

Pfanz H, B Beate andL Bohumir 1996. How do acidic air pollutants affect photosynthesis of tree leaves and bark collenchyma? Damage and avoidance strategies. In 17th International meeting for specialists in Air pollution effects on forest ecosystems. Firenze, sept. 14-19, Italy.

Polle A and H Rennenberg 1993. Significance of antioxidants in plant adaption to environmental stress In:Plant adaption to Environmental stress(Eds L Fowden,T Mansfield and J Stoddart).Chapman and Hall pp, 263-273.

Polle A, K Chakrabarti, W Schurmann and H Rennenberg 1990. Composition and properties of hydrogen peroxide decomposing systems in extracellular and total extracts from needles of Norway spruce [*Picea abies* (L.) Karst]. Plant Physiol., **94**, 312-319.

Polle A, M Eibleier and H Rennenberg 1994[a]. Sulphate and antioxidants in needles of scots pine (*Pinus sylvestris* L.) from three SO_2 polluted fields sites in eastern Germany. New Phytol., **127**, 571-577.

Polle A, T Otter and F Seifert 1994[b] . Apoplastic peroxidases and lingification in needles of Norway spruce (*Picea abies* L.) Plant physiol., **106**, 53-60.

Polle A, T Pfirrmann, S Chakrabarti and H Rennenberg 1993. The effects of enhanced ozone and enhanced carbon dioxide concentrations on biomass, pigments and antioxidative enzymes in spruce needles (*Picea abies* L.) Plant, Cell and Environ., **16**, 311-316.

Prasad M S V and J A Inamdar 1990. Effect of particulate pollution on *Pisum sativum*. J. Phytol. Res., **3**, 71-77.

Price A, P W Lucas and P J Lea 1990. Age dependent damage and glutathione metabolism in ozone fumigated Barley: A leaf section approach. J. Expt.Bot., **41**, 1309-1317.

Puccinelli P, M Anselmi and M Bragaloni 1996. Perioxidases. A possible marker of air pollution on trees in urban environment. In 17th International meeting for specialists in Air pollution effects on forests ecosystems, Firenze, Sept 14-19, Italy.

Rabotti C and A Ballarin-Denti 1996. Biochemical responses to abiotic stress in beech (*Fagus sylvatica* L.) leaves. In 17th International meeting for specialists in Air pollution effects on forest ecosystems, Firenze, Sept. 14 -19, 1996, Italy.

Raddi R, S Moricca, R Gellini and V Lorando 1992. Effects of natural and induced pollution on the leaf wax structure of three cypress species. Eur. J. For. Patho., **22**, 107-114.

Raitio H, J P Tuovinen and P Anttila 1995. Relation between sulphur concentrations in the scots pine needles and the air in Northermost Europe. Water, Air and Soil Pollut., **85**, 1361-1366.

Rantanen L,V Palomaki, A L Harrison,P W Lucas and T A Mansfield 1994. Interactions between combined exposure to SO_2 and NO_2 and nutrients status of trees:Effects of nutrient contents and uptake, growth needle ultrastructure and pigment. New Phytol., **128 (4)**, 689-701.

Rao M V and P S Dubey 1990. Biochemical aspects (antioxidants) for development of tolerance in plants growing at different low levels of ambient air pollutants. Environ. Pollut., **64**, 55-66.

Rao M V and P S Dubey 1993. Response of hydrogen peroxide scavenging system in two soybean cultivars Exposed to SO_2:Experimental avidence for the detoxification of SO_2 by enhanced H_2O_2 scavenging components. Environ. Pollut., **82**, 99-105.

Rautenkranz A A F, L Li, F Machler, E Martinoia, J J Oertli 1994. Transport of ascorbic and dehydro ascorbic acids across protoplast and vacuole membranes isolated from barley (*Hordeum vulgare*, L; ev Gerbel) leaves. Plant Physiol., **106**, 187-193.

Rennenberg H 1984. Fate of excess sulphur in higher plants. Ann. Rev. Plant physiol., **35**, 121-153.

Rennenberg H and A Polle 1994. Metabolic consequences of atmospheric sulphur influx into plants.In: Plant responses to the Gaseous Environment. Molecular metabolic and physiological aspects. (Eds R G Alscher and A R Wellburn). Chapman & Hall. pp, 165-180.

Rieder M 1989. The cuticles of conifers: Structurre, composition and transport properties. In:Forest Decline and Air pollution. A study of spruce (*Picea abies*) on Acid soils. (Eds Schulze E D, O L Lange and R Oren). Springer Verlag, Berlin pp, 157-192.

Saastomoinen T and T Holopainen 1989. Needle and root responses of small P*inus sylverstris* seedlings exposed to SO_2 and stimulated acid rains. Scand J. Res., **4**, 273-284.

Saha D 1998. Tolerance levels of certain tree species towards industrial air pollution. Ph D Thesis submitted to The M S University of Baroda, Baroda, India.

Saha D, J G Gavali and N S R Krishnayya 1999a. Differential response of tropical trees towards industrial air pollution. Tropical Ecology **40** (2), 289-297.

Saha D, J G Gavali and N S R Krishnayya 1999b . Studies of epicuticular wax in some trees exposed to industrial air pollution. Phytomorphology. **49** (3), 277-281.

Saha D, J G Gavali and N S R Krishnayya 2000. Stomatal studies of few tree species growing in industrial area. Advances in Plant Sciences. **13** (1), 209-212.

Sanders G E, L S Karby, M R Ashmore and J Fuhrer 1995. Estrablishing critical levels for the effects of air pollution on vegetation. Water, Air and Soil Pollut., **85**, 189-200

Sandmann G and H G Gonzales 1989. Peroxidative processes induced in Bean leaves by fumigation with sulphur dioxide. Environ. Pollut., **56**, 145-154.

Savada S, Y Hazama, T Hayakawa and T Totsuka 1989. Simulation effects of atmospheric ethylene on tree leaf shedding. Environ. Pollut., **61**, 173-185.

Scaefer D and W A Reiners 1990. Throughfall chemistry and canopy processing mechanisms. In: Acidic Precipitation Vol 3: Sources, Depositon and canopy Interactions. (Eds. S E Lindberg, A L Page and S A Norton), Springer verlag, New York, pp 241-284.

Schmieden U and A Wild 1994. Changes in levels of a - tocopherol and ascorbate in spruce needles at three low mountain sites exposed to ozone and Mg^{2+} - deficiency. Z Naturfrrsch., **49**, 171-180.

Schmieden U and A Wild 1995. The contribution of ozone to forest decline. Physiol. Plant., **94**, 371-378.

Schmeieden U, S Schneider and A Wild 1993. Glutathione status and glutathione reductase activity in spruce needles of healthy and damaged trees at two mountain sites. Environ. Pollut., **82**, 239-244.

Schwenn J D, B Depka and H H Hennies 1976. Assimilatory sulfate reduction in chloroplasts evidence for the participation of both stromal and membrane bound enzymes. Plant Cell Physiol., **17**, 165-176.

Sellden G and H Pleijel 1995. Photochemical oxidant effects on vegetation - Response in relation to plant strategy. Water, Air Soil pollut., **85**, 111-122.

Sergeichik S A and A A Sergeichik 1993. Nitrogen metabolism and metal accumulation in tree leaves in relation to resistance to air pollution. Vestsi Akad, Navuk Belarusi Ser., Biyal Navuk, **2**, 32-38.

Sgherri C L M, C Pinzino and F Navari Izzo 1996. Sunflower seedlings subjected to increasing stress by Water deficit: Changes in $O_2^{\cdot}$ production related to the composition of thylakoid membranes. Physiol Plant., **96**, 446-452.

Sharma G K 1992. *Bougainvillea glabra* L. Cuticular response to environmental pollution. ***Geobios*** **19**, 239-242.

Sharma G K 1996. Cuticular dynamics and environmental pollution in *Cercis canadensis* (redbud) in North

American deciduous forest. In 17 th International meeting for specialists in Air pollution effects on Forests ecosystems, Firenze, Sept 14 - 19, Italy.

Shaw P J A, M R Holland and N M Darrall 1993. The occurance of SO_2 - related forliar symptoms on Scot pine (*Pinus sylverstris* L.) in an open-air forest fumigation experiment. New Phytol., **123**, 143-152.

Sheppard L J 1994. Casual mechanisms by which sulphate, nitrate and acidity influence frost hardiness in red spruce: Review and hypothesis . *New Phytol.*, **127**, 69-82.

Shukla J, V Pandey, S N Singh, M Yunus, N Singh and K J Ahmad 1990. Effect of cement dust on the growth and the yield of *Brassica campestris* L. Environ. Pollut., **66**, 81-88.

Silvennoinen R, T Jaskelainen, K Nygnen, J Huttunen and P Parkkine 1995. Temporal and spatial environmental classification of pine reflectance spectra. Environ. Sci.Tech., **29**, 1456-1459.

Smeulders S, A Gorissen, N N Joosten and J A Van Veen 1994. Effects of short-term ozone exposure on the carbon economy of mature and juvenile Douglas firs [*Pseudotsuga menziesii* (Mirb.) Franco]. *New Physiol.*, 99, 1267-1270.

Smirnoff N 1996. The Function and metabolism of ascorbic acid in plants. Annals of Bot., **78**, 661-669.

Smith I K, T L Vierheller and C A Thorne 1989. Properties and functions of glutathione reductase in plants. Physiol. Plant., 77, 449-456.

Solvik S, A Siegmund, G Kindermann, R Riebeling and A Balaza 1995. Stomatal SO_2 uptake and sulphate accumulation in needles of Norway spuce stands (*Picea abies*) in central Europe. Plant and Soil, **168-169**, 405-419.

Speeding D J, I Ziegler R Hampp and H Ziegler 1980. Effect of pH on the uptake of 35S-Sulfur from sulfate, Sulfite and sulfide by ***Chlorella Vulgaris*** Z. Pflanzenphysiol, **97**, 205-214.

Steubing L, A Fangmeier, R Both and M Frankenfeld 1989. Effects of SO_2, NO_2 and O_3 on population development and morphological and physiology parameters of native herb layer species in a Beech forest. Environ. Pollut., **58**, 281-302.

Sturaro A, G Parvoli and L Duretli 1993. Plant tree bark as a passive sampler of polycyclic aromatic hydrocarbons in an urban environment. *J. Chromatog.*, **643**, 435-438.

Sutinen S, L Skarby, G Wallin and G Sellden 1990. Long term exposure of Norway spuce, [*Picea abies* (L.) Karst] to ozone in open-top chambers. II Effects on the ultrastructure of needles. New Phytol., 115, 345-355.

Szarek G 1995.Sulphur and nitrogen input to the Ratanica forested catchments Carpathian foorhills, southern Poland. Water, Air and Soil pollut., **85**, 1765-1770.

Takahama U, S Veljovic-Ionanovic, U Heber 1992. Effects of the air pollutant SO_2 on leaves. Inhibition of sulphite oxidation in the apoplast by ascorbate and of apoplastic peroxide by sulphite. *Plant Physiol.*, **100**, 261-266.

Terry G M, N J Strokes, P W Lucas and C N Hewitt 1995. Effects of reactive hydrocarbons and hydrogen peroxide on antioxidant activity in cherry leaves. Environ. Pollut., **88**, 19-26.

Tiwari S and S Bansal 1993. Effects of NO_2 pollution on *Mimusops elengi* Linn. Asian J. Pl. Sci., **5**, 43-46.

Turunen M and S Huttunen 1991.Effects of stimulated acid rain on the epicuticular wax of scots pine needles under northerly conditions. Can. J. Bot., **69**, 412-419.

Van Den Driessche R and C Langebartales 1994. Foliar symptoms, ethylene biosynthesis and water use of young Norway spruce (*Picea abies* (L) Karst.) exposed to drouhgt and ozone. Water, Air and Soil pollut., **78**, 153-168.

Van Hove L W A, O Van Kooten, E H Adema, W J Vredenberg and G A Pieters 1989. Physiological effects of long-term exposure to low and moderate concentrations of atmospheric ammonia on poplar leaves Plant, Cell Environ., **12**, 899-908.

Van Hove L W A, M E Bossen, M G J Mensik and O Van Kooten1992. Physiological effects of a long term exposure to low concentrations of NH_3,NO_2 and SO_2 on Douglas fir(Pseudotsuga menziesii). Physiol. Plant., **86**, 559-567.

Veljovic-Jovanovic S, W Bilger and U Heber 1993. Inhibition of photosynthesis, acididification and stimulation of zeaxanthin formation in leaves by SO_2 and reversal of these effects. Planta, **191**, 365-376.

Wahid A, R Maggs, S R A Shamsi, J N B Bell and M R Ashmore 1995. Air pollution and its impacts on wheat yield in the Pakistan Punjab. Environ. Pollut., **88**, 147-154.

Weckx J E J and H M M Clijsters 1996. Oxidative damage and defense mechanisms in primary leaves of *Phaseolus vulagaris* as a result of root assimilation of toxic amounts of copper. Physiol. Plant., **96**, 506-512.

Weigel H G, G Halbwachs and H J Jager 1989. The effects of air pollutants on forest trees from a plant physiological view. J. Plant Diseases and Protection, **96**, 203-207.

Wellburn A R 1988. Air pollution and acid rain. The biological impact, Longman publ., New york.

Wellburn AR1990.Why are atmospheric oxides of nitrogen usually phototoxic and not alternative fertilizers? New Phytol., **115**, 395-429.

Wiltshire J J J, C J Wright, M H Unsworth and J Criagon 1993. The effects of ozone episodes on autumn leaf fall in apple. New Phytol., **124**, 433-437.

Wingsle G and J E Hallgren 1993.Influence of SO_2 and NO_2 exposure on glutathione, superoxide dismutase and glutathione reductase activities in Scots pine needles. J. Expt. Bot., **44**, 463-470.

Woodbury P B, J A Laurence and G W Hudler 1994. Chronic ozone exposure alters the growth of leaves, stems and roots of hybrid *Populus*. Environ. Pollut., **85**, 103-108.

Wolfenden J and T A Mansfield 1991. Physiological disturbances in plants caused by air pollutants.In: Acidic deposition, Its nature and Impact,(Eds. F T Last and R Watting), Proc.of the Royal Soc.of Edinburgh 97B, pp, 117-138.

Yunus M., A K Dwivedi, K Kulshreshtha and K J Ahmad 1985. Dust loading on some common plants near Lucknow city. Environ. Pollut., **9**, 71-80.

Yunus M, N Singh and M Iqbal 1996. Global status of air pollution: An overview. In:Plant respose to Air pollution, (Eds. M Yunus and M Iqbal). John Wiley & Sons Ltd., Chichester, pp 1-34.

Zoettl H W and R F Huettl 1991. Management of nutrition in forests under stress. Kluwer Academic publ., The Netherlands.

Zundel T, O Rentz, R Dorn, A Jattke and M Wietschel 1995. Control techniques and strategies for regional air pollution control from energy and industrial sectors. Water, Air and Soil pollut., **85**, 213-224.

13

Shelter-Cum-Bio-Energy-Belt Along the Coast of Andhra Pradesh India

P.H.V. Vasudeva Rao[1], *S. Venkateswara Rao[2], E. Amminedu[3] and N. Subba Rao[4]

INTRODUCTION

India has 5700 km long peninsular coast; considering 25 Km land ward distance as the width of the coastal belt, nearly 0.15 million sq. km. is the area of the belt constituting 4.57% of the total 3.28 million sq. km geographical area of the country. Nearly 25% of the total population of 104 crores lives near the coast and derive its livelihood from coastal resources. Wind borne sterile sands form shifting sand dunes, expanding ravines, saline soils and/or marshy soils pose threat to the coastal regions. Cyclonic surges cause irreparable loss to the coastal regions. The much needed eco-conservancy can be achieved at least, in part by raising of shelterbelt of vegetation along the coast. Shelterbelt in the vegetational context is defined as the rows of plants of all the three tiers, herbs, shrubs and trees raised normal to the direction of wind flow. They are grown in the high wind-velocity prone areas to reduce, break or obstruct the wind speed from the windward to leeward side and mitigate the impact of cyclonic surges. Shelterbelts are environmentally and economically benign as they improve edaphic conditions, microclimate and recover landscape.

A proper management of shelterbelt would reduce the pressure on forests for fuel, timber and on water for irrigation for growing trees on main land. The significance of stabilisation of coastal dunes with shelterbelts need not be over emphasized as India has extensive tracks of coastal sandy soils sometimes extending up to 30 kms into the land.

Even though in the humid tropical coast, wind erosion is partly circumvented by the atmospheric moisture, the other meteoro-edaphic factors such as high temperatures, big particle

[1& 3] Department of Geo-Engineering, Andhra University, Visakhapatnam-530 003, A.P., India.
[2 & 4] Department of Geology, Andhra University, Visakhapatnam-530 003, A.P., India.

size of the soils, poor soil capillarity, poor water holding capacity and loose soil texture help the eolian sand transport landwards. This wind action proceeds till it encounters an obstacle or wind velocity is reduced. The relation between coastal eolian sand dunes, wind velocity and vegetation has been studied in India (Reddy, 1979; Singh, 1982) but not in relation with shelterbelt. World Meteorological Organisation (WMO) in the report on Agricultural Meteorology (1964) and Sturrock (1988), an international authority on shelterbelt, stressed the need for investigations on shelterbelts. The effect of wind breaks on microclimate has been discussed by Baymi (1976) Heisler & De Walle (1988), Mcnaughton (1988) and Wight (1988). But shelterbelts have scarcely been studied in India.

METHODOLOGY

Twenty stations situated in six districts were selected along the coastline of Andhra Pradesh for studying wind velocities, soil types and shelterbelt aspects (Fig. 1). Vizianagaram, West Godavari and Guntur Districts were not covered as they have negligible coastline. The wind velocities were measured on windward and leeward sides and in the midst of shelterbelt simultaneously with portable anemometers. On the downwind side velocities were measured at 3-5 m, 80-100 m and 300 m away from the shelterbelt. On the upwind side wind velocities were measured only at 10 - 12 m from the shelterbelt because of the close proximity of the shelter to the surging waves. Length, number of rows and espacement of the trees noted. The wind speed on the lee is expressed as the percentage of wind in the open or simply the ratio of wind speed due to shelter to that in the open (WMO, 1964). The general physiography, meteorology and pedology of the coastline are discussed with reference to the data of the last 10 years. The plant species commonly occurring along the coast are recorded. The secondary data on district-wise frequency of cyclones along West and East Coast of India was collected. Dry biomass parameters and the energy potential of shelterbelt trees are presented and discussed.

PHYSIOGRAPHIC FEATURES

Andhra Pradesh lies on the East Coast of India within 12°37′ and 19°54′ N.lat. and 76°46′ and 76°46′ and 84°50′ E.long. The coastal zone stretches from Kalingapatnam (Srikakulam District) in north to Pulicot (Nellore District) in the south along a narrow strip, which broadens in the middle along Godavari - Krishna deltas (upto 80 sq. km. inland). Precipitation, local topography, wind velocity and direction, humidity, sand and salt spray effect the coastal biosphere. The relatively uniform Andhra Pradesh coastal climate falls under sub-humid to semi-arid type as per unviersal concept of Thornthwaite (1948) which was later specifically applied to the Andhra Pradesh coast (Subrahmanyam, 1983).

Minimum temperature of 12°C was recorded in the northern district Srikakulam, whereas maximum of 42°C was recorded in central district of Krishna. The average annual rainfall varies from 1000 mm to 1250 mm. Heavy to very heavy rainfall climaxing in the cyclones commonly occurs during the post-monsoon from September to November due to northeast monsoon. Mean monthly wind speeds for a period of 10 years are given in the Table-1. A minimum wind speed of 4.4 km/h was recorded in October in the southern district Prakasam and a maximum of 17.0 km/h was observed in May in northern most district, Srikakulam. But during cyclonic period (October - November) a wind speed of 70 to 170 km/h has been recorded. The coast line of

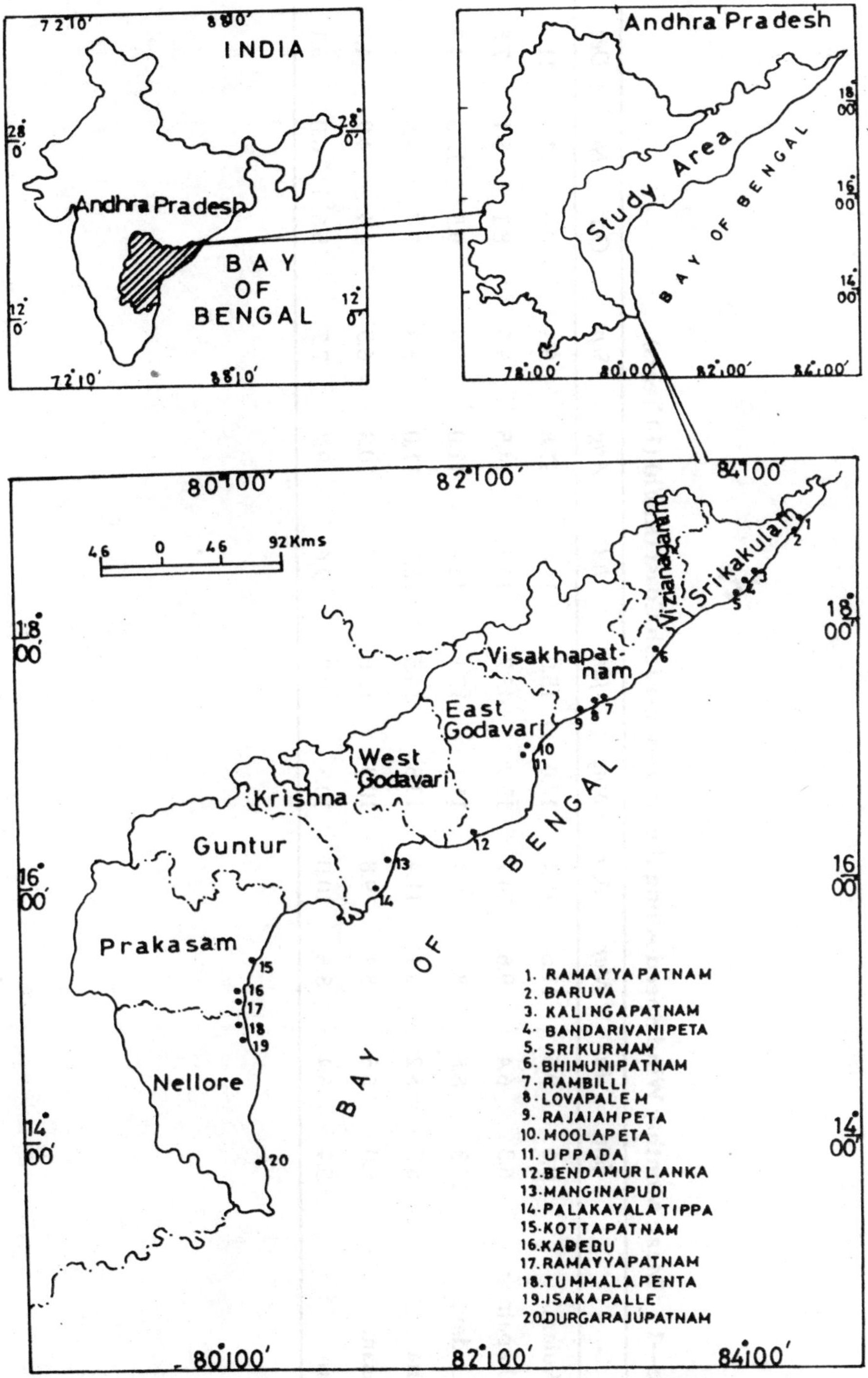

Fig. 1 Some of the Cyclone prone areas selected for shelterbelt studies.

Table—1 Mean Monthly Wind Speed Along the Andhra Pradesh Coast (Km/h) (10 Years)

Districts	*Jan*	*Feb*	*Mar*	*Apr*	*May*	*Jun*	*Jul*	*Aug*	*Sept*	*Oct*	*Nov*	*Dec*
Srikakulam	9.6	9.5	11.6	15.1	17.0	15.0	14.0	11.8	9.1	10.0	11.2	11.3
Visakhapatnam	6.3	6.4	9.6	14.7	15.9	13.7	16.5	14.5	9.5	8.1	8.1	7.5
East Godavari	10.3	8.8	8.3	9.1	11.1	12.1	12.3	11.0	8.6	9.5	12.0	4.3
Krishna	8.7	8.2	9.1	11.4	13.8	14.5	12.8	11.0	8.8	7.7	8.8	9.1
Prakasam	5.0	6.5	8.4	9.8	10.3	11.6	10.7	10.3	6.5	4.4	4.9	4.7
Nellore	5.4	6.4	8.6	10.0	10.3	9.8	9.6	9.2	7.7	5.6	6.0	6.1

northern districts of Srikakulam and Visakhapatnam has a very highly raising topography (above 200 msl) whereas the southern coast, i.e., Krishna and Prakasam districts can be found at less than 6 m above msl indicating that the water line has raisen relative to the land. The coast in the southern districts at some places has very marshy to semi-marshy soils, favouring fast water spread during high tides and cyclones. The extreme coast is sparsely covered by shrubs and grasses. The estuarine ecosystem of Godavari and Krishna rivers especially of the former harbours mangrove vegetation.

NATURAL CALAMITIES

The violent winds occurring during tropical cyclones along the coast adversely effect the coastal population, industries and maritime activities. The frequency of land falling of cyclones is five times more frequent in east coast (Bay of Bengal) than in the west coast (Arabian Sea) of India (Table-2). Out of the storms developing in the Bay more than 70% approach the East Coast in the post-monsoon (October - November) season and 30% in the pre-monsoon season (May - June). Andhra Pradesh, experiences the highest landfalling of cyclones among all the coastal states. Srikakulam, Krishna and Nellore districts in the state experience cyclones more frequently than the other coastal districts.

Singh (1982) opines that the cyclonic storms are inevitable in India and the resultant hazards on tropical agriculture are not preventable through the windbreaks. Nicholas (1988) argues only theoretically that the probability of cyclonic storms that occurr in any given year is sufficiently very low. He mentions that in any case most cyclonic storms are powerful enough to demolish a windbreak along with the crops. But the present authors feel that because of the immense losses, the "opined infrequency" has to be considered as more frequent. Even otherwise the data on frequency of cyclones on the East Coast of India show that a total of 418 medium and severe cyclones occurred during the last 90 years. This is a very high frequency for a coast length of 2,440 km. Reddy (1979) has reported that shelterbelts and plantation have lessened the effect of devastating cyclone in 1977 in coastal Andhra Pradesh.

SOILS

Visakhapatnam and Srikakulam districts have wide coastal alluvial soils and sandy soils. The former has isolated hills/hillocks projecting into the sea or into the coast. Along the Godavari estuary the soils are coastal alluvial often followed towards inland by saline and saline alkaline soils. Less frequently, wide strips of saline and saline alkaline soils are present along the coast followed by deltaic alluvial soils as in Krishna district. In Guntur and Nellore districts, medium to deep black soils are present contiguous to saline soils. Red loamy soils mixed with red sandy soils and less frequently with gravelly soils are found 1 to 2 kms away from the coast along the belt.

More than 40 per cent of farm land along the coast suffer from severe ecological constraints as soil erosion, compaction, water logging, salinity, acidity and alkalinity. This influence is more in Srikakulam, Krishna and Nellore districts.

As per the land capability classification, the sandy shore belt of 0.5 to 6 kms width is regarded, as poor land, not because of failure of efforts but due to lack of efforts for its

Table—2 Cyclone Land Fall in Various Coastal States in India (1891-1990)

West Coast		*East Coast*		*Coastal Andhra Pradesh*	
States	*No. of Cyclones*	*States*	*No. of Cyclones*	*District*	*No. of Cyclones*
Gujarat	20	West Bengal	33	1. Srikakulam	14
Karnataka	3	Orissa	45	2. Vizianagaram	Nil
Kerala	4	Andhra Pradesh	75	3. Visakhapatnam	5
Maharashtra and Goa	9	Tamilnadu	38	4. East Godavari	8
				5. West Godavari	1
				6. Krishna	16
				7. Guntur	2
				8. Prakasam	10
				9. Nellore	19
Total	**36**	**Total**	**191**	**Total**	**75**

reclamation. Most of the coastal Andhra Pradesh has moderately good soil adjacent to the poor land. The north coastal districts have good or moderately good land even along the extreme shore. West Godavari and Krishna districts have very good land along the shore; low lying areas experience frequent inundation and water-logging, leading to soil and water salinity and saline swamps due to cyclones. These hydro-meteorological features are responsible for the slightly to median alkaline followed by slightly acidic soils in the central coastal Andhra Pradesh. However, the extreme northern and southern coastal Andhra Pradesh regions have neutral soils with shifting acidity.

Sand dunes and sand sheets are of common occurrence along the coast. High atmospheric humidity, uniformly medium temperature and higher rainfall stabilise dunes of coastal sandy soils. The opposite extreme of the same meteorological components lead to destabilization of desert sands. Even during the high velocity winds, sand shift occurs only to a limited extent in the extreme shore region because of soil moisture. Where the coastal sandy soils extend more than 0.25 kms and is without any vegetation, the sand shift is severe covering the roads, buildings, crops etc. In majority of the locations in the East Coast, red soils are found at a superficial depth of 0.75 to 1.5 m at 100 m away from the coast rendering them more productive.

STRAND VEGETATION

Majority of the coastal area of Andhra Pradesh is covered by cultivable lands, with occasional patches of protected/reserved forests of residual type, at some places covered by scrub and grasses, probably indicating poor management of the soil. Flora of coastline spreading lengthwise is called as a 'strand flora/strand vegetation'. In Andhra Pradesh coast the strand area is discernible into three sub-types depending upon the type of substratum as sand, rock and coral strand. The natural strand vegetation has only the lower most tier, i.e., herbs and creepers, which also can grow inland in versatile meteoro-edaphic conditions. These plants exhibit psammophytic, xerophytic and lithophytic characters. They are herbs like *Ipomea biloba, Citrullus Colosynthis, Spinifex squarrosus, Tephrosia purpurea, Cynodon dactylon, Paspalum vagnatium, Trachys muronata, Perotis indica, Coix Lachryma-lagopus, Alueropus lagopoides, Jatropha Curcas etc.* The trees that are commonly found naturally in the psammic conditions are *Pandanus* sps., *Prosopis juliflora* and the raised *Casuarina equisetifolia, Eucalyptus tereticornis, Glyricidia saprim, Cocos nucifera* etc. It is note worthy that these species are efficient as shelterbelt constituents at five different tiers as (1) creepers forming a vegetation carpet (2) herbs (3) shrubs (4) big shrubs / small trees and (5) trees. The advantages of perennial herbaceous barriers for water conservation and protection of soils and crops in increasing soil temperature in winter and decreasing soil temperature in summer and functioning as green mulch to evaporation rate have been emphasized by Black and Aase (1988).

SHELTER AND WIND VELOCITIES ALONG THE ANDHRA PRADESH COAST

The status of shelterbelt and its effect on wind velocities along the coast of Andhra Pradesh is discussed with reference to 20 cyclone prone locations situated in six districts (Table-3). The shelterbelt throughout the coast has only *Casuarina equisetifolia*.

1. **Srikakulam District:** Srikakulam, the northern most coastal district of Andhra Pradesh, is predominantly rural and the most backward region in the state, experiencing the

Table—3 Shelterbelt and Wind Velocity Along the Andhra Pradesh Coast

S. No.	Field Sites	Width of the shelter belt (m)	Spacing (m)	Wind velocity and percentage of reduction: Near shore (wind ward in (km/h)	Midst of Shelterbelt: In Km/h	Midst of Shelterbelt: % of reduction	Immediately Behind Shelterbelt: in Km/h	Immediately Behind Shelterbelt: % of reduction	80-100 m Behind shelter-belt (km/h)	300-400 m behind shelterbelt: In Km/h	300-400 m behind shelterbelt: % of reduction
Srikakulam District											
1.	Ramayyapatham	42	2.0	8.54	5.52	34.00	5.48	35.00	–	–	–
2.	Baruva	–	–	–	–	–	–	–	10.51	7.32	30.35
3.	Bandarivanipeta	–	–	–	–	–	–	–	10.96	7.78	29.02
4.	Kalingapatnam	–	–	–	–	–	–	–	12.38	8.75	29.33
5.	Srikurmam	276	2.0	10.1	3.55	64.86	6.63	34.36	–	–	–
Visakhapatnam District											
1.	Bheemunipatnam Mangamaripeta	38	2.0	3.16	0.75	76.27	0.99	68.78	–	–	–
2.	Rambilli	250	2.0	8.91	1.07	88.00	0.89	90.02	8.04	3.57	55.6
3.	Rajayyapeta	120	2.0	4.03	0.06	99.52	0.04	99.01	4.80	2.52	47.5
4.	Lovapalem	250	2.0	5.42	0.01	99.82	0.35	93.35	11.93	4.27	64.21
East Godavari District											
1.	Uppada	–	–	–	–	–	–	–	10.30	6.64	35.53
2.	Mulapeta	100	2.0	–	–	–	–	–	20.00	16.08	19.65
3.	Vodalarevu	100	1.5	3.18	0.47	85.23	4.53	–	6.34	6.7	–

(Contd...)

(Table—3 Contd...)

S. No. Field Sites	Width of the shelter belt (m)	Spacing (m)	Wind velocity and percentage of reduction							
			Near shore (wind ward in (km/h)	Midst of Shelterbelt		Immediately Behind Shelterbelt		80-100 m Behind shelter-belt (km/h)	300-400 m behind shelterbelt	
				In Km/h	% of reduction	in Km/h	% of reduction		In Km/h	% of reduction
Krishna District										
1. Manginipudi	350	1.0	4.68	3.06	34.62	4.67	0.22	7.62	5.66	25.72
2. Paramahamsalapuram	300	1.5	–	–	–	–	–	21.22	12.76	39.87
Prakasam District										
1. Kothapatnam	350	1.5	5.1	–	–	1.65	67.65	4.71	4.23	10.20
2. Karedu	400	2.0	9.22	0.79	91.44	1.78	80.70	8.91	4.29	51.86
3. Ramayyapatnam	200	2.0	9.00	0.14	98.45	1.29	85.67	8.90	5.03	43.38
Nellore District										
1. Tummalapenta	25	1.5	2.74	1.60	5.06	–	–	–	–	–
2. Isakapalli	–	–	–	–	–	–	–	5.14	4.12	19.84
3. Durgarajupatnam	250	1.5	5.78	0.78	87.03	2.47	52.27	–	–	–

highest frequency of landfalling of cyclones (Table 2). Five locations, two with and three without shelterbelt were selected. All the five locations were severely effected during the cyclone of 1990. (Ramayyapatnam, Baruva, Bandaravanipeta Kalingapatnam and Sri Kurmam). The impact of shelterbelt was not felt to its full potentiality as it had wide gaps created due to improper management. The location 3 has 200 - 250 m wide swampy shore with an area of about 70 ha. The region is a highly saline and prone to inundation hence not suitable for raising traditional shelterbelt species. Ramayyapatnam and Baruva have shelterbelts of 42 and 27.8 m width respectively. The spacing of the trees in shelterbelts is 2 m. Wind velocity is reduced to 34 and 35% in the midst of and behind shelterbelt in the 1st location and the corresponding reduction for the 5th location were 64.8 and 34.3%. The percentage of reduction in the wind velocity for locations 2, 3 and 4,without shelterbelt it is only 20 to 29%. The soils in 1, 2, 4 and 5 locations are predominantly sandy and the location 3 has more clay content.

2. **Visakhapatnam District:** Even though Visakhapatnam is the fastest developing city in the country, it is one of the most backward districts because of the predominance of water-scarce rural and tribal areas. The district has 1,119 ha of coastal sandy soils occasionally extending 2 to 3 km inland constituting 1% of the total geographical area of the district. Water-logged/marshy land of saline and saline alkali soils covering an area of 156 ha is present. The coastline is broken by a number of baldheaded lands, the important one being the Yarada hill or Dolphin's Nose that sustains the natural harbour. The other important hills are Kailasagiri, Rushikonda, and Narasimha hill. Coastal alluvial soils or occasionally red loamy soils and deltaic alluvial soils, lie adjacent to the extreme belt of sandy soils/marshy lands. It changes to red loamy and red sandy soils further inland. These extreme and sub-coastal soils though considered unsuitable for cultivation of conventional crops, the edaphic limitations can be over come through afforestation with ergophytes (energy plantation), fruit trees and fodder crops. However, to the north and south of Visakhapatnam City, moderately good land is available in certain patches for conventional crops.

   Four locations were selected (Bheemunipatnam, Rambilli, Lovapalem and Rajayyapeta) to study the shelterbelt aspects in this district. The reduction in wind velocity due to shelterbelt varies from 68 to 99%. The extreme case of reduction is due to very low initial value of 4 km/h, which was reduced to 0.04 km/h on the lee due to shelterbelt. No variation in temperature and humidity is recorded between the windward and leeward sides probably because of free movement of air. The wind velocity at the shelterbelt-break-regions also recorded gradual reduction from the shore inland wards. But this reduction is very low when compared with reduction on the leeward side of the shelterbelt showing that the greater reduction is undoubtedly due to shelterbelt. In all the locations wind velocity reduction is greater on the lee than in the midst of the shelterbelt. It is observed that the vegetation behind sparse shelterbelt is also sparse. The protective impact of shelterbelt is also evident at certain locations, as paddy could be cultivated successfully on the leeward side of the shelterbelt.

   In certain region along the coast, a natural growth of *Borassus flabellifera* (toddy palm) and *Prosopis juliflora* in dense stands is observed.

3. **East Godavari District:** Mulapet and Uppada near the Kakinada Bay and Odalarevu in the southern part of the district were selected for the study. Mulapeta, situated near the estuarine region, has a swampy shore. A shelterbelt of 100 m wide is present only on one side of the Godavari canal joining the sea. The other bank of the canal is swampy area of nearly 40 ha without any vegetation. During 1986 cyclone, the shelterbelt and crops behind it were severely damaged. The damage could have possibly been mitigated, if shelterbelt were to be present on the other side of the canal. Uppada, without shelterbelt, does not suffer cyclonic loss of human lives because of concrete buildings. A slow and deadly process of soil erosion occur because of high tides and salt spray on the crops continuously. The soils are very fertile and hence food crops are grown in large areas. Shelterbelt along the Godavari banks is very thin and sparse and along the seashore it is hardly 30 m wide. Therefore, these areas suffered the cyclonic impact to very great extent in 1986. Hence, the existing shelterbelts need further strengthening in view of the high productivity of the agricultural crops.

4. **Krishna District:** Mangenapudi and Paramahamsalapuram, the sites most susceptible to cyclonic effect, were selected to obtain the data regarding shelterbelt aspects. Paramahamsalapuram is situated 2.50 km away from the sea with the intervening area being swampy and saline A 300m wide shelterbelt of casuarina was raised after 1977 cyclone around a swampy area with ecologically specialized groups, mangroves, halophytes and certain xerophytes. Even though a reduction of 39.87% was recorded from windward side to leeward side, it is not sufficient to mitigate the cyclonic impact. The situation demands further strengthening of the existing shelterbelt and reclamation of the vacant space lying as a wasteland. Mangenapudi has 350 m wide shelterbelt but it is subjected to illegal felling and hence the effect of shelterbelt is not appreciable.

5. **Prakasam District:** The selected sites Kothapatnam, Karedu and Ramayyapatnam have a considerable wide shelterbelt. In general, the percentage of reduction in wind velocity amidst and behind the shelterbelt is appreciable with the values ranging from 67 to 98. The reduction is more amidst shelterbelt than on the lee. This lesser reduction behind the shelterbelt is attributable to sudden outgress into an open area, where aerodynamic pressure is less than the turbulence created amidst the shelterbelt. However, no direct relationship between width of the shelterbelt and reduction in wind velocity could be traced. But at 300 to 400 m on the lee of shelterbelt, an increasing reduction in wind velocity with increasing width of the shelterbelt is observed.

   As all the three villages are situated on the banks of brackish Buckingham canal, the substratum tends to be saline. Conventional crops, such as groundnut, pulses coconut and the non-conventional plantations of *Eucalyptus* and *Acacia* are grown behind the shelterbelt. The vegetation behind the shelterbelt appears healthy. No crops were observed for long distances in the shelterbelt break regions.

6. **Nellore District:** The proven vulnerable cyclone sites - Tummalapenta, Isukapalli and Durgarajupatnam were selected for the study. The brackish Buckingham canal

influences the sub-surface and ground water and soil hence the flora is mainly xerophytic. The crops grown in the area are affected by the salt-laden winds from the sea and soil and water salinity because of Buckingham canal. Isukapalli has no shelterbelt and remaining ones have very sparse, narrow shelterbelt. The lesser growth in shelterbelt of *Casuarina* is probably due to soil salinity. Therefore, it is essential to raise certain salt tolerant species as *Tamarix aphylla, Prosopis juliflora, Pandanus* spp

DISCUSSION

Wind speed reduction windward and leeward of wind breaks and shelterbelt has been widely reported (Van Eimern *et. al.*, 1964; Baier, 1971). Bates (1944) concludes 'the values of the protection afforded, increase as the square of the length until the latter is 24 times the height'. According to Naegeli (1953), an optimum ratio of length to height of the shelterbelt is to be achieved. Blenk and Trienes (1955) propose that the protective role of shelterbelt becomes effective, if the belt is twice as long as height. Wherever the shelterbelt occurs along the coast of Andhra Pradesh, the minimum continuous length is 400 m with a tree height of 7 to 20 m (ratio of length to height varies from 20 to 55) thus exceeding critical safety levels suggested by all the above investigators. These earlier scientists hold that the wind velocity is reduced by windbreaks on the windward side over a distance which is 5 times of height of wind break and on the leeward side over a distance of 15-20 times the height of the windbreak. The results of present investigation more or less coincide with these observations. Reduction of wind velocity immediately on the lee is maximum (99.01% for the 3rd location in Visakhapatnam district) with the shelterbelt of 10 to 12 m height and 120 m width. In the 2nd and 4th locations the corresponding reductions are 90 and 88.0% respectively. The reduction in wind velocity at 80 to 100 m on the lee is more in location 4 than in 2 and 3 probably because of greater width and adequate height. A similar trend is also observed at 300 to 400 m behind the shelterbelt in three locations. Thus, on the basis of standards mentioned above, efficiency of shelterbelt is appreciable as revealed by the percentage of reduction in wind velocity in the present survey. It should be mentioned that the present survey records effect of shelterbelt on normal wind velocities but not on the storm wind velocities. When the windward velocities are low, the reductions in the velocities on the lee probably may not be as clear as they are when the windward velocities are high. However, some reduction is clearly observed in the present survey in the former conditions.

Heisler and De Walle (1988) have also recorded measurable reductions in wind speed on the lee to a distance as far as 50 times the height (h) of shelterbelt. Even though no significant correlation between distance on the lee, measured in terms of height of shelterbelt, and reduction in wind velocity could be traced along Andhra Pradesh coast, a general reduction in velocity with increase in distance could be noticed. The conclusion of Heisler and De Walle (1988) that single row tree breaks can provide wind reductions as great as multiple row windbreaks sounds rather far fetched, because, larger the number of rows in a shelterbelt, greater is the reduction on the downwind. It is observed in the present investigation and also earlier researchers (Baier, 1971; 1937; McNaughton, 1988; Naegeli, 1953, Sturrock, 1988). Heisler and De Walli (1988) seem to contradict their own conclusion as they mentioned that the 'efficiency of successive windbreaks is less than that of the first row of windbreak.' The statement reveals that the successive windbreaks definitely have reducing impact on wind speed. The wind in wider shelterbelt travels longer distances encountering larger number of obstacles with reduction in wind velocity at every obstacle. Therefore, wider the shelterbelt, greater is the reduction in wind velocity when height is optimum. It shows multiple rows have far greater effect than single row trees.

McNaughton (1988) coined the term 'quiet zone' to designate a triangular region bounded by a line from the barrier top to the ground at about 8 h, behind the long windbreak. The 'zone of influence' of Naegeli (1953), 'dead calm area' of Kreutz (1956) and 'quiet zone' of McNaughton (1988) overlap each other, probably, with minor differences in length of the zones on the lee, experiencing reduced wind velocities. They also more or less overlap the 3 - 5 m distance on the downwind in the present investigation.

On the basis of above discussion, the shelterbelt, in coastal context, can be defined as a specific afforestation strategy, wherein, stress tolerant, salt and sand-spray resistant, deep rooted tree species, adaptable to coastal meteoro-edaphic conditions, are raised at the required spacing, normal to the wind direction to reduce wind velocity on the leeward side, and also reduce sand shift, salt spray, and also sea water surges. These windbreaks of vegetation are called as permeable windbreaks as wind permeates through inter-arboreal continua. The reduction of turbulent exchange processes associated with the wind speed reduction from windward through shelterbelt to leeward and changes in wind direction result in a chain of reactions. They modify other elements such as microclimate, humidity, evapo-transpiration, air and soil temperature.

CHARACTERISTICS OF SHELTERBELT SPECIES

Shelterbelt species especially along the coast should be ecologically psammophytic or halophytic or of mangrove type (in general xerophytic) depending on the climatic factors. They must be profusely branched and physiologically efficient to thrive on nutrient imbalance/deficient soils because the coastal environment is a conglomeration of above conditions. Moreover, the plants should be tolerant to soil and water salinity and also to eolian salt and sand spray.

The leaf exomorphy and morphometry of the canopy play a significant role in sieving off the salt spray and sand particles. Commonly broad-leaved species are not preferred for coastal shelterbelt as the salt spray and sand particles get deposited on the laminae. The laminae suffer scorching effect because of insolation and the thermophilic property of salts and silica which leads to defoliation. Fast growing species, with dense spreading canopy with profuse deciduous minute leaves but with equally efficient foliage regeneration, quick coppicing capacity and deep efficient soil-binding root system, are preferable as shelterbelts. These characters are found in *Casuarina equisetifolia* with scaly leaves and numerous, densely ramified and cylindrical thin phylloclades. The plant is inherently immune to eolian silica deposition because of its inherent silicification of its epidermis. It has a well-spread deep root system. It is interesting to note that this species is widely cultivated as shelterbelt along the sandy soils of the East Coast under tropical semi-arid conditions. The hardy *Prosopis juliflora*, another ideal shelterbelt species has a densely branched canopy, small deciduous leaves and frequent leaf fall. It is found growing wildly in sandy and swampy/sandy soils of the West Coast of India. It spreads so fast that its removal becomes problematic. But this species does not grow as tall as that of *C. equisetifolia*. As the concept of shelterbelt is based upon the relation between the height of the shelterbelt and the distance up to which the wind velocity is reduced on the leeward side, *C. equisetifolia* with a greater height is more efficient and suitable to the East coast experiencing frequent cyclones. *P. juliflora*, with a less height, is considered suitable to the West Coast experiencing lesser frequency of less severe cyclones. It is ecologically versatile and inherently adaptable to various edaphic such as sandy soils, saline areas, swampy regions, and locations with industrial pollutants. *C. equisetifolia*, however is not suitable to saline soils and waterlogged soils.

Tamarix aphylla and *Pandanus* sps. are other species adaptable to sandy soils, marshy soils and swamps. *Pandanus* spreads fast with the help of stilt roots and can mitigate the impact of salt-laden wind and this has the potentiality to form a dense shelterbelt, especially along the extreme coast. The fast and widely spreading aerial stilt roots of *Pandanus* not only bind soil but also mitigate the velocity of surging waters. These species can serve as first line of defence. Its leaves are broad succulent and leathery and hence tolerate salt, sand spray and solar radiation. But it is yet to be tried as a shelterbelt species

The hardy fast growing species *Acacia auriculiformis, A. tortilis, A. nilotica, Eucalyptus terticornis, E. canaledulensis, Salvadora persica* etc., can be grown as second and successive lines of defence. They provide fuel and pulp. Conventional food or horticultural crops can be raised on the wind shadow lee. Thus, use of a shelterbelt can be of two-fold: eco-conservancy and economy. Sturrock (1988) emphasized the role of complimentary species in the shelterbelt in the disposition of some of the wind energy before reaching the taller species. The significance of trees with flexible branches as shelter constituents has also been stressed by Sturrock (1988), as they store and release the energy breaking down the larger eddies into smaller ones. *C. equisetifolia*, with rigid primary and secondary branches and flexible phylloclades of the last order can be considered as a semi flexible species, whereas *Tamarix* sps. and *P. juliflora* are rigid. These rigid or semi-flexible species followed by flexible species, such as *E. teriticornis* and *A. auriculiformis* would constitute a near ideal shelterbelt.

Along with tree species it is advisable to raise floor/carpet vegetation of herbs/grasses on the sand dunes. A fully developed herbal cover form a dense felt that controls the sand shift, resulting in dune stabilisation. The herbal cover protects the soil against insolation and reduces evaporation. The herbs adaptable to these coastal eolian sand dune environment are *Ipomea biloba; Spiniffex squarrosus, Aristida* sps, *Antigonon leptopus, Cenchrus ciliaris, C. setigerous, Cynodon dactylon* etc., and the legumes *Tephrosia purpurea, Indigofera tinctora, Cassia auriculata* etc. The herbs check the wind at surface level and trap oncoming supplies of sand. They help in enriching the organic content through their dead and decaying organic matter as they mostly undergo progressive degeneration of older parts. These herbs should be raised at least during three successive monsoons, because the aerial parts of these herbs initially succumb to insolation above and thermophilic sand below. The root system regenerates and produces new shoots after the onset of monsoon.

Shrubs like *Calotropis procera, Lantana camara, Jatropha Curcass, Agave americana* etc., are grown as the second tier of vegetation as they are good sand binders. It is interesting to note cashew (*Anacardium occidentale*) when grown on sandy soils, branches profusely, possess wide and dense canopy and the branches more or less droop to the soil level providing protection to the substratum against insolation, reducing evapo-transpiration, which favour natural undergrowth of herbs and grasses.

SHELTER AS A BIO-ENERGY BELT

According the Sturrock (1988) in both temperate and tropical regions the realisation that shelter and timber are compatible objectives is slow. The tropical climate of India is more prone to cyclonic weather and hence shelterbelt is an obligatory aspect in eco-conservancy. Cabron (1965) advocated the compatibility of shelter to timber productionin Britain. Similarly Bagley (1976) pointed out the neglect of timber potential of windbreaks. Chavasse (1984) in his review shows that efficient management to produce optimum yield of wood from shelter could make a substantial contribution to farming income and national productivity in addition to shelter value. The fuel, timber, and climatic importance of shelterbelt are incorporated in 'shelter cum bio-energy belt'. It has high utility value for the coastal belt of Andhra Pradesh, which is predominantly rural. The species selected and their total dry biomass yield along the Andhra Pradesh coast is given in the Table-4 (Vasudeva Rao et. al., 1990). Espacement is very important especially, when the thinning is to be done because the initial and final spacings should efficiently serve the shelterbelt and fuel wood purposes. *Casuarina equisetifolia* is more efficient as a source of fuel wood, even though it has lesser calorific value than *Prosopis juliflora*. But the former has a greater yield of biomass. Its higher efficiency as a shelterbelt specific exomorphism indicates its effectiveness as shelter cum bio-energy belt species. *P. Juliflora* because of its adaptability and profuse litter fall, remarkable copping capacity, reasonably good yield and high calorific value is also a good shelter cum bio-energy belt species. All the four species mentioned in Table-IV can meet the fuel needs of the villages when they are raised with definite strategy on the basis of quantum of the local demand. Therefore, an efficient management to get optimum yield of wood from shelter would make a substantial contribution to farming income and national productivity in addition to shelter value. The ecological, economic and social benefits due to shelterbelt along the coast are given in figure 2.

SUMMARY AND CONCLUSIONS

Subtropical semiarid coast line of Andhra Pradesh experiences highest land falling of cyclones and is predominantly rural and agro-based. Therefore, a coastal shelterbelt of plants is one of the effective eco-conservancy measures for economic and ecological sustainability. The swampy coastline of Krishna district and the open sandy shores of Nellore and Srikakulam districts suffer cyclonic impacts more frequently. Coastal physiography of other districts and the well-grown shelterbelt are probably responsible for mitigating cyclonic impact.

Shelterbelt in all the surveyed coastal areas of Andhra Pradesh is found to be effective in reducing the wind velocities from windward to leeward sides. The effect of shelterbelt in staggering rows is more effective than in parallel rows. The morphological characteristics and canopial topography of shelterbelt species are discussed. The definition of the shelterbelt is modified with reference to coastal environment. The efficiency of shelterbelt species in meeting the fuel needs of the local rural people has been emphasised on the basis of the biomass yield and calorific value

Table—4 Height and Dry Biomass Yield of Certain Fuel Wood Species of Shelterbelt Importance at the End of of 5 Years*

Plant name	*Spacing*				*Calorific value (Kilo cal/kg)*	*Density of wood (gms/cubic cm)*
	1.0 x 1.0 meters		*1.5 x 1.5 meters*			
	Height (m)	*Dry biomass weight (t/ha)*	*Height (m)*	*Dry biomass weight (t/ha)*		
Casuarina equisetifolia	10.5	120	10.0	60	5000	0.781
Eucalyptus tereticornis	12.0	78	14.0	72	4800	0.680
Acacia auriculiformis	13.0	90	12.0	70	4800	0.750
Prosopis juliflora	5.3	88	5.2	60	5200	0.700

***Source :** Vasudeva Rao, et. al., 1990.

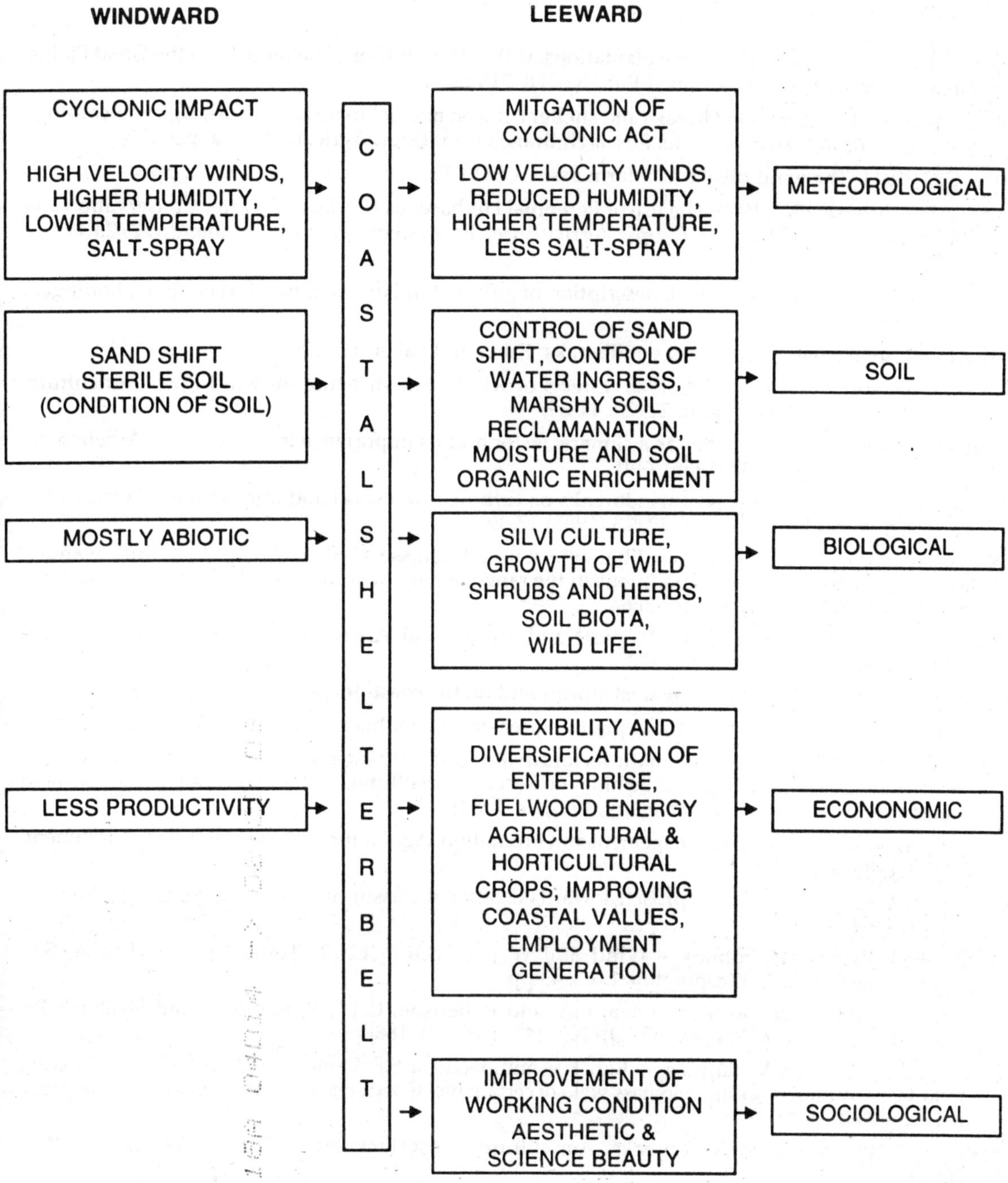

Fig. 2 **Ecological, Economic and Social Benefits due to Shelterbelt along the coastline**

REFERENCES

Bagley, W.T., 1976. Multipurpose tree plantations: R.W. Tinus (Editor), Shelter belts in the Great Plains. Great Plains Agricultural Council Pub. No. 78, 218 pp.

Baier, R.W. 1971. Effects of wind breaks and shelter belts on microclimate and plants, Agrometeorology section, Plant Research Institute, Research Branch Can. Dept. Agric.m Ottawa, pp. 173

Bates, C.G., 1944. The wind breaks as a farm asset. USDA Farmers' Bull. No. 1405, 22 pp.

Black, A.L. and Aase, J.K. , 1988. The use of Perennial Herbaceous Barriers for Water Conservation and the Protection of Soils and Crops. Agriculture, Ecosystems and Environment, 22/23 (1988) 135-148.

Blenk and Trienes, 1955. Wind break description of airflow Ein Film, Institute f. Wiss. Film, bottingen, Film No c 704 (1955)

Cabron, J.M., 1965. Shelter belts and windbreaks. Faber and Faber, London.)

Heisler, G.M. and De Walle, D.R., 1988. Effects of windbreak structure on wind flow. Agriculture, Ecosystems and Environment, 22/23: 41-69.

Kreutz, W., 1956. The wind shelter as a climatic factor and its importance in Agriculture, Arbeiten de; Deutsch. Landw. Ges. 37 (1956) 7-40.

Mcnaughton, K.G., 1988. Effects of windbreaks on turbulent transport and microclimate. Agriculture, Ecosystems and Environment, 22/23 (1988) 17-39.

Naegeli, W., 1953. Untersuchungen iiber olie winverbaltnisse in Bereich von Schilbrohrwanden (Investigations on the wind conditions in the range of narrow walls of reed). Mitt. Sehweiz. Anst. Forstl. Versurchswegen, 29: 213-266.

Nicholas, C.D., 1988. Plantings in Tropical and Subtropical Areas. Agriculture, Ecosystems and Environment, 22/23 (1988) 465-482.

Reddy, C.V.K., 1979. Shelter belts against storms and on the coast. Indian For. 105: 720-726.

Singh, S. P., 1982. Some thoughts about planning coastal shelter belts. Indian For., 108: 402-407.

Subrahmanyam, V. P., 1983. Agroclimatic factors in the Southern States and Orissa as seen from data of meteorological stations. Regional Seminar on Recent Agricultural Production Trends in the Southern States and Orissa, Andhra University, Waltair (1983).

Sturrock, J.W., 1988. Shelter: Its Management and Promotion. Agriculture, Ecosystems and Environment, 22/23 (1988) 1-13.

Thornthwaite, C.W., 1948. An approach towards a rational classification of climate, Geogr. Rev., 38 (1948) 55-94.

USDA, 1982. Erosion Handbook – Wind and Water, Section EH-2 (Texas edition). USDA, Soil Conservation Service, Washington, D.C., 43 pp.

Van Eimern, J., Karechon, R., Razumova, L.A. and Robertson, G.W., Windbreaks and Shelter belts. WMO Technical Note No. 59 (WMO)- No. 147. T. P. 70), 188pp.

Vasudeva Rao, P.H.V., B.V. Nageswara Rao, E. Ammineedu, Y.S.S. Rameswara Rao and K. Sai Prasad., 1990. Canopy reluctance and biomass yield of certain fuel wood species. World Renewable Congress, Reading, U.K. p. 23-28.

Wight, B., 1988. Farmstead windbreaks. Agriculture, Ecosystems and Environment, 22/23 (1988) 261-280.

World Meteorological Organisation., 1964. Windbreaks and Shelter belts. Report of a Working Group of the Commission for Agricultural Meteorology. WMO Technical Note No. 59 ((WMO No. 147, T.P. 70).

14

Solid Waste Disposal Practices: Implications on Groundwater Pollution

V.V.S. Gurunadha Rao, P. Nandakishore**, K. Srinivasa Reddy** & S. Sankaran

ABSTRACT

Solid waste disposal has become a problem as nature of wastes are becoming complex in nature. The municipal and environment protection authorities are facing problems of how safe disposal of a particular waste can be achieved while protecting the environment. Landfills are the most common means of disposing of municipal refuse and garbage and sludge from industrial wastewater treatment facilities. Rainfall percolates through the landfill and generates leachates. The existing municipal solid waste disposal sites around Hyderabad city are not engineered landfills. One of the Municipal Solid Waste landfill covers an area of 19 ha and receives about 800 MT/day of municipal solid waste. The thickness of garbage varies from 2 – 8 m. On the basis of hydrological data of 20 years, leachate to be generated within the dumpsite is computed using the HELP (Hydrological Evaluation of Landfill Performance) model. Another landfill site covers 14.5 ha and receives approximately 350 MT/day of municipal waste. The leachate generated from the two landfills has been computed. Suggestions are made for carrying out integrated hydrogeological, geophysical and numerical modeling studies for designing engineered landfills for a hazard free solid waste disposal, while protecting the groundwater regime.

INTRODUCTION

Quantity of waste has increased immensely with increasing industrialization, technical development and economic growth in all major cities. The nature and composition of waste have

* National Geophysical Research Institute, Uppal Road, Hyderabad–500007 India

** EPTRI, Survey No. 91/4, Gachibowli, Hyderabad–500019 India changed over decades, reflecting industrial and domestic practices in developing countries. Many types of waste materials are produced by society, of which domestic, commercial, industrial, mining and radioactive wastes probably are the most notable. The waste can be regarded as non-hazardous and hazardous. Waste may take the form of solids, sludge, liquids and gases or any combination thereof. Depending on the source of generation, some of these wastes may degrade into harmless products, whereas others may be non-degradable and/or hazardous, thus posing health risks and environmental problems if not managed properly. A further problem is the fact that deposited waste can undergo changes through chemical reaction, resulting in the production of dangerous substances. Disposal of wastes at appropriate locations is one of the fundamental tasks of environmental protection.

As waste products differ considerably from one another, the storage facilities they require also differ. Solid, non-reactive, immobile inorganic wastes can be disposed off at above ground disposal sites. To provide long-term isolation from the environment, high toxicity non-degradable wastes should be disposed off underground if they cannot be burnt. The best method of disposal is determined on the basis of the type and amount of waste, on one hand, and the geological conditions of the waste disposal site on the other.

The primary task of the site exploration, which follows, is to determine the geological and hydrogeological conditions. Their evaluation provides the basis of the models used to test the reaction of the system to engineering activities. The chemical analysis of rocks, may help to yield information about its origin, and hence about the future development of site. At the same time, the leaching capacity of the water is determined which allows a prediction to be made of the reactions between wastes and soil or rock. If groundwater must be protected, from highly mobile toxic or very slowly degradable substances present in wastes, impermeable liners may be used to inhibit the infiltration of leachate into the surrounding ground. In terms of waste disposal by landfill, a landfill is environmentally acceptable if it is correctly engineered. Unfortunately, if it is not constructed to sufficiently high standards, a landfill may have an adverse impact on the environment.

IMPLICATION

Surface water or groundwater pollution may occur due to leaching action of wastes by rainfall. Consequently a physical separation between waste, on the one hand and groundwater and surface water on the other, as well as an effective surface water diversion drainage system are fundamental to design. Domestic refuse is a heterogeneous collection of almost anything (i.e. waste food, garden rubbish paper, plastic, glass, rubber, cloth ashes, building waste, metals etc.) much of which is capable of reacting with water to give a liquid rich in organic matter, mineral salts and bacteria, namely leachate. Leachate is formed when rainfall infiltrates a landfill and dissolves the soluble fraction of the waste, including the soluble products obtained as a result of chemical and biochemical processes occurring within decaying wastes. The organic carbon content of waste especially is important, because this influences the growth potential of pathogenic organisms.

Any assessment of the state of a landfill and its environment must take into consideration the substances present in a landfill, their mobility at present and in the future, the potential pathways along which pollutants can travel and the targets potentially at risk from the substances involved. As far as the location of a landfill is concerned, site selection and project extent, finance, construction materials and site rehabilitation have to be studied. Important requirement in planning a landfill site is to establish exactly, by survey and analysis, the type, nature and quantities of waste involved.

MUNICIPAL LANDFILLS

Landfill is the most common means of disposing of municipal refuse, ashes, garbage and sludge from municipal and industrial wastewater treatment facilities. Precipitation that infiltrates into the waste dump can mix with liquid already present in the waste and may leach compounds from the solid waste. Leachate from the landfill can percolate from the site to the water table and could cause groundwater contamination. Leachate often contains high concentration of dissolved organic substances, resulting from the decomposition of organic material, such as vegetable matter and paper. The concentration of chemically reduced inorganic substances such as ammonia, iron and manganese, varies accordingly to the hydrology, chemical and physical conditions at the site. As an example, Brown et al (1972) calculated that 1000 m^3 of waste can yield 1.25 tonnes of potassium and sodium, 0.8 tonnes of calcium and magnesium, 0.7 tonnes of chloride, 0.19 tonnes of sulphate and 3.2 tonnes of bicarbonate. Barber (1982) estimated that a small landfill site with an area of 1 ha located in southern England could produce up to 8 m^3 of leachate during November to April per day for a rainfall of 900 mm/year, assuming that evaporation was close to the average for the region and runoff was minimal. Gray et al (1974) considered that the major criterion for the assessment of a landfill site as a serious risk was the presence of toxic or oily liquid waste, although sites on impermeable strata often merited a lower risk depending on local conditions.

When leachate from a landfill mixes with groundwater, it forms a plume that spreads in the direction of flow through the processes of advection and dispersion. The movement of contaminants in the landfill is facilitated by

- Prevalence of humid climate in the region
- Shallow water table
- Thin soil cover over the permeable bedrock
- Highly permeable rocks such as gravels
- Greater density of the source of contaminants

The volume of leachate produced is a function of the amount of water percolating through the refuse. Land disposal of solid waste in humid areas is more likely to produce large volume of leachate than land disposal in arid regions. The vadose zone in arid regions may receive little or no groundwater recharge. Under such conditions solid waste disposal is not likely cause much groundwater contamination.

Initially potential life of a landfill is estimated based on the waste survey, forecast. Its distance from the proposed waste catchment area is assessed. The character of the material, which it has to accommodate, influences the design of a landfill. The ideal landfill site should be acceptable based on hydrogeological conditions, should not be posing a potential threat to water quality, should be free from running or static water and should have a sufficient store of material suitable for covering each individual layer of waste.

It should be situated at least 200 m away from any residential area. The position of water table is important, as it determines whether wet or dry tipping is involved.

The position of water table also determines the location at which flow is discharged. The hydraulic gradient determines the direction and velocity of the flow of leachates when they reach the water table, and also influences the amount of dilution that leachates undergo.

Barber (1982) identified three classes of landfill sites based upon hydrogeologic criteria. When assessing the suitability of a site one of the principal considerations is the ease with which the pollutant can be transmitted through the substrata and the distance it is likely to spread from the site. Consequently, the primary and secondary permeability of the formations underlying a potential landfill area are of major importance. Generally, the third category, in which the leachate is contained within the landfill area is preferred.

Because all natural materials posses some degree of permeability, total containment only can be achieved if an artificial impermeable lining is provided over the bottom of the site. However, there is no guarantee that clay, soil-cement and plastic liner will remain impermeable permanently. Thus the migration of the contaminant from the landfill site into the substrata will occur eventually, although the length of time is uncertain.

In some instances, the delay will be sufficiently long for the pollution potential of the leachate to be effective. One of the methods of tackling the problem of pollution associated with landfills is by the dilution and dispersal of the leachate. Otherwise, leachate can be collected by internal drains within the landfill and conveyed away for treatment. One of the difficulties in predicting the effect of leachate on groundwater is the continued change in the characteristics of the leachate as the landfill ages.

Leachate may be diluted where it gains access to surface water runoff or groundwater, but this depends on the quantity and chemical characteristics of the leachate, as well as the quantity and quality of the receiving water.

DEGRADATION OF WASTE IN LANDFILLS

Leachate is formed by the action of liquids, primarily water within a landfill. The generation of leachate occurs once the absorbent characteristics of the refuse are exceeded. Initially, the decomposition of waste is aerobic. As decomposition proceeds, the oxygen in the waste rapidly becomes exhausted, and the conditions become anaerobic. Initially, complex organic materials are broken down into simple organic substances typified by various acids and alcohols. The nitrogen present in the organic material tends to get converted into ammonium ions, which are soluble readily and give rise to significant quantities of ammonia in the leachate.

The sulphate in the landfill may be reduced bio-chemically to sulphides. These sulphides tend to remain in the landfill as highly insoluble metal sulphides. During stages of landfill, dissolved salt content may exceed 10000 mgl^{-1} with relatively high concentration of sodium, calcium, chloride, sulphate or iron, whereas as a landfill ages, the concentration of inorganic materials usually decreases. Second stage of anaerobic decomposition involves the formation of methane. Methanogenic bacteria use the end products from the first stage of anaerobic decomposition to produce methane and carbon dioxide.

ATTENUATION OF LEACHATE

Physical and chemical processes are involved in the attenuation of leachate in soils. These include precipitation, ion exchange, adsorption and filtration. In the reducing zone, where organic pollution is greatest, insoluble heavy metal sulphides and soluble iron sulphides are formed. In the area between the reducing and oxidizing zones, ferric and inorganic hydroxides are precipitated. Clay minerals and humic materials in soil increase ion exchange capacity. Adsorption is susceptible to changes in the pH value as the pH affects the surface charge on a colloidal particle or molecule. Hence, at low pH values, the removal rates due to adsorption can be reduced significantly.

However, the removal of pollutants by the ion exchange and adsorption can be reversible. For example, after high-level pollution from a landfill subsides and more dilute leachate is produced; soil and rock may release the pollutants back into the leachate. Pathogenic microorganisms, which may be found in a landfill usually do not travel far in the soil, because of the changed environmental conditions.

SURFACE AND GROUNDWATER POLLUTION

Leachates contain many contaminants that may have a deleterious effect on surface water. If, for example leachate enters a river, oxygen is removed by decomposition of the organic compounds in the leachate. In cases of severe organic pollution, the river may be completely depleted of oxygen, with disastrous effects on aquatic life. When dissolved oxygen levels reduce below 2 mgl^{-1}, most fish cannot survive. The principal inorganic pollutants which may cause problems with leachate are ammonia, iron, heavy metals and to a lesser extent, chloride, sulphate, phosphate and calcium. Ammonia can be present in landfill leachates up to several hundred milligrams per liter. Ammonia is toxic to fish at concentration of 2.5 – 25 mgl^{-1} and may alter the ecology of the river.

Leachate containing ferrous iron is objectionable in a river, because ochreous deposits are formed by chemical and biochemical oxidation of the ferrous compounds to ferric compounds. The turbidity caused by the oxidation of ferrous iron can reduce the amount of light, leading to decrease in the number of flora and consequently fauna. Heavy metals are also toxic to river organisms. Physically leachate affects the river quality in terms of the suspended solids, color, turbidity and temperature. The suspended solids, color and turbidity reduce the light intensity in a river which can affect the food chain. Decrease in photosynthetic activity by plants reduces the oxygen replenishment in a river.

If a leachate enters the phreatic zone, it mixes and moves with the groundwater. The organic carbon content in the leachate leads to an increase in the biochemical oxygen demand (BOD) of the groundwater, which may increase the potential for the reproduction of pathogenic organisms. Organic matter is stabilized slowly, because the oxygen demand may deoxygenate the water rapidly, and usually no replacement oxygen is available. Under anaerobic conditions, metals such as iron and manganese may dissolve in the water causing further problems. If the groundwater has a high buffering capacity, the effects of mixing of acidic or alkaline leachate with the groundwater are reduced. The velocity of groundwater flow is another factor. A high velocity gives rise to more dispersion in the direction of flow and less lateral dispersion, so that the leachate plume forms a narrow cone in the direction of groundwater flow. Low groundwater velocity leads to a wider plume.

Ideally, a groundwater-monitoring program should be established prior to tipping, and should continue at least up to 20 years after closure of the site. The number, location and depth of monitoring wells depend on the particular site. Monitoring should include sampling from the wells and analysis for pH, chlorides, dissolved solids, organic carbon and the concentration of any particular hazardous waste that has been tipped in the landfill.

The most serious effect of a leachate on groundwater is mineralisation. Much of the organic carbon and organic nitrogen is biodegraded as it moves through the soil. Heavy metals may be attenuated due to ion exchange onto clay minerals. Mineralisation is brought by inorganic ions, such as chloride. The concentration may only be reduced by dilution. Continued input of inorganic ions into potable groundwater eventually means that groundwater becomes unfit for drinking.

HAZARDOUS WASTE

Co-disposal, that is mixed with very much larger quantity of 'non-hazardous' domestic waste are buried in disused quarries, clay pits and other convenient holes in the ground. The objective of co-disposal are to absorb, dilute and neutralize liquids, and to provide a source of biodegradable materials to encourage microbial activity. In other words, co-disposal makes use of attenuation processes in a landfill to minimize the impact of hazardous waste on the environment.

Current state of the art hydrogeologic and engineering practice for determining the locations of municipal waste landfill sites calls for careful hydrogeological and geophysical analysis of several alternative sites in order to identify low permeability formations, which have low groundwater contamination potential. Since in hard rock areas the secondary porosity like joints and fractures play a significant role in the groundwater movement, care should be taken during the site selection so that area is devoid of such structural features.

LANDFILL DESIGN

Landfills must be designed in a appropriate way considering the following aspects

- It should minimize the formation of leachate and also the amount of leachate escaping from the landfill
- Leachate may also be collected and treated in situ
- It should not interfere with the water table in the region
- Clay minerals attenuate the movement of contaminants because of ion exchange reactions and adsorption properties
- By constructing a landfill in the unsaturated zone the movement of the leachate is greatly reduced.
- Periodic monitoring of the water discharging from the under drains is needed to determine the impact of portion of the leachate that drains through the liner.

The movement of moisture in a landfill represents unsaturated flow through homogeneous porous media. The applicability of this theory has been successfully demonstrated in the laboratory studies. The risk of pollution by leachate is directly proportional to the quantity and type of chemicals found at the site and to the precautions taken to prevent its gaining access to groundwater. Pollution of groundwater due to landfills can best be avoided by proper site selection and interposing a physical barrier which will prevent the leachate from reaching the groundwater table.

It is desirable in most cases to construct landfills in deep water table areas. Most of the municipal solid waste disposal sites in the country are based on natural attenuation process of leachate. A natural attenuation landfill is one that relies totally on natural processes to attenuate any leachate originating from it. Such landfills should be well above the water table to promote maximum attenuation of leachate in the vadose zone.

EXAMPLE MUNICIPAL LANDFILLS STUDY

Twin cities of Hyderabad and Scunderabad and surrounding urban agglomerations approximately produce 2200 metric tones of municipal solid waste everyday. Few existing municipal solid waste disposal sites are not selected on a scientific basis and mostly were selected based on wasteland available and proximity to the habitations. The refuse is dumped in heaps above the ground level ranging from 1.5 to 8 m high. All the dumpsites are open and have not been engineered. Indiscriminate selection of dumpsites are causing severe problems in terms of unhygienic conditions in the surroundings, resulting in contamination of air, surface and ground water. The waste disposal sites selected with the twin cities comprises of country rocks like granite gneisses and intrusive bodies like dolerite dykes. The granite present in the area is pink and gray granite with grain sizes ranging from fine to coarse with porphyritic texture. One site is situated on plain land while the other is located on a shallow buried valley fill in the Hyderabad area.

Municipal Solid waste (MSW) landfill in Autonagar is located in an area of 19 ha and receives approximately 800 metric tons of municipal solid waste everyday. The area is located on the right bank of Musi River about 2 km away. The quantity of solid waste dumped so far is approximately 450000 tones (EPTRI, 2001). The thickness of garbage at this site varied from 2 to 8 m. The normal annual rainfall around Hyderabad is about 750 mm. The leachate generated within the Municipal Landfill is computed using average annual rainfall data of past 20 years at Begumpet airport, Hyderabad and considering the hydrogeological characteristics viz., permeability of different layers within the landfill, their moisture holding capacity, and total water availability from rainfall for leachate generation in the HELP (Hydrological Evaluation of Landfill Performance) model. The model output shows that about 40% of the average annual rainfall is available for evapo-transpiration (mostly soil evaporation only) and about 10% of rainfall for soil moisture storage within the landfill. Thus the average quantum of leachate that is being generated from the Municipal Landfill at Autonagar has been worked out to be 71000 m^3/year. The model considers the hydraulic conductivity of local formation as well as the solid waste dump. The chemical analyses of leachate collected from dumpsite showed high values of TDS, COD, BOD, NO_3 and other chemical constituents. The groundwater quality in the surrounding areas is found deteriorating and one of the reasons could be attributed to the presence of open MSW landfill.

The second MSW landfill site at Golconda is spread in an area of 14.5 ha and receives about 350 metric tons of municipal waste everyday. The area serves the western part of the study.

A tank is situated in the downstream area of the dumpsite. Risk due to leachate seepage is very high in post monsoon period because of presence of shallow water table. The amount of leachate that could be generated is calculated from HELP model as ~ 58000 m^3/year. Water quality analyses indicate similar chemical constituents at the site.

CONCLUSIONS

The selection of landfill site for a municipal waste requires a study of the geological and hydrogeological conditions. Domestic waste can be tipped at dry sites on granular material, which has a thickness of at least 15 m. Water supply sources should be located at least 0.8 km away from the landfill site. As far as potentially toxic waste is concerned, it is best to contain it, Gray et al (1974) suggested that such sites should be underlain and confined by at least 15 m of impermeable strata, that any source abstracting groundwater for domestic purpose should be 2 km away. The topography of site should be such that runoff can be diverted from the landfill. Containment can be achieved by an artificial impermeable lining placed on the base of the site. Still some leachate will eventually leak into the subsurface. In some instances, this will be sufficiently long for of pollution to create problems.

Groundwater monitoring system must be designed by understanding the hydrodynamics of contaminant propagation, the nature of the controlling flow mechanism (vadose zone, saturated flow) and the aquifer characteristics (permeability, porosity). At least three well fields could be monitored one in the up gradient for background water quality, one just close to the source and other one little away from the source in the down gradient. Under favorable conditions resistivity method can be used to determine the boundaries of a plume of contaminated groundwater. Vertical electrical soundings are used to determine top and bottom of contaminant plume. Resistivity profiling has to be carried out to determine the lateral extent of the contaminant plume. In this way a quantitative assessment of groundwater contamination from landfills can be made.

REFERENCES

Barber, C., 1982. Domestic waste and leachate. Notes on water research No.31. Water Research Centre. Medmenton, UK.

Brown, R.H., Konophyantsev, A.A., Ineson, J & Kovalersky, V.S., (eds.) 1972. Groundwater studies. An international guide for research and practice, studies and reports in hydrology, 7 UNESCO, Paris

Environmental Protection Training and Research Institute (EPTRI), 2001. Environmental status around Autonagar and Golconda municipal solid waste dumpsites in Hyderabad city, pp. 85. EPTRI, Hyderabad.

Gray, D.A., Mather, J.D. and Harrison, I.B. 1974. Review of groundwater pollution for waste sites in England and Wales with provisional guidelines for future site selection., Quarterly Journal of Engineering Geology, 7, p.181-196.

HELP Model, Unsat Suite, Waterloo Hydrogeologic Inc, Waterloo, Canada.

15

Environmental Impact Assessment Studies of Groundwater Regime Around a Proposed Alumina Plant in Doragarha, Rayagada District, Orissa

V.V.S. Gurunadha Rao,
B.A. Prakash & R.L. Dhar

INTRODUCTION

An alumina plant is proposed to be established in Doragarha village, Rayagada District, Orissa, India by Ms. Indal – Tata group. The alumina plant produces fine red-mud residue with liquor slurry containing Sodium Carbonate (Na_2CO_3) and traces of Caustic soda (NaOH). The concentration of Na_2CO_3 in the red mud slurry may reach 3500 mg/l. The slurry is pumped to the disposal area and is spread in layers on pre-graded drying beds. Upon reaching a target dryness, each layer will be covered with a fresh layer. The water drained from the red mud stacking will be collected in ponds for recycling. Seepage of Na_2CO_3 rich liquor from the holding ponds or even through initial phase of red mud spreading itself, if significant, may result in contamination of groundwater regime. Detailed geophysical and hydrogeological investigations were carried out in the micro-watershed around Doragarha for estimating the aquifer geometry and aquifer parameters respectively (Fig. 1). A groundwater flow and mass transport model was prepared using visual MODFLOW and MT3D Computer software to assess the impact of the likely migration of the Na_2CO_3 plume from the red mud ponds towards the Baraha nadi, a perennial stream flowing in northwest of the area.

* National Geophysical Research Institute, Hyderabad- 500 007 India

The study area is located in a Khondalitic terrane of Eastern Ghats. Charnokites and granites are also present in the surrounding areas. The Khondalite suite of rocks is represented by quartz-feldspar-garnet-sillimanite gneisses (Khondalite), quartzite and granulite. The Khondalite rocks are predominant on hill ranges. The topography is highly undulating with a number of small hills and mounds with steep slopes interspersed with intermontane narrow valleys. A number of streamlets with dendritic and sub-parallel drainage are observed around the plant area. High density of drainage due to undulating topography and steep slopes of hillocks (except the red mud ponds area) restrict infiltration and eventual recharge to groundwater. As such, substantial part of rainfall goes as surface runoff to the Barha nadi, a perennial stream. The base flow in the stream is expected from groundwater effluence.

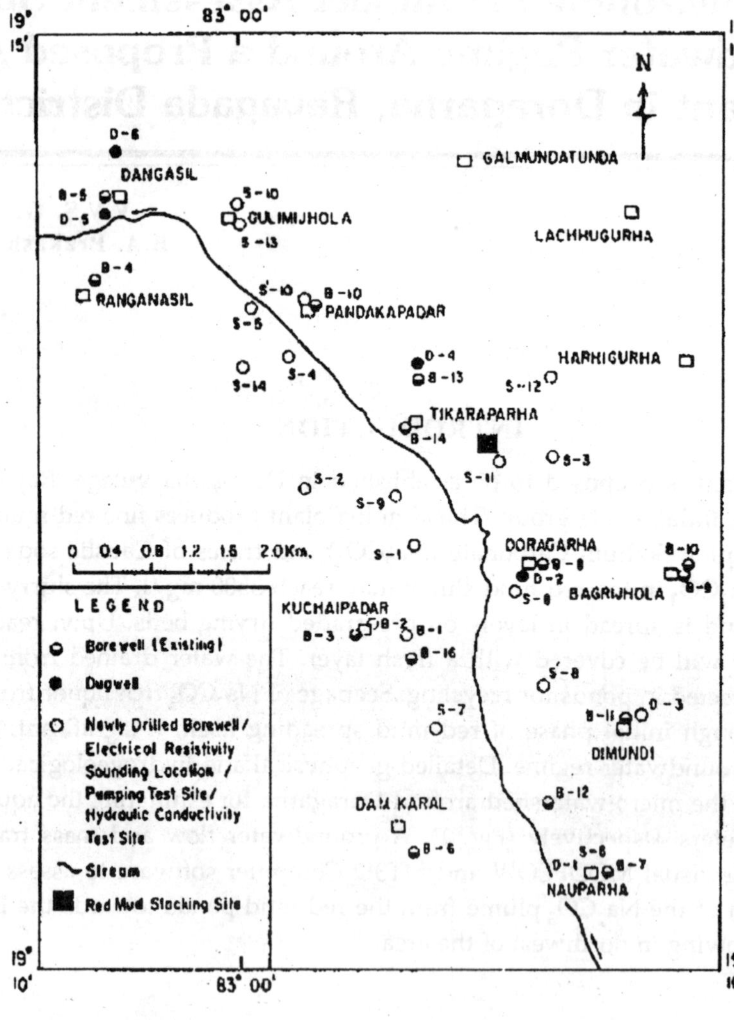

Fig. 1 **Location of proposed red-mud ponds of allumina plant around Doragarha, Rayagada district, Orissa**

CHARACTERIZATION OF HOST HYDROGEOLOGICAL SYSTEM

Groundwater occurs under water table condition in weathered zone and semi-confined conditions in deeper fractures. The thickness of weathering varies from 4-8 m. The weathered zone in Khondalite is generally more porous due to high quartz content, compared to other rock types, except in kaolinised zones. The felspathic gneis gives rise to clayey weathered zone, while Charnokites by virtue of their texture are massive in nature and are more resistant to weathering, as such the thickness of the weathered zone is less.

The over burden is predominantly clayey and its thickness close to the stream is about 20 m with some patches of deeper weathering at places. The clay thickness varies from a few meters to a few tens of meters. The subsurface (particularly below water table) may be quite impermeable as indicated by predominant clayey matrix in the lithologs of wells (Fig. 2). This is also confirmed by very low permeability values of about 1 m/day or even less which are determined from pumping tests. The water level configuration of March 1994 indicates substantial groundwater effluence to the Baraha nadi.

Near surface hydraulic conductivity measurements were carried out at four sites in the vicinity of the water collection pond using soil core samples. The hydraulic conductivity of soil cores ranges from 0.02 to 0.1 m/hr. The infiltration rates from near surface in-situ measurements were found to be lower than those from the laboratory measurements which may be attributable to the fact that the core might have got disturbed during transit. In-situ hydraulic conductivity measured at water collection pond is 0.006 m/hr. The lithologs of wells indicate that the sub-surface (particularly below water table) may be quite impermeable as evinced by predominantly clayey matrix. This is confirmed by very low transmissivity values obtained from the pumping tests conducted on the drilled wells. Generally, the transmissivity values estimated from pumping tests are of the order of 10 m^2/day and even lower (as low as 2 m^2/day). The recovery in the pumping wells is very slow (Dhar et al, 1994).

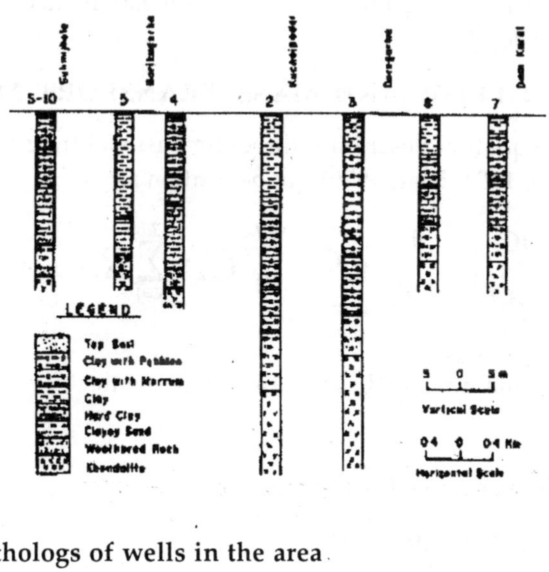

Fig. 2 Borehole lithologs of wells in the area

The permeability of the unsaturated zone is primarily controlled by the grain size distribution. However, around the red-mud stacking area, the surface exhibits cracks, which could influence the infiltration process. The development of cracks may result in compaction of the soil resulting in lowering the permeability of the small crack-free blocks. The overall infiltration rate could, however, increase. The cracks could also serve as preferred pathways for infiltration. However, the cracks seem to be confined to top soil and may close at a shallow depth. As far as the laboratory and in-situ permeability measurements are concerned, the sampling has been carried out at visibly non-cracked portions of the ground.

The water level measurements made on a number of observation wells have been utilized to prepare regional groundwater level contours of March 1994. The water level configuration indicates that the general groundwater flow direction is towards Baraha nadi. Stream flow measurements at Doragarha show a base flow of the order of 0.5 m^3/sec in Barha nadi during the lean period from January to May. The groundwater effluence to Barahanadi between Nuaparhar and Dangasil is of the order of 0.1 m^3/sec during the same period. Also the groundwater effluence of 0.1 m^3/sec from such a small area towards base flow in Barha nadi may be a result of the fact that the deeper portion of the overburden may be having low permeability and porosity. Shallow water table condition seen in the area could be due to the same reason.

Substantial part of the contaminant after percolating down to water table at shallow depths from the red mud ponds is likely to migrate laterally westwards to Barha nadi. There could be some preferred paths for groundwater flow such as cracks in the clay or fractures in the bedrock, which may also control the migration of contaminant. Electrical resistivity profiling carried out has not indicated presence of any major lineament/fracture in the area. The resistivity sounding studies indicate that the bedrock in the northern part of the red-mud stacking area may be quite shallow (about 6 m) and have the order of 30 m in the southern part. As mentioned earlier, the bedrock in the area is Khondalite. Being a hard rock, its permeability is expected to be extremely low and negligible at greater depths.

GROUNDWATER FLOW AND MASS TRANSPORT MODELING

The partial differential equation describing three-dimensional transport of contaminants in groundwater (Javandel, et. al, 1984, Bear, 1979) can be written as

$$\frac{\partial C}{\partial t} = \frac{\partial}{\partial x_i}\left[D_{ij}\frac{\partial C}{\partial x_j}\right] - \frac{\partial}{\partial x_i}(v_i C) + \frac{q_s}{\theta}C_s + \sum_{K=1}^{N} R_k \quad \text{............(1)}$$

where

C concentration of contaminants dissolved in groundwater

t time

x_i distance along the respective Cartesian co-ordinate axis

D_{ij} hydrodynamic dispersion coefficient

V_i seepage or linear pore water velocity

q_s volumetric flux of water per unit volume of aquifer representing sources (positive) and sinks (negative)

c_s concentration of sources or sinks

θ porosity of the porous medium

R_k chemical reaction term

Assuming that only equilibrium controlled linear or non-linear sorption and first order irreversible rate reactions are involved in the chemical reactions, the chemical reaction term can be expressed as (Grove and Stollenwerk, 1984)

$$\sum_{k=1}^{N} R_k = -\frac{\rho_b}{\theta}\frac{\partial \bar{C}}{\partial t} - \lambda\left[C + \frac{\rho}{\theta}\bar{C}\right] \quad \text{......................(2)}$$

where

ρ_b the bulk density of the porous medium

$\bar{C}$ the concentration of contaminants sorbed on the porous medium

λ the rate constant of the first-order rate reactions

rewriting

$$\frac{\rho_b}{\theta}\frac{\partial \bar{C}}{\partial t} = \frac{\rho_b}{\theta}\frac{\partial C}{\partial t}\frac{\partial \bar{C}}{\partial C} \quad \text{.............(3)}$$

We can rewrite equation(1) by substituting eqs. (2) and (3) as

$$\frac{\partial C}{\partial t} = \frac{\partial}{\partial x_i}\left[D_{ij}\frac{\partial C}{\partial x_j}\right] - \frac{\partial}{\partial x_i}(v_i C) + \frac{q_s}{\theta}C_s - \frac{\rho_b}{\theta}\frac{\partial \bar{C}}{\partial C}\frac{\partial C}{\partial t} - \lambda\left(C + \frac{\rho_b}{\theta}\bar{C}\right) \quad \text{......(4)}$$

Rearranging terms we get

$$R\frac{\partial C}{\partial t} = \frac{\partial}{\partial x_i}\left[D_{ij}\frac{\partial C}{\partial x_j}\right] - \frac{\partial}{\partial x_j}(v_i C) + \frac{q_s}{\theta}C_s - \lambda\left(C + \frac{\rho_b}{\theta}\bar{C}\right) \quad \text{..........(5)}$$

where R is called the retardation factor, defined as

$$R = 1 + \frac{\rho_b}{\theta}\frac{\partial \bar{C}}{\partial C} \quad \text{..................(6)}$$

Equation (5) is the governing equation underlying the solute transport model.

The transport equation is linked to the flow equation

$$v_i = -\frac{K_{ii}}{\theta}\frac{\partial h}{\partial x_i} \ldots\ldots\ldots\ldots\ldots(7)$$

where

K_{ii} a principal component of the hydraulic conductivity tensor

h hydraulic head

The hydraulic head is obtained from solution of three dimensional groundwater flow equation through MODFLOW software (McDonald and Harbaugh, 1988)

$$\frac{\partial}{\partial x_i}\left[K_{ii}\frac{\partial h}{\partial x_j}\right] + q_s = S_s\frac{\partial h}{\partial t} \ldots\ldots\ldots\ldots\ldots\ldots(8)$$

where S_s is the specific storage of the porous material.

The numerical approaches for solving the mass transport equations are based on computer-based particle tracking methods. They are approximate forms of the advection - dispersion equation (5) as a system of algebraic equations or alternatively simulating transport through spread of a large number of moving reference particles. These numerical approaches deal with variability of flow and transport parameters (hydraulic conductivity, porosity, dispersivity etc.).

Velocity values are computed by applying Darcy's equation using calculated hydraulic heads and porosity values. For steady state flow, the water level configuration of March 1994 was considered and the groundwater flow equation was solved once and thereby a single velocity field was determined for the mass transport simulation for all times. Dispersion is accounted for in the particle motion by adding to the deterministic motion as a random component, which is a function of the dispersivities. The mean concentration for each grid block is calculated as the sum of the mass carried by all the particles located in a given block divided by the total volume of water in the block (Zheng, 1990)

SIMULATION OF HOST HYDROGEOLOGICAL SYSTEM

On the basis of the available information, the following boundary conditions have been simulated in the model:

- The eastern and northeastern boundary has been considered as inflow boundary through which later flow enters the aquifer system. Initial estimate of inflow has been arrived from flow net analysis.
- Barahanadi has been considered as the outflow boundary because of steep groundwater gradient towards the stream and its effluent nature. The outflow has been distributed over the nodes representing the stream.

- Amount of red mud slurry to be discharged into the red mud ponds will be about 4000 m^3/day corresponding to 2700 tons of red mud.
- About 15-20% of this quantity may leak into the aquifer system. About 800 m^3/day of water is assumed to infiltrate into the aquifer system from two ponds and red mud stacking area. Thus each node was assigned an input of 400 m^3/day.
- It has been assumed that the infiltrated effluent with a concentration of 3000 mg/l would be reaching the groundwater regime just below the ponds.

The grid map showing the simulated model domain, model boundaries, pumping wells and red mud ponds have been shown in Figure 3. The permeability distribution simulated in the groundwater flow model varies from 0.75 – 2 m/day (Fig. 4). The phreatic aquifer thickness varies from 3 to 6 m only. The thickness has been determined from detailed geophysical investigations. The thickness of weathered zone is varying from 2- 6 m and is underlain by a fracture zone in charnokite basement. Groundwater withdrawal from individual dug wells from the phreatic zone is ranging from 25 to 75 m^3/day and total withdrawal in the watershed is about 500 m^3/day. The area has highly rugged topography and steep slopes and much of rainfall water goes as surface runoff and as such recharge to the groundwater regime could be very low (about 50 mm/year).

The groundwater flow model was prepared and the head distribution was computed using visual MODFLOW (Guiger & Franz, 1996 and McDonald & Harbaugh, 1988). Comparison of the computed and observed water levels during March, 1994 at the observation wells was carried out during the calibration of the flow model and the matching seems to be very good. The groundwater velocity field in the entire model domain was computed using hydraulic head gradient and porosity values. The porosity of the Khondalitic formations was assumed to be about 0.15. The computed water level contours along with computed velocity vectors indicate a predominant groundwater flow towards Baraha nadi (Fig. 5). The average groundwater velocity is about 0.005 m/day (Dhar et al, 1994).

SIMULATION OF MASS TRANSPORT MODEL

The concentration of Na_2CO_3 and traces of caustic soda in the red-mud ponds is likely to be as high as about 3500 mg/l. When the liquor is separated from the fine mud residue in the red mud stacking area it may have slightly higher concentration. However, adsorption by red mud itself may reduce the concentration of the liquor in red mud ponds. Thus a concentration of about 3000 mg/l for the infiltrated contaminant has been assumed at the water table of the phreatic aquifer for mass transport simulation. The longitudinal dispersivity of 100 m, horizontal transverse dispersivity of 0.1 m and vertical dispersivity of 0.001 was assumed based on literature values. The background initial concentration of Sodium Carbonate is assumed to be about 50 mg/l in the natural groundwater water. Using the above dispersion parameters and computed velocity field obtained from the MODFLOW, the contaminant migration from the red mud ponds has been computed for 30 and 50 years (Figs. 6 & 7) using MT3D (Mass Transport in Three Dimensions) model (Zheng, 1990). The contaminant plume with highest concentration of 3000 mg/l is confined to the red mud ponds only due to relatively low permeability of the formations covering red mud ponds. However, because of the natural hydraulic gradient and continuous loading conditions in the red mud ponds over 50

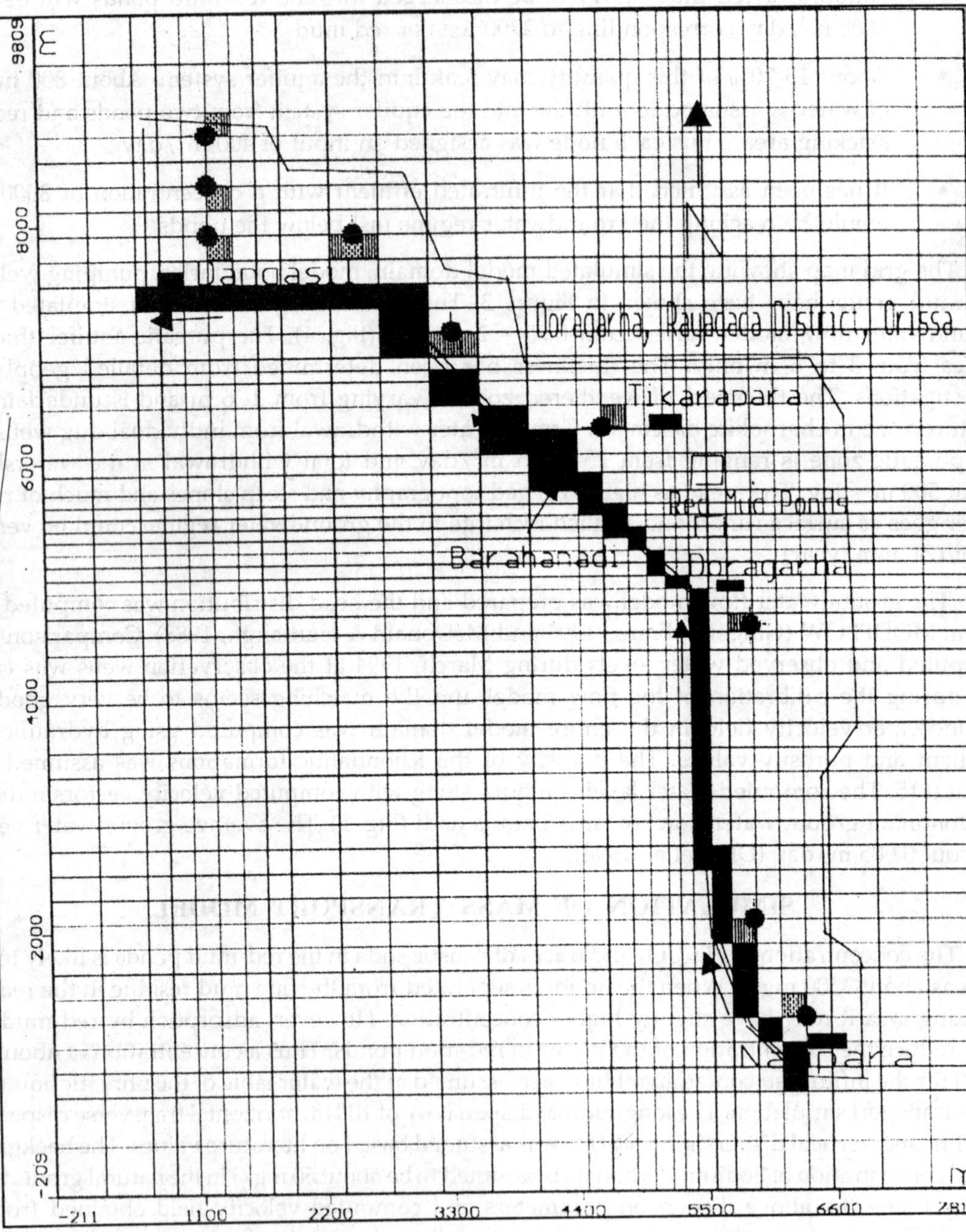

Fig. 3 Grid map with boundary conditions of the simulated model domain

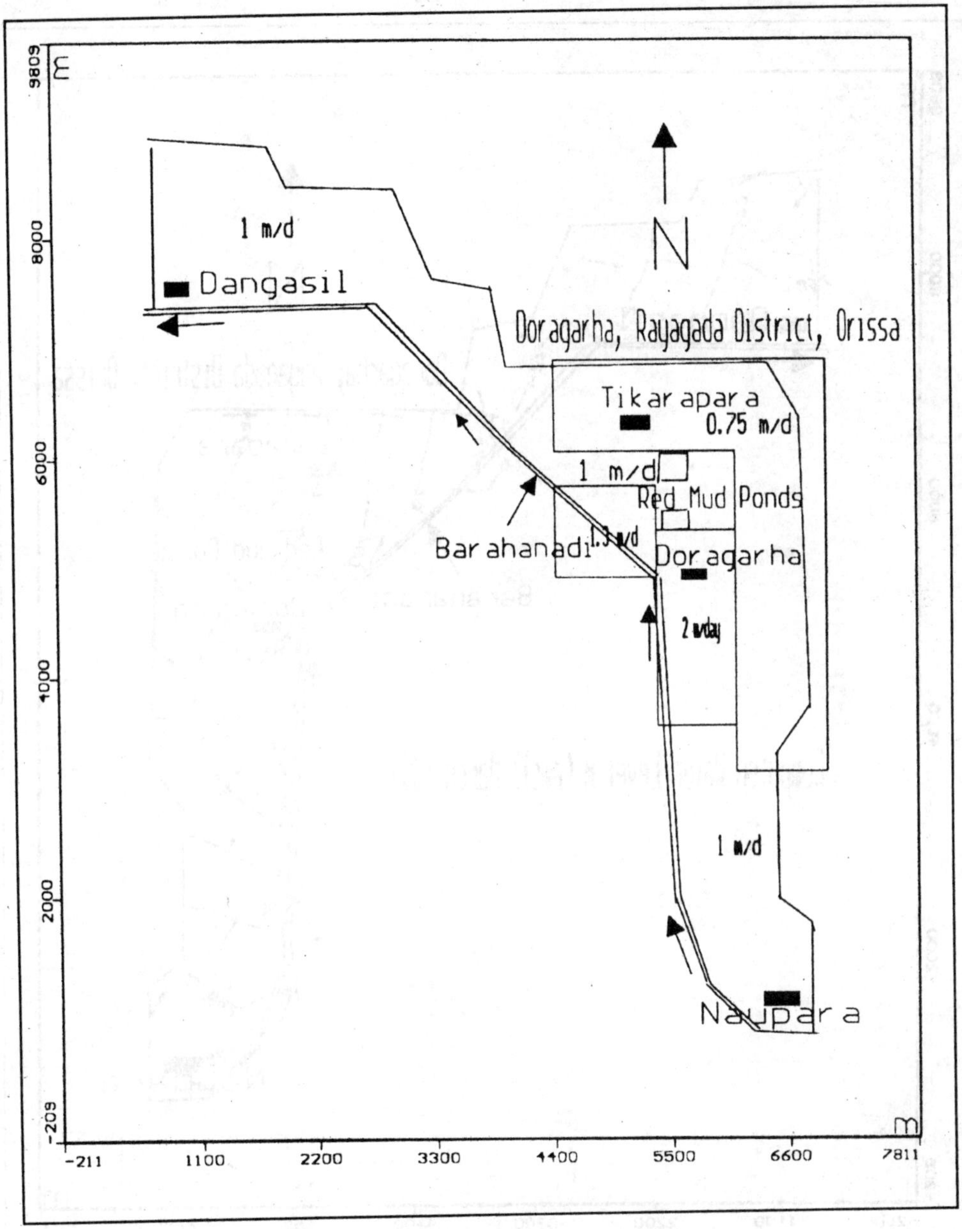

Fig. 4 **Permeability distribution (m/day) of the simulated mode domain**

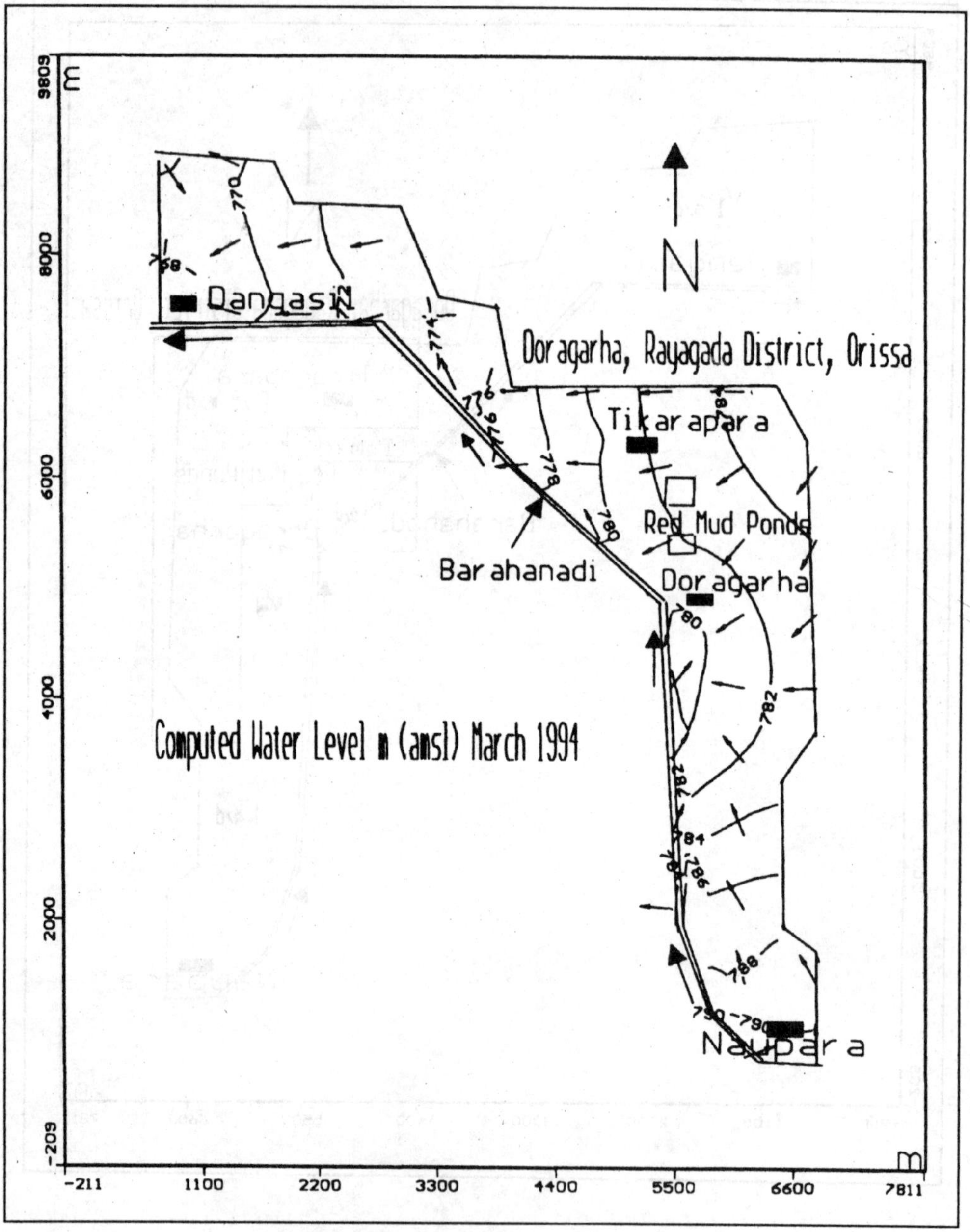

Fig. 5 Computed water level in m (amsl) and groundwater velocity vectors

year period, the computed Na_2CO_3 concentration in groundwater would attain 100 mg/l near the Barahanadi after 50 years. In the flow and mass transport simulations no preferred pathway for groundwater flow has been considered. If such a flow path exists the contaminant may reach Barahanadi even earlier than what has been predicted.

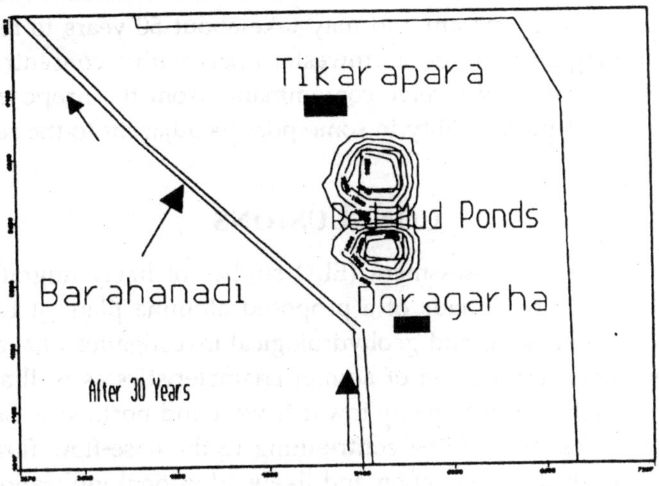

Fig. 6 **Computed concentration of Na_2CO_3 plume (mg/L) after 30 years of red-mud stacking and storage in ponds**

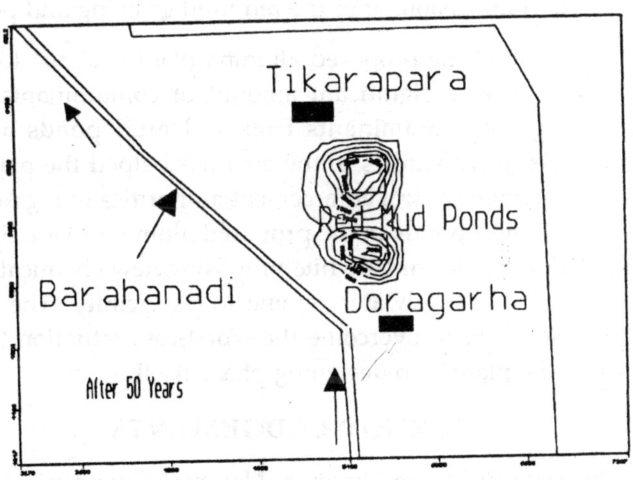

Fig. 7 **Computed concentration of Na_2CO_3 plume (mg/L) after 50 years of red-mud stacking and storage in ponds**

DISCUSSION OF SIMULATION MODEL RESULTS

The groundwater flow is mostly towards west and northwest and a substantial amount of groundwater effluence is expected to be contributing to the base flow in the Barahanadi. At the present rate of groundwater withdrawal and likely additional infiltration expected from the proposed red mud stacking and seepage from red mud ponds, the groundwater table would rise, which in turn may cause an increase in base flow to the Barahanadi. The contaminant plume is likely to migrate towards the stream and may take about 50 years to reach the stream near Kuchaipadar and adjoining areas in the northwestern part with a concentration of 100 mg/l. It may be likely that the stream may receive contaminants from the proposed red mud stacking even earlier because of high permeability in some pockets adjacent to the red mud ponds in the area.

CONCLUSIONS

Environmental Impact Assessment (EIA) studies of likely migration of contaminants likely to be infiltrating from red ponds of a proposed alumina plant at Doragarha have been carried out. Integrated geophysical and geohydrological investigations have been carried out in Doragarha watershed for determination of aquifer characteristics as well as its properties. The groundwater in the watershed flows mostly towards west and northwest. Substantial additional groundwater effluence is expected to be contributing to the base-flow flow in Barahanadi. At the present rate of groundwater abstraction and likely additional infiltration from the red mud stacking and ponds, the groundwater table would rise, in turn, may drive enhanced base flow in Barahanadi. The mass transport model predicted that the contaminant plume with a concentration of 100 mg/l would reach the Barahanadi in about 50 years time. It is likely that the Barahanadi may receive contaminants even earlier through preferred pathway flow. It is suggested that the monitoring network for measurement of water level and water quality should be installed along with the commissioning of the red mud stacking and ponds.

In this context, as the life of the proposed alumina plant is about 40 years, the Barahanadi, would not be expected to receive significant amount of contaminants through groundwater effluence from the seepage of contaminants from red mud ponds and stacking area. The environmental Impact Assessment studies in the area has helped the plant authorities to obtain clearance from the environmental control enforcement authorities in regard to likely groundwater contamination from the red mud ponds in the proposed alumina plant. Similar such EIA studies are required to be carried out before hand while proposing new chemical industrial units, which are likely to contaminate the groundwater regime in its vicinity. The studies will provide a foresight for the planner in order to overcome the worst-case situation likely to emerge during the life of the plant while the plant is in designing phase itself.

ACKNOWLEDGEMENTS

We are extremely grateful to the Director, National Geophysical Research Institute for allowing us to publish the paper. Engineers India Limited (EIL), New Delhi, sponsored the study. We wish to thank Mr. K. Shankar for assistance in preparation of the paper.

REFERENCES

Anderson M.P., and Woessner, W.W., 1992. Applied groundwater modeling simulation of flow and advective transport. Academic press. San Diego, CA., U.S.A.

Bear, J., 1979. Hydraulics of groundwater. McGraw Hill, NewYork, pp.210.

Burnett, R.D., and Frind, E.O., 1987. An alternating direction Galerkin techniques for simulation of groundwater contaminant transport in three dimensions, 2-Dimensionality effects. Water Resour. Res. 23 (4), p.695-705.

Dhar, R.L., Gurunadha Rao, V.V.S., Gupta, C.P., Subrahmanyam, K., Prakash, B.A.,

Ramachandra, Y.M. and Hodlur, G.K., 1994. Geohydrological and solute transport modeling studies for preliminary assessment of the impact of red-mud stacking on aquifer system at Doragarha, Rayagada, District, Orissa.

Grove, D.B. and Stollenwork, K.G., 1984. Computer model of one-dimensional equilibrium controlled sorption processes. U.S. Geol. Survey Water Resources

Investigations Report 84-4059, pp.58.

Guiger, N. and Franz, T. 1996. Visual MODFLOW: Users Guide. Waterloo Hydrogeologic, Waterloo, Ontario, Canada.

Javandel, I., Doughty, C. and Tsang, C.F., 1984. Groundwater transport: Handbook of Mathematical models. American Geophysical Union, Water Resources Monograph 10, pp.228.

Konikow, L.F. and Bredehoeft, J.D., 1978. Computer model of two dimensional solute transport and dispersion in groundwater. U.S. Geol. Surv. Water Resour. Invest. Book 7, Chapter C2, pp.90.

McDonald, J.M. and Harbaugh, A.W., 1988. A modular three-dimensional finite-difference groundwater flow model. Techniques of Water resources Investigations of the U.S. Geological Survey Book.6, pp.586.

Zheng, C., 1990. MT3D, a Modular three-dimensional transport model for simulation of advection, dispersion and chemical reactions of contaminants in groundwater system prepared for the U.S. Environmental Protection Agency.

16

Groundwater Environs of Hubli Dharwad Municipal Corporation Area, Karnataka

*S.C.Puranik

INTRODUCTION

Water is fundamental to all life and is one of the most valuable resources for mankind. The continuing growth and concentration of population and industries in urban and suburban areas in recent years has caused many problems. The concentration of large number of people results in increased demand for water which sometimes exceeds locally available quantum. Many industries/factories and bulk consumers who normally depend on municipal supplies are now resorting to manage their requirement from alternative sources. They are now relying on groundwater which they can tap independently. The present paper discusses groundwater environs of Hubli-Dharwad Municipal Corporation (HDMC) area, Dharwad District, Karnataka State.

The HDMC area lies between 15°18′25" to 15°30′47" north latitude and 74°57′37" to 75°11′0"east longitude (Fig.1). The aerial extend of the HDMC is about 150 sq.kms. Next to Bangalore, the HDMC is the largest Municipal Corporation in Karnataka State. The HDMC includes two urban areas viz. Hubli and Dharwad. These two urban areas are surrounded by suburbs and villages. The total population in the area is over 0.8 million. Because of rapid urbanisation the area is experiencing scarcity of portable water. Presently the twin cities are receiving water from the Malaprabha reservoir which is about 50 kms. away. Till last year the water supply was on alternate days but during 2001 and presently the supply is once a week. Apart from this supply, people are harvesting water from more than thousand bore wells. Only a few bore wells are managed by the corporation authorities.

* Department of Studies in Geology, Karnatak University, Dharwad 580 003 India

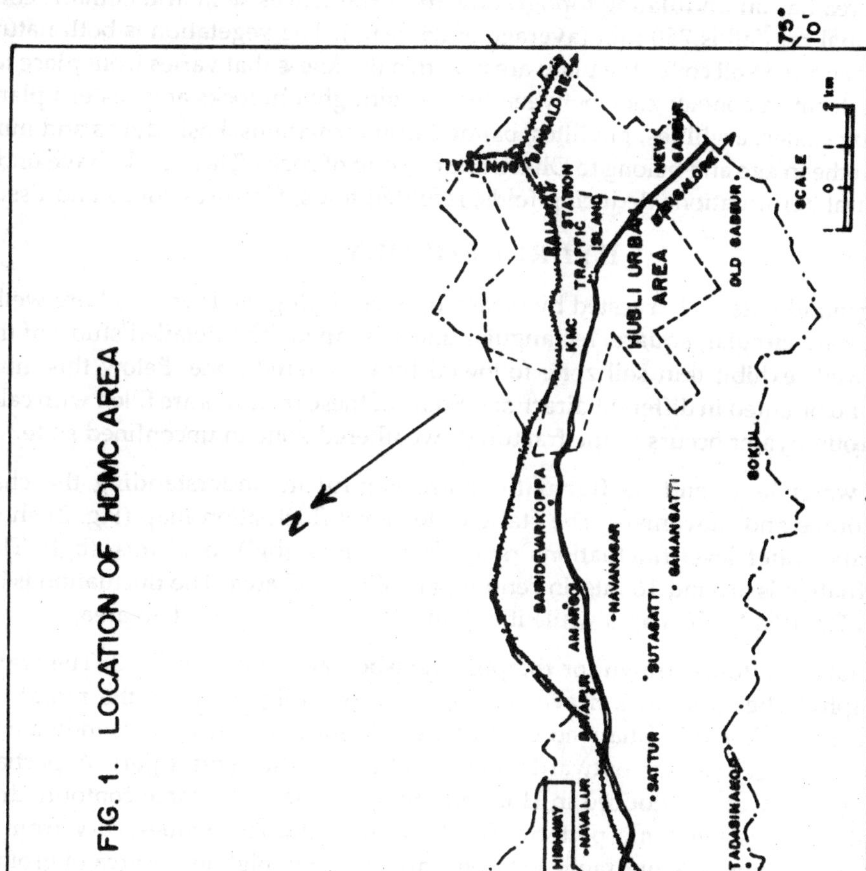

FIG. 1. LOCATION OF HDMC AREA

The study area has an undulating topography and experiences semi arid climatic condition. The average annual rainfall is 750 mm (average of 25 years). The vegetation is both natural and afforested. Black and red soil cover the plain areas with a thickness that varies from place to place. The HDMC area forms a contact zone between the western ghat hillocks and eastern plains. The area is covered by shales, argillites, phyllites, banded iron formations, basic dykes and numerous acidic veins of Archean age and belong to Dharwarian group of rocks. These rocks have undergone repeated structural deformations deducing folds, refolded folds, fractures, joints and fissures.

HYDROGEOLOGY

Normally groundwater is harvested by means of open or dug wells and by bore wells. Most of the open wells are circular, square, rectangular and D shaped. The detailed study of the cross section of dug wells exhibit thin soil zone followed by weathered zone. Below this, numerous fractures are found oriented in different directions. Some of these fractures are filled with calcareous material. The groundwater occurs in the fractured/weathered zone in unconfined state.

The static water level and its fluctuations are helpful in understanding the change in groundwater storage and movement. The static water level fluctuation map (Fig. 2) shows that the depth to static water level fluctuations range from 2 mts. (bgl) to 18 mts. (bgl). The static water level fluctuation is around 16 mts. in central part of HDMC area. The fluctuation is between 6 and 18 mts. in Dharwad urban area while it is 2 and 10 mts. in Hubli urban area.

The water table contours drawn for the post monsoon are shown in Fig.3. The water table follows topography. The flow lines drawn to these contour (Fig.3) shows the radial outward flow of groundwater as found in Dharwad urban area indicating presence of groundwater mound and the area of recharge. Similar outward flow is noticed in the central part. A partial radial inward flow of groundwater is noticed in Hubli urban area. The water table contours drawn for pre monsoon also shows the same pattern. The hydraulic gradient values vary from 0.00725 mts/kms to 0.025 mts/kms. The hydraulic gradients are generally high in the area of groundwater recharge and low in groundwater discharge area.

The water table fluctuation is shown in Fig. 4. The fluctuation is more in the Dharwad urban area i.e.2 to 16 mts and 1 to 3 mts in Hubli urban area. In order to know the behavior of water table fluctuation with the surface water bodies a water table fluctuation profile is drawn along X and Y (Fig. 4) and shown in Fig.5. The amplitude of the profile is more at three places which coincide with the areas of the surface water bodies(tanks) of Kellegeri, Navalur and Unkal. The approximate radius of influence on water table fluctuation of Kellegeri, Navalur and Unkal tanks are 1.4, 1.9 and 2 kms. respectively.

Based on the data of bore wells (yield, depth and casing) it is observed that 83% of the well yield is less than 2.52 lps of water and 17% of the wells yield between 2.52-11.36 lps. The depth range of 85% of wells is between 50 and 80 mts. Thus the economic depth of drilling is 90 mts. 78% of bore well have a casing of 20 to 50 mts. indicating presence of loose/collapsable subsurface material upto a depth of 40 to 45 mts.

The subsurface conditions are studied with the help of vertical electrical sounding (V.E.S). This study indicates that subsurface conditions are inhomogenous/anisotropic. This is indicated by multi layered system with varying resistivity and thickness. The resistivity of 1st and 2nd

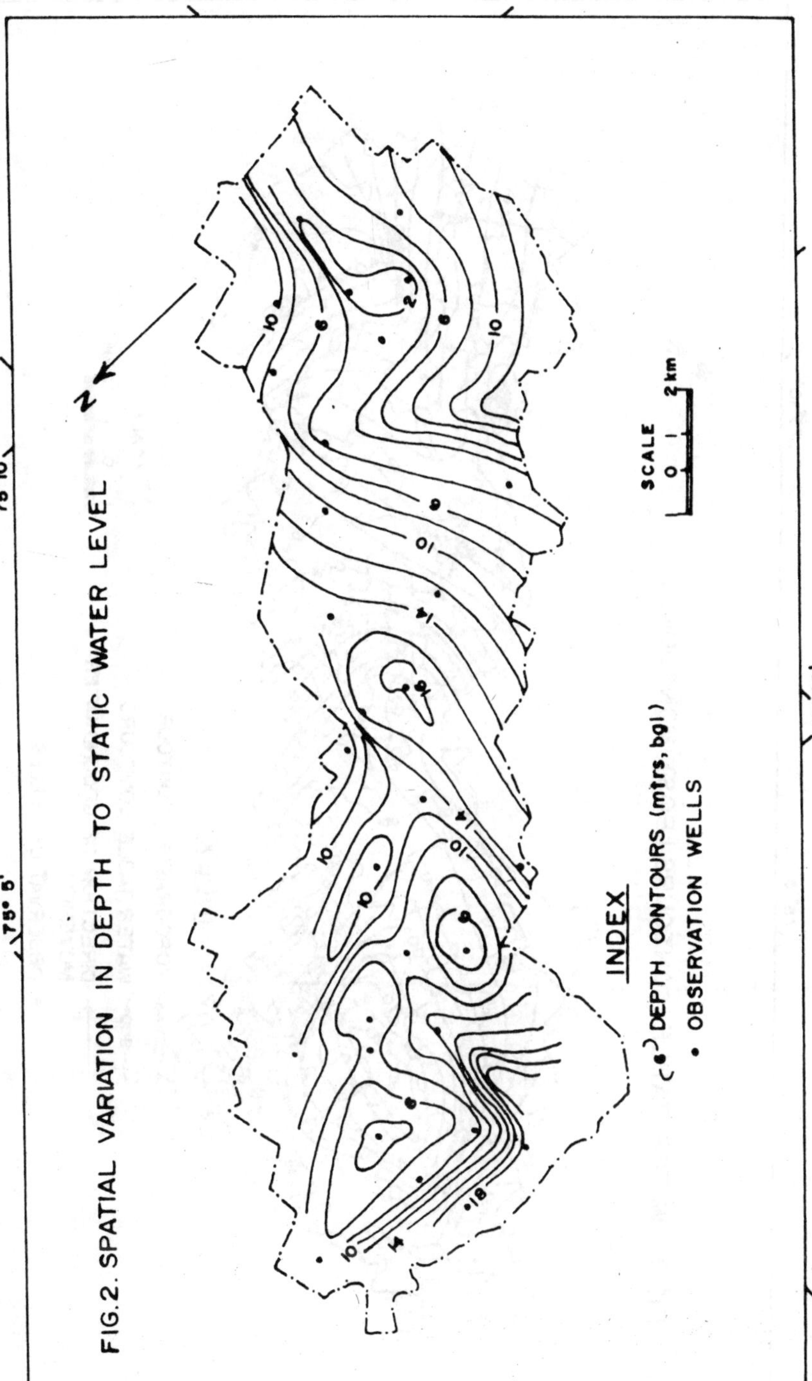

FIG.2. SPATIAL VARIATION IN DEPTH TO STATIC WATER LEVEL

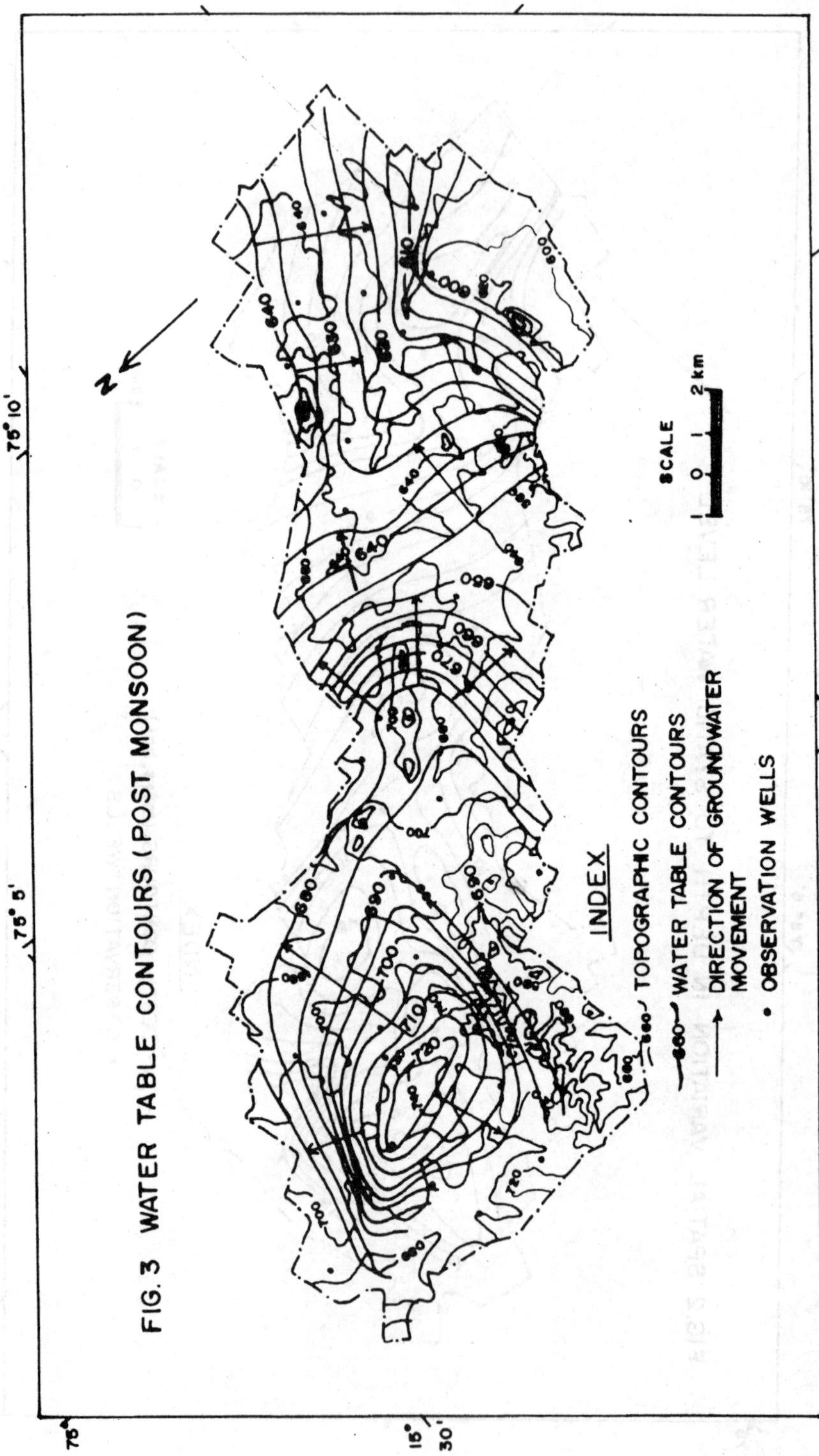

FIG. 3 WATER TABLE CONTOURS (POST MONSOON)

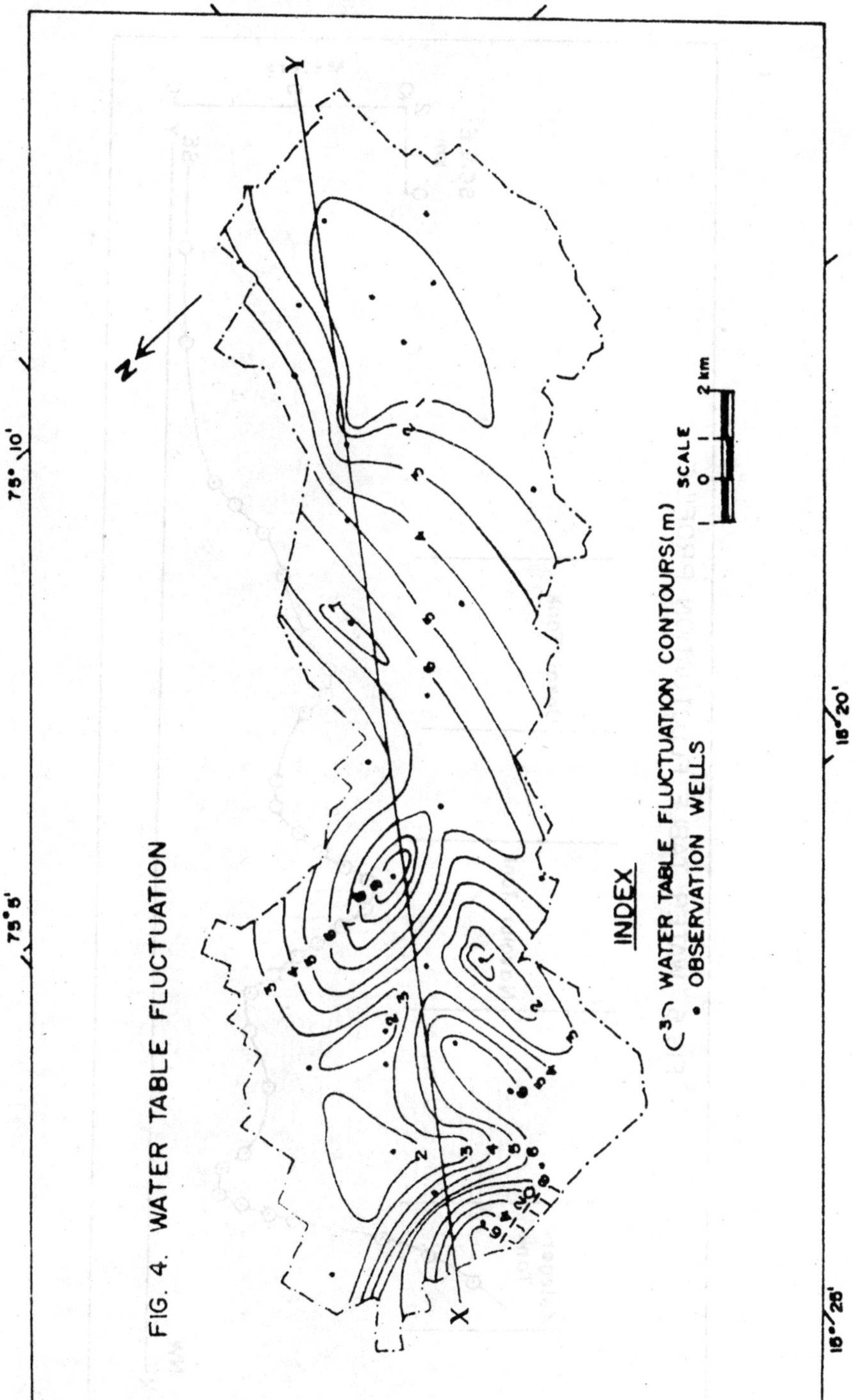

FIG. 4. WATER TABLE FLUCTUATION

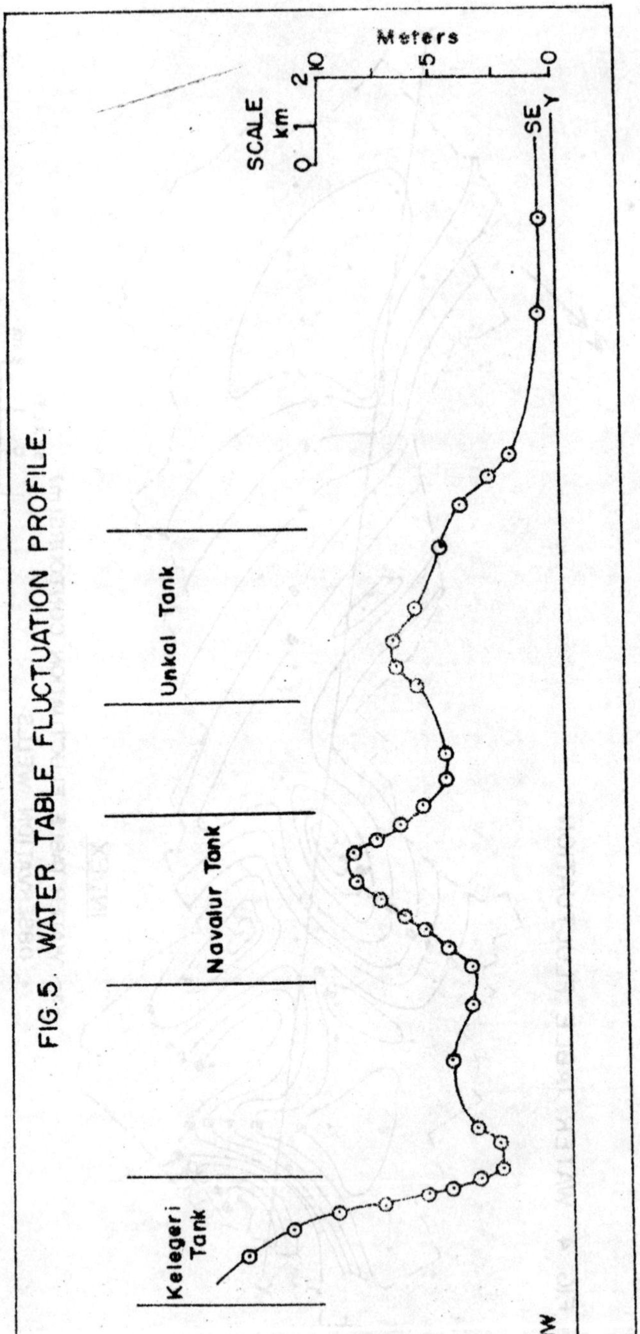

FIG. 5. WATER TABLE FLUCTUATION PROFILE

layers correspond to dry shale/soil, weathered phyllites/shales. The third and fourth layers resistivity correspond to fractured rock, sometimes hard rock/basement. The asymptotic behavior of the VES curves corresponds to hard rock/basement. The first layer resistivity varies from 3.5 ohm.mts to 580 ohm.mts and thickness from 0.2 to 2.98 mts. The second layer resistivity varies from 5.2 ohm.mts to 132 ohm.mts and thickness from 1.82 to 45.72 mts. The third layer resistivity varies from 7 ohm.mts to infinite and thickness from 1.09 to 91.4 mts. Except for few locations, fourth layer resistivity attains infinite value.

GROUNDWATER QUALITY APPRAISAL

In groundwater assessment studies, evaluation of quality of groundwater is as important as quantity. The study of groundwater quality involves a qualitative and quantitative assessment of the constituents therein and their relation to industrial, agricultural and domestic uses. A systematic assessment of groundwater quality of HDMC is performed by dividing the whole area into Hubli urban area, Dharwad urban area and Suburban area. The quality of groundwater of HDMC are studied under two heads

1. Quality Evaluation Part—I : Involves overall study of HDMC area.
2. Quality Evaluation Part—II : Involves impact of sewage course.

Quality Evaluation Part—I

The groundwater samples were collected from open and bore wells as shown below.

- 18 open well samples and 29 bore well samples from Hubli urban area.
- 18 open well samples and 23 bore well samples from Dharwad urban area and
- 12 open well samples and 15 bore well samples from Suburban area.

The chemical constituents like Ca, Mg, Na and K as major cations and HCO_3+CO_3, Cl and SO_4 as anion are estimated. EC (electrical conductivity) and pH values are determined in field. The minimum, maximum and average of above said constituents are given in Table 1.

Quality Evaluation Part—I reveals that most groundwater is potable in the study area. However a few groundwater samples show high concentration of Ca, Cl, TDS, TH, NO_3 and K. These are from Trovihakkal, Kumbar Oni, Jangalipet, Karjgi Oni, Tayammanagar, Yellapur Oni, Karki Basaveshwarnagar and Dhakappa Circle areas of Hubli urban area and Hosayellapur, Akhanigalli, Gulaganjikop, Mallapur, Maratha Colony areas of Dharwad urban area. In the urban areas, about 20% of groundwaters exceed the permissible limits (ISI, 1983) of Ca, 12% of TDS, 4 to 5% of Cl and 26 to 34% of TH. In Suburban area 11% of groundwaters exceed the limits of Ca and 7% exceed the limits of TDS, of drinking water standards. Almost all the groundwaters of the study area are suitable for different live stock except a few samples which do not meet quality requirement for poultry in their TDS contents. The groundwaters of suburban area are suitable for agricultural purposes. However, a few samples show salinity hazards. The groundwaters of the study area do not meet the quality requirement of TDS and TH for textile industries and of TDS, TH, Ca and Cl for canned, dried and frozen fruits and vegetables and of TH for leather tanning. A few samples exceed permissible limits of Cl specified for soft drinks and bottling.

Table—1 Minimum,Maximum and average values of chemical constituents from HDMC area

	HUBW			HUOW			DUBW			DUOW			SUBW			SUOW		
	Min	*Max*	*Avg*	*Min*	*Max*	*Avg*	*Min*	*Max*	*Avg*	*Min*	*Max*	*Avg*	*Min*	*Max*	*Avg*	*Min*	*Max*	*Avg*
pH	6.0	7.6	6.9	6.2	7.9	7.3	6.6	8.0	7.07	6.8	7.8	7.37	8.7	7.4	7.01	7.0	8.1	7.49
EC	711	4930	1845	335	3151	1258	305	4980	1511	203	3659	1152	157	3354	1054	419	2541	1149
Ca	62	63	198	29	432	105	46	501	158	41	353	120	13	392	116	66	232	108
Mg	16	147	60	15	97	49	8.0	341	55	5.0	114	32	7.0	135	43	15	90	41
Na	33	250	131	17	230	115	12	258	89	24	306	99	16	177	67	12	210	96
K	1.0	85	6.0	1.0	26	6.0	1.0	275	36	2.0	75	12	1.0	150	23	1.0	80	10
HCO_3 $+CO_3$	110	410	276	100	420	278	120	420	24	110	310	239	70	400	258	160	450	286
Cl	120	1235	425	853	774	261	64	1086	320	42	1086	249	28	717	195	71	738	265
SO_4	1.0	200	93	3.0	194	73	Tr	228	61	1.0	228	53	1.0	190	46	5.0	150	72
TDS	455	3155	1184	227	2016	805	195	3187	967	130	2342	731	100	2148	674	268	1626	735
TH	264	2146	734	146	1476	463	154	2590	619	134	1350	431	62	1536	452	184	948	446
CR	0.58	7.41	2.59	0.57	3.88	1.65	0.39	6.54	2.02	0.58	5.69	1.54	0.28	3.35	1.13	0.55	3.62	1.46

HUBW-Hubli urban bore well
HUOW-Hubli urban open well
DUBW-Dharwad urban bore well
DUOW-Dharwad urban open well
SUBW-Suburban bore well
SUOW-Suburban open well

* All values are in mg\l. except pH. Tr-Traces

There are four chemical types(Piper,1944) of groundwater in the study area (Quality Evaluation Part—I). They are "No one cation-anion pair exceeds 50% type" "Non-carbonate alkali exceeds 50%," "Carbonate hardness exceeds 50%" and "carbonate alkali exceed 50%" types. Most groundwaters of the study area belong to Ca+Mg-Na+K cation facies (Back,1966) followed by less dominating Na+K-Ca+Mg cation facies. The dominating anion facies is $Cl+SO_4-HCO_3$, followed by a less dominating $HCO_3-Cl+SO_4$, facies.

Quality Evaluation Part—II

Another set of water samples are collected to know the impact of urban sewage on quality of groundwater to assess the degree of quality deterioration of groundwater along and across the sewage course in Hubli urban area. 22 groundwater samples are collected and studied for physico-chemical characters, organic matter content and a few trace elements. The results are given in Table 2.(minimum, maximum and average values). The Table also contains minimum, maximum and average values of the trace elemental analyses.

Quality Evaluation Part-II reveals that wells located in the vicinity of sewage course have high content of TDS (upto 3423 mg/l), Cl (upto 1400 mg/l) Ca (upto 805 mg/l),Na (upto 650 mg/l) and K (upto 270 mg/l), indicating the quality deterioration in this area. These groundwaters are free from pollution by organic matter, as indicated by DO and COD concentrations which are within the permissible limits. Trace/minor elemental concentrations such as Fe, Mn, Cu, Ni and Zn are within the permissible limits for drinking water. However, a few samples exceed the limits of Pb and Cd which are toxic elements.

Chemical types (Piper,1944)of groundwater along the sewage course (Quality Evaluation Part—II) are "Non-carbonate alkali exceeds 50%" and "No one cation-anion pair exceeds 50%" types. The chemical facies (Back,1966) are Na+K-Ca+Mg which dominate over Ca+Mg-Na+K cation facies. $Cl+SO_4-HCO_3$ anion facies dominate over $HCO_3+Cl-SO_4$ facies.

Seasonal variation in the quality of groundwater is shown by increase in EC and Na and decrease in Ca, Mg, TH and SO_4 concentrations from pre-monsoon to post-monsoon seasons. Concentration of other constituents remained almost same. Most groundwater show rock dominance character (Gibbs,1970). A few samples, however, have an affinity towards evaporation dominance in Hubli urban and Dharwad urban areas.

A comparison of chemistry of groundwater (Quality Evaluation Part I and II) of Hubli urban area show that there is an increase (37%) of primary salinity type which indicates the increase in alkalis and strong acids in the groundwater samples from quality evaluation part II. Similarly there is a decrease in secondary salinity and secondary alkalinity type of groundwater samples from quality evaluation part II. The salinisation is more in the groundwater in the vicinity of sewage courses and/or in old city blocks. Similarly there is an increase (67%) in Na+K-Ca+Mg cation facies and also in $Cl+SO_4-HCO_3$ anion facies (17%) in the quality evaluation part II groundwater samples.

Most of the groundwater samples are corrosive in nature. The corrosivity value ranges from 0.57 to 7.41 in Hubli urban, 0.39 to 6.54 in Dharwad urban and 0.28 to 3.62 in Suburban area groundwater samples.

Table—2 **Minimum, Maximum and Average values of chemical constituents of samples along sewage course**

	pH	*EC*	*Ca*	*Mg*	*Na*	*K*	HCO_3 $+CO_3$	*Cl*	*SO4*	*TDS*	*TH*	*CR*
Minimum	6.6	839	12	14	140	Tr	300	78	14	537	135	0.31
Maximum	7.8	5348	805	108	650	270	520	1400	227	3423	2300	6.70
Average	7.1	2236	151	50	363	27	412	445	70	1431	570	1.87

	Fe	*Mn*	*Cu*	*Ni*	*Zn*	*Pb*	*Cd*	*COD*	*DO*
Minimum	0.052	0.001	0.0	0.015	0.085	0.067	0.004	0.0	0.3
Maximum	1.136	0.176	0.026	0.047	0.968	0.868	0.019	41	10
Average	0.279	0.028	0.007	0.026	0.220	0.146	0.010	11.81	6.30

*All values are in mg\l. except pH.

Tr - Traces

Table—3 Landuse/Land cover statistics

(Areas in Hectares)

Category	1930-31	1975-76	1988	2000
1. Builtup lands	1197.00	2925.00		
(a) Central business areas	-	-	683.25	710.00
(b) Residential/Industrial areas	-	-	4475.00	6266.00
2. Uplands	2104.00	675.00	475.00	452.00
3. Water bodies	582.75	362.50	337.50	304.00
4. Gardens	-	575.00	1300.00	1300.00
5. Wet land/Marshy area	-	-	781.25	746.00
6. Agricultural/Open lands	14116.25	13462.00	9950.00	8222.00
Total area in hectares	**18000.00**	**18000.00**	**18000.00**	**18000.00**

*The actual area of HDMC is considered in calculating land cover patterns for all years.

LAND USE PATTERN AND GROUNDWATER QUALITY

Urbanisation and industrialization cause social, economical and physical problems. The use and development of land and water resources are significant aspects of urbanisation.

The land use and land cover analysis of HDMC area has been carried out for a period of 70 years, from 1930-31 to 2000 in order to study their variation in space and time of landuse in HDMC area. Toposheets are used to identify land use and land cover features during 1930-31 and 1975-76 and satellite imageries of 1988 and 2000 are used to demarcate land use and land cover features. The areas in hectares of different categories of land use pattern is given in Table 3.

The land use and land cover analysis of HDMC area shows that there is an increase in the area of built upland and decrease in the areas of upland, water bodies, agriculture/open lands.

The TDS, Ca and Cl contents of groundwater are higher in both the urban areas than in suburban areas/ newly developed residential areas. The groundwater of Hubli and Dharwad urban areas show more affinity towards chloride apex than the groundwater of suburban areas, on Piper's Trilinear Diagram(1944). The two zones formed by plots of groundwater of the urban areas show more affinity towards secondary salinity than suburban area samples. There is a change in chemical facies of groundwater in Hubli urban area (Quality Evaluation Part -II) from Ca+Mg-Na+K to Na+K-Ca+Mg i.e. from alkaline earth salts to alkali salts. A few groundwater samples from Hubli and Dharwad urban areas show an affinity towards evaporation dominance field on Gibbs diagram (1970) indicating the increase in TDS content. This is because of contaminated groundwater rather than evaporation.

All the above observations support that there is relation between quality of groundwater and the land use pattern i.e. the groundwater quality is deteriorating in both the urban areas. The groundwater quality is good in suburban/newly developed residential areas.

REFERENCES

Back, W. (1966) : Hydrochemical facies and groundwater flow in northern part of Atlantic Coastal Plain, U.S.G.S. prof. paper , 498 A , pp. 442.

Gibbs, R. J. (1970) : Mechanisms controlling world water chemistry, Science , V. 170, pp. 1088 - 1090.

ISI (1983) :Indian Standard Specification for drinking water, IS - 10500 - 1983. Indian Standard Institution, New Delhi.

Piper, A.M. (1944) : A graphic procedure in the geochemical interpretation of water analyses. , Trans. Am. Geophys. Union. V. 25, No. 6, pp. 914 - 928, Washinton D C.

17

Detection and Measurement of Ultra-trace Levels of Mercury and Copper by Genetically Engineered Microbial Sensors

S.K. Dubey*, David S. Holmes,
G.R. Rao and S. Gangolli

INTRODUCTION

Many traditional analytical methods such as atomic absorption spectrometry (AAS), atomic emission spectrometry (AES), mass spectrometry (MS) and high pressure liquid chromatography (HPLC) have been used to measure metals, metalloids and organic compounds (Beaty, 1988) but these analytical methods involve time-consuming and often destructive sample preparation and require highly expensive and sophisticated instruments. Though they enable precise quantitation of total chemical or ion concentration, they fail to provide information on what fraction of the total metal is in free form and hence bioavailable. This type of information, vital in developing bioremediation and biodegradation strategies, can readily be obtained by the use of biosensors.

What are biosensors? A biosensor is a detection or measuring device containing a biological component which senses and responds to a target analyte in a quantifiable manner. In recent years, extensive research work has appeared on the use of a wide-range of biological components which provide the molecular recognition element for the biosensor (Rechnitz, 1988). These biological components include chemoreceptors, antibodies/antigens and biocatalyst such as isolated enzymes, DNA sequences, metabolites, whole cells (microorganisms), p1ant and animal tissues. Holmes and his associates have developed several bioluminescent biosensors that are

* Laboratory of Bacterial Genetics and Environmental Biotechnology Department of Microbiology, Goa University, Taleigao Plateau, Goa–403 206, India.

based on the ability of genetically engineered bacteria to emit visible light in response to target analytes viz. Cu, Hg, PCB (Holmes et.al. 1993a,b; Gangolli et.al. 1993; Dubey et.al. 1993). Biosensors have got unique attributes such as their sensitivity, and specificity to the target analyte. Such attractive characteristics have made these biosensors an important tool in pharmaceutical and other biotech industries involved in production of health care products, food processing and environmental monitoring of toxic substances.

In last two decades use of lux genes as a bioreporter has received immense attention and popularity due to their important and valuable attributes as compared to traditional gene reporters such as CAT, LacZ, GUS, etc. (Burlage et.al. 1990; Corbisier and Silver, 1992; Holmes et.al. 1993a, b; Heitzer et.al. 1994; Loessner et.al. 1996; Iizumi et.al. 1998; Wahlstrom and Saris, 1999).

There is a pressing demand for a simple and sensitive analytical technique to detect bioavailable levels of metals such as Hg, Cd, Cu, Zn, As, Pb, Sb and organometals viz. Methyl Hg, Tetra-ethyl Lead (TEL), tributyl-tins (TBTs) directly which is present in the aquatic environment. Development of biolumінsecent microbial sensors for biomonitoring of such environmental pollutants present at ultratrace levels have proved to be an important ground breaking research in the field of environmental biotechnology and genetic engineering.

In order to accomplish this work, it is of vital importance to understand the mechanisms that microbial cells employ to regulate metal responsive operons and hence metal ion resistance. A greater understanding of metal ion resistance mechanism has made it possible to customize biosensors using several different genetic regulatory elements specifically responsive to those analytes viz. metals, organo-metals and metalloids.

It is interesting to note that microorganisms specifically bacteria respresent a wide genetic diversity and possess metal resistance determinants either on chromosomal genome or on plasmids or transposons (Ohtake et.al. 1987; Silver and Misra, 1988; Cervantes and Ohtake, 1988; Cervantes and Silver, 1992; Lebrun et.al. 1994; Silver and Phung, 1996).

The mercury and copper biosensors developed, offer many advantages: their exquisite specificity to the target metal, high sensitivity, easy detection and measurement. The presence of a metal (Hg or Cu) is readily revealed by the bioluminescence produced by the biosensor microorganism. It can be seen by naked eye, measured by luminometers and also documented using CCD camera.

In this report we describe the ability of these genetically engineered microbial biosensors to detect ultratrace quantities of Hg or Cu ions which makes them ideal tools to detect these toxic metals in contaminated environmental samples, industrial effluents, sludges and clinical samples such as saliva, blood and urine.

MATERIALS AND METHODS

Plasmids and Bacterial Strains: Plasmids pDU1003 (gift from Prof. Simon Silver) and pUCD615 (gift from Prof. C. I. Kado) were used to construct the mercury biosensor. Plasmids pPA223 (gift from Prof. B.T.O. Lee) and pUCD615 were used to construct the copper biosensor. *E.coli* JM101 (Stratagene, U.S.A.) was used for transformation experiments. Transformants serve as biosensors for specific metal analytes viz. Cu and Hg.

Growth of Biosensor Cells: Biosensor cells were grown at 30°C in Luria-Bertani (LB) medium or on LB solid medium (solidified with 1.5% agar) in the presence of 100 µg/ml of ampicillin. Mercury biosensor cells grown to log phase were immobilized in LB-agar matrix to prepare biodiscs and biofilms. These biofilms with embedded biosensor cells were also tested for their viability and analyte detection performance.

Measurements of Growth and Bioluminescence: Bacterial cell growth was measured in Spectronic 20 (Bosch & Lomb). Light emission from bacteria grown in liquid culture was recorded in a BioOrbit luminometer (Model 1250) in millivolts (Geiselhart *et al.* 1991) or in MGM Optocomp-I luminometer in relative light units (RLUs). Light emission from bacteria grown on solid media was recorded using an intensified CCD camera and an Argus-10 image processor. The images were analyzed with an LG-3 video capture board (Scion, Inc. USA) and the NIH image software (public domain) in a Macintosh II-CX computer (Fig. 1).

VISUALIZATION OF BIOLUMINESCENCE
BY CCD CAMERA

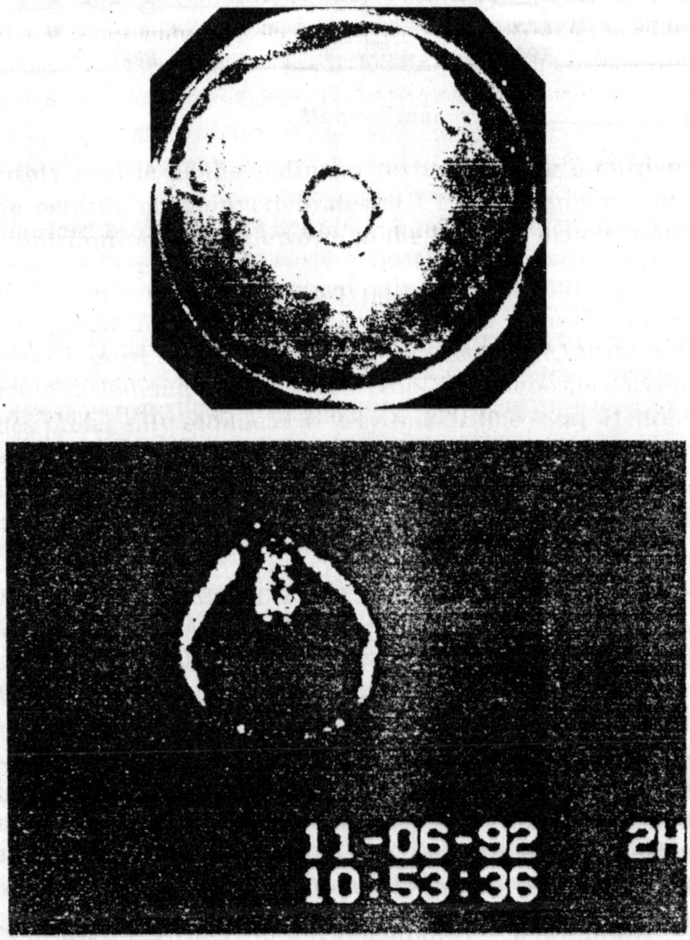

Fig. 1 A. Visualization of bioluminescence by CCD camera.

B. Real organizational set up of CCD camera, image intensifier, image processor, video monitor, VCR and personal computer (PC).

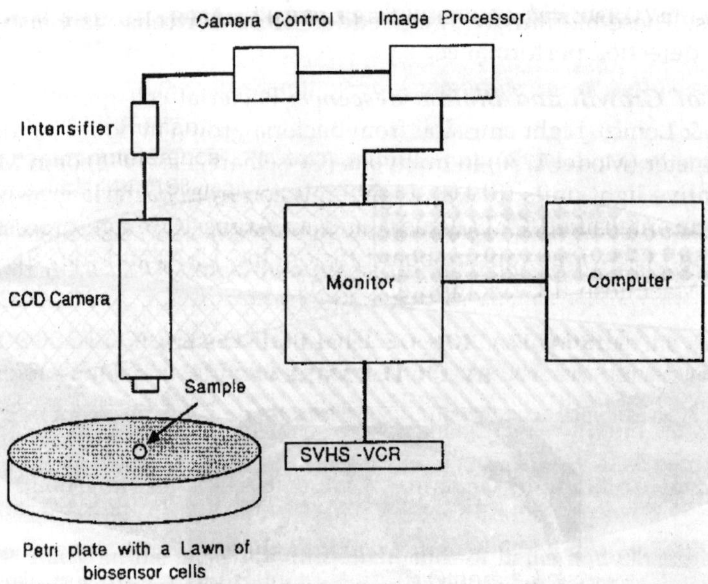

Fig. 1 C. Diagrammatic sketch of instrumentation set up to record bioluminescence.

RESULTS

Construction of Biosensor Organisms: Lux structural genes from *Vibrio fischeri* were fused with genetic regulatory elements from *Escherichia coli* and *Serratia marcescens* , which respond to copper and mercury ions, respectively (Fig. 2). Cells of *E.coli* JM101 were then transformed separately with these constructs. The transformants (biosensors), *E.coli* JMI01/pCULUX and *E.coli* JM101/pSD30 emit light in response to copper and mercury ions, respectively. Simplified diagrammatic representation of these constructs and the products of five *lux* structural genes are shown in Fig. 2A. The biochemical mechanism of bioluminescence is also shown in Fig. 2B.

GENES AND PROTEINS INVOLVED IN METAL INDUCED LIGHT PRODUCTION

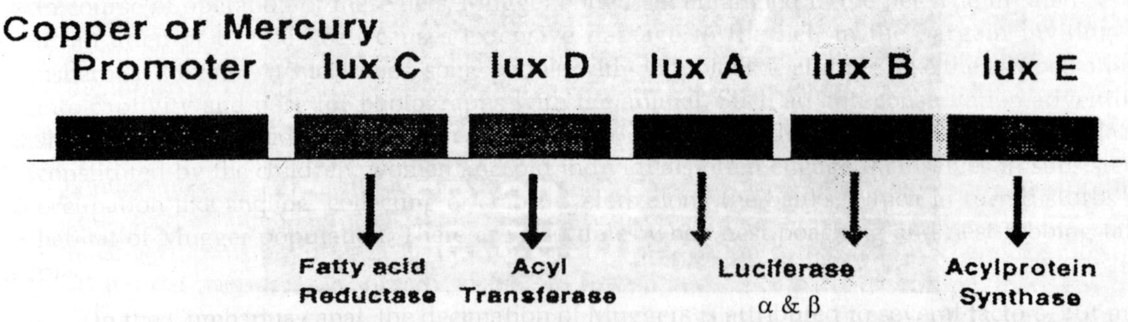

Fig. 2 A. Diagrammatic representation of the fusion of metal responsive regulatory elements upstream of the *lux* genes. They activate the *lux* genes resulting in bioluminescence in response to Cu and Hg ions.

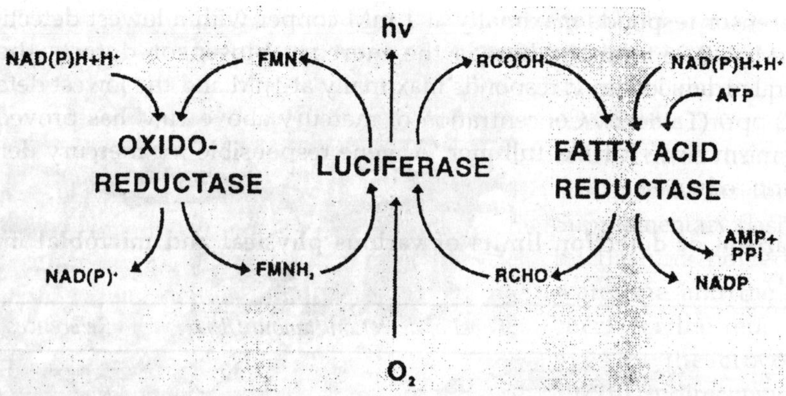

Fig. 2 B. Biochemical mechanism of bioluminescence.

Sensitivity and dynamic range of biosensors: The biosensors grown in LB broth responded to increasing concentrations of Cu and Hg. They both produced similar maximal light output. It is in the range of 10^5 to 10^6 RLUs (Relative light units) per ml of cell culture. 10^6 RLUs correspond to approximately $4x10^3$ photons/sec/cell. For both the biosensors the light output peaks in their midlog growth phase. The dynamic range, limit of detection and specificity are shown in Table 1.

Table—1 Dynamic range, limit of detection and specificity of copper and mercury biosensors@

	Cu biosensor	*Hg biosensor*
Concentration showing max. light output	1 mM	1 µm
Dynamic range	1 µM - 1 mM	0.01 nM – 1.0 µM
Lowest detection limit	1 nmol/ml	0.01 pmol/ml*
Specificity to:		
Target metal - Cu	>95.5%	0
Target metal - Hg	0	100%
Related cations (0.1 mM)		
Cobalt	0.5%	0
Cadmium	0	0
Nickel	0	0
Zinc	0	0

@ Data based on luminescence measured in BioOrbit luminometer.

* Using MGM Optocomp -I luminometer detection of 0.2 to 2 fg/ml (tested in culutre, biodisc) was achieved (for comparison with other methods, see Table 2).

The copper biosensor responds maximally at 1 mM copper with a lowest detection limit of 1 nmole/ml (1 μM = 0.06 ppm), whereas the mercury biosensor detects the metal at concentrations several orders lower. It responds maximally at 1μM and the lowest detection limit is 0.01 pmole/ml (2 ppt) (Table 2). Concentration of mercury above 1μM has proved lethal for biosensor microorganism since it lacks full mer 'A' gene responsible for mercury detoxification mediated by mercuric reductase.

Table—2 Comparison of detection limits of various physical and microbial methods for mercury

Method	*Level of detection (amount/L)*	*References*
Physical		
ICP-AES	10-20 μg (10-20 ppb)	Molders (1990)
Graphite- AAS	1-5 μg (1-5 ppb)	Molders (1990)
Hydrid- AAS	0.5 μg (0.5 ppb)	Molders (1990)
ICP-MS	0.1 μg (0.1 ppb)	Molders (1990)
MCA-90-AAS	0.01 μg (10 ppt)	Moffett (1992)
Cold vapor-AAS	0.001 μg (1 ppt)	Molders (1990)
Mercury biosensors		
E. coli C600 (pGL4)	0.2 μg (200 ppt)	Molders (1990)
E. coli JM101 (pSD 30) (biodisc)	2.0 pg (.002 ppt)	Present study
E. coli JM101 (pSD 30) (biofilm)	2.0 pg (.002 ppt)	Present study
E. coli JM101 (pSD 30) (culture)	0.2 pg (.0002 ppt)	Present study

ICP-AES: Inductively coupled Plasma-Atomic Emission Spectrometry

ICP-MS: Inductively coupled Plasma-Mass Spectrometry

AAS: Atomic Absorption Spectrometry

MCA: Mercury Concentration Accessory

Specificity of Biosensors: The mercury biosensor emitted virtually no light in presence of metals like Cd, Co, Ni, Zn and Cu. The background emission of light by the copper biosensor in the absence of added copper is only 0.5% of the fully activated level. The low background emission is probably due to the uptake of trace quantities of Cu contaminating the reagent grade chemicals used in preparing LB medium. In presence of Cd^{2+}, Ni^{2+}, Zn^{2+}, and Co^{2+} ions, the copper biosensor emits 0.5-1% of the level of light induced by Cu^{2+} (Table 1). Its response to Cu^{2+} and Cu^{1+} is almost the same.

Applicability of Biosensors: Both copper and mercury biosensors are very specific and sensitive to their target analytes. The copper biosensor can detect even copper ions released from penny and also from malachite ore (Fig. 3). In addition it can also detect trace amounts of copper

Visualizing Release of Copper

Penny Visualized in it's own light:

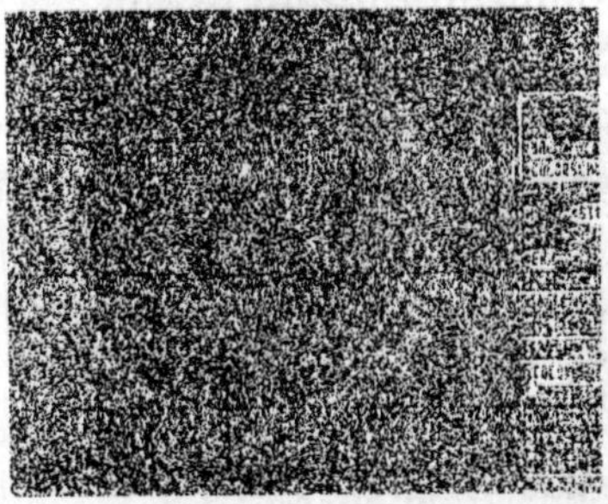

Fig. 3 A. Detection of copper released from a penny using copper biosensor.

Light from Malachite ore.

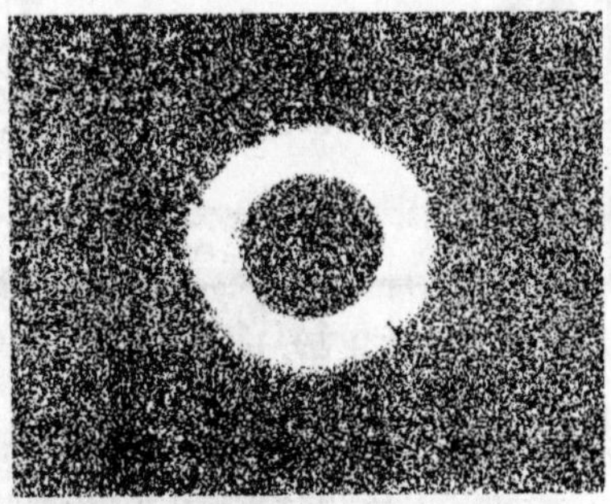

Fig. 3 B. Detection of copper released from copper ore, malachite Petri plates were prepared with 20 ml of Luria-Bertani agar and surface-inoculated with 100 µl of overnight grown culture of *E.coli* JMI01/pCulux. A piece of malachite ore or a penny was placed in the centre of the agar plate. The plates were then incubated at 30°C.

ions present in mine tailings (Fig. 4). The mercury biosensor even responus to soil contaminated with mercury (Fig. 4).

The biosensor cells were grown in LB agar and poured into petri plates to develop biofilms and biodiscs. By direct spotting of $HgCl_2$ on these plates trace quantities of mercury (down to 0.5mg) can be detected. The biodiscs and biofilms prepared from LB agar embeded biosensor cells can detect as low as 2 fg Hg^{+2} /ml when the induced bioluminescence was measured in a highly sensitive Optocomp-1 luminometer. The sensitivity is further increased by 10-fold by studying the effect of mercury on growing cultures. The detection limits achieved by mercury biosensor cells and those by atomic absorption method are compared (Table 2). The mercury biosensor cells also detect ultratrace level of mercury present in water, saliva of dental patients and mercury released from dental fillings (Figs. 5 and 6).

DISCUSSION

Though there are several reports of non-*lux* reporter genes, but in few cases only *lux* genes have been used to develop specific biosensors to detect specific metal analytes (Holmes et.al.

DETECTION OF COPPER AND MERCURY FROM ENVIRONMENT

Detection of Mercury in Soil

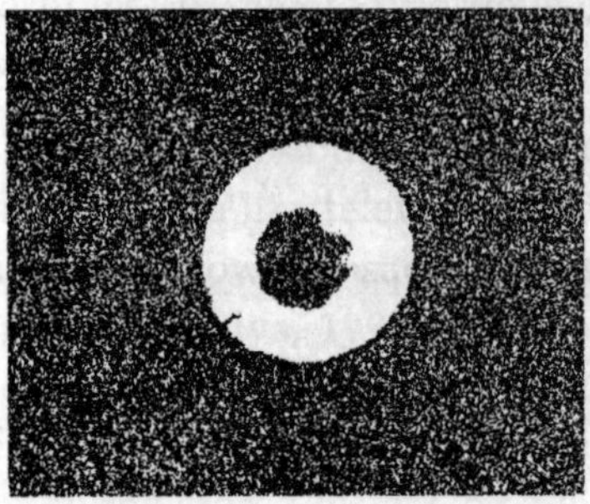

Fig. 4 A. **Detection of mercury in soil sample by mercury biosensor.**

Detection of Copper in Mine Water

Fig. 4 B. **Detection of copper in mine tailing (water) by copper biosensor.**

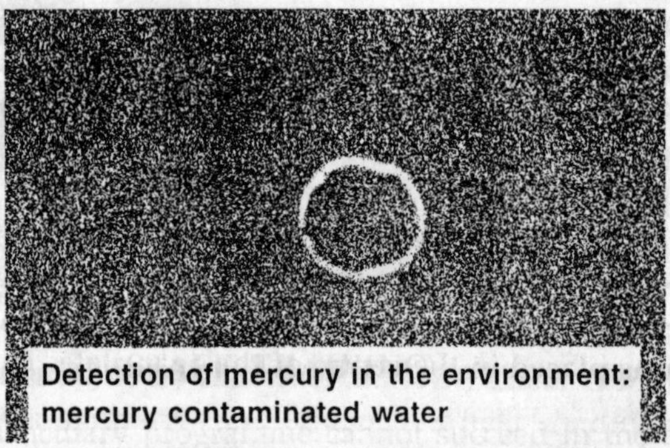

Fig. 5 **Detection of mercury in mercury contaminated water sample by mercury biosensor *E.coli* JMlOl/pSD30. LB agar plates were surface-inoculated with the biosensor microoganism, and sample was placed in the centre of agar plate with lawn of biosensor cells. Light is seen as a halo around the test material.**

1993a). The biosensors reported here can detect mercury and copper ions in the concentration ranges that are useful for environmental monitoring. Besides, they appear to be sensitive and specific for their target metals and do not suffer interference from relatively high concentrations of potentially competing metal ions. Their specificity may be attributed to their construction strategy since metal responsive genetic determinants (genes) are fused upstream to the *lux* reporter genes. This is accomplished by transcriptional fusion technology.

The ease of detection of light by naked eye and its quantitation by luminometers makes this type of biosensors especially valuable and attractive. They could be effectively used to monitor metal ions in a non-destructive fashion and almost in real time. The method also permits multiple or continuous image recording on a single sample over a period of time through a CCD camera.

Visualization of Mercury in Dental Samples

Extracted tooth with fillings

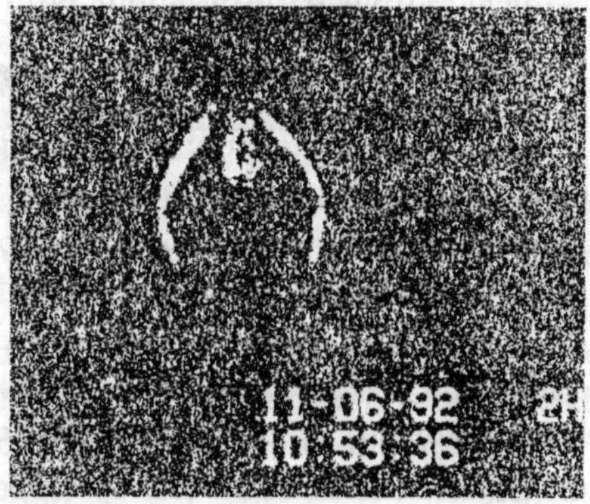

Saliva of a person with dental filling

Fig. 6 Detection of trace quantities of mercury by mercury biosensor *E.coli* JMlOl/pSD30 in dental amalgam (A) and saliva sample (B). Light halo was observed around dental filling showing a small inhibitory zone whereas in case of saliva sample all biosensor cells in contact with saliva were glowing.

The bioluminescent zones captured as images are amenable for quantification, using image analysis software from NIH, U.S.A.

The traditional analytical methods do not distinguish the bioavailable component in the total chemical concentration, which is vital information for two reasons. It is this form which is responsible for the toxicity and needs bioremediation. The biosensors developed in the present study are capable of detecting the bioavailable mercury in soil, water, dental fillings (amalgam), saliva and copper released from malachite ore, penny and also in mine water.

The mercury biosensor shows exquisite specificity and responds to ultratrace levels of mercury. The lowest level of mercury detection achieved has far surpassed (by 500 to 1000-fold) the sensitivity reported for cold vapor atomic absorption spectrometry. We have successfully evaluated the viability of biofilms and biodiscs developed by immobilization of biosensor cells in LB agar matrix. Therefore, these microbial sensors also possess immense potential for commercialization as portable device to monitor copper and mercury ions in environmental as well as clinical samples.

ACKNOWLEDGEMENTS

Authors gratefully acknowledge financial support from office of Naval Research, U.S.A.. Authors also feel obliged to Dr. Simon Silver, Dr. B.T.O. Lee and Dr. C.I. Kado for their plasmids and *E.coli* strains as gifts.

REFERENCES

1. Beaty, R.D. 1988. Concepts, instrumentation and techniques in atomic absorption spectrophotometry, Perkin Elmer Corporation, USA.
2. Burlage, R.S., G.S. Sayler and F. Larimer 1990. Monitoring of naphthalene catabolism by bioluminescence with nah-*lux* transcriptional fusion. **J. Bacteriol.** 172:4749-4757.
3. Cervantes, C. and H. Ohtake 1988. Plasmid determined resistance to chromate in *Pseudomonas aeruginosa*. **FEMS Microbiol. Lett.** 56: 173-176.
4. Cervantes, C. and S. Silver 1992. Bacterial chromate resistance and reduction. **Plasmid** 27: 65-71.
5. Corbisier, P. and S. Silver 1992. Overview of bacterial heavy metal resistances. **In**. Proceedings of Conference on the bioremediation of sediments held in New Jersey, USA.
6. Dubey S.K., S.S. Gangolli and D.S. Holmes 1993. Development of biosensors to measure metal ion bioavailability in environmental samples. **J. North East. Geol.** 15: 88-94.
7. Geiselhart L., M. Osgood and D.S. Holmes 1991. Construction and evaluation of a self luminescent biosensor. **Ann. New York Acad. Sci.** 646:53-60.
8. Gangolli, S. S., S.K. Dubey and D.S. Holmes 1993. Development of biosensors for copper and mercury., **In**. Proceedings of 93rd General meeting of American Society for Microbiology held in Atlanta, Georgia, U.S.A.
9. Heitzer, A., K. Malachowski, J.E. Thonnard, P.R. Bienkowski, D.C. White and G.S. Sayler 1994. Optical biosensor for environmental online monitoring of naphthalene and salicylate bioavailability with an immobilised bioluminescent catabolic reporter bacteria. **Appl. Environ. Microbiol.** 60: 1487-1494.
10. Holmes, D.S., S.K. Dubey and S.S. Gangolli 1993a. Genetically engineered biosensors for environment monitoring, food testing and health applications. **In**. Proceedings of Biochemical society of chile.
11. Holmes, D.S., S.K. Dubey and S.S. Gangolli 1993b. Development of biosensors to measure metal ion bioavailability in mining and metal wastes. **In**. Torma, A.E., M.L. Apel, and C.L. Brierley (Eds). **Biohydrometallurgical technologies** pp. 659-666. The Minerals, Metals and Materials Society, Philadelphia, U.S.A.
12. Iizumi, T., M. Mizumoto and K. Nakamura 1998. A bioluminescence assay using *Nitrosomonas europaea* for rapid and sensitive detection of nitrification inhibitors. **Appl. Environ. Microbiol** 64: 3656-3662.
13. Lebrun, M., A. Audurier and P. Cossart 1994. Plasmid borne cadmium resistance genes in *Listeria monocytogenes* are similar to Cad'A' and Cad'C' of *Staphylococcus aureus* and are induced by cadmium. **J. Bacteriol.** 176:3040-3048.

14. Loessner, M.J., C.E.D. Rees, G.S.A.B. Stewart and S. Scherer 1996. Construction of luciferase reporter bacteriophage A511:: *lux* AB for rapid and sensitive detection of viable *Listeria* cells. **Appl. Environ. Microbiol**. 62:1133-1140.

15. Moffet, J. 1992. The routine measurement of ultratrace levels of mercury. **Am. Environ. Lab**. 10/ 92: 18-24.

16. Molders, H. 1990. German Patent application. W090/08836.

17. Ohtake, H., C. Cervantes and S. Silver 1987. Decreased chromate uptake in *Pseudomonas fluorescens* carrying a chromate resistance plasmid. **J. Bacteriol.** 169:3853-3856.

18. Rechnitz, G.A. 1988. Biosensors. **Chem. Engg. News.** 66:24-36.

19. Silver, S. and T.K. Misra 1988. Plasmid mediated heavy metal resistances. **Ann. Rev. Microbiol**. 42: 717-743.

20. Silver S. and L.T. Phung 1996. Bacterial heavy metal resistance: New surprises. **Ann. Rev. Microbiol.** 50:753-789.

21. Wahlstrom, G. and P.E.J. Saris 1999. A nisin bioassay based on bioluminescence. **Appl. Environ. Microbiol.** 65:3742-3745.

18

Ecological History, Contemporary Status and Conservation Future of Marsh Crocodile *Crocodylus Palustris* in Goa

***Manoj R. Borkar**

ABSTRACT

In India, the Mugger crocodiles (*Crocodylus palustris, LESSON*) with a widespread distribution had adapted to a wide range of environmental conditions once upon a time. Today it is confined to few rivers, reservoirs, man made tanks and even irrigation canals. Goa is one of the few states which has a wild population of Muggers. Based on reliable memory of living witnesses, as also on the perusal of historical records, it is known that the Muggers had infested the estuaries, rivers and the riparian forests of Goa until about four decades ago. Habitat alterations and loss, through the post-liberation times has decimated the reptiles to a point of no return; and today the only surviving wild population of Muggers in Goa is confined to a small stretch of mangrove fringed and tide flushed canal of Cumbarjua.

The reptile is vulnerable to a variety of anthropogenic pressures. Today it is a victim of apathy and indifference of statutory agencies in the state and is reduced to a "biocommodity", unsustainably exploited by the eco-tourism entrepreneurs. Their numbers are declining at an alarming rate and in the last census conducted in 1993 only 14 individuals were counted **(Borkar, et al., 1993 b).** The Mugger, which defended the Goan territory from invaders during the pre-Portuguese times, is now defenceless against its degraded habitat in its very own land.

It is time for stakeholders at all levels of this precious resource to reflect on the situation and initiate remedial action. This paper makes a scrutiny of all the pertinent issues concerning the Muggers and its habitat in Goa. It also highlights a unique eco-theological approach of

* Bio-diversity Research Cell, Carmel College for Women, Nuvem-Goa, India— 403 604 carmel@goatelecom.com

conservation of the reptile practiced by the eco-tribes of Goa **(Borkar, 1992).** The paper also offers suggestions for conservation and sustainable management of the Muggers in Goa. The data presented here is a part of first status report on the Mugger of the Cumbarjua canal of Goa.

HISTORICAL BACKGROUND

Ancestors of present day crocodiles probably appeared on the earth during the Triassic era. They have adapted to exploit waters and land, and thus possess advantage of utilizing resources from both the life zones.

India has three species of crocodiles, the Mugger *(Crocodylus palustris)*, the salt-water crocodile (*Crocodylus porosus*) and the Gharial (*Gavialis gangeticus*) **(Daniel, 1983).** These species have a specific habitat distribution; the Muggers are encountered throughout the lowlands; the salt-water crocodile on the coastal northeastern parts and the Gharial was restricted to the more inland northern rivers. Reports and travelogues provided by hunters and travelers visiting India in late 1800 and early 1900 had similar picture to project about their well being **(Shortt, 1921).**

In fact, the dam construction projects that were carried out between 1900 and 1950 had exterminated the crocodiles and irreversibly altered the habitat. It is believed that remnants of some of the *Crocodylus palustris* survived and these formed the nuclei of the small present day population in reservoirs. In the early seventies reports of Zoological Survey of India, the Bombay Natural History and the Madras Snake Bank Trust drew attention of the people to the severely depleted status of all these species due to over hunting, habitat destruction and egg eating by natural predators and some tribes.

The government of India in recognition of this alarm then requested the services of the FAO for devising the remedial measures **(Biswas, 1970; Anon, 1972; Whitaker, 1974; Whitaker & Daniel, 1978; Gupta & Srihari, 1990).** Protective legislation began in the early 1970's and management programme was envisaged with technical help from the FAO & UNDP. The national programme was principally focused on rehabilitating the wild population through captive breeding in 9 states. Consequently the states of Tamilnadu, Orissa, Andra Pradesh, Uttar Pradesh, West Bengal, Madhya Pradesh, Bihar & Rajasthan have been most active in the field of crocodile management. A conspicuous feature of the FAO programme was to help create protected areas for crocodiles and promote *in situ* conservation of this valuable reptilian resource. The managerial base for this long-term programme was later established by creating the Central Crocodile Breeding & Management Training Institute at Hyderabad.

Extensive work has been done on the biology and ecology of the Mugger crocodile in India and elsewhere. Data on life history of *Crocodylus palustris* has been reviewed by **Whitaker (1977b)** The species has a roughly triangular range. In the west, they extend up to Sarba River, in Iran, eastwards; in Pakistan also their position is precarious. Even the once famous captive assemblage at the sacred Muslim shrine of Karachi, the *Mugger Pir*, is now reduced to a few individuals. In Nepal small isolated population of Mugger is found in several tributaries of the Ganges, at the base of the Himalayas. In Bhutan and Bangladesh too, a few Muggers have been reported. But the most dramatic number of Muggers has been encountered in Sri Lanka, concentrated in two of its Natural Park, Yala & Whilpattu.

In India, the Mugger crocodiles *(Crocodylus palustris)* with a widespread distribution had adapted to a wide range of environmental conditions once upon a time. Today it is confined to few rivers, reservoirs, man made tanks and even irrigation canals.

Goa is one of the few states that have a wild population of Muggers. The reptile has been protected under wild life protection Act that came into force on 9th September 1972 and was made applicable in Goa with effect from 16th July 1973. Unfortunately, this reptile has been neglected in this state for an unpardonable length of time. Review of the historical literature reveals that the crocodiles have been living in Goa since antiquity **(Decada Segunda Livro V).** Although no systematic study had been carried out on the crocodiles of this region on earlier occasions, it is a paradox that their long-standing existence in this state and particularly in the Cumbarjua canal is well documented. Besides their antique status, their role in territorial defence could probably be one of the earliest examples of bio-warfare. Available literature suggests that forty years prior to the Portuguese conquest on Goa, the Muslim rulers of this land had used crocodiles for defending their territorial boundaries. **(Borkar & Mallya 1995b).** On retrospective temporal scale this dates back to Adilshah's rule in Goa.

The original report also highlights the controversy of their origin in that place, either that the nature has produced them there or that they might have been brought in for the purpose of defending the island. A similar opinion is held by **Deraniyagala (1936)** on crocodiles of Sri Lanka, who writes *"in all probability it was this reptile which was so petted by the Portuguese soldiers at Malwara, Colombo, Kalutara and other river forts, and Kayman's gate in Colombo perpetuates the memory of their former abundance"*. Thus the existence of wild Muggers in Goa since antiquity stands confirmed.

This report also has a reference to the island of Tiswadi starting from the east pass called Banastari, that entire island was infested by the *" lizards of waters"* with horns of cattle filling their mouths. This was the only document that makes a casual reference to the crocodiles of Cumbarjua canal.

Village elders in Goa narrate exciting accounts of the river banks occupied by these *'basking monsters'* even as recently as four decades ago. The reptile was such a menace then, that hunters were invited to shoot them. Crocodile skulls, teeth, skin and other artifacts adorning the private collections of some elite families in Goa are a testimony to the hunting pressures on the Muggers in the bygone days. The quickened pace of development in the post liberation days caused further fragmentation of their habitats and an accompanying decline in their numbers.

There are also confirmed reports of crocodiles from other parts of Goa such as Quepem, Aldona, Panchwadi, Amona, Madkai, Chorao, Thivim and Carambolim. It is very interesting and unusual that Crocodile have also been kept as pets in the Varca village.

It also deserves to be stated that Whitaker's consultancy report on the manmade lake and marsh land of Ciba Geigy Santa- Monica plant in Goa, does mention the existence of wild crocodiles in the Zuari River and the Cumbarjua canal. **(Whitaker, 1991).** The report records that the crocodiles inhabiting this manmade lake are Muggers and imply that the resident population of the canal constitutes a single species population of *Crocodylus palustris*.

However their long-standing existence in Goa failed to arouse interest in the scientific community and no systematic study on their status could be accomplished until about a decade ago **(Borkar & Mallya 1993 a)**. So also there was an acute lack of authentic information on their taxonomic position **(Untawale, 1990)**. Crocodiles occasionally sighted by fisher folks in Goa have been mysterious animals indeed. The mystique surrounding these reptiles had been precipitated by the exaggerated accounts of people living in their range.

Despite the varied environmental conditions that it has been adapted to, it is a fact that it generally prefers to remain closely associated with a source of water, which could either be pond, lake, river or even estuary. Muggers have found a niche in many riverine as well as estuarine ecosystems of Goa. It is rather surprising that Muggers have adapted very well to the saline waters of the Cumbarjua canal, which is unlikely to have happened anywhere in its entire Indian range. Following a catastrophic decline in their numbers due to indiscriminate hunting and allied environmental pressures, the Muggers are now rare in Goa and are virtually restricted to the Cumbarjua canal which is fringed by a luxuriant mangrove cover **(Plate 1)**. Perhaps this is the last natural pocket for the assemblage of these wild species and therefore the future integrity of this canal and its ecology must be guaranteed in the context of the environmental pressures that have been identified for the species in Goa.

Plate 1 Luxuriant mangroves and exposed mudflat of Cumbarjua canal during a low tide.

Crocodilian environment is extremely fragile and is subject to a variety of pressures, some of which are natural and others manmade. Most unfortunately, the reptiles in general have failed to gain sympathy and support of people as a result of their unfamiliar and secluded life style. Therefore it is not very surprising that these reptiles are still regarded as being in league with evil powers **(Borkar, 1993 c)**. Bible has made a reference to the crocodile as being an evil, fire-spitting creature known as *Leviathan*. Similarly *Vishnu Purana* has a story of *Gajendra Moksha* in which the crocodile has been depicted as having attacked Lord Indra's vahan, an elephant called *Aairavath*. Much has not been changed in this respect of human attitude, because contemporary man still confronts these animals with little understanding. In this study the attitude concept of the people towards the crocodile in Goa has been studied using standard procedures **(Kellert, 1976; 1980)**

STUDY AREA

The state of Goa (15^{0}48'00"N & 74^{0}20'54"E) lies on the central west coast of India, with length of 105 km^2 and an altitudinal range from sea level to 1,022 meters. The state of Goa is

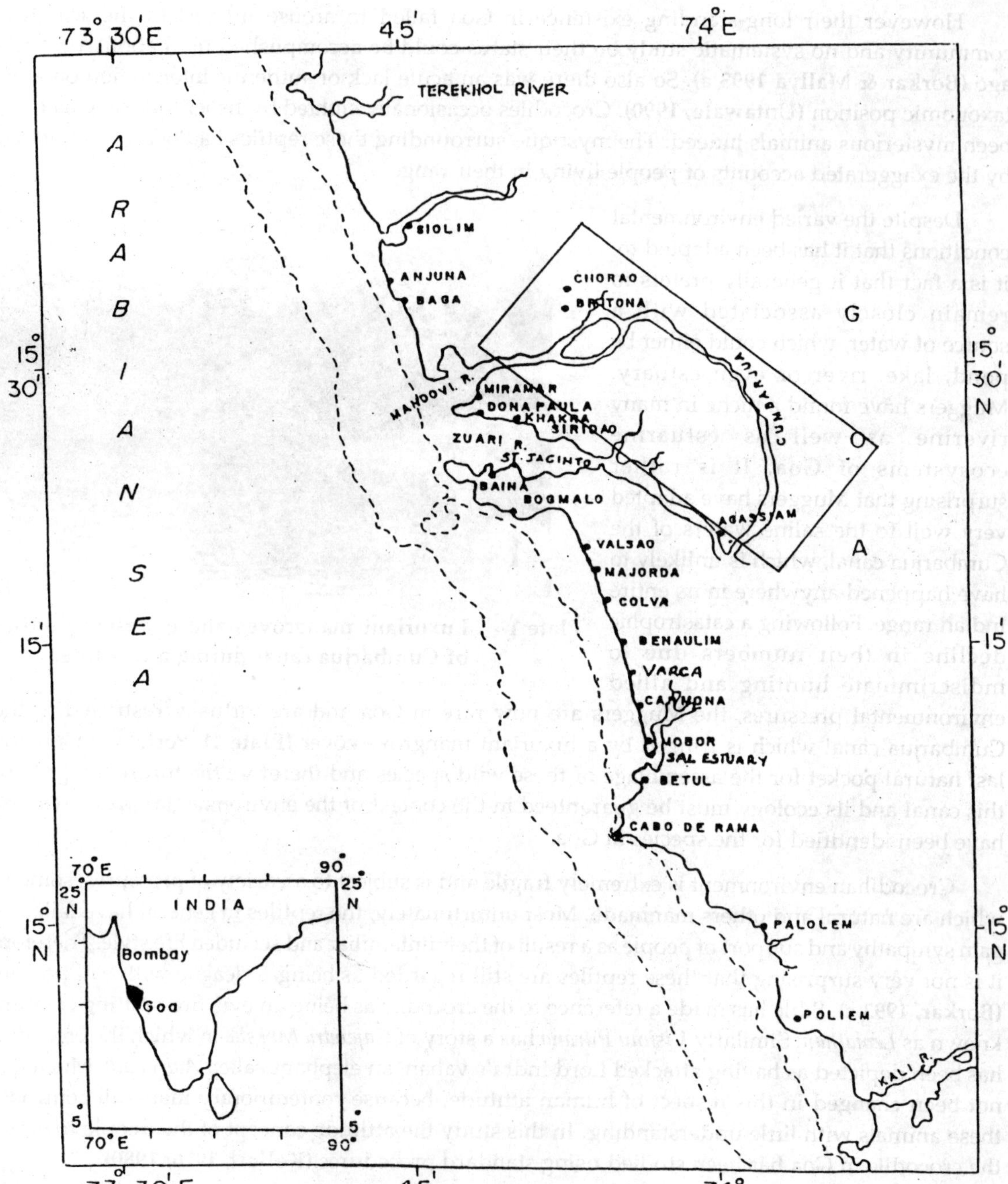

Fig. 1 **Details of the area under study**

bounded on the North by Sindhudurg district of Maharashtra, on the East by Belgaum district, on the South by Karwar district of Karnataka and on the West by Arabian Sea. All its rivers have origin in Sahyadris and are navigable throughout the year. All these rivers flow westward into the Arabian Sea. Between the rivers and the sea are the estuaries and the bays that pass through the hinterland.

The estuaries of Goa are fringed by a luxuriant mangrove cover. Cumbarjua canal, the site of this investigation alone has more than 15% of the mangrove cover of the state. The dominant species here are *Rhizophora mucronata, Sonneratia alba, Avicennia officinalis, Kandelia candel* and *Bruguiera gymnorrhiza*. The canal connecting the Mandovi and Zuari estuaries has navigable waters measuring about 17 km in length **(Plate. 2)**. The canal has an average depth of 3 meters. The salinity over here varies widely between 0.5 parts per thousand to 35 parts per thousand.

Plate 2 Navigable canal waters: A barge carrying iron–ore.

MUGGER CENSUS

Knowledge of the numerical strength of a species is useful in designing the conservation programme for that species. The density of population has spatial and temporal frames of reference. It is an established fact that density of populations is not static but rather a dynamic character. It is therefore essential to arrive at realistic as well as statistically confident figures to know the population trends. Then only would one come to know whether the population is increasing, stable or on a decline and accordingly appropriate management may be recommended.

The dynamics of population could be followed only if enumerations are accomplished periodically. Once the status of the population becomes clear one could arrive at an agreement on number of individuals of the species that is in a good balance with the environment, without violating the carrying capacity of their actual habitat.

Wildlife population enumeration is a difficult task, considering the adversities encountered in adopting the methodologies. Also it needs to be emphasized that the procedures that look simple on the paper, take an entirely different turn when transferred to the actual field situation.

METHODOLOGY FOR MUGGER ENUMERATION

The entire length of the canal along with its associated creeks was scanned using a motorboat maneuverable in shallow water. **(See map)** The boat was run in the core waters equidistant from

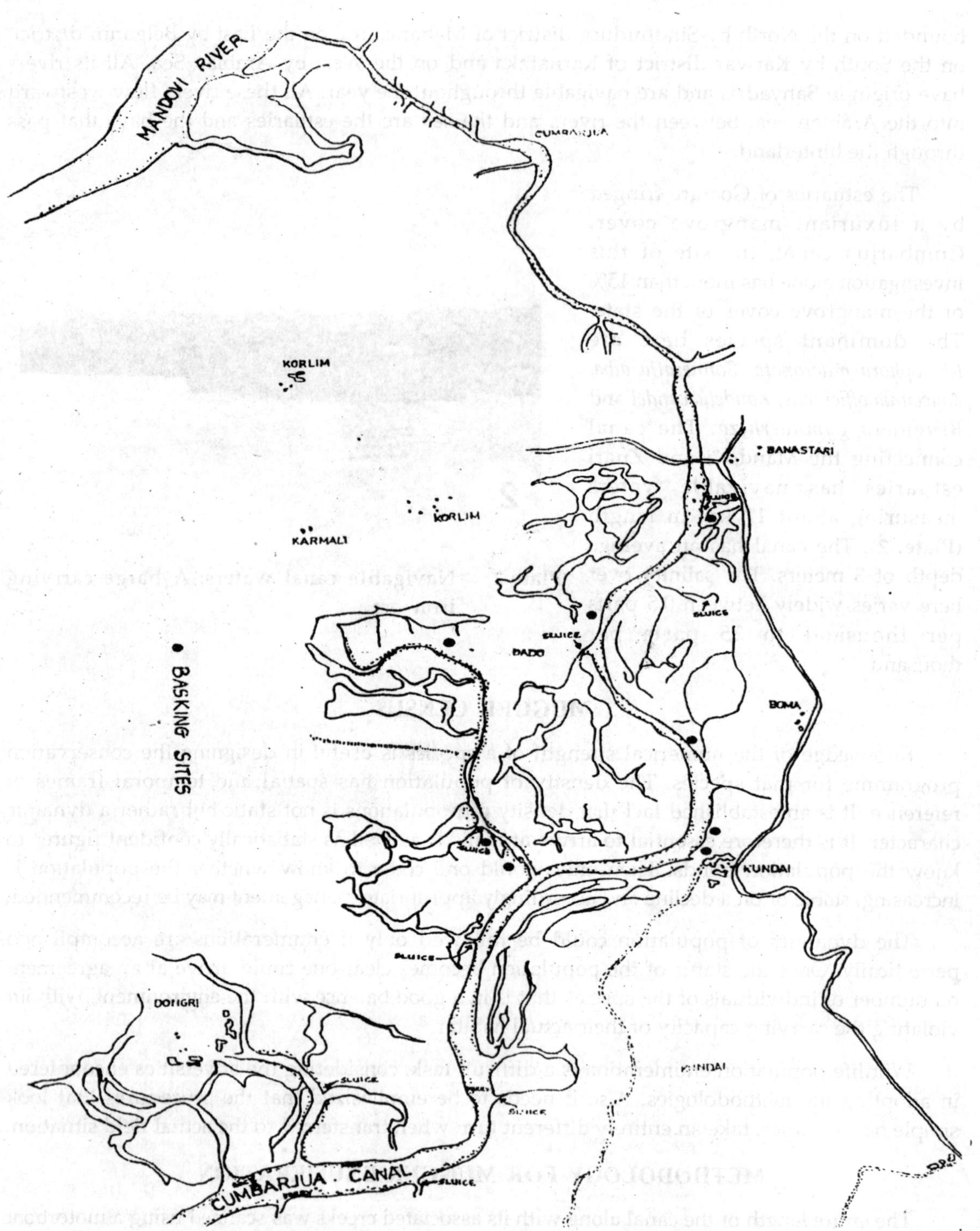

Fig. 2 Toposheet of the Cumbarjua canal showing basking sites of the muggers.

Plate 3 The croc has left imprints of its claws on the wet yet compact silt.

both the banks of the canal with moderate speed, to avoid disturbing the shoreline by propeller generated waves. The population was enumerated in relative as well as absolute terms through day counts. Supplementary flash counts were made by flashing powerful torches on the waterline and the bunds at night to check for the reflected glow from the eyes of the crocodiles **(Whitaker, 1977b).** Indirect evidences such as claw imprints on the exposed mudflats were also taken into account **(Plate 3).**

Having sighted a crocodile, its position was marked on the topography sheet provided by the forest department with the eventual aim of understanding the distribution pattern of the resident population and preferred basking sites of this reptile **(Plate 4).**

Plate 4 A *'basking Monster;'* an adult croc on a dyke.

Relevant information was collected from the locals and the fishermen in particular; whose knowledge of crocodilian ecology can surpass that of a qualified herpetologist. The collected information was then processed to verify the causes of numerical decline, human attitude etc **(Borkar et al., 1993 b).**

SOME OF THE HIGHLIGHTS OF THE PRESENT STUDY ARE AS FOLLOWS

The distribution pattern of the Mugger population in this canal was found to be rather patchy, the greatest concentration being found between the Banastari bridge and the Karmali creek. Adult Muggers preferred to reside along the main canal **(Plate 5)** where as the juveniles were mostly confined to the remote creeks **(Plate 6).** Most of the Muggers were found to bask perpendicular or diagonal to the canal waters.

The extensive survey done by us, based on Kellert's Scheme **(Kellert, 1976; 1980)** which provides insight into the attitudes of the resident public along the canal, who are an integral

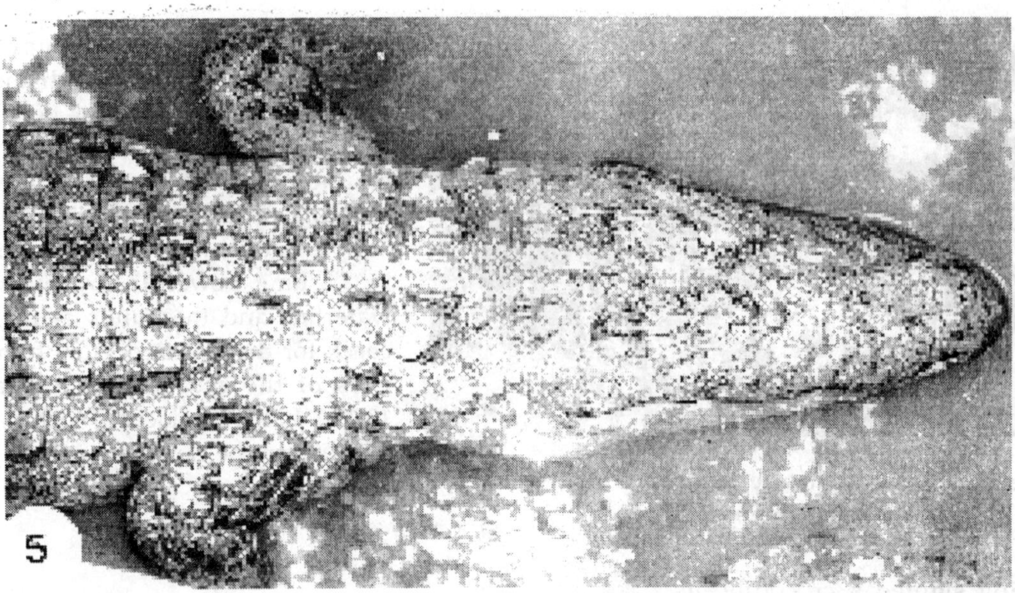

Plate 5 **An adult Mugger partially submerged in waters: Note the four occipitals behind the head.**

Plate 6 **A Mugger juvenile among "breathing roots"**

part of the reptilian environment. Our survey of Boma and other crocodile habitats in Goa, where fisher folks form an inseparable parts of crocodilian environment, reveals that fisher folks have an element of hatred regarding this animal as they feel that their means of livelihood has been threatened by the crocodile, which competes for the fish. However these views have been contradicted by investigators, who have found that waters inhabited by crocodiles also have a good fishery potential. In fact they are known to be performing an ecological role of predator as well as scavenger. Such a myth has been conclusively ruled out by gut content analysis.

It is interesting to ponder over people's reluctance to change their unscientific preconceptions even after being confronted with sound scientific evidence. Even educated individuals are a victim of such selective reading and retention. A case in point is the Mugger's image as a man-eater. Within reliable memory of living witnesses to crocodile's past glory, that the author has met, the attack has never been actually witnessed but only heard of. The narrations, always being laced with exaggeration. Only one man was seen with scars of crocodile bites on the thigh, when as a child he was attacked by Mugger. This could well be an act of miss-predation by the Muggers. However the fortunate victim survived to narrate this horrific experience. In the recent past a much hyped account of a boy having been attacked and eaten by crocodile appeared in local newspapers. The body of this ill- fated 5 years old boy was found floating in a small rain fed laterite quarry pit in Ponda taluka, with his soft parts eaten away, strengthened the image of the reptile as a man-eater though for a short time. The post mortem of the victim however did not corroborate this story. On the contrary forensic experts absolved the crocodile as being the cause of boy's death, implying that the reptile was only scavenging on the body.

As indicated by the survey, level of education and gender appeared to be the most consistent social predictor of an individual's view. Women were inclined towards moralistic attitude whereby they expressed concern for the welfare of the crocodile, while a few women showed indifference. Naturalistic, utilitarian, dominionistic and aesthetic attitudes were more characteristic of men than women. Degree of education and positive attitude were parallel. It was realized from the analyses of our findings that the support among the general public for the crocodile was largely emotional rather than knowledge based.

The findings of the attitudinal survey have implications in the management of the species, as the concerned agencies have to deal with rather larger interest groups with varied interests. It is only through incorporation of this knowledge of attitudes, that wildlife professionals can provide responsible stewardship.

Mangroves of Cumbarjua, which is yet another important resource that is neglected and mismanaged is in need of immediate attention. In a recent Conservation Assessment and Management Plan Workshop under the aegis of Bio-diversity Conservation Prioritisation Project India, a total of 60 taxa of mangrove plants were assessed according to the New IUCN categories. It was a disheartening find of this assessment that most of the taxa were under threat; habitat loss, human interference and trade being the major causes. **(See pie chart I)**. Socio-economic factors of the resident population influence the extent of pressure on the mangrove forest. The decrease in fish catch in the recent years has forced people to fell the mangroves forest for firewood. **(Whitaker, 1977a; Bustard & Choudhary, 1980; Whitaker, 1982; Borkar et al., 1993 b; Borkar, 1993 c)**. Dependence of crocodiles on mangroves was well demonstrated on the Malabar coast of south-west India in the early 1900's, where destruction of mangroves for increase in

PIE CHART I: THREATS AFFECTING MANGROVES

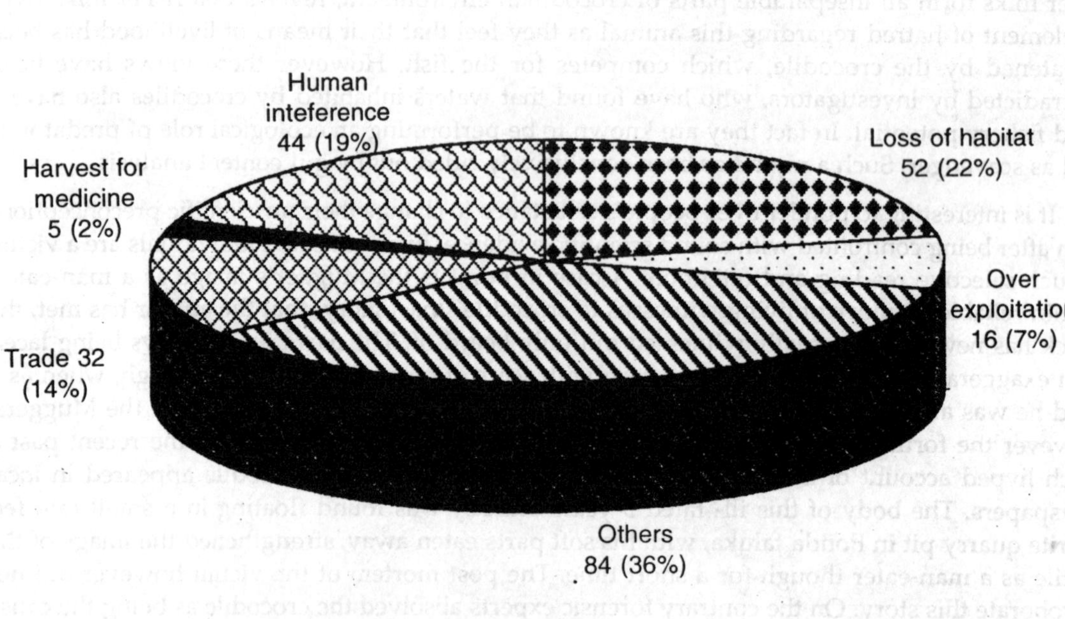

Number of mangrove species assessed = 60

Number of threatened mangrove species = 58

Source: C.A.M.P Workshop on Indian mangrove Ecosystem (1998)

human needs limited the chance of survival of *Crocodylus porosus*. There is no reason to disbelieve that mangroves are indispensable for *Crocodylus palustris* also.

Traditional fisheries that lack an aspect of organization pose a serious threat to the Muggers of this canal. The *'Khutani'*, which is a traditional gear of the locals, is the most common death trap for the Crocodiles of this canal besides the *'Hari'*, which is a bigger version of the *'Khutani'*. In course of operation of these nets, Muggers often get entangled in the net. The ill-fated reptile in an attempt to free itself causes extensive damage to the net, in the bargain inviting the fishermen's wrath. At such times, some people with dominionstic attitude take the trapped animal into captivity and pose for photographs with the animal. Such an anti-conservation adventures should be discouraged and not given any publicity. The less productive segment of the population constituted by the children, women and old individuals often engage themselves in subsistence occupation like angling, collecting of crabs, oysters along the banks, which in turn disturbs the habitat of Mugger population. There are also times when nest poaching and nest robbing takes place.

In the Cumbarjua canal, the decimation of Muggers is attributed to several factors, but most commonly accidents, hunting and trade take a significant toll of these animals **(Borkar, et al., 1993 b; Borkar, 1994 a)**. There are few cases reported, wherein the reptiles often get caught in the barge propellers that navigate through this canal.

In Goa crocodile killers are usually a group of itinerant fishermen and nomadic tribes. Information regarding poaching statistics is difficult to obtain unless a good rapport is built with the resident population. The decimation pressure on the animal exerted by the local people is mounting. The *'converted Gavdas'* of Goa rank foremost among the crocodile decimating tribes of Goa. Once upon a time when the Mugger populations were undisturbed, spearing and shooting were common methods of hunting, but baited hooks are more common now.

It is often found that when people volunteer to give information on poaching, the element of exaggeration cannot be ruled out. Various nomadic tribes establish makeshift camp by the canal. For example *Piddukkars* or *"Bead sellers"* of Karnataka, poach the Muggers.

Apart from the cash value attached to this reptile, the Mugger also has an ethno-medical value. Locally known as *'ons'*, the crocodile fat, which on melting yields oil, is deemed to have curative properties against rheumatism. There are reports that the liver and spleen cure bronchitis, gall bladder has remedy for cataract and crocodile eggs are fine nutritional package, complete with vitamins and minerals. Such medicinal values attributed to crocodiles put additional pressures on this species, which may not be reported from the non-Indian range of this animal.

The Mugger has not been emancipated from the hazardous fallout of unplanned developmental activities. This reptile has a potential to support wildlife tourism. In the lure of sighting moving crocodiles, the foreigners are often exploited by the unskilled guides who charge them heavily. Such unplanned wildlife tourism generates seasonal employment, but is a serious disturbance to this reptile, considering the fact that they are often disturbed by oars. The construction related earthmoving operations also result in heavy sedimentation and siltation of the waters. The vulnerability of Cumbarjua canal for effluent discharge is obvious, taking into account number of industrial units as well as human settlements situated along its banks. The fish processing units discharge enormous quantities of organic wastes, which lower the oxygen content of its water. In this context, it is alarming to note that many such fish canning units dump enormous quantities of spoilt fish, besides the byproducts of canning industries. So also pesticide manufacturing unit also has its effluent discharge point in this water body. Similarly significant quantities of non-degradable waste like plastic containers are found floating. Faecal contamination is another common type of organic pollution. Proximity of the agricultural land to the canal enhances the possibility of agricultural runoff rich in pesticides, during the monsoons. The possibility of the transfer of the pesticides along the food chain and biomagnifications cannot be entirely ruled out as being hypothetical. If such pollutants continue to be dumped hap hazardously, they would naturally reduce the potentialities of the ecosystem to support a rich biological spectrum, of which crocodile is a part.

A comprehensive Management Plan For Rehabilitation Of Muggers In Goa: Some Key Issues & Approach:

- One of the most chronic problems in wildlife management is lack of public support which is due to basic failure of wild life biologists to articulate their arguments effectively. Major weakness in modern wild life managers is in the areas of communication and public relations. Such communication gap often strengthens the mistaken impression, that principles of wild life management are guided by environmentalism or preservationist philosophy that is opposed to economic development.

- The conventional conservation policies and practices had their limitations, as their historical roots have been anchored in the colonial era and therefore are ineffective and defunct in the present context. Wild life managers simply cannot refuse to take cognizance of dependence of local people on natural resources, as also their traditional resource management practices and customs. It is a deliberate indifference and ignorance of their eco-friendly cultural ethos that has made these people hostile to the modern wild life protection programmes
- Include grass root level stakeholders of these *'ecosystem goods'* and decentralize wild life management to make the way for participatory conservation. The urgent need of the hour is to explore the scope for co-management of conservation issues. **(Saberwal, et al., 2001)**
- Classical exclusionary logic of wildlife management, which harps on reducing human consumptive use of resources within their boundaries, may not hold good so far the crocodiles are concerned. On the contrary serious ecological problems may stem from such exclusionary approach, traditionally adopted by conservationists in India. Myth of pristine wilderness needs to be replaced by a policy that permits animal man co-existence.
- Publicize the value and usefulness of these reptiles so that they can be protected. Sustained management of Indian Crocodiles is intimately linked with their acceptance by the local people, who are integral part of few remaining habitats of crocodiles.
- To ensure the long-term integrity of mangroves, which are the important component of crocodiles environment, strict legislations should be implemented. Judiciously managed commercial utilization of this valuable silvi-agri-aquacultural complex shall prevent their continuous deterioration and ensure the future of the crocodiles.
- There is also an urgent need for collection of data pertaining to human dependence on mangroves and related socio-economics **(BCPP, 1998) (See table 1)**. Providing alternative fuels may reduce mangroves felling for fuel and fodder, for which participation of coastal populace in mangrove silviculture is of vital importance.
- Banning of fishing is undesirable and would be difficult to achieve. However, compromises may be made and fishing may be controlled and restricted in the key areas of the Cumbarjua canal. Chambal National park, Bhitarkanika Sanctuary and Satkosia Gortge Sanctuary are examples of places, where fishing has been restricted or banned in the interest of crocodile. **(Bustard & Choudhary, 1980)**.

CROCODILE RANCHING

Crocodile ranching is the most ideal and economically viable industry that can be established on a small scale with simple inputs. **(Whitaker, 1986)**. By making available a certain percentage of their stock for restocking, the scheme seems necessary for boosting the wild population. Ranching can have reciprocal benefits to the conservation of wild habitat.

Table—1 Some important mangrove taxa of the Cumbarjua canal, the threats affecting them and their conservation status as per IUCN categories.

SPECIES	*THREATS*	*IUCN*
Acanthus ilicifolius	Damming, Human interference, Harvest, loss of habitat	EN
Avicennia officinalis	Cattle grazing, Damming, Human interference, Harvest for food, Harvest for timber, Harvest, Loss of habitat, Over exploitation, Trade	EN
Bruguiera gymnorrhiza	Human interference, Harvest for timber, Harvest, Harvest for medicine, Loss of habitat, Over exploitation, Trade	CR
Derris heterophylla	Human interference, Harvest for medicine, Loss of habitat, Trade	EN
Excoecaria agallocha	Human interference, Harvest, Loss of habitat, Over exploitation, Trade	VU
Kandelia candel	Human interference, Harvest, Loss of habitat, Pollution	EN
Rhizophora mucronata	Human interference, Harvest for timber, Harvest, Loss of habitat, Over exploitation, Trade for parts, Trade	EN
Sonneratia alba	Cattle grazing, Climate, Damming, Human interference, Harvest for timber, Harvest, Loss of habitat, Over exploitation, Trade	EN
Sonneratia caseolaris	Climate, Human interference, Loss of habitat, Over exploitation, Trade	EN

*IUCN Red List Categories: EN-Endangered, CR-Critically endangered, VU - Vulnerable

Source: C.A.M.P Workshop on Indian mangrove Ecosystem (1998)

Ranching would be easier to accomplish if the tribals are employed in the collection of eggs and juveniles. **(Whitaker, 1979)**. Employing tribals could give them cash incentives thereby discouraging their participation in nest poaching and egg robbing. It is strongly felt that such ranching programmes must be taken up on experimental basis. Feasibility of starting a ranch at Bondla Sanctuary of Goa may be evaluated. Such project shall require the State Forest Department to recruit officers for training. The author and his co-workers have carried out studies on sex determination, health conditions and biometry of the captive Muggers of Bondla Zoo, the only place in Goa where Muggers are exhibited **(Borkar and Mallya, 1995 a)**.

ECO-TOURISM AND PUBLIC EDUCATION

The canal with its rich biodiversity has a tremendous potential to support tourism, the Mugger obviously being the main attraction. **(Borkar, 1992; Borkar et al., 1993 b; Borkar, 1993 d and D'Souza, 2001)**. This opinion is meaningful if one examines the current trend of foreign tourists patronizing this quasi-sanctuary. However it is regrettable that there are no organized efforts to tap this potential for revenue generation. Boat trips to the canal could be arranged by a collaborative exercise of the forest department and the tourism department. Such venture can serve a dual purpose. Substantial revenue can be generated by minimizing the disturbances arising out of *'unscientific Crocodile safari.'* Simultaneously the poaching activities could be checked, since the formerly less frequented portions of the canal that were *'Hunters Heavens'* would no longer offer privacy for hunting.

For successful eco-tourism, adequate publicity is essential, and the same can be achieved through mass media like television, radio and newspaper. Certain agrarian tribes residing along the canal have bestowed a sacred status on the Mugger and accord reverence to them. **(Borkar, 1992; Borkar & Mallya, 1993 e; Borkar, 2001 b).** Such conservation oriented, folk celebrations could also be of anthropological interest to people. The tourism department should display posters of this ceremony with relevant information at the tourist encounters. The author has prepared a video film on the crocodiles of Goa, which also covers the folk celebrations of the tribals residing in the Mugger range. This film can act as a useful tool in generating public interest and awareness. Apart from this, the author has also published several articles with photo features on the life of Muggers in Goa and allied aspects; in local and national dailies as well as in popular and scientific magazines, with the understanding that public education is of vital importance in execution of any conservation programme.

But one should not forget that wildlife tourism should not be left entirely in the realm of private entrepreneurship and that the activity must be monitored by the trained field personnel of the forest department. Such joint venture would prevent wildlife tourism from deteriorating into routine business without an element of conservation. So also it should be mandatory that a certain percentage of the revenue generated be set aside for supporting conservation programmes, thus repaying economic value of these reptiles.

STAFF TRAINING

Even the best-planned sanctuary programme cannot succeed in the absence of a cadre of well-trained management personnel. Monitoring should be given the first priority for a critically threatened Mugger population. The forest department should undertake such surveys in collaboration with competent bodies like the Madras Crocodile Bank and the State Government should fund the same.

ENVIRONMENT IMPACT ASSESSMENT

Development activities like Seismic blasting, dam and bridge construction, diversion, canals and concrete embankments and the consequential earth moving operations are detrimental to the crocodilian habitat and make the problem of habitat conservation very complicated. These activities may either prevent nesting or affect the hydrobiology of the canal negatively, reducing the productivity of the canal. Therefore environment impact assessment prior to the execution of any such work is necessary. In this regard it is pertinent to note that the Cumbarjua canal has been declared as a 'No Development Zone'

UPHOLDING TRADITIONAL CONSERVATION FOLK-RELIGIOUS RESPONSES

Ecological responses in India are growing at political level and are also being strengthened at cultural levels. Cultural response to eco-crisis is primarily achieved through religious beliefs and myths. The rich cultural heritage of Indian tradition has provided a unified code of conduct that promotes mutual co-existence, however it may also be realized that religion not only provides a code of spiritual but also environmental conduct that permits a sustainable and ecological use of our natural resources. Such conservation oriented religious responses constitute eco-theological basis of conservation.

Tradition of animal worship has originated in the primitive fear and need to propitiate the animal that posed threat to men or served him **(GRZIMEK 1984)**. In many parts of the world from Indonesia, Africa and India, man's fear of crocodiles has sublimed in religious rites and beliefs. In Karachi, crocodile pool having 50 large Muggers is given the status of priests, and pilgrims too honour them. The crocodiles in a pond called *'Magar Pir'* near Karachi are worshipped. and fed with mutton by the devotees who return content with their blessings **(Anon, 1988)** In ancient Egypt, crocodiles, the God of river was propitiated by sacrificing virgins in the Nile River. Muggers at Peer Khan Jahan Ali Pond, in Bangladesh are considered holy and blessings are believed to be granted by placing cash offerings on the croc's tail **(Cox, 1993)**. Crocodiles have always been an important symbol, in many areas, sacred deities, for the swamp and river people of New Guinea Island **(Whitaker & Whitaker, 1998)**. In Hindu religion also crocodiles have gained deep sense of respect through myths. It is displayed as the mount of the Varuna (water) and river Yamuna. The sun temple of Konark in Orissa has a huge stone crocodile worshipped by people. In Gujarat three tribal communities worship *'Mugger Dev'*, a double-headed crocodile carved in wood and a goat is sacrificed to propitiate this god. The most revered goddess of Gujarat, *'Khodiyar Mata'* also has a crocodile as her mount **(Plate 7)**.

Plate 7 "Khodiyar Mata": A revered deity of Gujarat with Crocodile as its mount.

In Goa, such a religious practice goes on in Bhoma and Dhurbhat Wadi in Ponda Taluka. **(Borkar, 1992; Borkar, 1993 c; Borkar & Mallya, 1993 e; Borkar, 1994 a; Borkar, 1995 c)** Locally known as *"Mannge Thapnee"*, the ceremony is held on the new moon day of *'Pausha'* month of Hindu calendar, when a crocodile dummy is prepared by scooping the silt from the inundated paddy fields **(Plate 8)**. Incidentally this day also coincides with the commencement of threshing of the harvested paddy. The scutes and eyes are made by planting clam shells; teeth are made with small straight sticks. The crocodile is decorated with flowers and vermilion. A live chick is then sacrificed in the pit of the dummy followed by a community payer. The offerings of puffed rice and jaggery are then shared by the devotees. The devotees are the agrarian tribals, the Hindu Gawdas.

In Bhoma village along the Cumbarjua canal a goat is sacrificed after the ceremonial procession and its meat is distributed to all the villagers as *'prasad'*. Thus where animism is practiced as in the case of the crocodile worship in Goa, it is not possible to exploit nature in a mood of indifference to the feelings of natural objects and a consequent indiscriminate exploitation, is prevented.

8

Plate 8 "Mannge Thapnee": A decorated crocodile replica made out of Mangrove silt.

In this regard it is tempting to establish a parallel between the *'Gumatji'*, an aboriginal tribe of Arnhem Land in Australia's tropical far north and Hindu Gawdas of Goa. The *'Gumatji'* people believe since time immemorial, that crocodile is their spiritual father. The *'Mannge Thapnee'* has its roots in the belief that the sea was appeased by the agrarian settlers of Goa by worshiping the Mugger, after large tracts of agricultural lands were inundated by the saline waters **(Borkar, 2001 b).** There is some phonetic resemblance between the konkani name for the crocodile 'Mannge', the Marathi 'Magar' and 'Mongar' a term in Okian language of New Guinea.

However it is alarming to note that these cultural practices are slowly disappearing, with modernization creeping in and subsistence occupation which once supported such communities, rapidly loosing out.

In this ultimate analysis of this religious response there is sufficient reason to believe that the religious dogmas of the least anthropocentric religions like Hinduism have a positive side, in that they affirm their followers commitment to respect all forms of life and promote conservation of biological resources.

SUGGESTIONS FOR FUTURE CONSERVATION OF MUGGERS IN GOA

It is a matter of speculations on how the original agrarian settlers of this land perceived the Mugger. Did they look at this reptile as source of meat? Did ethno-medicinal values get consolidated as a corollary of croc meat consumption? Did they view the Mugger as threat to their life or live-stock? Such a primitive fear eventually manifesting itself into a very curious practice of the crocodile worship is still prevalent in Goa. It is difficult to find answers to many of these questions, though not impossible; except through integration of minds and approaches of anthropologist and zoologist.

Wise decisions about resource use and manipulation depends on quality research. There is a growing acceptance of the fact that ignorance of science is an unjustifiable excuse for environmental abuse. It has been increasingly realized that all important decisions on management of natural resources should have an element of research. It is for this reason that scientific collection of database is necessary.

It may also be realized that the viability of a natural resource and its sustainability depends on the dynamics of its renewal **(Borkar & Mallya 1994b)**. Furthermore such resource renewal dynamics fluctuates on temporal and spatial scales. Hence it becomes prudent to have periodic assessment of the resources before drawing conclusions and expressing opinions. Such a monitoring would be all the more necessary if the resource were limited.

It is axiomatic that if we do not maintain the habitat that wild life requires, we cannot maintain wildlife either. Therefore it implies that wild life manager has to seek co-operation of other resource managers within the habitat of this target species. This simply means that intersectoral resource management is an important ingredient of wild life management **(Peek, 1986)**.

Habitats are subject to adverse manipulation that favours other interests. In such a case we need to understand the specific habitat requirement of species, plus the effects of artificially produced and natural trends in habitat change. In the time tried concepts and ideas that are associated with habitat retention and management, manipulation is the basic underpinning.

A wildlife manager is not exempt from the tendency of general propositions. The consequence of relying on minimum information, or using one population to guide management polices or a similar yet geographically different population can do a lot of harm. Such a *"management by hindsight"* is however not very uncommon and can invite problems in public relations and reflect inadequacy of reasonable database for the population in question. Therefore we must recognize the basically local nature of wild life management. On spatial and temporal scales, if the population of Mugger in Goa, is to be effectively managed, there is no compromise on generating information within the limitations of time and budget, and this has to be periodically updated. This calls for a periodic assessment of Mugger population in Goa, as also an environment audit of the habitat. At the same time it is not wise to ignore the historical records, written and other wise, to minimize the chances of duplicating inefficient management.

Natural resource messages must be presented to the public in such a way that they will be understood and accepted. The time of concentrating on biological side of natural resources management is over. It is now time to concentrate on the sociological principles of human management so that our public will allow us to apply sound biological principles **(Gilbert, 1964)**.

Against this background training in social sciences should be a core component of wild life training. Most wild life management problems start out as a biological problem but eventually become people's problem. Though this has been adequately recognized, contemporary wild life managers continue to stockpile scientific evidences for better management, ignoring elements of human ecology and behaviour. **(Shea, 1948; Peterle, 1961)**.

Having convinced that conventional policy and practice is inappropriate and ineffective in the present time, it becomes imperative to take a cognizance of the dependence of local communities on the resource of that area to be conserved, as also to give a serious thought to their conservation oriented knowledge and practices. If this is not done the people will feel alienated and turn hostile to wild life protection program. Fortunately there is an increasing recognition of the fact that the local communities and the allied stakeholders must be centrally involved in the conservation.

At the same time it must also be realized that such eco-tribes lack the resources to tackle threats and ecologically sensitive issues at regional scales. Whereas the statutory agencies of the state lack the requisite site-specific knowledge to implement management plan. There is therefore a need to forge a partnership in conservation involving local resource users and the statutory agencies, ensuring equal participation to both.

A matter of great concern is the non- investment of the revenue earned from wild life tourism back into conservation programmes. Against this background one of the resolutions passed at the XXI meeting of the Indian Board of Wildlife, held on January 21st 2002 is that tourism revenue should be ploughed back into conservation. The profit though small, is quite substantial to support the conservation programmes. A case in point is a crocodile safari accredited by State Forest Department of the Goa Government that charges Rs. 950/- per tourist for a boat trip to the Cumbarjua canal. The tourist season commences from the month of October and extends up to May. It is a paradox that not even a single rupee earned through this wildlife enterprise is returned back for the management or the conservation of crocodile and its fragile mangrove habitat, nor does the benefit accrue to local communities along the Cumbarjua canal.

CONCLUSION

The Muggers of Goa, now largely confined to the Cumbarjua canal are extremely vulnerable to a lot of environmental pressures. It is difficult to say with confidence whether such pressures can be nullified by their reproductive capacities and recruitment to the stock. This lacuna in our knowledge on their breeding ecology needs to be filled with urgency.

It is not enough to look at the crocodile as an eco-tourism resource alone, but its ecological role also needs to be highlighted and impressed upon the general public. The reptile has had a glorious past in the state, but the modern times have taken a toll. It is a collective responsibility of the intellectually elite, regulatory agencies and the common man to ensure that the pressures on the Mugger are alleviated and it is prevented from being pushed to a point of no return.

ACKNOWLEDGEMENTS

This study was funded by World Wide Fund for Nature, India, Goa State Office. The author is indebted to Dr. A.G. Untawale, former advisor to Biological Oceanographic Division of the National Institute Of Oceanography, Goa for his constructive criticism and constant encouragement. Technical assistance for field investigation was provided by the State Forest Department, Government of Goa.

Miss Neelam K. Komarpant, J.R.F., Bio-diversity Research Cell, Carmel College for Women, has been kind enough to assist in the preparation of this manuscript. Also placed on record is the gratitude towards the Principal of Carmel College for Women, Nuvem Goa for allowing the author to pursue his research despite a tight teaching schedule.

REFERENCES

Anon, 1972. Circular from the Bombay Natural History Society. ***Mimeo, Postal Survey Of Crocodilians.***

Anon, 1988. Letters to the Times Of India, TOI/Press, 1988, P.Wilkins, ***The Times,*** 22nd August 1846.

Bio-diversity Conservation Prioritization Project (BCPP) India, 1998. Endangered species project: ***Conservation Assessment and Management Plan (C.A.M.P.) Workshop*** on Indian Mangrove Eco-system (Ed. by Rao, et al.,), a publication of the Zoo outreach Organization, Coimbatore.

Biswas, S., 1970. A preliminary survey of Gharial in the Kosi River. ***Indian Forester,*** 96: 705-10.

Borkar, M.R., 1992. Eco-theological basis of Crocodile Conservation in Goa. Seminar on ***'Biodiversity of Goa'*** held on 16th February 1992, organized by Botanical Society of Goa (BSI) and All Goa association of Zoologists (AGAZ), Abstract No. 29.

Borkar, M.R. and Mallya, M., 1993 a. Project Crocodile- Country Profile. ***Wildlife Institute of India-News letter.*** 8 (1-4): 5

Borkar, M.R., Mallya, M., Christopher, S. and Phatak, S., 1993 b. The status of the Mugger Crocodile (*Crocodylus palustris,* Lesson), in the Cumbarjua canal of Goa. ***A report submitted to W.W.F. Goa Division.***

Borkar, M.R., 1993 c. ***Crocodiles.*** Environmental Education kit on Bio-diversity of Western Ghats, jointly prepared by WWF Nature India and International Institute For Rural Reconstruction, Manila Philippines,

Borkar, M.R., 1993 d. 'Sacred yet scientific'. ***Gomantak times,*** 20th April 1993.

Borkar, M.R. and Mallya, M., 1993 e. Eco-theological basis of Crocodile conservation in Goa. ***Biology Education,*** 9(4): 297-298.

Borkar, M.R., 1994 a. Conservation future of the Mugger (*Crocodylus palustris,* Lesson) in Goa with special reference to the Cumbarjua canal. ***Master's thesis submitted for Post Graduate Diploma in Ecology & Environment*** to the Indian Institute Of Ecology and Environment, New- Delhi.

Borkar, M.R. and Mallya, M., 1994 b. Study of Environmental Pressures and Resource Renewal With Reference to the Mugger Crocodiles of Goa. Nat. Symp. on ***Eco-environmental Impact And Organism Response,*** held on 15-17th October 1994, organized by P.G. Department of Zoology and Botany, Govt. Vidharba Mahavidyalaya, Amaravati. Abstract No. 15.

Borkar, M.R. and Mallya, M.K., 1995 a. Sex determination, health conditions and biometric study of captive Muggers of Bondla Zoo, Goa. ***ZOO'S PRINT,*** 10 (7): 39-42,

Borkar, M.R. and Mallya, M., 1995 b. Crocodiles in territorial defense in pre-Portuguese Goa. ***HORNBILL,*** 1995(1): 20-22.

Borkar, M.R., 1995 c. ***Crocodile Specialist Group Newsletter,*** 141B, 3-4.

Borkar, M. R., 2001 a. Crocodile is God. In: ***Exploring Goa,*** A teacher's handbook on Goa's environment. Published by Center for Environmental Education. CEE, Ahmedabad.

Borkar, M. R., 2001 b. Crocodile Worship: A Herpeto- eco-theological tradition of Goa's Agrarian past and its Contemporary Relevance, UGC sponsored Seminar on ***Eco-Spirituality - Re-***

discovering Ecology in our Religion and Culture, held on 6-7th December 2001, Organized by Environment Protection Club and Bio-diversity Research cell, Carmel College for Women, Nuvem, Goa.

Borkar, M. R., 2001 c. Community knowledge of lesser known fauna with special reference to Marmugoa, Salcete talukas of Goa. A report submitted to the Goa Foundation under ***National Bio-diversity Strategy and Action plan, Programme (NBSAP),*** New - Delhi.

Bustard, H.R., and Choudhary, B.C., 1980. Conservation future of the salt water Crocodile (*Crododylus porosus*, schneider) in India. ***J. Bombay Nat. Hist. Soc.,*** 77 (2): 201-215.

Cox, J., 1993. Holy Crocodiles. ***Crocodiles specialist Newsletter 141 B,*** 2919 Colony Road, Charlotte, NC 28211, USA.

Daniel, J.C., 1983. ***The Book of Indian Reptiles***. B.N.H.S. publication.

Decada Segunda Livro V Da Asia De Joao Barros Capitulo-I. The Crocodiles and Alligators.

Deraniyagala, P.E.P., 1936. A new Crocodile from Ceylon. ***Ceylon J. Sci., Sect. B. Zool. and Geol.*** **19**: 279-86.

D'Souza, C. A., 2001. Prospects and Problems of Eco-tourism in Goa - An evaluation of a novel enterprise ***Master's thesis submitted for Post Graduate Diploma in Ecology & Environment*** to the Indian Institute Of Ecology and Environment (IIEE), New- Delhi.

Gilbert, D.L., 1964. ***Public Relations in Natural Resources Management,*** Burgess Publishing Co., 426 South 6th Minneapolis, 227.

Grizmek 1984. Grzimek's ***Animal life Encyclopedia*** **6**: Edited by Grzimek. Van Nostrand Reinhold Co. 137

Gupta, R.C. and Srihari, P., 1990. On the burrow utilization by the Indian Marsh Crocodile (*Crocodylus palustris* Lesson), at Bhorsaindan Crocodile Sanctuary, Haryana. ***J. Environ. Biol.,*** **11**(2): 169-178.

Kellert, S.R., 1976. Perception of animal In American Society. ***Trans. N.A. Wildl. and Nat. Res. Conf.,*** **41**: 533-546.

Kellert, S.R., 1980. Public attitude towards critical wildlife and natural habitat issues. ***Natural Technical Information Service. Accessation. No. PB-80-138332, 139.***

Peek, M. J., 1986. ***A review of Wild Life Management*** Prentice Hall, Englewood Cliffs. New Jersey 07632.

Peterle, T.J., 1961. The hunter – who is he? ***Trans. N. Amer. Wild. Conf.,*** **26**: 254-266.

Saberwal, V., Rangarajan, M. and Kothari, A., 2001. ***People Parks and Wildlife towards co-existence,*** vol. **14** Orient Longman Ltd. Publication, New Delhi.

Shea, J.P., 1948. A new approach to farmer sportsmen Co-operation. ***Trans. N. Amer. Wildl. Conf.,*** **13**: 163-169.

Shortt, W.H.O., 1921. A few hints on Crocodile shooting. ***J. Bombay. Nat, Hist. Soc.,*** **28**: 77.

Untawale, A.G., 1990. Marsh Crocodile in Goa. ***Personal communication.***

Whitaker, R., 1974. ***Gharial Survey Report.*** Madras Snake park Publication No. **1**: 1-16.

Whitaker, R., 1977 a. Notes on the status of Gir Crocodiles ***J. Bombay Nat. Hist. Soc.,*** 75 (1): 224-227.

Whitaker, R., 1977 b. Growth rate of *Crocodylus palustris* ***J. Bombay Nat. Hist. Soc.,*** 1: 231-232.

Whitaker, R. and Daniel, J.C., 1978. The status of Asian Crocodilians. ***Tiger paper,*** 5: 12-17.

Whitaker, R., 1979. Crocodile egg collection in Tamil Nadu. ***Indian Forester,*** **105** (92): 1-5.

Whitaker, R., 1982. Crocodile's conservation in India. ***The Zimbabwe Science News,*** **16** (9): 210-215.

Whitaker, R., 1986. Crocodile Ranching: Animal husbandry opportunity for tribal people ***Zoo's Print,*** 1: 8

Whitaker, R., 1991. Mugger Lake and Marsh land – Ciba Geigy Santa Monica Plant- Goa. ***Consultancy Report.***

Whitaker, R. and Whitaker, Z., 1998. Wildlife Adventures In New Guinea In: ***Crocodile Fear,*** Orient Longman Ltd. Publication, Chennai. 34

Wild Life Protection Act, 1972, Natraj publication, Dehradun. In technical association with WWF, TRAFFIC INDIA.

19

Carambolim Lake an Ideal Freshwater Wetland of Goa

*A.B. Shanbhag,
Sonali D. Borges
& Renu Walia

Most ancient civilizations blossomed and flourished on the banks of rivers such as Indus, Nile, Tigris and Euphrates, depending entirely on them for their daily needs. Not withstanding this stark reality and confronted with an ever-increasing demand on the space for living, promotion of aquaculture/agriculture and industries around the cities/ metropolis, man unmindfully ventured in to the transformation of low-lying waterlogged fringe areas. Till the early sixties these fringe areas had little or no status even in ecological domain, as they did not fit in to either of the 2 broad categories of ecosystems – aquatic and terrestrial, recognized till then. Since, presence or absence of water made all the difference between the 2 categories, these fringe areas that did not fit with either of the systems came to be considered as wastelands. It is only in late sixties that progressive commercial minded American farmers flooded the low-lying areas in their ranches and post-harvest fields to attract aquatic birds, especially migratory ducks with an intention of promoting game hunting. It was soon realized that whenever such areas were kept in a continuously flooded state, their productivity decreased and so did their potential to attract the birds (Baldassare & Bolen, 1994). This phenomenon opened the floodgates of scientific inquiry/ research into the ecology of these shallow waters, thus bestowing upon them a separate identity as new set of ecosystems, 'the wetlands', distinctly different from the earlier two categories of ecosystems. In the year 1971, wetland conservation got an added dimension when ornithologists and bird conservationists met at Ramsar in Iran and collectively stressed the need to conserve them as a prerequisite strategy for conserving the waterbirds in general and migratory birds in particular. Since then general awareness as to the multifarious benefits

* Department of Zoology, Goa University, Goa 403 206, India

rendered by the wetland ecosystems is increasing and countries around the globe are realizing the need to check their loss rather than transform them for 'single purpose' commercial ventures. Thus a host of academic institutions, research institutes, government and nongovernmental organizations are continuously engaged in understanding the ecology of wetlands- the unique ecotones between the well-established aquatic and terrestrial ecosystems and evolving improved versions for their management.

In this background, the chapter attempts to throw light on the multifarious utilities, the orientation of research inquiries and conservation strategies for maintenance of functional integrity of freshwater wetlands in general, through our findings on the ecology of Carambolim Lake (CL), a principal fresh water wetland of Goa.

WETLANDS DEFINED

A thorough survey of the existing literature reveals the absence of a 'single, indisputable, ecologically sensible' definition on wetlands (Vijayan, 1986). However, from a wider perspective, the broadly accepted definition adopted by Ramsar Convention (1971) reads- "wetlands are the areas of marsh, fen, peatlands or water, whether natural or artificial, permanent or temporary, with water that is static or flowing, fresh, brackish or salt, including the areas of marine water the depth of which at low tide does not exceeds 6 meters".

LOCATION, PHYSIOGNOMY AND WATER REGIME OF CL

Goa, second smallest state of the Indian Union on the western board of the peninsula, receives an average 3000 mm rainfall through southwest monsoon, which normally lasts from June to September. Until its liberation from the Portuguese colonial yoke in the year 1961 the agricultural/ horticultural practices of the state principally depended upon rain fed water. However, a community based micro level minor-irrigation programme based on the impoundment of water from surface runoff of the receding rains and/or surface springs in the low lying areas by earthen bunds existed in the state even prior to the colonial rule. This practice was further strengthened and improved by the Portuguese through local community management called 'communidades'. The water in these shallow depressions not only catered the irrigation need of the paddy fields in the adjoining area during the *Kharif* season but also yielded revenue through the sale of naturally grown fish harvested late in summer. Goa has numerous such freshwater wetlands, CL being prominent amongst them.

CL (73^0 55'N and 15^0 30'E) is at a distance of 12 km from Panjim, the capital city of Goa. Its total area of 70 ha, comprises of a central, chiefly water-laden 40ha surrounded by grassland. The western embankment of the lake serves as the base for the recently laid Konkan Railway track. The lake boundaries are fringed on 3 sides by coconut and mango trees. Except for the central watercourse, the lake-bed is flat with the greatest depth of 2.75m at the sluice gate point at the southwestern side of the lake. The lake is fed entirely by rainwater and surface runoff and holds 20,34,212 m^3 water when flooded. The maximum volume of water attained by the end of August remains high during post monsoon and starts depleting precipitously through winter and reaches its lowest limit by late summer (Fig. 1). During late summer/early monsoon, the entire lake is reduced to a single zigzag watercourse running diagonally across the lake.

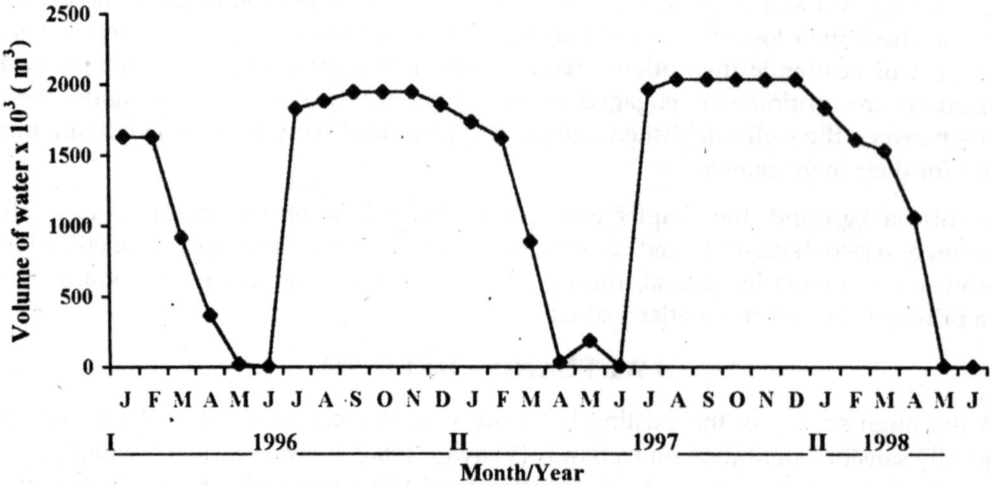

Fig. 1 **Monthly variations in the volume of water at CL. Semipermanent nature of the lake due to the sluice gate operated water regime can be appreciated.**

PHYSICOCHEMICAL CHARACTERISTICS OF THE LAKE WATER

Water as a medium plays a significant role in maintaining the integrity of the entire ecosystem by internal balance of its physical and chemical characteristics (Reid and Wood, 1976). This in turn decides its ability to support life forms, patterns of aquatic biota as well as productivity. Therefore, a host of interacting physical and chemical characteristics in the form of hydrological factors discussed in the following passages contribute to the trophic structure of the system, their total biomass and aquatic food web (Hutchinson, 1957).

Water pH is considered to be the reflection of many chemical and biological processes that take place in natural waters. The changes in pH are known to primarily indicate metabolic changes in CO_2-CO_3/ HCO_3 content, reflecting phytoplankton production and photosynthetic activity. Therefore, it is believed that pH can be used as an index of production influencing habitat diversity (Minns, 1989). Conductivity of water is known to indicate dissolved and suspended organic matter that provides large pool of energy for the organisms that utilize these components. Hence, conductivity has been used as the index of nutrient status of the water body significantly affecting macrophytic distribution (Crowder *et al.*, 1970). Waters with conductivity values more than 200µmho are considered eutrophic (Rawson, 1960). The total alkalinity of water is due to Ca++ and Mg++ ions, and high levels of calcium have been considered to be the signs of eutrophic state of water bodies (Sahai and Sinha, 1969). Dissolved O_2 (DO) is a measure of photosynthesis in surface waters. Another common gas, free CO_2 in water has been observed to be inversely proportional to DO. Changes in levels of these gases are known to indicate nature of biological activities and the nature of water quality (McManus *et al.*, 1996). Nutrients such as total nitrogen and phosphorus have been attributed with a key role in eutrophication of the water bodies (Shapiro, 1973). Thus, it is clear that the physico-chemical parameters as a whole, portray the general health of the water body, its productivity and pollution status.

Water temperature at CL, which ranged from a minimum of 24^{0}C in Jan. to a maximum of 38.6^{0}C in June, closely followed the broad pattern of ambient temperature (Fig.2), clearly reflecting the inherent character of shallow water bodies truly fitting well with the definition of wetlands. Water pH was very close to neutrality oscillating within a narrow limit of mild acidity ranging from 5 in March to 6.8 in August. Average pH during monsoon was 6.67 while that during summer was 5.58. According to pioneer limnologists (Hutchinson, 1976; Rump and Krist, 1988) "every case where water is neither very acidic nor very alkaline, it may justly be assumed that pH is regulated by carbon dioxide-carbonate- bicarbonate system". Thus it is obvious that CL waters are not affected by inflow of severely acidic/alkaline effluents of domestic or industrial nature. Further, the pH of the lake waters ranging from 6.10 to 6.63 coincides with the pH known to be conducive for aquatic life (Trivedy and Gurdeep Raj, 1992). Hence, the lake has a great potential for rich aquatic biota.

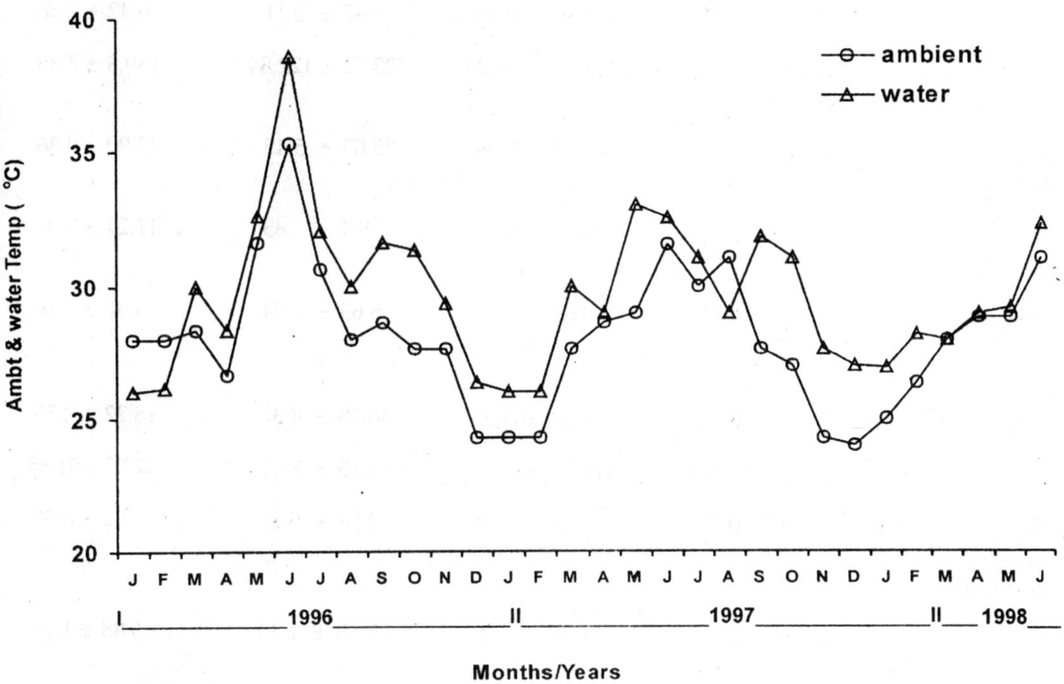

Fig. 2 Comparative profile of water temperature and that of ambience at CL on monthly basis. A close follow up between the two, reflecting shallowness of the lake can be noted.

The conductivity was highest in May and lowest in October (Fig.3). Seasonwise, it was maximum during summer and minimum during post monsoon (Table 1). Chloride values showed a close parallel. Electrical conductivity of water is known to depend on concentration and dissociation of ions as well as temperature. A close correlation between conductivity and chlorides is considered to be characteristic feature of wetlands (Briggs *et al.*, 1985). Organic waste is known to be one of the main contributors for the chloride content of flooded bodies and hence is known

Table—1 Seasonal Variations in the Physico-chemical parameters of waters of Carambolim Lake

Sr. No.	Physico-chemical Parameters	Winter (Dec. Feb.)	Summer (Mar.-May)	Monsoon June-Sept.)	Post-Monsoon (Oct.-Nov.)
1.	Ambient temperature (°C)	24.70 ± 0.67	28.59 ± 0.27	30.30 ± 0.86	26.63 ± 1.35
2.	Water temperature (°C)	25.60 ± 0.10	29.88 ± 0.20	32.06 ± 0.18	29.8 ± 1.65
3.	pH	6.38 ± 0.17	5.58 ± 0.23	6.47 ± 0.11	6.42 ± 2.56
4.	Conductivity (μmho)	198.43 ±16.49	760.35 ±369.03	323.72 ± 122.84	130.8 ± 7.36
5.	Turbidity (NTU)	32.00 ± 2.51	42.87 ± 2.99	35.02 ± 5.45	23.90 ± 0.36
6.	Free carbon dioxide	8.76 ± 2.95	22.62 ± 0.59	8.4 ± 1.85	11.24 ± 3.01
7.	Dissolved oxygen	2.12 ± 0.41	3.68 ± 0.21	5.69 ± 0.94	3.97 ± 0.10
8.	Chloride	16.99 ± 0.64	98.98 ±63.92	36.28 ± 8.32	18.22 ± 0.35
9.	Alkalinity	109.17 ±53.41	170.01 ±46.10	99.15 ± 2.41	122.3 ±51.85
10.	Biological oxygen demand	3.93 ± 0.21	7.11 ± 1.16	2.78 ± 0.52	3.2 ± 0.50
11.	Total organic carbon	15.40 ± 0.64	26.98 ± 2.04	7.68 ± 1.38	6.88 ± 1.20
12.	Total hardness	29.20 ± 1.58	52.45 ± 1.58	22.81 ± 4.22	21.75 ± 2.01
13.	Phosphate	2.11 ± 0.72	4.29 ± 0.32	0.84 ± 0.33	0.06 ± 0.01
14.	Ammonical nitrogen	2.53 ± 0.47	5.49 ± 0.41	1.39 ± 0.52	3.60 ± 0.65

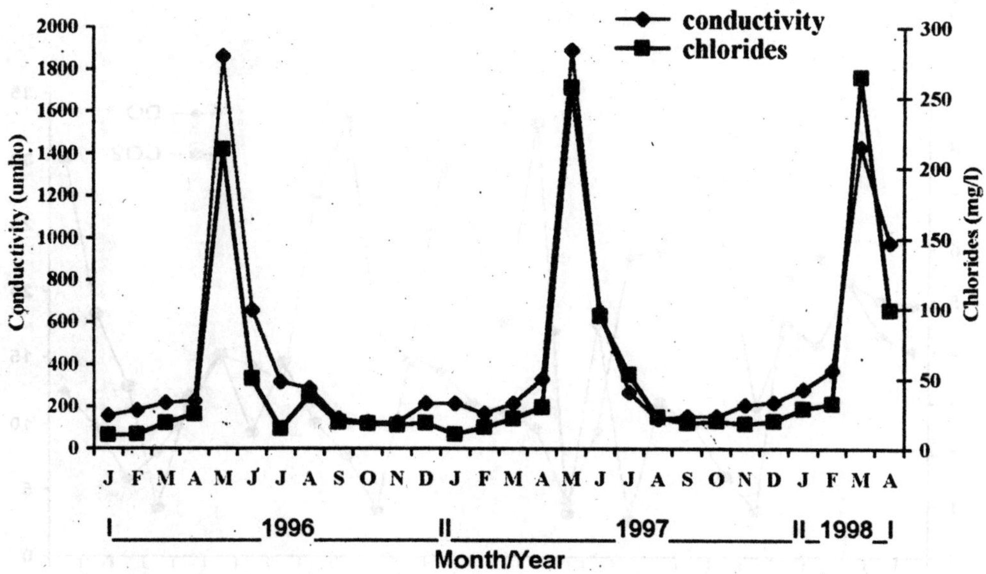

Fig. 3 **Monthly variations in the chlorides and conductivity of CL waters. A close correlation between the 2 parameters and the sharp peaks during summer owing to depleting water column and death/decay of organisms is obvious.**

to have a close correlation with eutrophication. Further chloride contents are also treated as the index of pollution through fecal contamination (Trivedy and Gurdeep Raj, 1992). At CL, the conductivity remained moderately high, close to 200μmho throughout most of the year shooting up drastically during summer. CL is a semi permanent water body, wherein bulk of water is released in late summer. This invariably led to the death and decay of aquatic vegetation and animals thereby naturally enriching nutrients in already depleted water column, in turn increasing conductivity. The defecation by large herds of grazing cattle during late summer must be adding to the dramatic shoot up of the chloride and conductivity of CL waters during the season. On the whole it is very clear that despite there being no overt pollution of any kind, the lake is on the threshold of eutrophication in terms of high chloride and conductivity through major part of the year.

DO was highest in July (8.5mg/L) and lowest in December (1.05mg/L) while free CO_2 was highest in April (30mg/L) and lowest in June (3.5 mg/L) (Fig.4). The peak levels of DO during monsoon could be primarily because of influx through rainwater. Secondary source could be the stabilized photosynthetic activity of aquatic autotrophs. The respiration of aquatic biota and decomposition of organic matter are the two major processes responsible for the FCO_2 in water (Hutchinson, 1957). Death and decay of macrophytes and other organisms due to depleting water column during summer, hastened by increased temperature must be the factor responsible for increased CO_2 levels during the season.

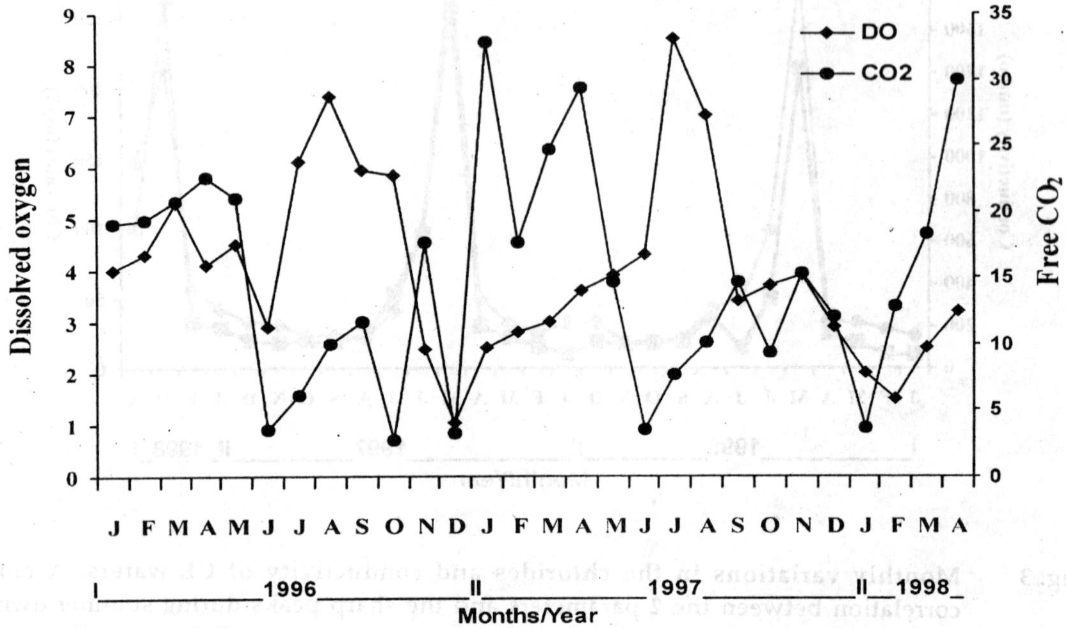

Fig. 4 **Monthly profile of the dissolved gases- O_2 and CO_2 of CL waters. The diagonally opposite trends of the 2 gases especially during summer and monsoon are noteworthy.**

Alkalinity, total hardness and calcium hardness showed identical seasonal profile at CL. All these parameters that were at their minimum during late monsoon and post monsoon, started rising through winter, reached peak values in summer and declined through mid monsoon (Fig.5). According to Spence (1964) alkalinity levels above 60ppm are the signs of nutrient richness in the water bodies. The alkalinity level as prescribed by WHO is 120mg/L. In this background, consistently higher alkalinity levels in the lake amount to categorize it to be eutrophic, and more so during summer.

Nitrates and phosphates, the principal nutrients meant for the growth and reproduction of autotrophs are considered to be good indicators of pollution and eutrophication. PO_4 in CL never exceeded the limit of 10mg/L (Fig.6). The summer high of PO_4 could be due to the decreased water column and increased organic enrichment. The high NO_3 content observed in late summer declined steadily during late monsoon and early winter. The growth of emergent and submergent vegetation during monsoon and early winter might be increasingly utilizing the nutrients in the water thereby keeping them under check. In tune with the concept that nitrogen deficiency is a characteristic feature of tropical waters (Khan and Chowdhury, 1994) its content at CL was within 5mg/L.

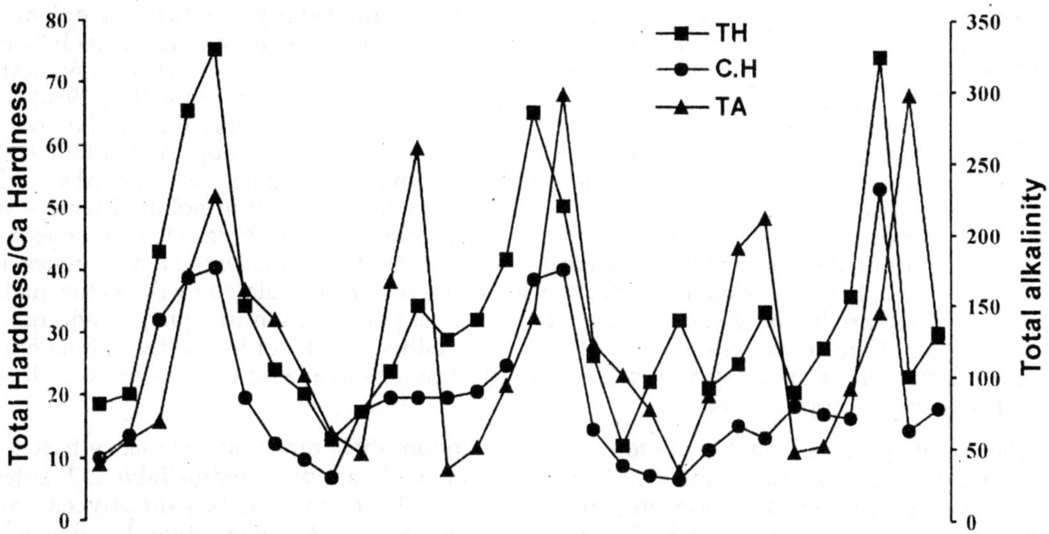

Fig. 5 Monthly changes in the total calcium and magnesium hardness of CL waters.

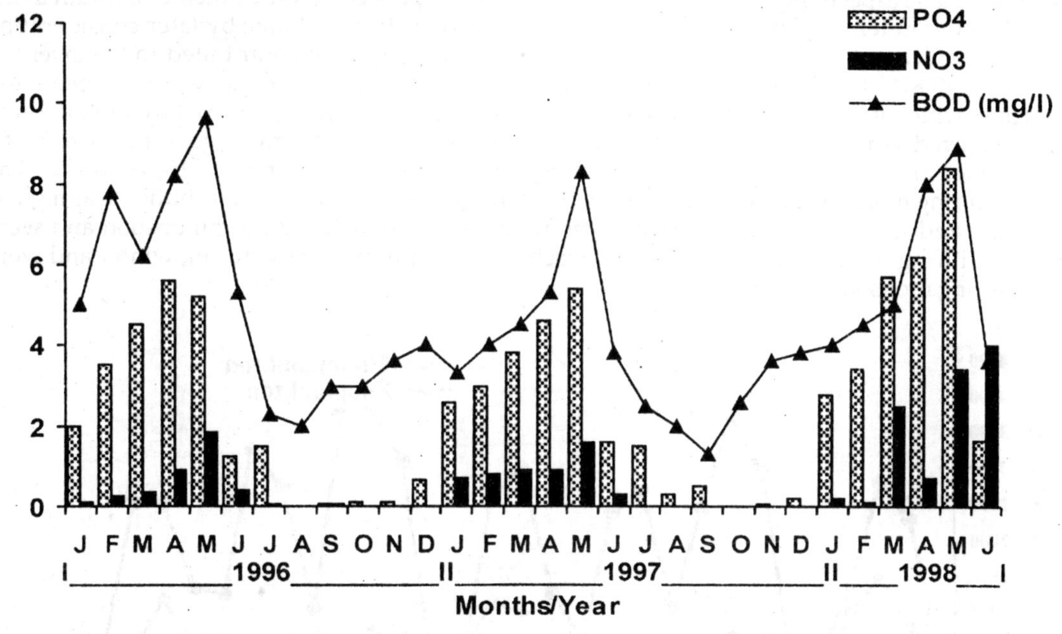

Fig. 6 The profile of the 2 principal nutrients- PO_4 and NO_3 in CL waters depicting the peak values during summer and a parallel with BOD.

PLANKTON AND PRODUCTIVITY

'Plankton' (Gr. Wandering or drifting) denotes a group of organisms – either plants or animals that passively float and drift with flow of tides, waves and currents. Therefore, planktonic community is a heterogeneous assemblage of plants and animals suspended in aquatic systems. Based on the mode of nutrition, the plankton is categorized as phytoplankton and zooplankton.

Freshwater, one of the important aquatic biomes for these potential group of which phytoplankton forms the major source of primary production, while zooplankton serves as important link between producers and higher forms. Composition and dominance of plankton groups have been suggested to be of indicator value in earmarking trophic status to the water bodies (Ilmavitra, 1980). The application of biological data for assessing the water quality was first reported by Kolkwitz and Marsson (1908). Nygaard in 1949 established 5 indices based on 4 major groups of plankton to get meaningful evaluation of the extent of pollution, which are now used routinely as Nygaard's trophic state indices to assess degree of eutrophication. Certain planktons grow in polluted waters and their presence characterizes the environment in which they live. Therefore, because of their tolerance to polluted environment they are treated as indicators. Palmer (1969) presented a comprehensive list of algal genera; based on which generic/species indices can be calculated to assess the quality of water. Primary productivity is one important biological phenomenon in aquatic environment upon which the entire heterogenic community depends either directly or indirectly. Estimation of primary production of an ecosystem by different techniques available enables us to estimate production at higher trophic levels.

The list of species of phytoplankton and zooplankton, their ranks of dominance based on cumulative picture of longterm analysis of net plankton at CL are provided in Table 2. Eighteen species of phytoplanktons belonging to 3 families- Chlorophyceae, Cyanophyceae and Bacillariophyceae were encountered. The top 3 ranks based on their cumulative densities were occupied by a Cyanophycean, *Oscillatoria formosa* and 2 Chlorophyceans- *Chlorella vulgaris* and *Spirogyra lineata* respectively. Of the 3 major groups, Chlorophyceans contributed maximum to the phytoplankton in terms of number of species and their density, weightage by later consideration being 52.4%. Cyanophyceans and Bacillarophyceans as a group each contributed to the extent of 27.73% and 19.87% respectively. There was a distinct seasonality in the density of phytoplankton as a group, which varied from 212ind/L in May to 1459ind/L in March (Fig.7a). Phytoplankton as a whole showed a distinct seasonality in its density with a perceptibly high range from 711-855ind/L in summer and post monsoon limits were reasonably low amounting to 310-389ind/L. The densities during monsoon and winter were nearly identical. Chlorophycean contribution was highest in summer and lowest during monsoon. Cyanophyceans dominated during monsoon and were minimum during winter. Bacillarophyceans reached a maximum density during winter and were lowest during monsoon (Fig.7b).

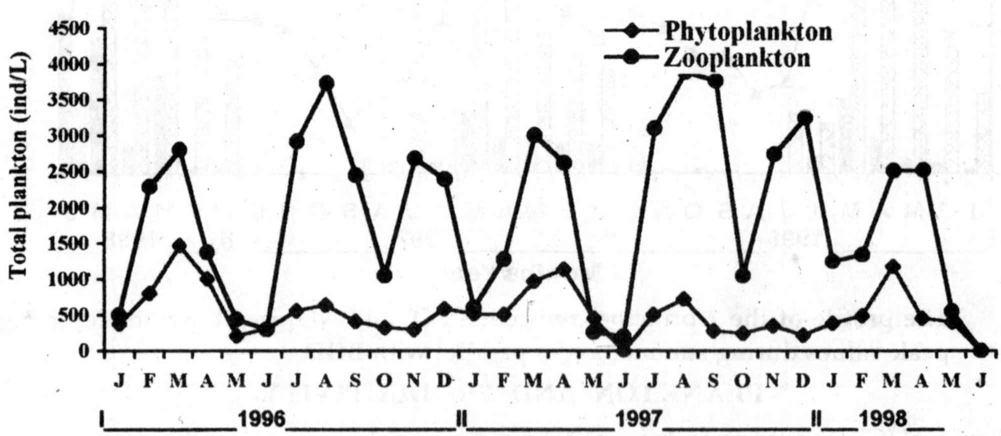

Fig. 7a

Table—2 Diversity and dominance ranks of planktons in the waters of Carambolim Lake.

Phytoplankton		Zooplankton	
Family/Species	**Rank of Dominance (1–18)**	**Family/Species**	**Rank of Dominance (1–20)**
Chlorophyceae		**Copepoda**	
Ankistrodemus fulcatus	11	*Cyclops bicuspidatus thomsi*	3
Chlorella vulgaris	2	*Mesocyclops leukarti*	4
Oocardium	14	*Naupliar larva*	1
Scenedesmus quadricauda	6	*Phyllodiaptomous thomsi*	8
Pediastrum duplex	7	**Rotifera**	
Pandornia morum	5	*Brachionous angularis*	20
Selenastrum gacile	8	*B. calyciflorus*	9
Spirogyra lineata	3	*B. caudatus*	6
		B. falcatus	13
Cyanophyceae			
Anabaena sp.	12	*Felina longiseta*	14
Microsystis aeruginosa	4	*Keratella tropica*	12
Oscillatoria formosa	1	*Lecane ploensis*	15
Spirulina sp.	15	*Monostylla bulla*	16
		Phylodina sp.	18
Bacillariophyceae			
Astrionella formosa	10	*Trichocera porcellus*	11
Fragillaris capucina	9	*Biatura affinis*	10
Navicula tuscula	18		
		Cladocera	
N. plea	13	*Daphnia carinata*	5
N. recta	17	*Lydgia ciliata*	2
Synedra ulna	16	*Moina brachiata*	7
		Hemiptera	
		Notonecta	19
		Ostracoda	
		Cypris sp.	17

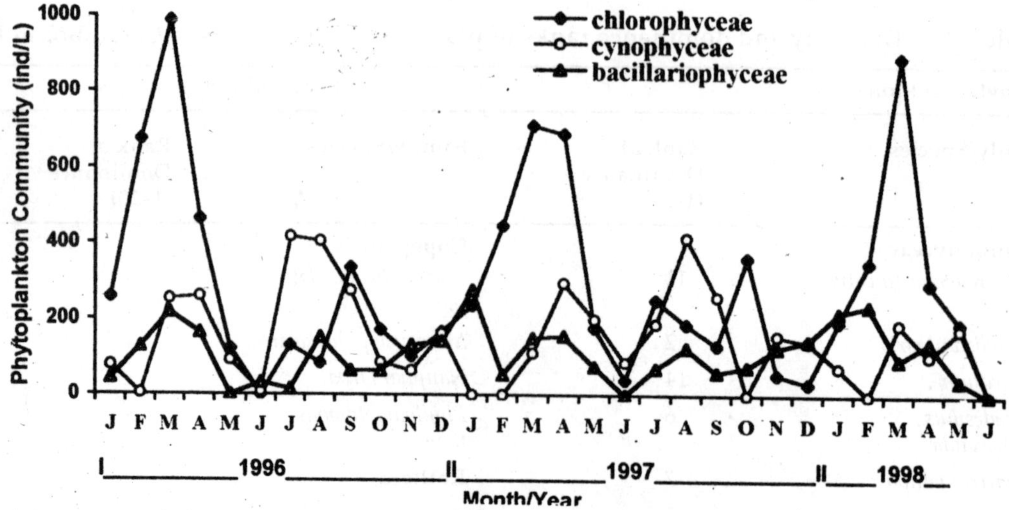

Fig. 7b

Zooplankton was represented by 11 species of rotifers, 4 species of copepods, 3 species of cladoceres, and a species each of insect and ostracod. On the basis of overall densities across the seasons, representation of copepods, rotifers and cladoceres by percentage was 37.02, 30.05 and 32.03 respectively. Zooplankton density ranged from 100ind/L in June to 3820ind/L in August. The density showed sharp and steep spasmodic peaks with intervening but brief depressions (Fig. 7c). The highest peaks were during monsoon followed by those of post monsoon and summer. During interphases between late summer/early monsoon, late monsoon/post monsoon

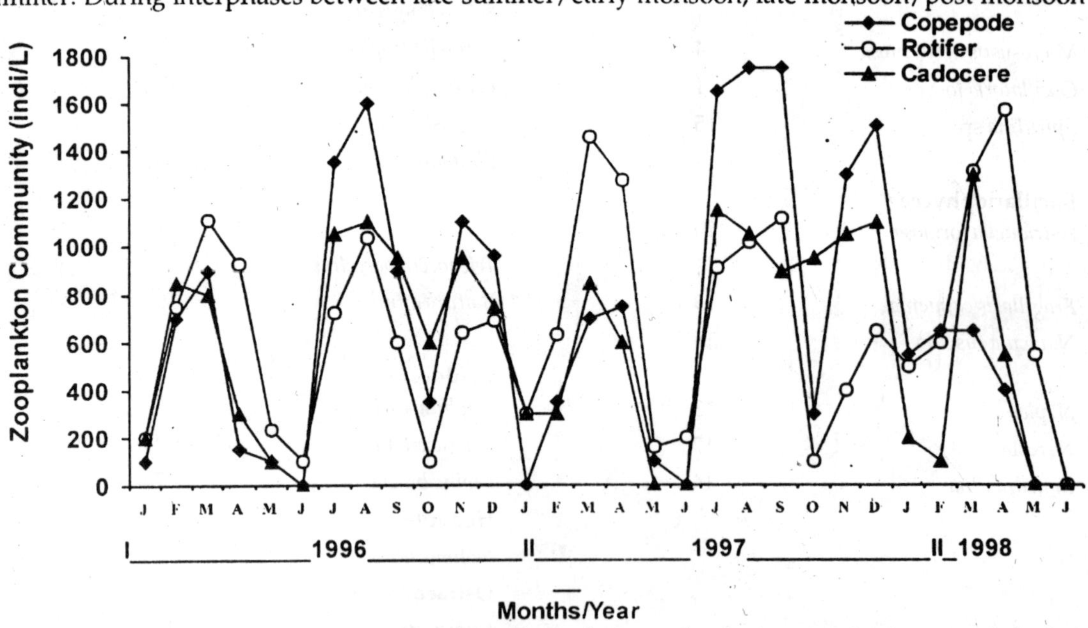

Fig. 7a,b&c: Monthly variations in the densities of phytoplankton, zooplankton and their 3 major groups each. A distinct gap between the densities of phytoplankton and zooplankton may be noted indicating adequate grazing pressure.

and winter, the density was the least. Most of the times the zooplankton density was 2-3 times higher than that of phytoplankton. Copepods showed maximum density during monsoon and minimum during summer, while rotifers were abundant during summer and minimum during post monsoon. Cladoceres were abundant during post monsoon and were least during summer. Thus, distinct seasonal abundances of zooplanktonic groups- rotifers, copepods and cladoceres during summer, monsoon and post-monsoon/winter were parallel with those of phytoplanktonic groups- chlorophyceans, cyanophyceans and bacillarophyceans.

The gross planktonic primary productivity studied by routine light and dark bottle method varied from a minimum of 0.04mgCm^2/day in July to 0.17mgCm^2/day in November (Fig.8). On seasonal basis it remained highest during post monsoon followed by summer, winter and monsoon respectively (Table 3). Community respiration also varied from 0.01mgCm^2/day in February to 0.012mgCm^2/day in November.

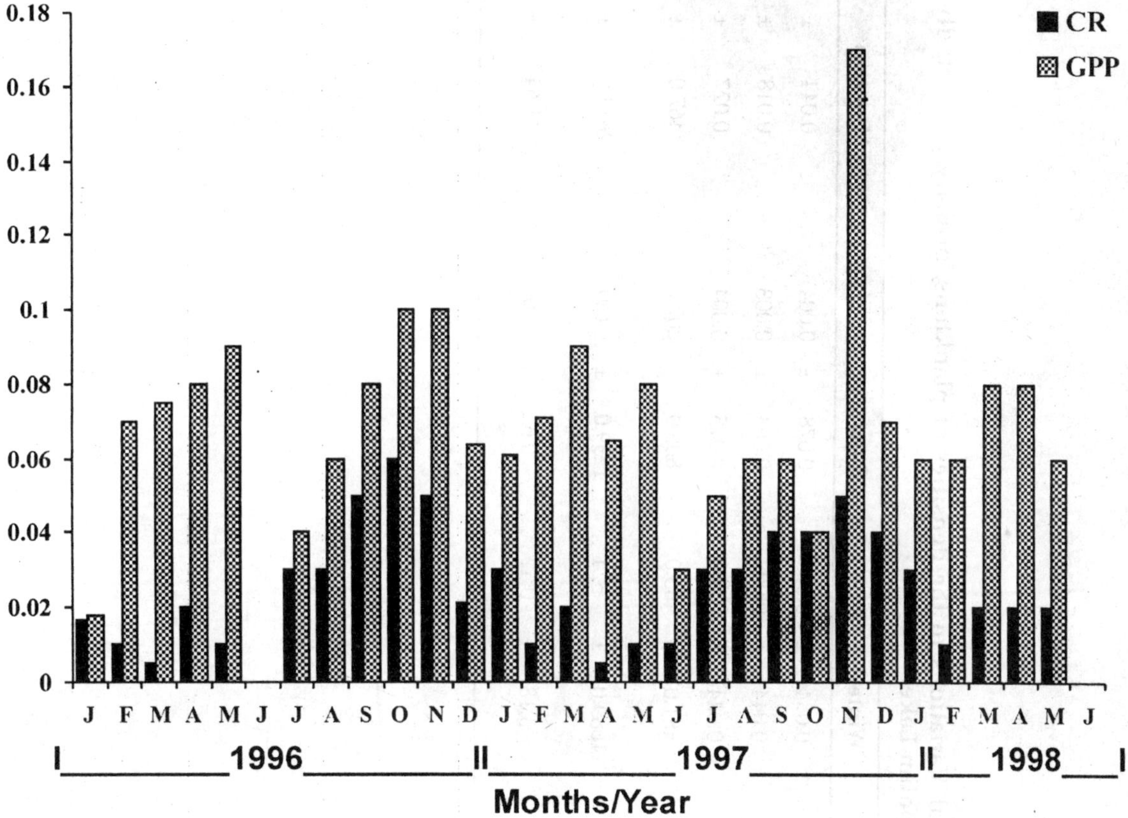

Fig. 8 **Monthly variations in the gross primary productivity and the community respiration in the CL waters.**

It is very clear that phytoplankton was abundant during summer a season characterized by high temperature, long sunshine hours and lake water with higher alkalinity, nitrates and

Table—3 **Seasonal Variations in the densities of planktons, primary productivity and the macrophytic biomass at Carambolim Lake.**

Productivity	Winter	Summer	Monsoon	Postmonsoon
GPP	0.064 ± 0.003	0.078 ± 0.004	0.044 ± 0.014	0.103 ± 0.025
NPP	0.044 ± 0.007	0.063 ± 0.003	0.018 ± 0.006	0.053 ± 0.005
CR	0.024 ± 0.006	0.015 ± 0.004	0.027 ± 0.009	0.050 ± 0.005
Phytoplankton (indi/I)	481.0 ± 65.0	806.0 ± 207.0	467.0 ± 49.0	350.0 ± 35.0
Zooplankton (indi/I)	1689.0 ± 524	1239.0 ± 691	2614.0 ± 794	1960.0 ± 702
Macrophyte biomass/m^2	3.18 ± 0.42	0.62 ± 0.00	1.64 ± 0.32	3.58 ± 0.15

phosphates. The minimum phytoplanktonic density during monsoon a season characterized by reverse meteorological and environmental conditions of aforesaid characteristics is also obvious. Association of green algae with shallow waters has already been emphasized (Happey-Wood, 1988). High nutrients and acidic pH have been shown to favor Chlorophyceans (Shapiro, *et al*, 1975). Hence, dominance of Chlorophyceans during summer is only but natural. Talling (1976) showed experimentally that the Cyanophyceans could utilize C at relatively higher pH, when most of it is available in the form of CO_3/HCO_3. Mukherjee and Pankajakshi (1995) reported accelerated growth of blue green algae where FCO_2 may be in low concentrations. Therefore, it is not surprising that Cyanophyceans at CL were dominant during monsoon when FCO_2 recorded was very low. Bacillarophyceans are known to prefer clear waters and low temperatures (Tilman, *et al*, 1986) which justifies their dominance during winter. The succession profile of algae with dominance of Chlorophyceae, Cyanophyceae and Bacillarophyceae during summer, monsoon and winter respectively is in agreement with similar findings by Ganpati and Sreenivasan (1968) in the reservoirs of Tamil Nadu. Under the scheme of Nygaards algal indices, the Chlorophycean index was never below the upper limit of eutropy. Likewise, Cyanophycean index was always in the eutrophic range while diatom index often crossed the limit of eutropy. Hence, CL can very well be categorized as an eutrophic lake. Zooplankton density surpassed phytoplankton density almost all through the year with a substantial gap. Therefore, according to the grazing theory (Kurasawa, 1975), the findings indicate enormous grazing pressure of zooplankton on phytoplankton at CL. Thus, descending vertical control of phytoplankton because of sustained grazing pressure in the lake is an encouraging sign indicating that the lake is still far from dystrophication (Goel, 1997).

Rotifers dominated the zooplanktonic domain in summer. According to Junk (1977), decaying macrophytes besides providing protection from predation also serve as a food source, promoting their population. Rotifers are also known for their reproductive potential which is reported to be dependent on temperature besides food availability (Edmondson, 1972). Thus, it is only logical that the soaring temperatures of summer in the region, shrinking volume of water leading to considerable decay of macrophytes and concentration of nutrients must have provided congenial environment for the growth and reproduction of rotifers in the lake resulting in their dominance in the season. Zooplanktons in general are known to avoid blue green algae because of their poor nutritional value, unmanageability and toxicity (Porter & Orcutt, 1980). Large filamentous Cyanophyceans are known to interfere with cladocers in more than one ways (Infante & Riehl, 1984), immobilizing their thoracic limbs and clogging their food-filtering apparatus. Considering the multifaceted antagonism by the cyanophyceans with cladocers the abundance of the later during winter/post monsoon, a period with less population of blue green algae is only a natural consequence. Although both copepods and cladoceres are raptorial omnivores, the former are found to be competitively inferior to the later (Kinsely & Geller, 1986). Secondly an inverse relation has been reported between copepods and chlorophyceans as the former do not ingest the later (Lampert & Taylor, 1985). Thus, the dominance of copepods during monsoon, a season with relatively less of green algae and free from competition with cladoceres seems to be justified. At CL, where a sharp distinction between dry and wet phases existed due to regulated water regime, GPP and phytoplanktonic species diversity were at their peak during post monsoon, immediately after the filling in phase of the lake. Peak zooplanktonic species diversity succeeded during summer with short lapse of about a season.

BENTHOS

Benthic community and its substratum constitute an indispensable and inseparable component in the cyclic dynamism of any aquatic system. Autotrophs and heterotrophs from the water column stretching across littoral, planktonic and nektonic domain on their death sink to the bottom and become the trophic inputs to the benthic system. Decay and putrefaction of these organisms brought about by saprobic benthos leads to the formation of detritus that provides the nourishment to benthic detritivores. Biochemical processes on the residual detritus results into the release of nutrients in the form of NH_3, NO_2, NO_3 and PO_4, setting forth the cyclic processes. The retention and resuspension of nutrients into the overlying water column depends on the physicochemical processes, pumping action of some of the rooted macrophytes and bioturbation owing to the activities of benthic organisms. On the other hand excreta of the benthic animals also adds to the nutrients in the system. Thus, one of the major components of benthos- the benthic macroinvertebrates provides an important link between primary production and higher trophic levels. Sediment texture in the form of grain size and composition of 3 main constituents- sand, silt and clay, not only have a significant influence on the water percolation/ retention, levels of nutrients, but also on the nature and extent of the macrobenthic community.

The sediments are inhabited by variety of macro invertebrates, principally annelids, arthropods and molluscs, all of them interact in some way or the other with their substratum (Brinkhurst, 1974). For any particular organism the nature of this interaction is determined by the manner in which food is obtained (trophic level), mobility of the organism and position in relation to sediment water interface life activities are carried on. Benthic forms are known to adopt different food habits such as suspension feeding, deposit feeding, browsing, carnivory, scavenging and parasitism. Primarily because of these varied food habits, different animal interaction such as amensalism, competition and mutualism are reported to exist in the benthic community, which leads to modulation, altering their structure and abundance.

Associations of macro invertebrate species or community have been extensively studied in biomonitoring of water quality, as water pollution is known to cause alteration in the structure of macrobenthic community. Use of benthic community in evaluating effects of pollution has been favored by many biologists because organisms on or within the substratum reflect the environmental condition of water prior to the time of sampling, whereas the analysis of water mass or plankton will comparatively reflect the condition only at the time of sampling. Though for quite sometime the pollution indicator species concept was in vogue and benthic animals like chironomids and oligochaetes were identified with polluted waters (Wiederholm, 1980), some other studies (Reish, 1979; Srivastava & Desai, 1997) which were more comprehensive and based on wide spectrum ecological analysis have not been fully supportive of the concept.

At CL, the sediment consists of 91% sand, 6.01% silt and only 2.99% clay. The sand, silt and clay were in higher proportion during monsoon, summer and winter (Table 4 and Fig. 9a). Seasonal variations existed in all the 3 principal nutrients of the sediment. The peak concentrations of P, N and C were 0.036mg, 0.022mg and 0.555mg per 100g of sediment during summer, post monsoon and winter respectively (Fig. 9b). The diversity of macrobenthic community was low and represented only by an annelid-bristle worm, a nematode and an insect-*Dytiscus* in the order of dominance(Fig.9c).

Table—4 Seasonal Variations in % composition of sediment texture, sediment nutrients and the density of benthic macroinvertebrates at Carambolim lake.

Productivity	Winter	Summer	Monsoon	Postmonsoon
Texture				
% Sand	86.37 ± 1.19	84.29 ± 1.45	95.96 ± 0.62	90.25 ± 0.42
% Silt	4.16 ± 0.63	13.13 ± 0.63	2.96 ± 0.31	4.88 ± 0.14
% Clay	10.99 ± 1.27	4.15 ± 0.17	1.33 ± 0.44	6.79 ± 0.3
Nutrients (mg/100g of sediment)				
P	0.026 ± 0.002	0.036 ± 0.001	0.005 ± 0.002	0.014 ± 0.002
N	0.019 ± 0.003	0.01 ± 0.001	0.012 ± 0.001	0.022 ± 0.001
C	0.555 ± 0.025	0.261 ± 0.039	0.039 ± 0.02	0.305 ± 0.045
BMI (ind/m^2)	72.91 ± 45.65	49.18 ±18.13	2.90 ± 1.17	43.9 ± 1.25

BMI—Benthic Macro-invertebrates

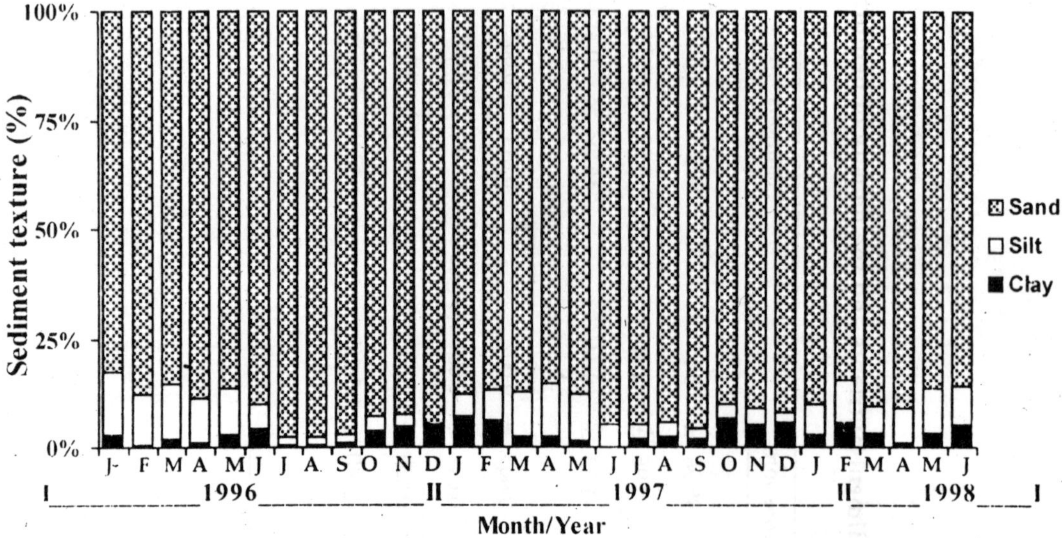

Fig. 9a

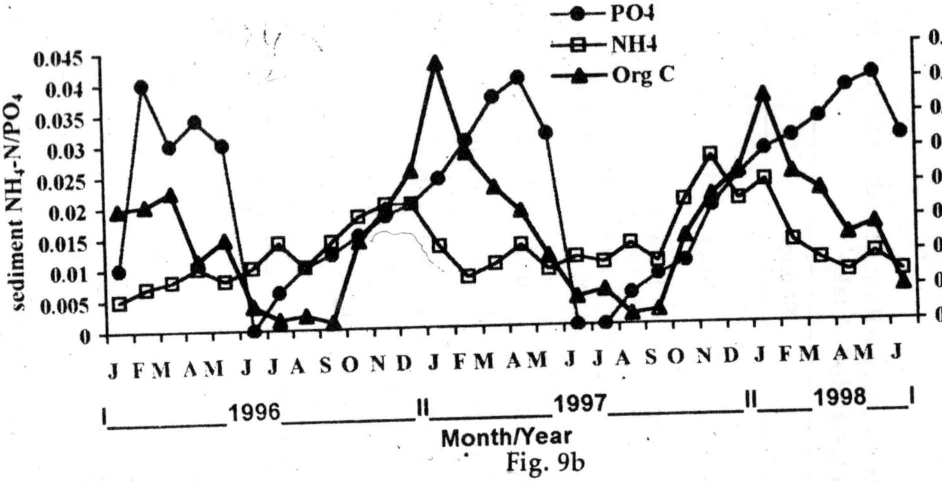

Fig. 9b

Sediment of CL can be classified, as silty sand as the sand component crossed the 80% mark all through the year, silt being the next predominant component. As the lake passes through a transient lotic state due to the opening of the sluice gate prior to the onset of monsoon, increased sand content must be due to the new sediment load brought in by the surface run off and also the local disturbances caused by the torrential rains during the season. Silt is primarily a product of partial decomposition of organic matter of plant and animal origin. Generally the waterlogged soils are deficient in oxygen because the pore spaces remain filled with water. On account of this limited supply of O_2 to the saprobic organisms the rapid decomposition of organic matter is decreased resulting into the high concentration of silt (Mohanty & Dash, 1982). Extensive hydrophytic vegetation is found dying and decaying during summer owing to the depletion of water column. It is evident that the increased silt in the lake is due to autocthonous input rather than allocthonous. The clay particles being less than 2μm in size remain suspended in water for

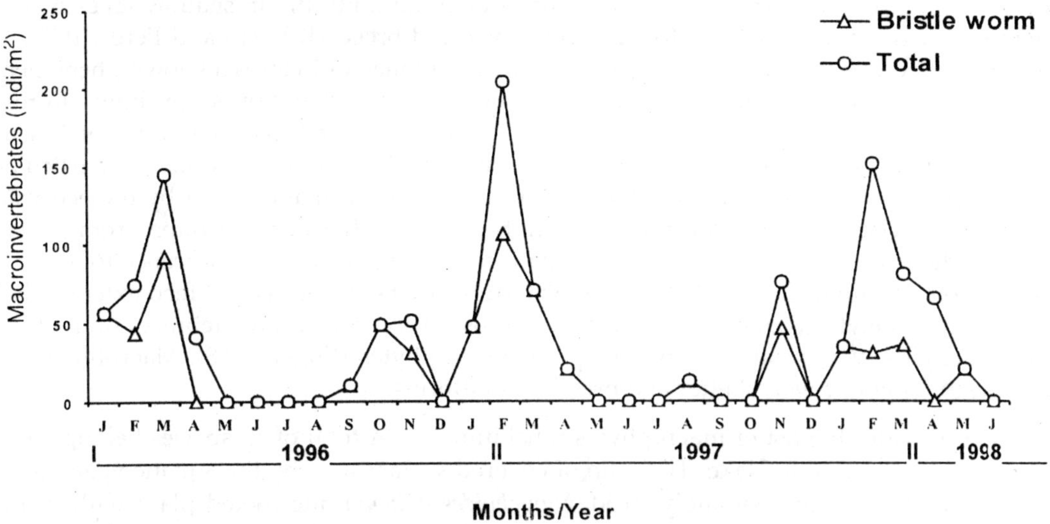

Fig. 9a,b&c: Monthly profile of the sediment texture principle nutrients of the sediment and the benthic macroinvertebrates at CL. Predominance of sand, silt and the major contribution of bristleworm to the benthos may be noted.

longer period and their settlement is promoted only when the turbulence is least and waters are tranquil. Thus it is natural that the highest clay content is recorded during post monsoon. The least level of clay at CL must be due to its annual transient lotic phase of the lake accompanied by monsoon torrents dislodging the previous years clay deposits. According to Reish (1963), benthic polychaetes prefer silty sediment to very fine or fine sand. Hence, it is obvious that the sediment at CL where silt is the second dominant constituent holds bristle worms as its dominant form. The benthic community of CL can be classified as polychaete type due to its dominance amongst the 3 different macro invertebrate species recorded in the lake.

MACROPHYTES

The term 'aquatic macrophytes', refers to macroscopic vegetation including macro algae, mosses, ferns and angiosperms that grow in aquatic and wetland habitats (Klaine & Ward, 1983). Based on their attachment to the substratum they are classified into 3 types namely rhizophytes, haptophytes and pleustophytes (Best, 1983). Haptophytes are attached to the substratum but do not penetrate it. They are therefore mainly limited to the bulk medium as nutrient source. Pleustophytes occur freely floating and obtain their nutrients in soluble form from ambient water through their shoots and often their roots. On the basis of the location with reference to the water column, and their link with the substratum, macrophytes are also categorized as submerged, rooted submerged, free floating, rooted with floating leaves and emergent types (Gopal *et al*, 1978).

Macrophytes distributed widely in tropical and subtropical aquatic systems are one of the prime producer groups of the waterbodies that receive radiant energy from the sun and convert it into chemical energy in the form of food, upon which most of the aquatic life depends. Therefore, a balance between biomass of macrophytes and consumers is essential to maintain the stability of the aquatic ecosystem. Macrophytes contribute greatly to the primary productivity and accelerate eutrophication. They are the centers of community life in shallow lakes wherein invertebrates, fishes and waterfowl forage, seek cover and breed (Kikuchi and Peres, 1977). An inverse relation between plankton production, nutrients and macrophytes is a known phenomenon (Mulligan and Baranowski, 1969). High levels of nitrogen and phosphorus are shown to result in the increased growth of plankton, whereas their low levels promote high yield of vascular plants. This interesting inverse relation between phytoplankton and macrophytes, particularly the submergents, leading to an alternative equilibrium has been made use of by the ecosystem managers to exercise biological control measures to control either of the groups promoting the other as per the need (Weisner *et al.*, 1997). Aquatic macrophytes are of significance to the life of waterfowl from multiple angles. Many of the waterfowl are herbivorous and feed extensively on submerged macrophytes (Lodge, 1991). Secondly most of the macro invertebrates preferred by waterfowl are closely associated with the aquatic macrophytes (Engel, 1988). Macrophytes also provide the birds with cover, shelter and breeding platforms.

Table 5 provides the list of macrophytes inhabiting CL. A total of 13 species belonging to 9 families were recorded in the lake. The ratio of covered surface area by the aquatic vegetation to clear water at CL was approximately 70:30. *Nymphoides indicum,* the rooted plant with floating leaves was the most dominant species representing 67% of the total macrophytic biomass. The emergent wild paddy, *Oryza rufipogon* which constituted 15.46% of the biomass, grew luxuriantly in sublittoral belts on all the sides except some areas on the eastern bank. *Hydrilla verticellata*, which was restricted to the deeper central region of the lake contributed to 12.46% of the biomass. The monthly variations in the standing biomass of the total aquatic macrophytes are provided in Fig 10 and the seasonal variations in the same are shown in Table 3. A well-defined cyclicity was noticeable in the total standing biomass of macrophytes in the lake through the season. It was minimum during summer (0.57kg/m^2). It rose to a moderate level of 1.57-1.7 kg/m^2 during monsoon but shot up glaringly during post monsoon and winter reaching a high of 2.87-2.05kg/m^2.

Once the aquatic vegetation takes a foothold in any tropical or subtropical waterbody, uninterrupted nutrient recycling sustains enormous standing plant biomass without many fluctuations. Hence, Westlake (1967) opined that an attempt to look for seasonal minima and maxima in aquatic macrophytic vegetation in tropical or subtropical waters is a futile exercise. Contrary to this opinion there was a drastic reduction in the standing biomass at CL during summer. This limitation on the standing biomass of hydrophytes at CL must be due to the annual fill-in and draw-down water regime, which renders the waterbody a semipermanent type. Negligible growth of littoral plants in the lake could be because of enormous grazing pressure by stray cattle. The gradient of the lake bottom and the depth of water column are considered to be some of the important physical attributes that decide the nature of hydrophytic community (Chambers, 1987). CL being a shallow lake harbored a good population of water lilies, the rooted macrophytes with floating leaves. Another reason for the dominance of rooted plants with floating leaves in the lake could be the higher silt content in the sediment as observed in Hoakersar lake of Kashmir (Kaul, *et al.*, 1978). Submergents dominated the brief period between initiation of filling-in phase and the establishment of stands of lilies. The loss of dominance status of submergents beyond postmonsoon and early winter could be due to the competition from lilies through coverage of surface water by their floating leaves and reduction in penetration of light, so also the foraging and destruction pressure by herbivorous birds like coots, which consisted of contingents of over 500 birds by mid winter.

Table—5 The structural composition of aquatic macropyte community at Carambolim lake.

Sr. No.	Family/Species
	Salviniaceae
1.	*Azolla pinnata* (free floating)
2.	*Salvinia* sp. (free floating)
	Cyperaceae
3.	*Cyperus iria* (emergent)
4.	*Scirpus* sp. (emergent)
	Ceratophyllaceae
5.	*Ceratophyllum demersum* (submergent)
	Ponterderitaceae
6.	*Hydrilla Verticelata* (submergent)[3]
	Menyanthaceae
7.	*Nymphoides indica* (rooted floating)[1]
8.	*Nymphoides cristata* (rooted floating)
	Poaceae
9.	*Oryza rufipogon* (emergent)[2]
	Onagraceae
10.	*Jussiaea ripens* (emegent)
	Trapaceae
11.	*Trapa natans*
	Lentibulariaceae
12.	*Utricularia aurea* (submergent)
13.	*U. exoleta* (submergent)

Note: ½/3 indicate in order of dominance of the 3 most dominant macrophytes.

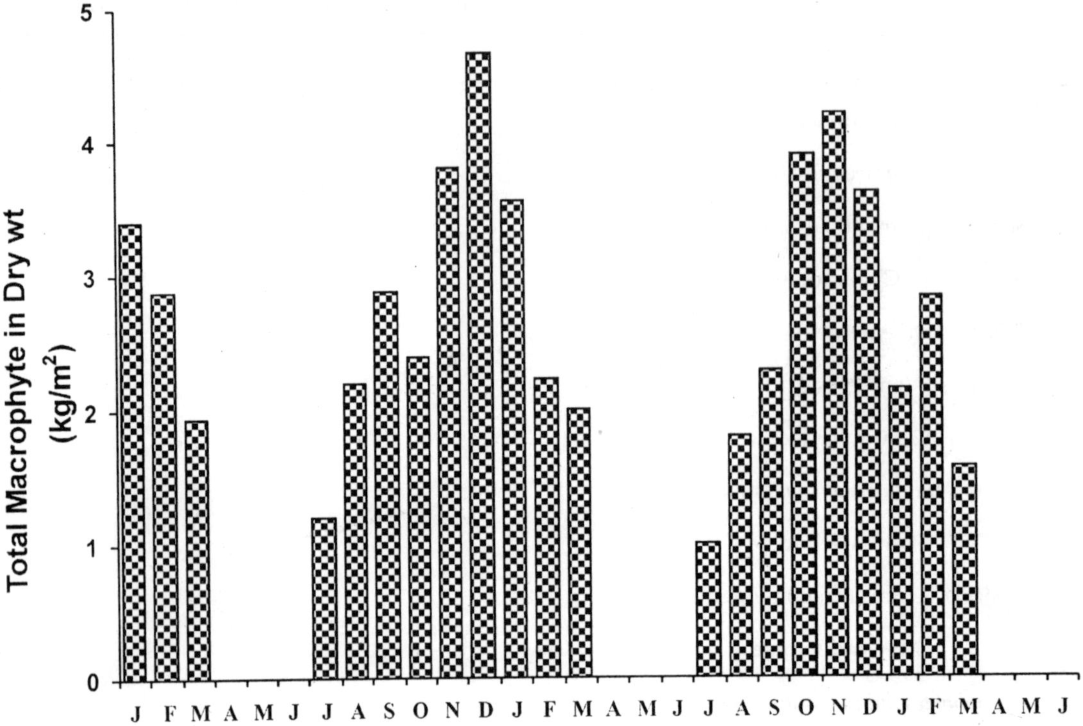

Fig. 10 **Variations in the aquatic macrophytic biomass at CL. The peaks coinciding with the impoundment of water during postmonsoon/winter can be appreciated.**

FISHERY POTENTIAL

Brackish water fish culture in the country has of late come under heavy clouds of uncertainty. This is primarily due to ingression of freshwater aquifers by saltwater and deterioration of soil fertility to the point of no return due to unmindful, greedy expansion of intensive brackish water pisciculture without any regard to the ecological norms. Therefore, it is important that human population diverts its attention towards freshwater pisciculture, at least extensive, if not intensive to cater to the ever-increasing demand for the animal protein. Techno-economic survey of erstwhile Union Territory of Goa, Daman & Diu, conducted immediately after liberation of the region from colonial rule stated that "….however, the detailed information about the sheet of water, their suitability for fish culture are not available" (Anonymous, 1964). The present study carried out after almost after the lapse of 3 decades seems to be unique in this regard from intensive as well as extensive considerations.

Two bagarids- *Aorichthys seghala* and *Heteropneustis fossilis* and two cyprinids- *Barbus amphibious* and *Rasbora daniconius* were prominent amongst the fishes that grew naturally in the impounded waters of the lake and were harvested at the end of summer. The later 3 species dominated the waters in that order. From totalitarian angle, the fish catch ranged from 24.03kg/ha to 31.89 kg/ha. Irrespective of the seasons and the species, on an average the fishes

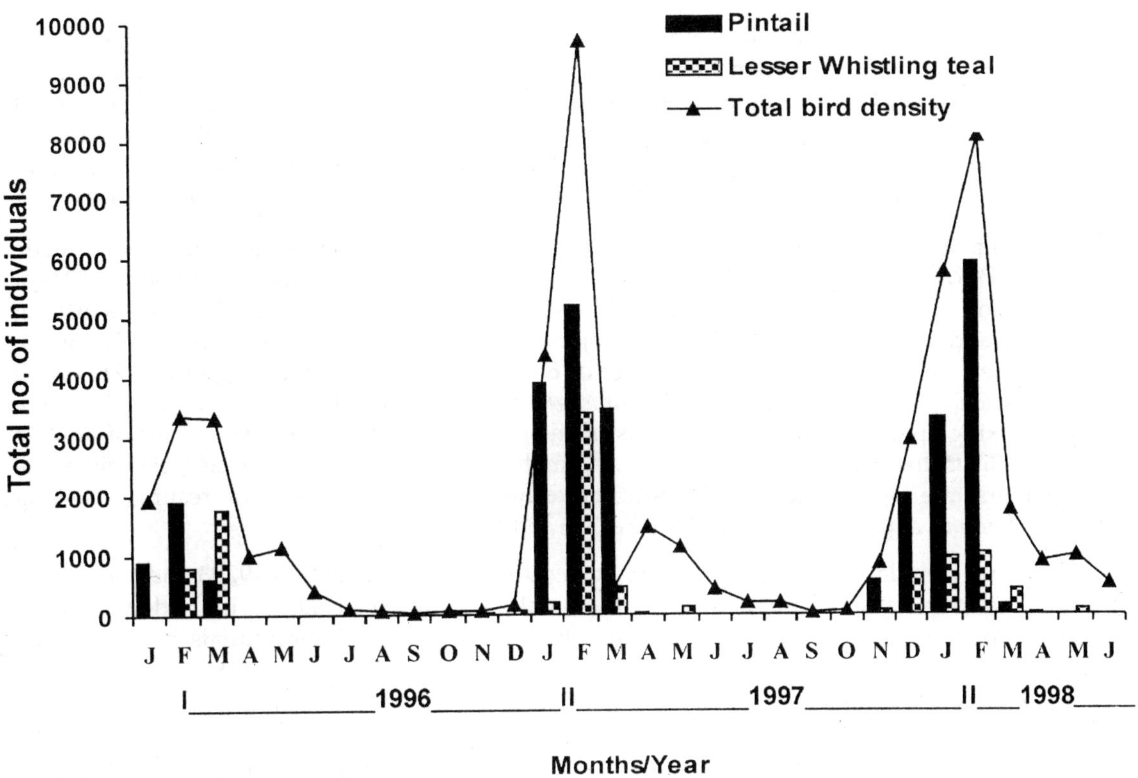

Fig. 11 Monthly variations in the bird density at CL. The major contribution by the 2 migratory species- *A. acuta* and *D. javanica* may be noted.

from the catches in the lake were 5.9cm long, 1.7 cm deep and weighed about 18g. The smaller size of the fish from the lake is to be viewed in the background of programmed impoundment and draw-down of the lake on annual basis, thereby providing a restricted period of about 6-7 months of growth before the total fish harvest. Besides extensive waterfowl foraging, predominant benthivorous catfish population must be responsible for the moderate benthic macroinvertebrate fauna in the lake.

AVIFAUNA

Birds in general and waterfowl in particular, despite being categorized as completely dependent, less dependent and opportunistic users of the wetlands (Vijayan, 1986) are inseparable faunistic component of the ecosystem. Martinez (1993) impressively outlined the multifaceted significance of birds in aquatic environment as 1) consumers at different levels of the trophic chain, 2) suppliers of organic matter, especially at the gregarious roosting sites and nesting colonies, 3) transporters/ interchangers of materials between aquatic and terrestrial environment, 4) modifiers of environment through utilization of aquatic macrophytes and bioturbation of sediment during feeding and locomotion, and 5) causal transporters of living organisms including pathogens across distant geographical areas. In addition, waterfowl generally occupy higher trophic levels, in most of the wetland food chains, and are highly susceptible to wetland contaminants and disturbance alike.

Hence, they are supposed to be good indicators of the general condition of wetland ecosystems. In India, most of the work on wetland birds is concentrated around a handful of larger wetlands, especially the Ramsar sites. As flight is the primary mode of locomotion in birds and distance is not a real limitation on their dispersal, they do use smaller and widely spaced satellite water bodies besides the major ones to meet their needs. Therefore, the need to study lesser-known waterbodies was stressed by the editors of the Directory of Indian Wetlands (Hussain and DeRoy, 1993).

Avifauna at CL was represented by 116 species, of which 38 species of aquatic birds made up the 98.50% of the population while 54 species of terrestrial birds contributed to only 1.50% of the population (Table 6). Among the aquatic birds, herbivores like Pintail ducks, Lesser whistling teals, coots and moorhens constituted the major group reaching to the level of 83.37%, while piscivores and insectivores were in the range of 6.76% and 8.67% respectively (Table 7). Pintail duck, *Anas acuta*, Lesser whistling teal, *Dendrocygna javanica* and Cattle egret, *Bubulcus ibis* were the top ranking species in the lake in terms of the overall population. Of 38 species of aquatic birds 28 utilized the site largely for feeding. A small population of only 7species, especially the large waders like Open billed Storks and White necked Storks used it for resting/ basking purpose. None of aquatic birds used the lake as a roosting site.

The bird population ranged from 19 individuals in September to 9703 individuals in February. From a seasonal point if view, the small population of 60-481 individuals during monsoon/ post-monsoon, shot up to the monthly average of 5602 individuals in winter and precipitated back to less than half the winter population during summer. Among the winter visitors, Pintails (67.13%), Lesser whistling teals (27.71%), coots (3.83%) and Blue winged teals (0.70%) were predominant ones. Pintails started arriving by December/January and reached the peak by February and receded through March/ April (Fig 11). Stray sightings numbering up to 10 were recorded till mid monsoon. In terms of initial arrivals the lesser whistling teals differed in schedules from year to year, from November to February, reached peak in February and left the site before April. Like Pintails, stray sightings of Lesser whistling teals were also not uncommon during monsoon/ post-monsoon months. Influx of Coots also occurred between November to February. But their major population stayed for a short period of 2 months during January/ February. Indian moorhen, Pheasant tailed jacana and Bronze winged jacana with young ones were routinely sighted, indicating the lake environs being used as a breeding site by the birds. Although there was a phenomenal decline in the avifaunal population during the construction and implementation phases of the Konkan railway, the numbers of birds visiting the lake has steadily increased since then (Walia & Shanbhag, 1999; Shanbhag, *et. al.*, 2001).

Nonbreeding wintering waterfowl are reported to feed mainly on aquatic plants to meet their energy requirements, in contrast to breeding waterfowl and young ones that largely depend on protein rich diet comprising of macro invertebrates (Tamisier and Boudouresque, 1994). Further aquatic plants- submergents and emergents with their vegetative parts, seeds, nuts and tubers are known to be extensively used by water birds as food. Therefore, it is but natural that CL with diverse aquatic plants such as *H. verticillata, U. aurea, Nymphea* sps. and *Oryza rufipogon* is primarily made use of as a feeding ground by over 80% of the water bird community. Van de Ven (1987) described southern Indian Subcontinent as the minor wintering ground of coot, *Fulica atra*. But the bird visited CL in good number, sometimes crossing the 500 mark. However, its stay was for a brief period of 2 months-January and February. The period coincided with the extensive growth of *H. verticillata*, which is known to be the choice food of the bird (Montalbano *et al.*, 1979).

Table—6 A systematic list of the birds of Carambolim lake with their resident status, abundance, habit and habitat.

Sr. No.	Order/Family/Species	Common name	Rank of Dominanc	Resident Status	Habitat	Habit/ Feeding type
1	2	3	4	5	6	7
	Podicipediformes/Podicipedae					
1.	*Podiceps ruficollis*	Dabchick	17	R	WL	D/I
	Pelicaniformes/Phalacrocoracidae					
2.	*Phalacrocorax fuscicollis*	Indian shag	49	R	WL	D/P
3.	*P. niger*	Little cormorant	13	R	WL	D/P
	Ciconiformes/Ardeidae					
4.	*Ardea cinerea*	Grey heron	23	R	WL	W/P
5.	*A. Purpurea*	Purple heron	23	R	WL	W/P
6.	*A. grayii*	Pond heron	4	R	WL	W/P
7.	*A. alba*	Large egret	21	R	WL	W/P
8.	*Bubulcus ibis*	Cattle egret	3	R	WL	W/I
9.	*Egretta garzetta*	Little egret	7	R	WL	W/P
10.	*E. intermedia*	Median egret	22	R	WL	W/P
11.	*E. gularis*	Reef heron	46	R	WL	W/P
	Ciconidae					
12.	*Anastomus oscitans*	Openbill stork	19	LM	WL	W/P
13.	*Ciconia episcopus*	Whitenecked stork	22	LM	WL	W/I
14.	*Leptopilos javanicus*	Leeser adjutant stork	40	LM	WL	W/P
	Threskiornithidae					
15.	*Threskiornis aethiopica*	White ibis	38	LM	WL	W/I
16.	*Plegadis falcinellus*	Glossy ibis	18	LM	WL	W/I
17.	*Platelea leucordia*	Spoonbill	23	LM	WL	W/I
	Anseriformes/Anatidae					
18.	*Dendrocygna Javanica*	Lesser whistling teal	2	LM	WL	S/H
19.	*Anas acuta*	Pintail	1	DM	WL	S/H
20.	*A. querquedula*	Garganey	12	DM	WL	S/H
21.	*A. clypeata*	Shoveller	28	DM	WL	S/H
22.	*A. Poecilryncha*	Spotbilled duck	16	DM	WL	S/H
23.	*Nettapus coromandelianus*	Cotton teal	14	LM	WL	S/H
24.	*Sarkidiornis melanotus*	Nakta	36	LM	WL	S/H

(Contd...)

1	2	3	4	5	6	7
	Falconiformes/Accipitridae					
25.	*Milvus migrans*	Pariah kite	20	R	SF	R/C
26.	*Haliastur indus*	Brahminy kite	34	R	SF	R/C
27.	*H.leucocyphus*	Pallas fishing eagle	49	LM	SF	R/C
28.	*Accipiter badius*	Shikra	48	R	SF	R/C
29.	*Spilornis cheela*	Crested serpent eagle	49	R	SF	R/C
30.	*Circus aeruginosus*	Marsh Harrier	46	DM	SF	R/C
	Gruiformes/Rallidae					
31.	*Amaurornis phoenicurus*	Whitebreasted waterhen	37	R	WL	W/H
32.	*Gallinula chloropus*	Indian moorhen	11	R	WL	W/H
33.	*Porphyrio porphyrio*	Purple moorhen	5	R	WL	W/H
34.	*Fulica atra*	Coot	6	LM	WL	S/H
	Charadriformes/Charadridae					
35.	*Hydrophasianus chirurgus*	Pheasant tailed jacana	9	R	WL	W/H
36.	*Metopidus indicus*	Bronzewinged jacana	8	R	WL	W/I
37.	*Vanellus indicus*	Redwattled lapwing	29	R	WL	W/I
38.	*Tringa totanus*	Redshank	39	LM	WL	W/I
39.	*T. erythropus*	Spotted redshank	24	LM	WL	W/I
40.	*T. stagnatilis*	Marsh sandpiper	30	DM	WL	W/I
41.	*T. glaroela*	Spotted sandpiper	32	DM	WL	W/I
42.	*T. hypoleucos*	Common sandpiper	31	DM	WL	W/I
43.	*Gallinago gallinago*	Common snipe	52	LM	WL	W/I
	Recurvirostridae					
44.	*Himantopus himantopus*	Blackwinged stilt	43	LM	WL	W/I
	Laridae					
45.	*Larus brunnicephalus*	Brownheaded gull	26	DM	WL	Sk/P
46.	*Sterna aurantia*	River tern	34	R	WL	Sk/P
47.	*Chlidonias hybrida*	Indian whiskered tern	33	LM	WL	Sk/P
48.	*Gelochelidon nilotica*	Gullbilled tern	36	LM	WL	Sk/P

(Contd...)

1	2	3	4	5	6	7
	Columbiformes/Colum bidae					
49.	*Columba livia*	Bluerock pigeon	+	R	SF	F
50.	*Streptopelia chinensis*	Spotted dove	38	R	SF	F
	Psittaciformes/Psittacid ae					
51.	*Psittacula krameri*	Roseringed parakeet	41	R	SF	F
52.	*P. cyanocephala*	Blossomheaded parakeet	41	R	SF	F
53.	*P. columboides*	Bluewinged parakeet	46	R	SF	F
54.	*Loriculus vernalis*	Lorikeet	46	R	SF	F
	Cuculiformes/Cuculida e					
55.	*Eudynamys scolopacea*	Koel		37	R	SF I
56.	*Rhopodytes viridirostris*	Small greenbilled malkoha	45	LM	SF	I
57.	*Centropus sinensis*	Coucal	41	R	SF	I
	Strigiformes/Strigidae					
58.	*Athene brama*	Spotted owlet	37	R	SF	I
	Coraciformes/Alcedinid ae					
59.	*Ceryle rudis*	Lesser pied kingfisher	39	R	WL.	Sk/P
60.	*Alcedo athis*	Small blue kingfisher	35	R	WL	Sk/P
61.	*Pelargopsis capensis*	Storkbilled kingfisher	45	R	WL	Sk/P
62.	*Halcycon smyrnensis*	Whitebreasted kingfisher	33	R	WL	Sk/P
63.	*H. pileata*	Blackcapped kingfisher	48	R	WL	Sk/P
	Meropidae					
64.	*Merops philippinus*	Blue tailed bee eater	25	R	SF	I
65.	*M. leschenaulti*	Chestnut headed bee eater	40	R	SF	I
66.	*M. orientalis*	Small green bee eater	16	R	SF	I
	Coracidae					
67.	*Coracias bengalensis*	Blue jay	25	R	SF	I

(Contd...)

1	2	3	4	5	6	7
	Pisciformes/Capitonidae					
68.	*Megalaima zeylonica*	Large green barbet	42	R	SF	F
69.	*M. viridis*	Small green barbet	38	R	SF	F
70.	*M. haemacephala*	Coppersmith	42	R	SF	F
	Piceidae					
71.	*Micropternus brachyurus*	Rufous woodpecker	42	R	SF	I
72.	*Dinopium benghalense*	Lesser goldenbacked woodpecker	46	R	SF	I
	Alaudidae					
73.	*Galerida cristata*	Crested lark	45	R	SF	I
	Hirudinidae					
74.	*Hirundo smithii*	Wire tailed swallow	23	R	SF	I
75.	*H. daurica*	Red rumped swallow	10	R	SF	I
76.	*Lanius excubitor*	Grey shrike	47	R	SF	I
77.	*L. vittatus*	Baybacked shrike	44	R	SF	I
78.	*L. schach*	Rufousbacked shrike	40	R	SF	I
79.	*Artamus fuscus*	Ashy swallow	45	R	SF	I
	Dicruridae					
80.	*Dicrurus adsimilis*	Black drongo	33	R	SF	I
81.	*D. paradiseus*	Ricket tailed	48	R	SF	I
	Sturnidae					
82.	*Sturnus pagodarum*	Brahminy myna	43	LM	SF	F
83.	*S. roseus*	Rosy postor	29	DM	SF	F
84.	*Acridotheres tristis*	Indian myna	28	R	SF	I
85.	*A. fuscus*Jungle myna	15	R	SF	I	
	Corvidae					
86.	*Dendrocitta vagabunda*	Indian tree pie	38	R	SF	I
87.	*Corvus splendens*	House crow	+	R	SF	Sc
	Pycnonotidae					
88.	*Pycnonotus jocosus*	Redwhiskered bulbul	38	R	SF	F
90.	*P. luteolus*	Whitebrowed bulbul	30	R	SF	F
91.	*P. priocephalus*	Gray headed	53	LM	SF	F
	Muscicapidae					
92.	*Turdoides striatus*	Jungle babbler	36	R	SF	I
93.	*Terpsiphone paradisi*	Paradise flycatcher	40	R	SF	I

(Contd...)

1	2	3	4	5	6	7
94.	*Prinia socialis*	Indian wren	41	R	SF	I
95.	*P. subflava*	Ashy wren	39	R	SF	I
96.	*Orthotomus sutorius*	Tailorbird	42	R	SF	I
97.	*Acroccphalus dumetorum*	Blyth's reed warbler	31	R	SF	I
98.	*Acrocephalus stentorus*	Great reed warbler	44	R	SF	I
99.	*Phylloscopus collybita*	Brown chiffchaff	46	R	SF	I
	Turnidae					
100.	*Saxicoloides fulicata*	Magpie robin	41	R	SF	I
101.	*Copsychus saularis*	Indian robin	32	R	SF	I
	Motacillidae					
102.	*Anthus novaeseelandiae*	Paddyfield pipit	31	R	SF	I
103.	*Motacilla flava*	Yellow wagtail	43	DM	SF	I
104.	*M. caspica*	Gray wagtail	44	DM	SF	I
105.	*M. alba*	White wagtail	44	DM	SF	I
106.	*M. maderaspatensis*	Large pied wagtail	26	R	SF	I
	Dicaeidae					
107.	*Dicaecum agile*	Thickbilled flowerpecker	42	R	SF	N
	Nectarinidae					
108.	*Nectarinia asiatica*	Purple sunbird	41	R	SF	N
109.	*N. zeylonica*	Purplerumped sunbird	41	R	SF	N
110.	*N lotenia*	Lotens' sunbird	43	R	SF	I
	Ploceidae					
111.	*Passer domesticus*	House sparrow	+	R	SF	G
112.	*Pteronia xanthocallis*	Yellow throated sparrow	50	R	SF	G
113.	*Ploceus philippinus*	Baya weaver	18	R	SF	I
114.	*Lonchura malabarica*	Whitethroated munia	27		SF	I
115.	*L. striata*	Whitebacked munia	29	R	SF	I
116.	*L. punctulata*	spotted munia	37		SF	I
117.	*L. malacca*	Blackheaded munia	43	R	SF	I

Note: + = Counts not maintained.
Resident status:— R = Resident, LM = Local migrant, DM = Distant migrant
Habitat:—WL = Wetland, SF = Scrub Forest
Habitat:—D = Divers, R = Raptors, S = Swimmers, Sk = Skimmers, W = Waders
Feeding type:— C = Carnivore, F = Frugivore, G = Graminivores, H = Herbivore, I = Insectivore, N = Nectivores, P = Piscivore, Sc = Scavengers.

Table—7 Percentage distribution of the bird community with special reference to their habitat, feeding habit and activity.

Bird Community	% Total Population	% Species Composition
Habitat		
Terrestrial	1.51	59.83
Aquatic	98.50	40.17
Habit		
Waders/Wading Birds	2.23	23.08
Swimmers/Divers	77.68	9.40
Skimmers	0.07	7.69
Feeding type		
Insectivores	8.67	48.72
Herbivores	84.37	11.11
Piscivores	6.76	17.09
Carnivores	0.19	5.13

The assemblage of larger number and diverse bird species, according to Pianka (1974) is an indication of less species competition due to diverse niche requirements. Though CL supports a large population of herbivorous birds, a closer look reveals that they are different from one another in terms of their fine food requirements. Male Pintails remained segregated in locations rich in tall emergent Oryza, which provided them with nutrient rich food along with better cover. Coots, though known to be herbivorous, choose Hydrilla, if available and also consumed macro invertebrates brought to the surface by the activity of other diving birds. On the other hand, moorhen, the surface forager is known to derive the benefit of food brought to surface by the agitative feeding habit of coot in its association. The female Pintails and Lesser whistling teals were found to be actively feeding on Nymphea leaves and flowers, primarily in the central region of the lake. Thus the finer niche variations within the broadly herbivorous species coupled with broad based interspecific interactions in the form of commensalism appear to be the key factors contributing to the better species diversity along with their abundance in the CL.

THREATS AND RECOMMENDATIONS

CL with its bustling biodiversity, of late appears to be in the grip of severe problems threatening its very existence. Some of the threats to the lake deserving special attention form managers and the well-meaning public alike are listed below.

WEED INFESTATION

As of today the most serious threat to CL is weed infestation in the form of a pteridophyte, Salvinia. The Salvinia weed has presently created a severe ecological imbalance in the lake (Borges

& Shanbhag, 2002). The increased input of organic nutrient into the lakes owing to surface run-off, washing of cattle, and night soil generated by the floating population and the wash down from soiled food package dumps along the edge of the railway station seem to facilitate the luxuriant growth of the weed in the lake. If an immediate plausible solution is not affected, it will not be long before the lake is converted into a single mass of Salvinia leaving little or no open spaces habitable for the waterfowl.

SILTATION

Ageing and the death of a neglected waterbody is a natural phenomenon. In the tropics, this process is hastened by the excessive silt generated within the water body and that imported through surface runoff. Therefore, lakes like CL need be desilted in a phased manner at regular intervals of 4-5 years so as to maintain the natural gradient of the lakebed and prolong its life in a healthy state.

HUMAN INTERFERENCE

CL is also faced with problems of encroachment and human interference. A buffer zone of 50-100m has to be demarcated and strictly maintained along the entire contour of the lake. This will go a long way in minimizing the problem.

REGULAR MAINTENANCE

The embankments of Carambolim Lake are plagued with breeches. These need be strengthened and widened so as to serve as trails for birdwatchers. A *machan,* if erected at a vintage point overlooking the lake would provide the many birdwatchers and naturalists a rewarding view of the lake. This may also promote ecotourism in the state.

REFERENCES

Annonymous 1964. **Technoeconomic Survey of Goa Daman and Diu**. National Council of Applied Economic research, New Delhi.

Baldassare G. A. and E. G. Bolen 1994. **Waterfowl Ecology and Management**. John Wiley and Sons, Inc., New York.

Best E. P. H. 1983. Effects of water pollution on freshwater submerged macrophytes. In. Varshney C. K. (Ed) **Water Pollution and Management Reviews.** South Asian Publishers, New Delhi.

Borges S. D. and A. B. Shanbhag 2002. Freshwater lakes in Goa buckling under the pressure of Salvinia weed menace. **Newsletter to Birdwatchers**. **41**:70-77

Briggs S. V., M. T. Maher and S. M. Carpenter 1985. Limnological studies of waterfowl habitat in south-western New South Wales II. Aquatic macrophyte productivity. **Aust. J. Mar. Freshw. Res. 36**: 707-715.

Brinkhurst R. O. 1974. **The Benthos of Lakes**. Macmilan, England.

Chambers M. R. 1987. The freshwater lakes of Papua New Guinea: an inventory and limnology review. **J. Tropical Ecol**. 3: 1-23.

Crowder A. A., J. M. Bristow, M. R. King and S. Vanderkloet 1970. The aquatic macrophytes of some lakes in south -eastern Ontario. **Naurastiste Can**. **104**: 457-464.

Edmondson W. T. 1972. Nutrients and phytoplankton in Lake Washington. **Am. Soc. Limnol. Oceanogr. Spec. Symp. 1**: 173-188.

Engel S. 1988. The shallow lakes initiative: restoring aquatic macrophytes in Wisconsin. **Wis. Dep. Nat. Resour. Res. Manage. Findings 15**: 4.

Ganapati S. V. and A. Sreenivasan 1968. Aspects of Limnology, primary production and fisheries in the Stanely Reservoir, Maharastra State. **Hydrologia 32**: 551-569.

Goel P. K. 1997. **Water Pollution: causes, effects and control**. New Age International Ltd., New Delhi.

Gopal B., K. P. Sharma and R. K.Trivedy 1978. Studies on ecology and production in Indian freshwater ecosystems at primary producer level with emphasis on macrophytes. In. Singh J. S. and B. Gopal (Eds.) **Glimpses of Ecology**. Inter Science Publications, Jaipur, India.

Happey–Wood C 1988. Ecology of freshwater planktonic green algae. In. Sandgren, C. (Ed) **Growth and Reproductive Strategies of Freshwater Phytoplanktons** pp 175-226 Cambridge Univ. Press. New York.

Hussain, S. A. and R. DeRoy 1993. **Directory of Indian Wetlands**. W. W. F., Indian and Asian Wetland Bureau.

Hutchinson G. E. 1957 and 1976. **A Treatise on Limnology Vol. I. Geography, Physics and Chemistry of Lakes. Vol. III. Aquatic Communities and Energy Relationships**. John Wiley and Sons, New York.

Ilmavirta V. 1980. Phytoplankton in 35 Finnish brown lakes of different trophic status. **Developments in Hydrobiology, 3**: 121-130.

Infante A. and W. Riehl 1984. The effect of Cyanophyta upon zooplankton in a eutrophic tropical lake (lake Velencica, Venezuela) **Hydrobiologia 113**: 293-298.

Junk W. 1977. The invertebrate fauna of the floating vegetation of Bung Borapet, a reservoir in central Thailand. **Hydrobiologia 53: 229-238.**

Kaul V., C. L. Trisal and J. K. Handoo 1978. Distribution and production of macrophytes in some water bodies of Kashmir. In. Singh J. S. and B. Gopal (Eds) **Glimpses of Ecology** Inter Science Publication, Jaipur, India.

Khan M.A.G. and S. H. Chowdhury 1994. Physical and chemical limnology of Lake Kapati, Bangladesh. **Tropical Ecology 35: 35-51.**

Kikuchi T. and J. M. Peres 1977. Consumer Ecology of Seagrass Beds : Seagrass Ecosystems and Scientific Perspective. In. Mc Roy and C. Hafferich (Eds) **Marine Science Vol. 4.** Marker Decler, Inc., New York.

Kinsely K. and W. Geller 1986. Selective feeding of four zooplankton species on natural phytoplankton. **Lagra 69**: 86-94.

Klaine S. J. and C. H. Ward 1983. The Use of Submerged Macrophytes for the Evaluation of Phytotoxicity. In. Varshney C. K. (Ed) **Water Pollution and Marine Reviews.** South Asian Publishers, New Delhi.

Kolkwitz R. and M. Marsson 1908. Oekologie der pflanzlinccken Saprobein. **Ber. Dtsch. Bot. Ges. 26:** 505-519.

Kurasawa H. 1975. Production of community – Japanese and inland water. Part - 9 zooplankton JIBP. Synthesis 10, 436. Tokyo Univ. Press.

Lampert W. and B. E. Taylor 1985. Zooplankton grazing in a eutrophic lake: implication of diel vertical migration. **Ecology, 66**: 68-82.

Lodge D. M. 1991. Herbivory on freshwater macrophytes. **Aquat. Bot.** 41: 195-224.

Martinez M. M. 1993. Birds and Limnology. In. Boltovaskoy A. and H. L. Lopez (Eds) **Conference on Limnology**. Institute of Limnology, La Plata.

McManus J., R. Callier, J. Dymend , C. Geeff Wheat and G. L. Larsen 1996. Spatial and temporal distribution of dissolved oxygen in Crater Lake, Oregon. **Limnol. Oceanogr. 41**: 722-731.

Minns C. K. 1989. Factors affecting fish species richness in Ontario lakes. **Trans. of Am. Fish Soc., 118**: 533-545.

Mohanty S. K. and R. N.Dash 1982. The Chemistry of Waterlogged soils. In. Gopal B. (Ed) **Wetland Ecology and Management.** National Inst. of Ecol. And International Scientific Publications. 389-396.

Montalbano F. I., S. Hardin and I.Hetrick 1979. Utilization of Hydrilla by ducks and coots in Central Florida. **Proc. Annu. Conf. South – East Assoc. Fish and Wildl. Agencies** 33: 36-42.

Mulligan H. F. and A. Baranowski 1969. Growth of phytoplankton and vascular plants at different nutrient levels. **Verh. Internet Verein Theor. Angew. Limnol. 17**: 802-810.

Mukherjee B. and G. V. N. Pankajakshi 1995. The impact of detergents on plankton diversity. **J. Environ Biol. 16**: 211-218.

Nygaard G. 1949. Hydrobiological studies on some Danish ponds and lakes II. The quotient hypothesis and some new or little known phytoplankton organisms. **Kgl. Danske. Viderskagernes Sklskab. Biol.Skirfter 7**: 1-293

Palmer C. M. 1969. A composite rating of algae tolerating organic pollution. **J. Phycol.** 5: 78-82.

Pianka E. R. 1974. Niche overlap and diffuse competition. **Proc. Natl. Acad. Sci., U. S. A. 71**: 2141-2145.

Porter K. G. and J. D. Jr. Orcutt 1980. Nutritional Adequacy, Manageability and Toxicity as Factors that Determine the Food Quality of a Green and Blue – green algae for Daphnia. In. Kerfoot, W. C. (Ed) **Evolution and Ecology of Zooplankton Communities.** Univ. Press, New England.

Rawson D. S. 1960. A limnological comparison of twelve large lakes in North Saskatchewan. **Limnol. Oceanogra. 5**: 195-211

Reish D. J. 1963. A qualitative study of the benthic polychaetous annelids of Bahia de San Quitin, Baja, California. **Proc. Nat. 3**: 399-436.

Reish D. J. 1979. Bristle worms (Annelids : Polychaeta). In. Hart C. W. and F. L. H. Samuel (Eds) **Pollution Ecology of Estuarine Invertebrates.** Academic Press, New York.

Reid G. K. and R. D. Wood 1976. **Ecology of Inland Waters and Estuaries**. Reinhold, New York.

Rump H. H. and H. Krist 1988. **Laboratory Manual for the Examination of Water, Waste Water and Soil**. VCH Publishers, New York.

Sahai R. and A. B. Sinha 1969. Investigations on bioecology of inland waters of Gorakhpur, India I : Bioecology of Ramgarh lake. **Hydrobiologia 34**: 344-447.

Shanbhag A. B., R. Walia & S. D. Borges 2001. The impact of Konkan Railway Project on the avifauna of Carambolim lake in Goa. **Zoos' Print J. 16 (6)**: 503 - 508.

Shapiro J. 1973. Blue–green algae – why they become dominant. **Science 179**: 382-384.

Shapiro J., V. Lamara and M. Lynch 1975. Biomanipulation : Ecosystem approach to lake restoration. In. Brezonik P. L. and Fox J. L. (Eds) **Proc Symp. On Water Quality Management Through Biological Control.** U. S. Environmental Protection Agency, Washington, 85-96.

Shrivastava N. P. and V. R. Desai 1997. Studies on the bottom macrofauna of Rihand reservoir (U. P.). **J. Environ. Biol. 18**: 325-331.

Spence D. H. 1964. The macrophyte vegetation on freshwater lakes, swamps and associated fen. In. Burne J. H. II. (Ed)**The Vegetation of Scotland**, Oliver and Boyd, London, 306-425.

Talling J. F. 1976. Water Charecteristics. **The Nile, Biology of an Ancient River**. Dr. W Junk B.V. Publishers. The Hague.

Tamisier A. and C. Boudouresque 1994. Aquatic bird population as possible indicators of seasonal nutrient flow at Ichkeul Lake, Tunicia. **Hydrobiologia 279/280**: 149-156.

Tilman D., R. Kresling, R. S. Sterner, S. S. Kitham and F. A. Jhonson 1986. Green, Blue-green and diatom algae – taxonomic differences in competitive ability for phosphorus, silicon, and nitrogen. **Arch. Hydrobiol. 106** : 473-485.

Trivedy P. R. and S. Gurdeep Raj (Eds) 1992. **Water Pollution**. Akashdeep Publishing House, New Delhi.

Van de Ven J. 1987. **Asian waterfowl 1987**. IWRB Report, Slimbridge, 103.

Vijayan V. S. 1986. On conserving the fauna of Indian wetlands. **Proc. Indian Acad Sci.**, B. Suppl: 91-101.

Walia R. & A. B. Shanbhag 1999. Status of the avifauna at Carambolim lake in Goa (India) prior to the implementation of Konkan Railway Project. **Pavo 37(1&2)**: 27 - 32.

Westlake D. F. 1967. Some effects of low -velocity currents on the metabolism of aquatic macrophytes. **J. Exp. Bot. 18**: 187-205.

Weisner S. E. B., J. A. Strand and H. Sandsten 1997. Mechanisms regulating abundance of submerged vegetation in shallow eutrophic lakes, **Oecologia**, 109: 592-599.

Wiederholm T. 1980. Use of benthos in lake monitoring. **J. Wat. Poll. Control. Fed.** 52: 537-547.

20

Studies on Ornitho-botanical Aspects of Anashi National Park in Western Ghats (India)

Dr. J.C.Uttangi

INTRODUCTION

India is endowed with abundant flora and fauna that nearly 4-5 percent of all known plant, animal and bird species on earth are found in the Indian Sub-Tropics. The main reason for this remarkable diversity of life forms in a single country like ours is the presence of African, Europian, Chinese and Indo-Malayan elements in the flora and bird fauna of India.Different kinds of climates have influenced to produce every major type of habitat. Genetically, India has become important for a number of food crops, forest trees including medicinal and aromatic plants as well as domesticated animals. However, during the past few decades, the country has witnessed serious depletion of its Biological Heritage, because of pressure exerted by the enormous growth of human and livestock population. Realising the importance for greater attention towards preserving natural habitats, a network of scientifically managed Protected Areas called as National Parks, Sanctuaries, Biosphere Reserves and representative samples of all significant Biographic Sub-Divisions have been established. This network of Protected Areas steadily increased from 20,000 square kilometers to 100,000 square kilometers during the period between 1970-1985. The number of National Parks increased from 19 to 53 and the Sanctuaries to 247. In the year 1973, the 'Project Tiger' was launched to save the species and its habitat in the dense forests. The programme at present is spread over nearly eleven states in our country. Different schemes involving total ecosystem preservation have been described as one of the great conservation success of modern times.

In the existing situation one of the most serious weaknesses experienced however, has been the quality of management and consideration for the local people residing in and around wildlife

15/12086 Mission Compound, Dharwad, Karnataka— 580 001 India reserves. In order to meet their needs, appropriate management system, law enforcement, education etc. have been suggested in the Action Plan. But, so far in our country a scientific assessment of all endangered and threatened species has not yet been fully achieved. A total of 135 threatened species of plants has been listed by the Botanical Survey of India. Studies and explorations are in progress to determine their status. So far as birds are concerned, experts have identified nearly 200 species as threatened out of the 2586 species found in the Oriental Region(Tim Inskipp & others, 1988). Birdlife International has listed nearly 76 species as threatened out of the 1250 birds found in the Indian Sub-Continent and about 14 species as restricted species in the state of Karnataka alone . To work out means for restoration of plants, animals and birds, specific measures are being taken up on National scale. More notable steps have been taken to review the National Forest Policy by launching an All India Survey of National Parks Sanctuaries and Biospheres Reserves. Conservation of endangered plants used by the tribal societies is given top priority. To fulfill these tasks, how far the Government and non-government agencies would come together is a question for which we have yet to wait for a satisfactory answer. Unless Law is respected and conservation action made a peoples movement it may not help to make any lasting contributions to the problem. Pressure on forest produce, is already acute. Continued land encroachment, grazing and poaching being continued are affecting long term conservation. Many patchy habitats need attention today. Therefore knowledge about floral and faunal conditions of every type of Protected Area or Park under management is very important.. Since disturbed areas have less number of species the undisturbed forest biotope like Anashi Core Area needs listing of all species of wildlife to understand the level of disturbance. For conservation, smaller areas with greater number of trees are better than larger areas with lesser number of trees.

Anashi National Park is located in Joida Taluka of Uttarkannada District of Karnatak in Western Ghats and commands a pristine semi-evergreen forest area of nearly 250 square kilometers. Programmes to study the flora and bird fauna has not been carried out. Gratifyingly, the U.K. based Oriental Bird Club(OBC), kindly sanctioned necessary funds to carry out a study of its bird fauna and flora, through a survey which was conducted between December 1993 and February 1994 and again between April 1994 and December 1994.

METHODS

A primary survey was undertaken to study and acquaint with conditions, location, limits and boundaries of the Park a preliminary with the assistance of the staff of the Forest Department. Different routes, including foot paths, extraction routes,fair-weather Jeepable roads cart-roads, available within the Park area and outside its boundaries were used during survey. Patchy habitats occurring within the boundaries were taken into account to estimate species diversity and status of bird population in these buffer zones. Habitat conditions of every locality surveyed for birds and plants were noted down including heights of hills. Trees species with flowers and fruits that attracted birds were also noted down and wherever possible they were photographed. Insects, butterflies, snakes and other animals seen during survey were noted down separately. Birds seen on road-sides were also taken into consideration. Soil and water samples were taken to study their physical and chemical status. Hamlets and villages occurring within the park limits were visited to enquire into the socio-economic problems of the people living there.

Bird survey was carried out throughout the day, mostly between 7.30 a.m. to 12 noon in the morning and in the evening between 3 p.m. to 6 p.m. Where it was found necessary to wait and watch feeding flocks of birds or individual species, different vantage spots were selected for the purpose. A list of forest trees, shrubs and climbers recorded from different areas of the Park have been furnished. Names of plants associated with bird species are summarized in Table 2. Wintering bird species within the Park area were listed during the months of December and January. Bird life and floral change during the monsoon were also noted down. Identification of bird species was based on both observation as well as specific calls. The compact Hand book of Birds of India and Pakistan by Salim Ali & Ripley (1987) was used as reference for determining the bird species. Wherever possible the 'transect' method of recording birds was also tried. For 'range' the field Guide to the Birds of southwestern India by R. J. R. Daniels has been used.

PHYSICAL FEATURES OF THE STUDY AREA

Anashi National Park (Fig 1) lies between latitudes 14° 54' and 15° 08' N and longitudes 74° 15' and 74° 31' E. The park lies within the forest reserves of Kumbarwada, Kundal, Anashi, Kadra and Ulvi in Joida taluka of Uttar Kannada district. It forms part of the Western Ghats (Sahyadri range) where hills have an average height of 700 meters.

The crest line ridges mark the water divide line between drainage that flows westwards (towards Arabian sea) and the rivers that flow eastwards over the Deccan plateau.In the higher terraces the vegetation improves in density and quality compared to the western spurs abutting the low level laterite plateaus. The Joida region with Anashi in the centre forms typical central belt where a contrast between hill-tops and ridges could be seen. They support a thinner vegetation but the Kaneri river valley existing there holds a luxuriant plant growth. Evergreen strands are a frequent feature but, the river banks almost invariably support the thickets of Bamboo growth. Towards Dandeli there is increasing dryness in landscape. In Anashi range there is no growth of Teak.In the south below Kadra there is Kali river to which several nalas, like Bendinala, Karadinala and Vakinala flowing though Dokreppa area join. The Kaneri river along the eastern border of the Park and along with the tributaries like Vakinala, Kattenala and Maradinala flow down to join the Kali river in the south.

The rocks exposed in this region are of Precambrian age.These include Dharwarian schists, Archean Gneisses (The Penincular Gneisses) and granites. The rocks of the Dharwar system include Quartzite, Banded Magnetite Quartazite, Quartz sericite schists, dolomite and basic igneous rocks. Ores of iron and manganese are the chief economic deposits. Bands of limestone are found in Supa. Limestone is also exposed in the Kali river at southern border of the park. Such mineral deposits are important for birds. One such bird that is seen regularly on the Ghat road is the Emerald Dove. Other mineral deposites in the Park area include bauxite, copper, clay, mica, mineral pigments, rock-crystal sand, stealite and dolerite. Occurrence of springs at foot hills in the park area are also found at a number of places. The average annual rainfall annually is about 2,741 mm. Water pools are rarely seen as the soil is porous. For this reason wild animals tend to migrate locally to lower areas which lie outside the Park boundaries.

The Department of Wildlife, Dandeli, is making efforts to construct small check-dams to stream for storage of water so as to enable the animals to stay within the Park area during dry season. At Maradinala point there is a 'Tiger Tank'. Attempts to conserve water in these water spots are being made by the Department of Wildlife, Dandeli.

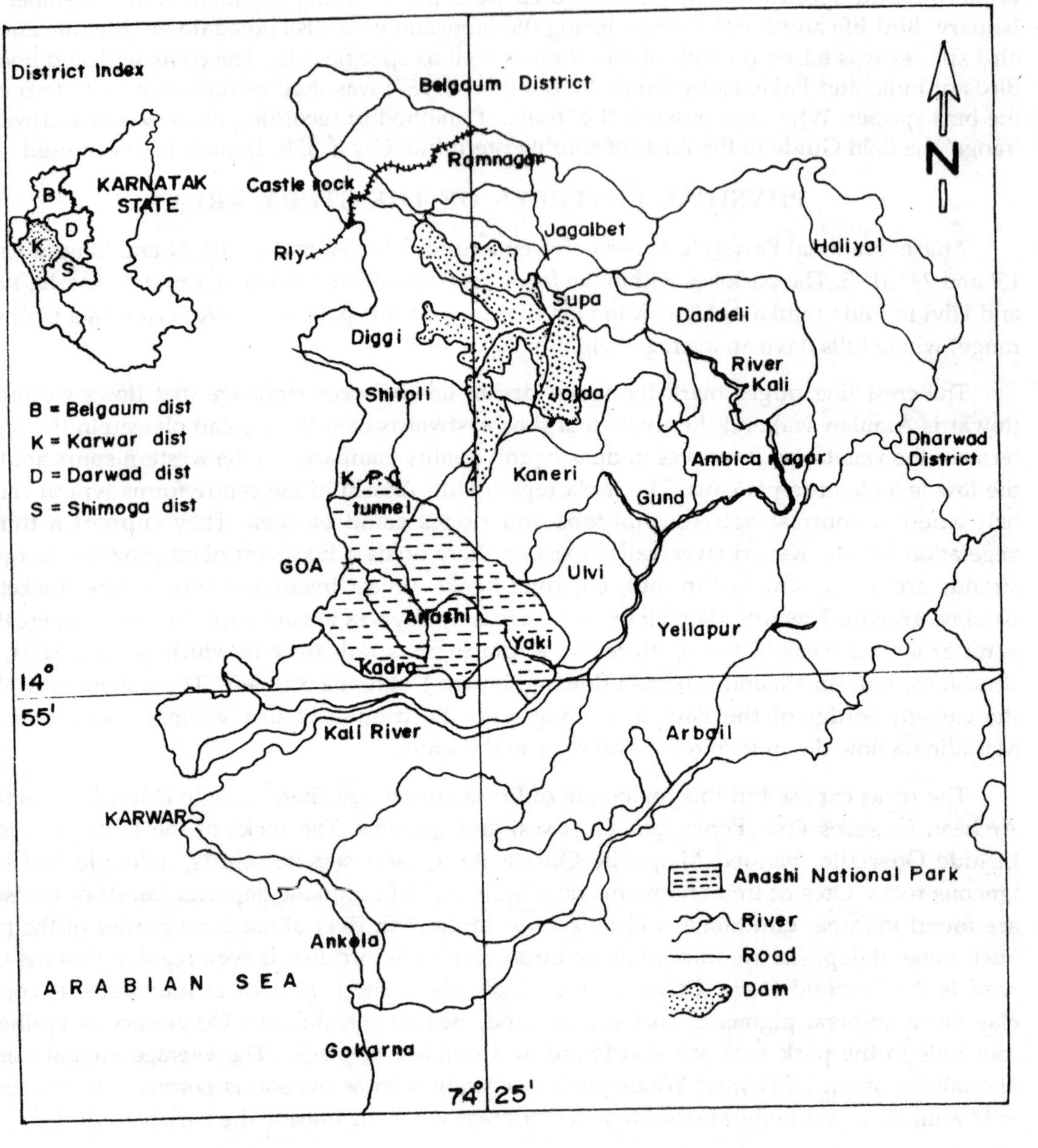

Fig. 1 Map showing location of Anashi National Park in Uttar Kannada District of Karnatak

The types of forests found in the Park area are semi-evergreen to evergreen and large patches of moist deciduous vegetation are also found. The timber is soft, the softness is indicative of the forest affinities with semi-evergreen type. A deciduous growth of Bamboo is also seen. The high forests in south-eastern region near Gund attract elephants and bisons. The forest found along river banks and slopes have timber of high quality and their dense canopies support rare birds like Yellow-Browed Bulbul, Babblers, Laughing Thrush, Fairy Bluebird and similar restricted types including the Malabar Whistling Thrush. The elevation of the central Ghat tract varies from 200 to 900 meters, and is characterized by high humidity all round the year along western side and drier along the central and eastern parts. Rainfall is quite adequate in the Western Ghats touching Anashi National Park. The variation in temperature and rainfall is not very significant. During April and May, the maximum daily temperature stands between 32°-33° degrees Celsius and the mean daily temperature is between 25°-26° degrees Celsius along western side of the Park. The mean monthly humidity varies between 60-86%. The northern part of the Park receives more rain than the southern part. The rainfall ranged between 77 to 126 mm. The pH of the soil ranged from 6.5 to 8.5 and the electric conductivity ranged from 0.8 to 1.2 mhos. in the study area.

The eastern border of the Park has a 'sacred grove' called ' Ulvi ' with a temple. The term ' Ulvi ' in Kannada has some reference to the whiteness of the stem of the tree called ' Bilimatti ' (*Terminalia paniculata*), the leaves of which have transparent 'pellucid' glands. Festivals are held regularly each year during 'sankranti' (14th January) which is attended by more than 1 lakh devoties. The traffic is found to disturb the movements of migratory birds along Ulvi-Anashi road in winter. This narrow asphalted road passing through the dense forest is ideal for birdwatching and other favourite pursuits. Table—1 gives the localities of observation. The list of plants associated with birds is summarized in Table—2. The diversity and status of avifauna is presented in Table—3

A FEW CONSERVATION MEASURES SUGGESTED FOR ANASHI NATIONAL PARK

1. The survival ratio of trees due to air pollution is most vital. The complex causes involved in such cases should be studied in detail. A study on the survival ratio of climax tree species will help to determine the health of the canopy. Actually, this is based on the defoliation done by 'leaf minor' and other noxious insects in summer. If the trees cannot tolerate the effect, they deteriorate. Any abnormal and unusual leaf fall (defoliation vigour) is attributed to the absence of insectivorous birds in the Park area. Some birds like tits are very sensitive to noise or traffic in the Park area. Therefore a special study to understand the relations existing between different species of plants, animals, insects, birds and butterflies becomes necessary.

2. Biotic conditions of Anashi National Park have degraded considerably due to logging operations and other forestry activities adapted massively in the past few decades. Their removal interferes with the life cycle of many lower animals, insects and birds, as well as, the recycling process of the rain forest. Large trees standing dead are regularly used as nesting habitats by birds like parakeets, woodpeckers, hornbills, barbets, blue jay and other hole nesting species.

TABLE—1

(a) Ulvi-cross to Kumbarwada

Code No.	Locality	Altitude (m)
1	2	3
1.	Amboli (Bamboo Brake)	509
2.	Chandrali (Spring)	508
3.	Matkarni (secondary forest)	600
4.	Ambadgalli (dense mixed jungle)	560
5.	Bombulsay (teak plantation)	564
6.	Kalasay (nala)	600
7.	Kondari (plantation)	644
8.	Tadshet (")	642
9.	Rumbadishet (")	569
10.	Godshet (dense mixed jungle)	680
11.	Davulli (Eucalyptus plantation)	642
12.	Kumbarwada cross (Christian memorial)	
	(b) Ulvi-cross to Anashi	581
13.	Pavala Halla (causeway river Kaneri)	486
14.	Khamshetadi (spring & sacred groove)	500
15.	Nargali (Katte Halla)	508
16.	Kodugali (dense forest)	600
17.	Vakinala (bank vegetation)	520
18.	Anashi (forest depot)	542
19.	Anashi Hill (dense vegetation)	
	(c) Kumbarwada-Anashi-Khadra	724
20.	Kaneri River (bridge point)	552
21.	Gundali (nala basin) west	838
22.	Malshet (teak plantation) west	700
23.	Nujji (kuneman locality) west	614
24.	Kuneman (dense mixed jungle) west	614
25.	Gavali (cultivated land) east, hill	700

(Contd...)

1	2	3
26.	Ajji (katte halla basin) east down-hill	622
27.	Dokreppa cross (dense mixed jungle) west	614
28.	Dudgali (hamlet) east	680
29.	Talapa (hamlet) west	732
30.	Moppai (dense mixed jungle) east to Bigundi	663
31.	Bigundi-kadpod-mayagini areas (east)	663
32.	Baradi (dense mixed forest) west	615
33.	Badapoli (dense mixed jungle) west	615
34.	Bhaki (hill with rocky knob) east/plantation	514
35.	Anashi (P.W.D.) east bungalow	542
36.	Anashi hill (dense forest) west	742
37.	Masoli west and kailwada east (rocky)	400
38.	Digali Nalla (dense bank vegetation) west	400
39.	Barpoli bridge (dense forest) east	552
40.	Hingewadi (ghat area with dense forest) W	515
41.	Pankadra (down hill) southern border	200
42.	Sakali Nalla (secondary forest) West	200
43.	Khadra forest bungalow	210
44.	Maradi (Maradi nalla bank side) south	
	(d) Kumbarwada to Kundal (north)	112
45.	Kaneri river cross (dense mixed vegetation)	552
46.	Jhalvadi (K.P.C. dam site) north	632
47.	Hale Kundal (tiger spot) north	616
48.	Kutavli hill	726
49.	Kundal (hamlet) north along Goa border	
	(e) Dokreppa to ambal	688
50.	Dokreppa cross point (dense vegetation)	552
51.	Malchepani (West)	600
52.	Tuppegali (West)	600
53.	Bendehalla (rounded rock bad)	640

(Contd...)

1	2	3
54.	Gali (hamlet) Goa border (West)	906
55.	Kumgali area north west	700
56.	Aspad area north west	680
57.	Ambal hill north west	927
58.	Ambal (hamlet) west along Goa border	
	(f) Barpoli cross to Maradi nala basin.	702
59.	Maradi nala basin (Calophyllum growth) east	620
60.	Bison tank (dense mixed jungle) east	620
61.	Chana Kunagi (dense) east	640
62.	Kailvada (hamlet) east	473
63.	Ambewadi (dense growth) east (sacred grove)	678
64.	Halyal border Bridge point (sign post) w.life	
	(g) Ghotegali-Bargadde to Jaddi basin	600
65.	Gotegali cross (along Sadashiv Ghad Rd.)	187
66.	Baragadde (secondary forest)	400
67.	Goyar (dense mixed jungle with bamboo)	410
68.	Kamargaon area (west along Goa border)	750
69.	Jaddi (nala basin along Goa border) n. west	
	(h) Adjoining boundaries	756
70.	Joida tank	500
71.	Kulgi tank	450
72.	Kumbarwada	500
73.	Gund (K.P.C. dam site to Kaveri river)	480
74.	Kalni (Gund road with bamboo growth)	500
75.	Jaggi halla with bamboo	550
76.	Potoli cross (dense evergreen growth)	500
77.	Kunda-kundi nalla along Gund Rd	400
78.	Karsingala nala	500
79.	Kalpai nursery	550
80.	Kalasai site (Ulvi road)	500

Table—2 Names of Plant Species Associated with Birds.

(a)	Terminalia bellerica	(b)	Eupatorium sp.
(c)	Glycosmis pentaphylla (Shrub)	(d)	Ficus nervosa
(e)	Mussaenda frondosa	(f)	Lasiosiphon eriocephalus
(g)	Myristica malabarica	(h)	Calophyllum tomentosum
(i)	Loranthus sp. (flowers)	(j)	Cedrela Toona (river)
(k)	Helicteres Isora	(l)	Salmalia malabarica
(m)	Lenna grandis (leaves shed)	(n)	Bauhinia malabarica
(o)	Flacourtia Montana (fruits)	(p)	Randia malabarica (flowers)
(q)	Grewia tiliaefolia (flowers)	(r)	Cassia sp. (creeper with pods on Terminalia)
(s)	Mucuna pruriens (creeper)	(t)	Terminalia arjuna
(u)	Careya arborea (fruits)	(v)	Ficus asperrima (fruits)
(w)	Xylia xylocarpa (flower)	(x)	Callicarpa lanata (flowers)
(y)	Terminalia paniculata	(z)	Syzygium cumini
(a′)	Dillenia pentagyna (fruits)	(b′)	Leea samboscina (shrub)
(c′)	Wendlandia Notoniana	(d′)	Saraca indica (flowers)
(e′)	Xylia dolabriformis	(f′)	Strychnos nux-vomica
(g′)	Pongamia glabra (river)	(h′)	Hopea Wightiana (river)
(i′)	Barringtonia acutangula (r)	(j′)	Trewia nudiflora (river)
(k′)	Flemingia strobilifera (shrub)	(l′)	Breynia patens (flowers)
(m′)	Allophylous Cobbe	(n′)	Rubia cordifolia
(o′)	Maba nigrescens (flowers)	(p′)	Madhuca nerifolia (river)
(q′)	Carallia integerrima	(r′)	Anthocephalus Cadamba
(s′)	Mallotus albus	(t′)	Odina Woodier (young leaves)
(u′)	Garcinia Cambogia	(v′)	Alstonia scholaris
(w′)	Dipterocarpus indicus	(x′)	Gnetum ulla (hanging fruits)
(y′)	Lagerstroemia lanceolata	(z′)	Alstonia scholaris
(a″)	Hydnocarpus laurifolia	(b″)	Osbekia truncata (shrub)
(c″)	Dalbergia tamarindifolia	(d″)	Calamus flagellum (palmae)
(e″)	Holigarna Arnottiana	(f″)	Albizzia odorattisima
(g″)	Calycopteris floribunda	(h″)	Artocarpus hirsute
(i″)	Mangifera indica	(j″)	Eugenia corymbosa (flowers)
(k″)	Vitex leucoxylon (fruits)	(l″)	Diosporys Melanoxylon
(m″)	Ochlandra Talboti (river)	(n″)	Aporosa Lindleyana
(o″)	Bombax Malabaricum (flowers)	(p″)	Macaranga tomentosa
(q″)	Mitragyna parviflora (flowers)	(r″)	Memecylon edule
(s″)	Ficus glomerata (fruits)	(t″)	Salix tetrasperma (river)
(u″)	Diospyros angustifolia	(v″)	Homonia riparia (shrub)
(w″)	Rauwolfia serpentina	(x″)	Vanda (orchid)
(y″)	Oberonia (orchid)	(z″)	Cymbidium (orchid)

Table—3 Table Showing Bird Species Occupied In Anashi National Park

Nos.	*Name*	*Code No. of Locality*	*Code No. of Associated Plants*	*Status*
1	2	3	4	5
1	*Phalacrocorax fuscicollis* (INDIAN CORMORANT)	46	Kaneri river K.P.C.dam site	RS
2	*Ardeola grayii* (POND HERON)	43, 44	g′	RS
3	*Buteo buteo* (COMMON BUZZARD)	69	a	W
4	*Aquila clanga* (GREATER SPOTTED EAGLE)	12,19,8	t″	W
5	*Ictinaetus malayensis* (BLACK EAGLE)	35	m	RS
6	*Spilornis cheela* (CRESTED SERPENT EAGLE)	75	Aerial	RS
7	*Falco amurensis* (AMUR FALCON)	68	o	W
8	*Falco subbuteo* (EURASIAN HOBBY)	71	y	W
9	*Gallus sonneratii* (GREY JUNGLEFOWL)	5,6,19,40,36	b″, v″	RS
10	*Pavo cristatus* (INDIAN PEAFOWL)	14	k″	RS
11	*Gallinula chloropus* (INDIAN MOORHEN)	70,71	Joida Tank	RS
12	*Amaurornis phoenicurus* (WHITEBREASTED WATERHEN)	70,71	Bamboo bush	RS
13	*Vanellus indicus* (REDWATTLED LAPWING)	62,28,29	Kali river bed	RS
14	*Sterna aurantia* (RIVER TERN)	43,17	Kaneri river	RS
15	*Treron pompadora* *(POMPADOUR GREEN PIGEON)*	13,50	o, v, k″	RS
16	*Treron phoenicoptera* (YELLOWFOOTED GREEN PIGEON)	36,37	v, s″	RS

(Contd...)

1	2	3	4	5
17	*Ducula badia* (*MONTAIN IMPERIAL PIGEON*)	13	h'',y',t	RS
18	*Columba elphinstonii* (NILGIRI WOOD PIGEON)	17	k', v''	E & RR
19	*Streptopelia orientalis* (ORIENTAL TURTLE DOVE)	35,36	Anashi road	RS
20	*Streptopelia chinensis* (SPOTTED DOVE)	72	e, z	RS
21	*Chalcophaps indica* (EMERALD DOVE)	40,41,43	Ghat road	RS
22	*Psittacula krameri* (ROSERINGED PARAKEET)	23,26	r, x, k''	RS
23	*Psittacula cyanocephala* (PLUMHEADED PARAKEET)	43	t, dried stem	RS
24	*Psittacula columboides* (MALABAR PARAKEET)	23	r. p, g''	E
25	*Loriculus vernalis* (VERNAL HANGING PARROT)	72	w', g'', i	RS
26	*Cuculus varius* (COMMON HAWK-CUCKOO)	68	y', t, c	RS
27	*Centropus bengalensis* (LESSER COUCAL)	73	c', n, f''	RS
28	*Bubo nipalensis* (*SPOTBELLED EAGLE OWL*)	71	h', j'	RS
29	*Bubo coromandus* (*DUSKY EAGLE OWL*)	79	l'', a'', p	RS
30	*Strix leptogrammica* (*BROWN WOOD OWL*)	8, 9	h', y', g''	RS
31	*Glaucidium radiatum* (*JUNGLE OWLET*)	63	Bamboo Brake	RS
32	*Caprimulgus macrurus* (*LARGETAILED NIGHTJAR*)	10,11	Forest floor	RS
33	*Zoonavena sylvatica* (WHITERUMPED SPINETAIL SWIFT)	35,36	Aerial	RS
34	*Tachymarptis melba* (ALPINE SWIFT)	32,33	Aerial	RS
35	*Apus pacificus* (FORKTAILED SWIFT)	78,79	Aerial	W
36	*Hemiprocne coronata* (CRESTED TREE SWIFT)	59,60	h, t, z	RS

(Contd...)

1	2	3	4	5
37	*Harpectes fasciatus* (MALBAR TROGON)	46	h′, y′	RS
38	*Ceryle rudis* (PIED KINGFISHER)	71,46	Kaneri river	RS
39	*Alcedo atthis* (COMMON KINGFISHER)	47,71	Vakinala	RS
40	*Halcyon smyrnesis* (WHITETHROATED KINGFISHER)	75,69,45,6	K.P.C.dam site	RS
41	*Merops leschenaultia* (CHESTNUTHEADED BEE-EATER)	54,55	w, j′ Bamboo	RS
42	*Merops orientalis* (SMALL GREEN BEE EATER)	18,22	n, g′, f″	RS
43	*Nyctyornis athertoni* (BLUEBEARDED BEE-EATER)	46	h′, y′, w′	RS
44	*Coracias benghalensis* (INDIAN ROLLER)	70,71	u′, t′,v′	RS
45	*Eurystomus orientalis* (DOLLARBIRD)	60	z′,Bison Tank	RS
46	*Ocyceros birostris* (INDIAN GREY HORNBILL)	69, 7, 8	b″, k′, v″	RS
47	*Ocyceros griseus* (*MALABAR GREY HORNBILL*)	24,30	o, p, v, a′	RS
48	*Anthracoceros coronatus* (*MALABAR PIED HORNBILL*)	16,17	v, o, a″, k″	RS
49	*Buceros bicornis* (*GREAT HORNBILL*)	16	a′, c″, u	RS
50	*Megalaima viridis* (*WHITECHEEKED BARBET*)	41,42,10	l, t, y′	RS
51	*Megalaima zeylanica* (BROWNHEADED BARBET)	37,38,39	c″, e″, g	RS
52	*Megalaima haemacephala* (COPPERSMITH BARBET)	79,80	s″	RS
53	*Picumnus innominatus* (SPECKLED PICULET)	65,66	k, y	RS
54	*Dinopium bengalense* (BLACKRUMPED FLAMEBACK)	14,18	l, v′, g″	RS
55	*Dinopium javanense* (COMMON FLAMEBACK)	3, 4, 6	y, x	RS

(Contd...)

1	2	3	4	5
56	*Dryocopus javensis* (WHITEBELLED WOODPECKER)	16,22	i	RS
57	*Hemicircus canente* (HEARTSPOTTED WOODPECKER)	68,69	x, n' on crown	RS
58	*Hirundo daurica* (REDRUMPED SWALLOW)	18	Aerial	RS
59	*Hirundo rupestris* (EURASIAN CRAG MARTIN)	17,18	Aerial	W
60	*Hirundo smithii* (WIRETAILED SWALLOW)	46,73	on wire	RS
61	*Delichon urbica* (NORTHERN HOUSE MARTIN)	78,79	Aerial	RS
62	*Lanius schach* (LONG-TAILED SHRIKE)	46,72	b', k'	RS
63	*Lanius cristatus* (BROWN SHRIKE)	67	n", p"	W
64	*Oricolus oriclus* (EURASIAN GOLDEN ORIOLE)	71	u', o"	W
65	*Oricolus xanthornus* (BLACKHOODED ORIOLE)	23,26	w, a	RS
66	*Daricrurus macrocercus* (BLACK DRONGO)	57,58	a', i', f"	RS
67	*Dicrurus leucophaeus* (ASHY DRONGO)	39,43	q', r'	W
68	*Dicrurus paradiseus* (GREATER RACKET-TAILED DRONGO)	61,62	t, m	RS
69	*Artamus fuscus* (ASHY WOOD SWALLOW)	49,50	t,(dried twig)	RS
70	*Acridotheres fuscus* (JUNGLE MYNA)	13,14,73	l, j	RS
71	*Gracula religiosa* (HILL MYNA)	14,17,18	t, y	RS
72	*Dendrocitta vagabunda* (RUFOUS TREE PIE)	57,58	k, w'	RS
73	*Corvus macrorhynchus* (LARGEBILLED CROW)	59,60	e", f"	RS
74	*Pericrocotus flammeus* (SCARLET MINIVET)	76,77	f", h"	RS

(Contd...)

1	2	3	4	5
75	*Pericrocotus cinnamomeus* (SMALL MINIVET)	54,62	q″, g″	RS
76	*Aegithina tiphia* (COMMON IORA)	2,64,72	i″, h	RS
77	*Chloropsis aurifrons* (GOLDFRONTED LEAFBIRD)	12,40,43	v″, b″	RS
78	*Iole indica* (YELLOWBROWED BULBUL)	13,17,27,53	d′, m″	RS
79	*Hypsipetes leucocephalus* (BLACK BULBUL)	1,39,40	z, p	RS
80	*Pomatorhinus horsfieldii* (INDIAN SCIMITAR BABBLER)	14,15	f″, b″	RS, NT
81	*Turdoides striatus* (JUNGLE BABBLER)	4,10	u, v″	RS
82	*Pellorneum ruficeps* (PUFFTHROATED BABBLER)	70,72	x′, n′	RS
83	*Garrulax delesserti* (WYNAAD LAUGHING THRUSH)	13,14	b′, r″	RS, NT
84	*Garrulax jerdoni* (GREYBREASTED LAUGHING THRUSH)	16,17	Ardesia solanacea	E & RR
85	*Muscicapa muttui* (BROWNBREASTED FLYCATCHER)	9,10	z, j′	W
86	*Muscicapa ruficaudata* (RUSTYTAILED FLYCATCHER)	37,38	e″, q″	W
87	*Ficedula parva* *(REDBREASTED FLYCATCHER)*	24,26	r″,m	W
88	*Ficedula nigrorufa* *(BLACK-AND-ORANGE FLYCATCHER)*	39,40	Maesa indica	E & RR
89	*Cyornis tickelliae* *(TICKELL'S BLUE FLYCATCHER)*	39	Melanostroma malabaricum	RS
90	*Eumyias thalassina* *(VERDITER FLYCATCHER)*	73	l	W
91	*Phylloscopus collybita* *(EURASTAN CHIFFCHAFF)*	53,54	Ixora coccinea	W
92	*Phylloscopus affinis* *(TICKELL'S LEAF WARBLER)*	68,69	m′	W
93	*Phylloscopus griseolus* *(SULPHURBELLED WARBLER)*	57,58	u	W

(Contd...)

1	2	3	4	5
94	*Phylloscopus trochiloides* (GREENISH WARBLER)	43,32,33	p	W
95	*Copsychus saularis* (ORIENTAL MAGPIE ROBIN)	46,47	Callicarpa lantana	RS
96	*Myiophonus horsfieldii* (MALABAR WHISTLING THRUSH)	43,44	Maradi nala	RS
97	*Zoothera citrinea* (ORANGEHEADED THRUSH)	66,67	Bamboo thicket	RS
98	*Turdus merula* (EURASIAN BLACKBIRD)	49,68	t	RS
99	*Parus major* (GREAT TIT)	27,28	u	RS
100	*Sitta castanea* (CHESTNUTBELLED NUTHATCH)	24,25	w′	RS
101	*Dendronanthus indicus* (FOREST WAGTAIL)	17,34	forest road	W
102	*Motacilla flava* (YELLOW WAGTAIL)	11, 15	forest road	W
103	*Motacilla maderaspatensis* (WHITEBROWED WAGTAIL)	70,71,72	Garcinia indica	RS
104	*Dicaeum agile* (THICKBILLED FLOWERPECKER)	6,7,19	j″, q″	RS
105	*Dicaeum concolor* (PLAIN FLOWERPECKER)	11,12,59	t′	RS
106	*Nectarinia zeylonica* (PURPLERUMPED SUNBIRD)	40,65,66	b, (nest)	RS
107	*Nectarinia minima* (CRIMSON-BACKED SUNBIRD)	74	i, p, j″,o″	E, T
108	*Zosterops palpebrosa* (ORIENTAL WHITE-EYE)	23,26	p, q, i″	RS
109	*Amandava formosa* (GREEN AVADAVAT)	43	open grass	V
110	*Caprodacus erythrinus* (ROSEFINCH)	39,38	t (dry twig)	W,NT

RS = Resident

E = Endemic

W = Wintering

RR = Range Restricted

NT = Near Threatened

T = Threatened

V = Vegrant

3. The people associated with agricultural activities have extended into the forest area through ' kumri ' cultivation which has reduced the status of many nearby virgin areas. Most tribals dwelling here are unwilling to move away elsewhere. Programmes to educate them about the need for conservation of natural habitats for survival of wildlife should be organized from time to time. The objective would also help in streamlining the management and administration of Protected Areas in conformity with Political Administration.

4. The proposed 250 sq. km area of Anashi National Park now stands reduced to a limited core area of only 104 sq. km, due to people's demands and political pressure. The Endemic and Range Restricted bird species and other threatened and near threatened species need enough protection from all kinds of external threats like mining, dam construction in the vicinity of the Park area or through it. Strict implementation of such legislation would help in maintaining Park's ecosystem diversity as well as species and genetic diversity. For such action the best approach would be to focus on land use pattern on habitats.

5. Percolation of water from, two water tanks viz; the Bison-tank and the Tiger-tank should be reduced using appropriate technology. The iron observatory needs replacement as it causes noise pollution.

6. Afforestation measures should include fruit and flower bearing trees. All open areas should be given priority.

7. A 'Butterfly Nursery' somewhere near Ajji, Nujji or at Anashi village may be established to boost up the ecological niche of breeding birds as well as to maintain the healthy growth of plants in the region.

8. A few semi-evergreen patches exist near Kumbarwada, Potoli, Gund and Ulvi. They should be included in the Park. This will indirectly help in maximizing the size of the bird population in the Park.

9. It is better to set up a ' Buffer Zone ' to the present ' Core Area 'of the Park.

10. A speed limit should be enforced on traffic along Sadashivgadh (Karwar) road which passes through middle of the Park Area to reduce disturbance to animals and birds.

11. Government should take up/award projects to study this pristine park which enables to plan conservation strategies.

ACKNOWLEDGEMENT

The author takes this opportunity to express his sincere thanks to the Council Members of the Oriental Bird Club (OBC), Slimbridge (U.K.) for granting necessary funds to carry out the survey. The author also thanks the OBC conservation Officer Mrs.Carol Inskipp for taking pain in obtaining special permission from the Department of Forest and Environment, Government of Karnataka, to carry out the said Avifaunal Survey in the National Park at Anashi. Thanks are also due to Dr. M.V. Shivalingaradhya, Reader in Botany, Karnataka University Dharwad for his kindness in doing the hard work of identifying all the plant species during survey. The author is specially indebted to Dr. G. R. Hegde, of the Department of Botany for his kindness in doing the hydrological studies and last but not least, thanks are also due to the Forest Staff including the A.C.P. and C.C.F. of the Wildlife Wing of Dandeli for their hospitality.

REFERENCES

1. Ali.S. (1979), The book of Indian Birds (BNHS) pp. 1-187
2. Ali S. & Ropley (1978), The concept Hand Book of The Birds of India and Pakistan (Oxford U. Press) New Delhi, pp. 733
3. Amit Mitra & Sumanta Pal (1994), The spitit of the Sanctuary (Down to Earth)
4. Ben King et al (1975), The field Guide to the Birds of South East Asia (Collins) London, pp. 449
5. Bingham C.T. (1905), The butterflies of India Vol. I, (F.B.I.) London pp.501
6. Daniel R.J. (1994), A Landscape Approach to Conservation of Birds (Jr. of Eclogy) Vol. VII, Pune, pp. 1-5
7. Daniel R.J. (1997), A Field guide to the Birds of South-Western India (Oxford U.Press) Bombay, pp. 195
8. Ganesh Pangare & Richards. C. (1993), Pastoralists at Cross Roads (Down to Earth) Dec. 15, New Delhi, pp. 25-27
9. Jon Charles Co. (1994) Animals and their Habitats are inseprable (Down to Earth)
10. Kamat S.V. (1985), Uttarkannada District (Gazetteer, Govt. of Karnataka) pp. 1-110
11. Kehimkar–Gay–Punetta (1992), Common Butterflies of India (Oxford U. Press), Bombay, pps.67
12. Mavinkurve G.R. (1965), Working plans for Joida Supa Forests, Karnataka North Division, pp. 1-44
13. Muraleedharan T.R. (1994), Mushroom Magic (Down to earth, May 31, New Delhi, page.47)
14. Martin B.P. (1987), "World Birds" (Guinness - USA) pp. 1-191
15. Numata M. (1993), A Historical Review of conservation Efforts in Japan (Jr. Environmental Awareness Baroda, Vol. 16, No.3, pp. 85-98)
16. Narendra K. (1993), The Broken Mirrow (Down to Earth) new Delhi, pp. 26-31
17. Rabb G.B. (1993), The evolution of Zoos, Auarium & Botanical Gardens in relation to Protected Areas (Jr. Environmental Awareness Baroda, Vol. 15, No.4, pp. 143-146)
18. Rajendran S. (1994), Villagers Unite to Conserve Resources (Down to Earth)
19. Richard Grimmet & C. Inskipp (1998), Birds of the Indian Sub-Continent (Oxford u. Press, Delhi, pp. 1-887)
20. Shanmukhappa. G. (1996), Working Plan of Unorganised Forests of Anmode – Anashi – Ulvi, Ranges (D.F.O) Belgaum, pp. 1-174
21. Singh H.R. (1985) An Introduction to animals Ecology (Chand & Co.) Vishal Publication, Delhi, pp. 1-301
22. Tim Inskipp & Others (1998), Birds of the Indian Sub-Continent (OBC). Pp-888

OTHER REFERENCES

1. Cooke T. (1958), Flora of the Presidency of Bombay Vol. I, II & III (Botanical Survey of India) Reprint.
2. Saldanha C.J. (1984), Flora of Karnataka Vol. I (Ox. & I.B.H) N. Del.
3. Wetzel R.G. (1975), Limnology (W.B. Saunders Co,) London.
4. Odum E.P. (1971), Fundamentala of Ecology (Toppan) Japan, pps. 557
5. Uppin S.P. (1987), Field guide to forest trees. Pp. 1-135
6. Talbot W.A. (1909) Forests flora of Bomaby and Sindh, Vol. I & II
7. Salim Ali. (1996) The Book of Indian Birds (B.H.S)
8. K. Gunathilaga – Raj (1998) Some South Indian Butterfiles (Banglore pp. 1-265)
9. Ghanshyam Saxena (1990) The Forest Crises (Natraj Publishers Dehradun pp. 1-177.S)

Marine Environment

21

Introduction

K.G.Hiremath

The ocean controls the climate which is the very foundation of livelihood of all life on the earth. It is therefore imperative that we identify and understand the numerous features of ocean. Effective and sophisticated utilization of ocean will promise prosperity and future development of mankind.

Man has turned to ocean to harness energy, food, medicine, fodder, fertilizers, fuel and many other industrial products. Easy accessibility, rich variety of organisms, communities and environmental gradients experienced by oceans, especially by coastal waters, make them favourable area for research, education, conservation and exploitation.

Oceans unfortunately are also considered as sinks to dump terrestrial wastes. Oceans have purified and rendered harmless all that had been discharged, but now they have begun to fail to cope with the powerful influx of anthropogenic wastes. Thus there is need, more than ever, to intensify the monitoring of oceans. Ocean monitoring involves a deeper analysis of theoretical and experimental aspects to determine the present ecological status and plan for future exploration. Environmental oceanography is the knowledge acquired by studies undertaken by physical, chemical, biological, geological and other emerging areas in oceanography.

Biological oceanography is most complex of all marine science. Since, they have to interact with specialists from other fields. The living communities have to be studied with respect to their diversity, density, population dynamics, community analysis, distribution, and behaviour patterns. This section has papers on biological and geological oceanography and marine corrosion and bifouling.

The paper on "Distribution of family Rhizophoraceae along the west coast of India" by Dr. Untawale has reviewed the distribution of mangrove along west coast of India. The study warns

* Head, Department of Botany, Dhempe College of Arts & Science, Miramar, Panaji, Goa-403001 India.

about degrading population of mangrove belonging to Rhizophoraceae. Reclamation, deforestation, cultivation and other developmental activities are severely affecting the density and diversity of this group of ecological important plants.

Phytoplankton accounts upto 95% of ocean primary productivity which justifies the importance given to phytoplankton research in oceanographic studies. Conventional oceanographic techniques are time consuming and cannot cover large areas. The launching of Sea W :FS in 1997 enabled instantaneous large-scale coverage of ocean for biological data. The paper by Dr. Neera Chaturvedi on "Application of Remote Sensing to Biological Oceanography : Indian context" presents a data collected by satellite sensors. The paper explores Arabian sea and Bay of Bengal for remote sensing of algal blooms, characterization of internal waves in the Andaman Sea, interrelationship between oceanographic parameters and finally inter annual and intraannual chlorophyll variability. It has defined an approach for developing bio-optically coupled models. Algal blooms account for one of the major carbon fluxes and are an interesting biological event taking place in the sea. The threat by harmful algal blooms to human health and fisheries is expanding significantly. The paper "Harmful Algal Blooms : Causes, Impact & Concerns" by Dr. Monika Ram give a general overview of blooms. It describes various diseases caused by toxic blooms and stresses the need by all nations to keep the line of communication and coordination open. The author also stresses the need for in depth understanding of bloom biology and ecology.

The ubiquitous nature and faster metabolic rate make the microbes an ideal tool for bioremediation. As discussed else-where, the bioremediation is one of the emerging areas in pollution control which can effectively supplement the physical and chemical measures. Dr. Ramaiah in his paper on "Environmental pollution detection and bioremediation by marine bacteria gives a few leads useful in appreciating the role of bacteria. The study suggests that bacterial strains endowed with the capacity to degrade aromatic compounds and tolerate toxic metals are not uncommon in polluted environment.

Molluscs have served as food and cheap source of protein for man from time immemorial. They have other uses such as medicine, ornaments and utility articles. They also play a destructive role as biofouling organisms. Their conservation and sustainable management is most important. These aspects have been comprehensively discussed by Dr. Modassir in her paper on "Conservation and sustainable management of molluscan resources along Goa coast, central west coast of India". The paper exhorts for greater awareness and research on biological information about clams, oyster and mussels.

The use of biological parameters to monitor the impact of industry and other anthropogenic changes is increasingly being appreciated. Benthic fauna have been recognized as bonafide community of environmentally disturbed ecosystems. They are the first casuality of such disturbances. A paper on "Response of benthic fauna to different pollutants : some case studies and relevance of benthos to environmetal impact assessment" by Dr. Ansari et.al. concludes that benthic fauna can be conveniently used as sensitive sentinels of pollution.

Deep seas are potential sources of metals such as Cu, Ni, Co, Mn, Fe and deep sea mining has been a subject of interest for over three decades. Dr. Sharma in his paper on "Environmental studies for deep seabed mining" discusses the status of deep sea mining, potential environmental effect. It also reviews environmental studies undertaken by different groups. The paper puts forth many questions that remain unanswered. The paper cautions about extrapolation of small scale experiments to assess the environmental impacts.

Marine corrosion and biofouling is a phenomenon which confronts marine structural engineers and marine scientists. Biofilm development on metal surface, conditions the metal surfaces which hasten the biofouling. A paper on "Biofilm development on metal surfaces in tropical marine waters" by Dr. Bhosle and Fraddry describes the stages of succession of marine microorganisms during the process of biofouling. The paper exhorts the need to understand the role of degraded and /or terrestrial organic matter in subsequent fouling.

The oceans are large, but are unable to sustain the burgeoning world's demand for fishes and other resources. Marine ecosystem has been severely damaged especially along the coastal regions. Ecologically sound planning of coastal development is of paramount importance. It is essential to make efforts to protect oceans which will be more beneficial to the future of mankind.

22

Distribution of Family Rhizophoraceae Along the West Coast of India

A.G. Untawale* and Thomas Mathew**

INTRODUCTION

The distribution of various mangrove species are considered to be the cumulative result of complex interaction of biological, chemical and geophysical environments now and in the past (Wells, 1983). Mangrove swamps develop only where coastal physiography and energy conditions are favourable. According to Walsh (1974) mangroves are extensive where there is a shore gradient and occupy a broader belt of shore line with a large tidal range. A sheltered habitat is essential for mangrove development and they normally flourish on coasts which are exposed to tides. It is believed that the mangrove vegetation also promotes sediment accretion and so one species prepare the way for another. Rosevear (1974), Banijbatana (1957), Kuenzler (1974) and Lipeng and Weixinem (1983) reported that salinity of seawater may greatly affect the morphological and physiological characters of mangrove plants. The mangrove flora of India comprises more than sixty species of plants belonging to 29 different families (Blasco, 1975). The family constitutes one of the major components of mangrove swamps in India.

In this paper an attempt has been made to outline the global distribution of Rhizophoraceae with special reference to the west coast of India. Various families of mangroves including Rhizophoraceae received attention of various workers like Kulkarni (1957), Patil (1957), Qureshi (1957), Puri (1957), Joshi and Shinde (1979), Untawale and Parulekar (1976). To find out the present status of Rhizophoraceae, several field trips in different seasons were conducted to 15 stations along the west coast during the period 1989-93. It is found that members of this family are not equally distributed along the coastal areas. Of the six species studied viz. *Rhizophora apiculata, R. mucronata, Bruguiera gymnorhiza, B. cylindrica, Ceriops tagal,* and *Kandelia candel,* it is

* 48, Nirmiti, Sagar Society, Dona Paula, Goa

** Dr. Thomas Mathew, Sada, Vasco Goa.

found that *R. mucronata* is the most dominant one and is of common occurrence except in Kerala and Gujarat coasts. *B. cylindrica* is not yet reported from Gujarat. *R. apiculata* and *B. cylindrica* occupies a second position in other regions while *B. gymnorhiza* and *Ceriops tagal* are restricted to very few places.

GLOBAL DISTRIBUTION

The distribution of the family Rhizophoraceae is well illustrated in Tables 1-4 and Figures 1-4. Members of Rhizophoraceae are widely distributed in tropical coastal environments with very limited extension into the sub-tropics due to frost and extended periods of near freezing temperatures (Tomlinson, 1986). Breteler (1977) studied the distribution of western species while Fosberg enlisted (1975) the eastern species. The works of Salvoza (1936), Savory (1953), Keay (1953), Stearn *et.al* (1959) and Jonker (1959) revealed the existence of three different species viz. *R. mangle, R. racemosa, R. harrisonii* on the West African, East American shores and Carribbean Islands. Salvoza (1936) reported various new species such as *R. brevistyla* from West coast of America, *R. samoensis* from Pacific coast of America and from west Pacific while *R. lamarckii* from New Caledonia. Ding Hou (1960) concluded that pacific region contains all species of *Rhizophora* except *R. racemosa*.

R. mucronata is the most widely distributed species in the old world tropics occurring from the coast of East Africa throughout and south eastern Asia to Pacific Islands (Ding Hou, 1960). Duke and Bunt (1979) noted *R. mucronata and R. lamarckii* from Australian east coast. A very limited distribution of *R. mucronata* is reported from Northern Australia, Queensland and eastern New Guinea and Solomon Island (Tomlinson, 1978). In the northern hemisphere it occurs in the Northern Phillipines and Indo China (Fig. 1).

R. stylosa, a species closely related to *R. mucronata* is found from Southern India (Tomlinson, 1986), Formosa, throughout Malaysia, New Britain, Solomon Islands, New Caledonia, Northern Australia, Fiji and Micronesia, Guam and Marshal Islands, (Ding Hou, 1960).

The distribution of *R. lamarckii* is restricted to New Caledonia, Solomon Islands, New Guinea, New Hebrides and isolated localities in Queensland (Fig. 1).

R. samoensis is reported from western pacific and *R. selala* from New Caledonia and New Hebrides *R. mangle* a western species is reported from Western Africa and Pacific coast of Tropical America (Tomlinson, 1986). *R. harrisonii*, an infertile hybrid between *R. mangle* and *R. racemosa*, occurs in West Africa as well as Atlantic coast of tropical America (Keay, 1953, Tomlinson, 1986). The distribution of *R. racemosa* ranges from western Africa to Northern South America (Fig. 1)

Genus *Bruguiera* has the widest range of all the mangroves (Fig. 2) and is reported from India, East Africa, Ceylon, Malay Archipelago, Micronesia, Polynesia (Samoa) Ryukys Islands, Western Australia and Queensland (Tomlinson, 1986).

The two species of *Ceriops* are widely distributed from East Africa, Madagascar, throughout tropical Asia, Queensland to Melanesia (New Caledonia and Solomon Island) (Tomlinson, 1986). The monotypic genus *Kandelia* occupies a narrow niche in the mangrove forest (Fig. 3), but found in many tropical countries of Asia including India, Myanmar (Burma), Thailand, Malay Peninsula, North West Kalimantan, Hong Kong, Taiwan and Southern Japan (Snedkar & Snedkar, 1977).

Table—1 Distribution of Rhizophoraceae along Gujarat and Maharashtra coast

Name of species	*Gujarat*												*Maharashtra*													
	1	**2**	**3**	**4**	**5**	**6**	**7**	**8**	**9**	**10**	**11***	**12**	**13**	**14**	**15**	**16**	**17***	**18**	**19**	**20**	**21**	**22***	**23**	**24**	**25**	**26**
R.m	-	-	-	-	-	+	+	+	-	-	-	-	-	+	+	+	+	+	-	+	+	+	+	+	+	+
R.a	-	-	-	-	-	-	-	-	-	-	-	-	-	-	+	-	-	-	-	+	-	-	+	+	-	-
B.g.	-	-	-	-	-	-	-	-	-	-	-	-	-	-	-	-	-	-	-	-	-	+	+	+	+	-
B.c.	-	-	-	-	-	-	-	-	-	-	-	-	-	+	+	+	+	-	+	+	+	+	+	+	+	-
C.t.	+	+	+	+	+	+	+	+	+	-	-	+	+	+	+	+	+	+	-	+	+	+	+	+	+	+
K.c.	-	-	-	-	-	-	-	-	-	-	-	-	-	-	-	-	-	-	-	-	-	-	-	-	-	-

Gujarat: (1) Kori creek (2) Jakhau (3) Mundra (4) Kandla (5) Noru (6) Bhaider (7) Chank (8) Pirotan Island (10) Ghogha (11) Alia bet (12) Hazira

Maharashtra: (13) Dahanu (14) Vasai (Bassein) creek (15) Malad—Manori creek (16) Diva—Mumbra (17) Bandra—Kurla complex (18) Versova (19) Elephanta Island (20) Thane creek (21) Vasishtri River (22) Ratnagiri (23) Vijaydurg (24) Devgad (25) Achra (26) Malvan

(R.m: *Rhizophora mucronata*; R.a: *Rhizophora apiculata*; B. g: *Bruguiera gymnorhiza*, B.c.: *Bruguiera cylindrica*; C.t: *Ceriops tagal*; K.c.: *Kandelia candel*)

* Locations of study area

- Source: Untawale (1984); Untawale & Wafar (1988)

+ Present, - absent).

Table—2 Distribution of Rhizophoraceae along Goa and Karnataka coast

Name of species	Goa						Karnataka													
	*28	*29	30	*31	*32	*33	*34	35	36	37	38	*39	40	41	42	*43	44	45	46	47
R.m	+	+	+	-	+	+	+	+	+	+	+	+	+	+	+	+	+	+	+	+
R.a	+	+	+	+	+	+	-	-	-	+	+	-	-	+	+	+	+	+	+	-
B.g	-	+	+	-	+	+	-	-	+	-	-	-	-	-	-	-	-	-	-	-
B.c	+	+	-	-	+	+	-	-	+	-	-	-	-	-	-	+	+	-	-	-
C.t	-	-	-	+	-	-	-	-	-	-	-	-	-	-	-	+	+	-	-	+
K.c	+	+	+	-	+	+	+	-	+	-	-	-	+	-	-	-	-	-	-	-

(28) Chorao Island, (29) Cortalim, (30) Ambora, (31) Talpona creek, (32) Zuari creek, (33) Mandovi creek, (34) Kali Nadi, (35) Hattikeri, (36) Kalinadi creek, (37) Gangavali river, (38) Agnashini, (39) Kumta river, (40) Sharavati (Honavar), (41) Saokadahung Hole, (42) Baindur Hole, (43) Coondapur (Kollur river), (44) Haladi river, (45) Sita river, (46) Kopalandi (Udipi), (47) Manjala Halla.

R.m: *Rhizophora mucronata*, R.a: *Rhizophora apiculata*, B.g: *Bruguiera gymnorhiza*, B.c: *Bruguiera cylindrica*, C.t: *Ceriops tagal*, K.c: *Kandelia candel*.

1. * Locations of study area.

2. + Present, - absent.

3. Source: Untawale (1984), Untawale & Wafar (1988)

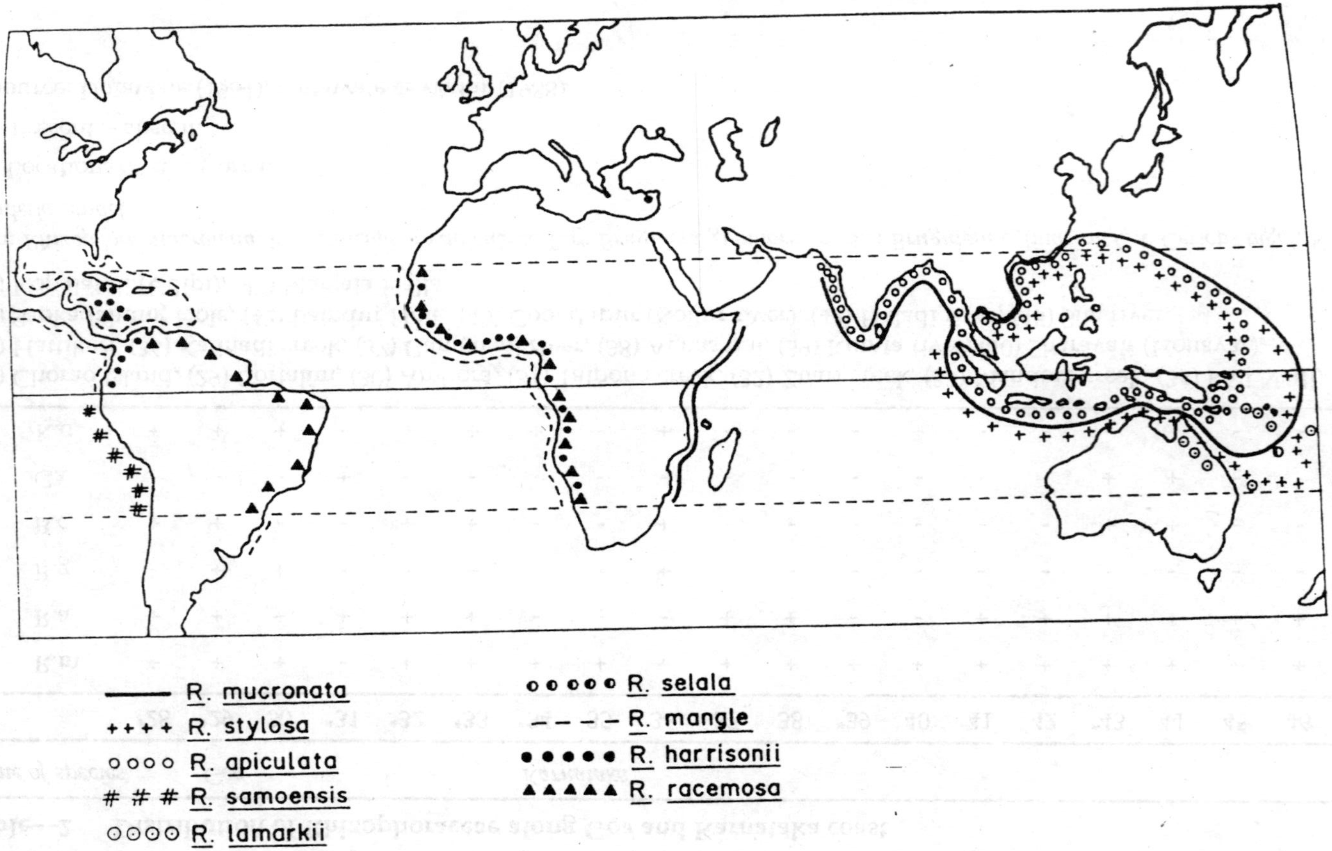

Fig.1 World distribution of *Rhizophora* species (After Ding Hou 1960, Vu Van Cuong 1964, Chapman 1970 & Tomlinson 1986)

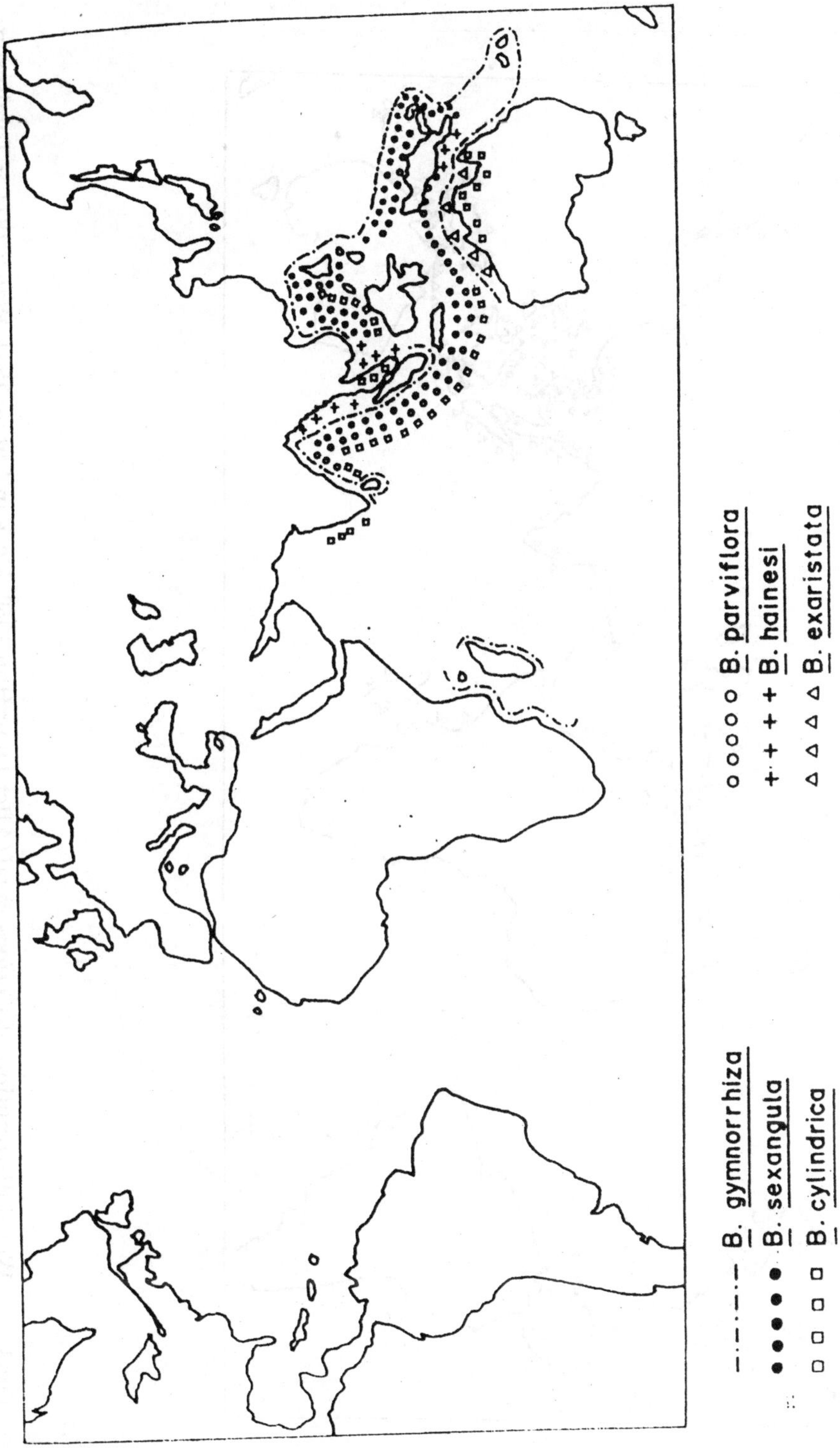

Fig. 2 World distribution of *Bruguiera* species (after Vu Van Cuong 1964, Chapman 1970 & Tomlinson 1986)

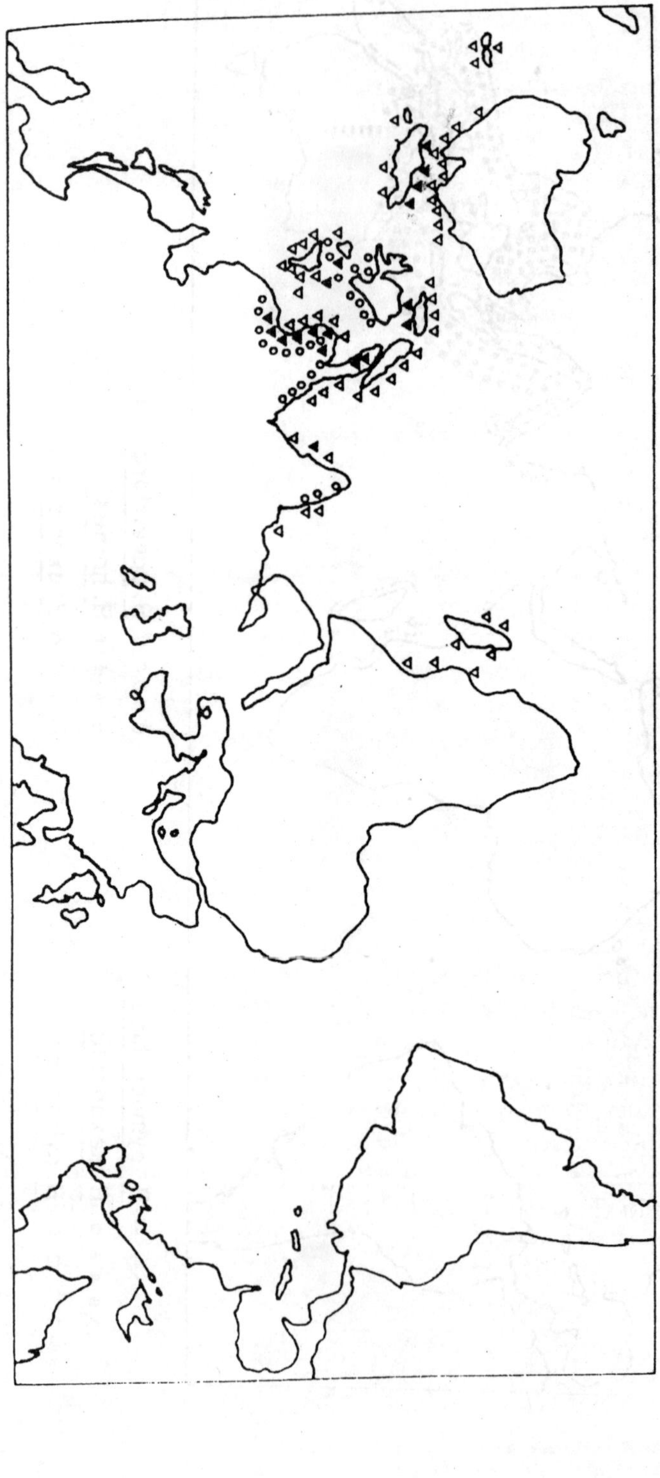

Fig.3 World distribution of *Ceriops tagal* (After Ding Hou 1960, Vu Van Cuong 1964, Walsh 1974 & Tomlinson 1986)

EAST COAST OF INDIA

The Ganges-Brahmputra delta complex houses the greatest halophytic formations in the world. Floristic researches of Gangetic delta began at a very early date by notable botanists like Roxburgh (1814), Clarke (1896) and Prain (1903). Blasco (1975) reported the occurrence of 9 species of Rhizophoraceae viz. *R. mucronata, R. apiculata, B. cylindrica, B. sexangula, B. gymnorhiza, B. parviflora, C. decandra, C. tagal* and *K. candel* from the east coast of India including Andamans and Nicobar Islands (Table 4). Mangroves of Cauvery delta has been investigated by Blasco, Caratini and Thanikaimoni of French Institute, Pondichery (Tissot, 1987). Lakshmanan and Jayalakshmi (1984) studied the floral biology of *Ceriops* species and reported the occurrences of *C. tagal* from the coastal areas of Krusadi, Rameshwaram Islands and Gulf of Mannar and *C. decandra* from Pichawaram Reserve Forest. The occurrence of *R. mucronata, R. apiculata* and *C. decandra* were reported by Tissot (1987) from Cauvery delta. Sidhu (1963) had studied the mangroves of East Godavari region.

WEST COAST OF INDIA

Gujarat

Blatter (1908, 1909) was the first to study the mangroves of Gujarat. The distribution of Rhizophoraceae is restricted (Table 1) to Gulf of Kuchchh, Mundra creek, Kandla creek and Navalakhi (Fig. 4,5 & 6). Untawale and Wafar (1988) had reported the presence of *R. mucronata,* and *C. tagal* from the islands like Nova, Banda, Chank Gandhra and Dhani. *R. mucronata* were reported from Pirotan Island, Ambla creek. Mathauda (1957) reported the presence of *R. mucronata, R. conjugata, C. candolleana, C. roxburghiana, K. rheedi,* from Gujarat coast. According to Untawale *et. al.* as a result of the threat from animals and the over exploitation many species like *C. candolleana (C. tagal), C. roxburghiana (C. decandra) K. rheedi, B. gymnorhiza and B. eriopetala* (Figs. 4,5,6) *have* been wiped out from Gujarat coast, (Untawle *et al* 1988).

Maharashtra

Maharashtra coast was explored by various workers like Blatter (1908), Sidhu (1963), Untawale *et. al.* (1980) to study the pattern of distribution of mangrove vegetation from its estuaries and creeks. About 62208 ha of mangrove area have been reported by Sidhu (1963b). But recently using remote sensing techniques, Untawale *et. al.* (1980) revealed 18 major estuarine systems approximately covering 17335 ha of mangrove vegetation. Species of *Rhizophora, Avicennia, Sonneratia, Bruguiera, Acanthus* and *Ceriops* are the dominant species along this region. In the beginning of this century Mumbai coast had a luxuriant mangrove vegetation (Blatter, 1905). Now it is available only in a few places like Vasai (Bassien) creek, Bandra creek and Thane creek (Table 1). The species like *R. mucronata, C. tagal (C. candolleana), K. rheedi (K. candel), B. gymnorhiza, B. cylindrica (B. caryophyllodides)* and *B. parviflora* had been reported from Bombay Presidency by Blatter (1908). Later in 1957, Patil reported the presence of *R. mucronata* from Versova creek, Bassien Creek, Ghodbunder and Mahim creek; *B. cylindrica* from Versova creek, Bassien creek and Mahim; *C. tagal* from Versova and Bassien creek. Navalkar (1956) had surveyed mangroves of Bombay and Salsette Islands and studied its detailed ecology. In Ratnagiti there is a well protected mangrove vegetation. Here even today there is a luxuriant growth of *R. mucronata,* while *R. apiculata* is not common (Fig. 4). In 1984 Untawale reported a moderate

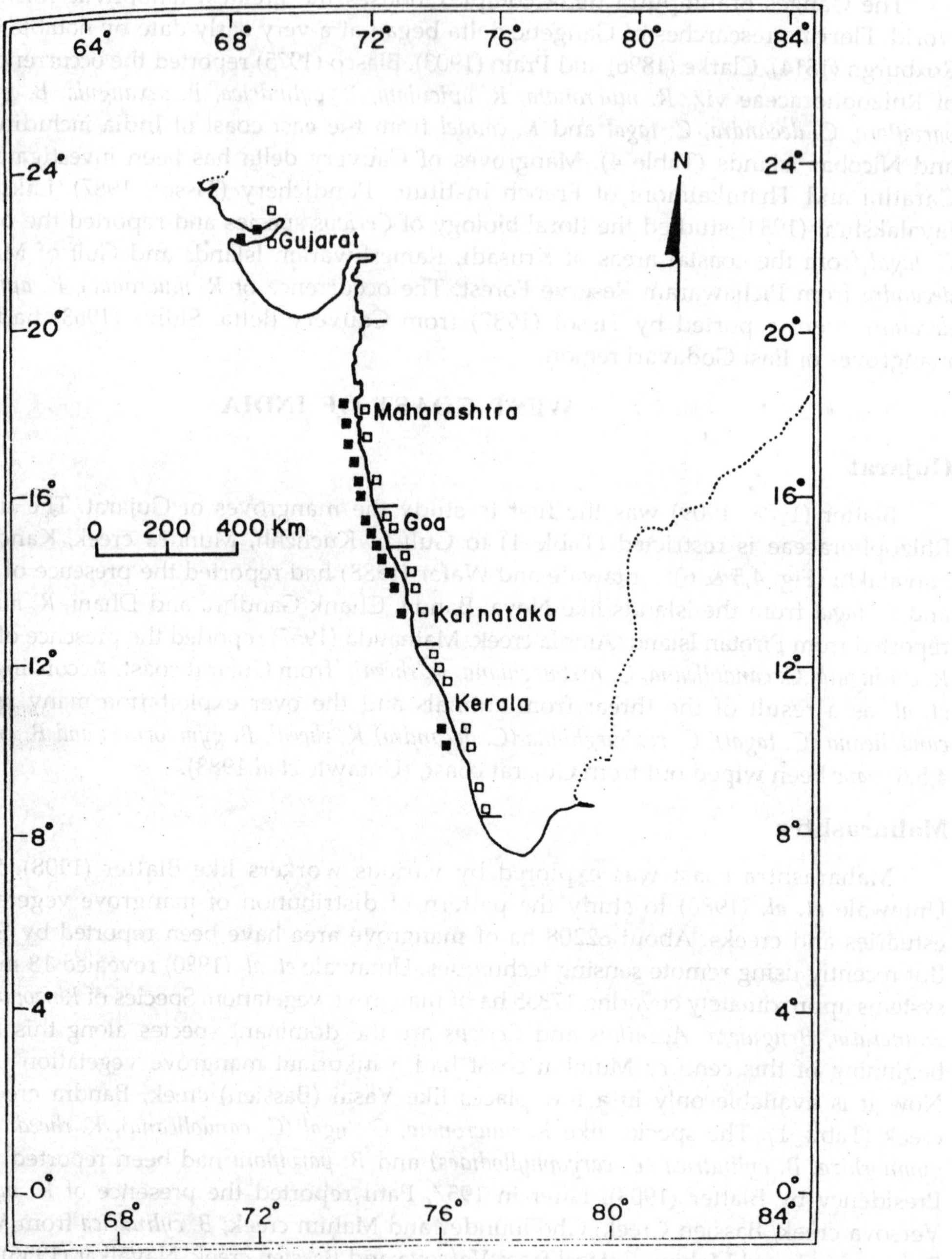

Fig. 4 **Distribution of *R. mucronata* (■) & *R. apiculata* (□) along the west coast of India**

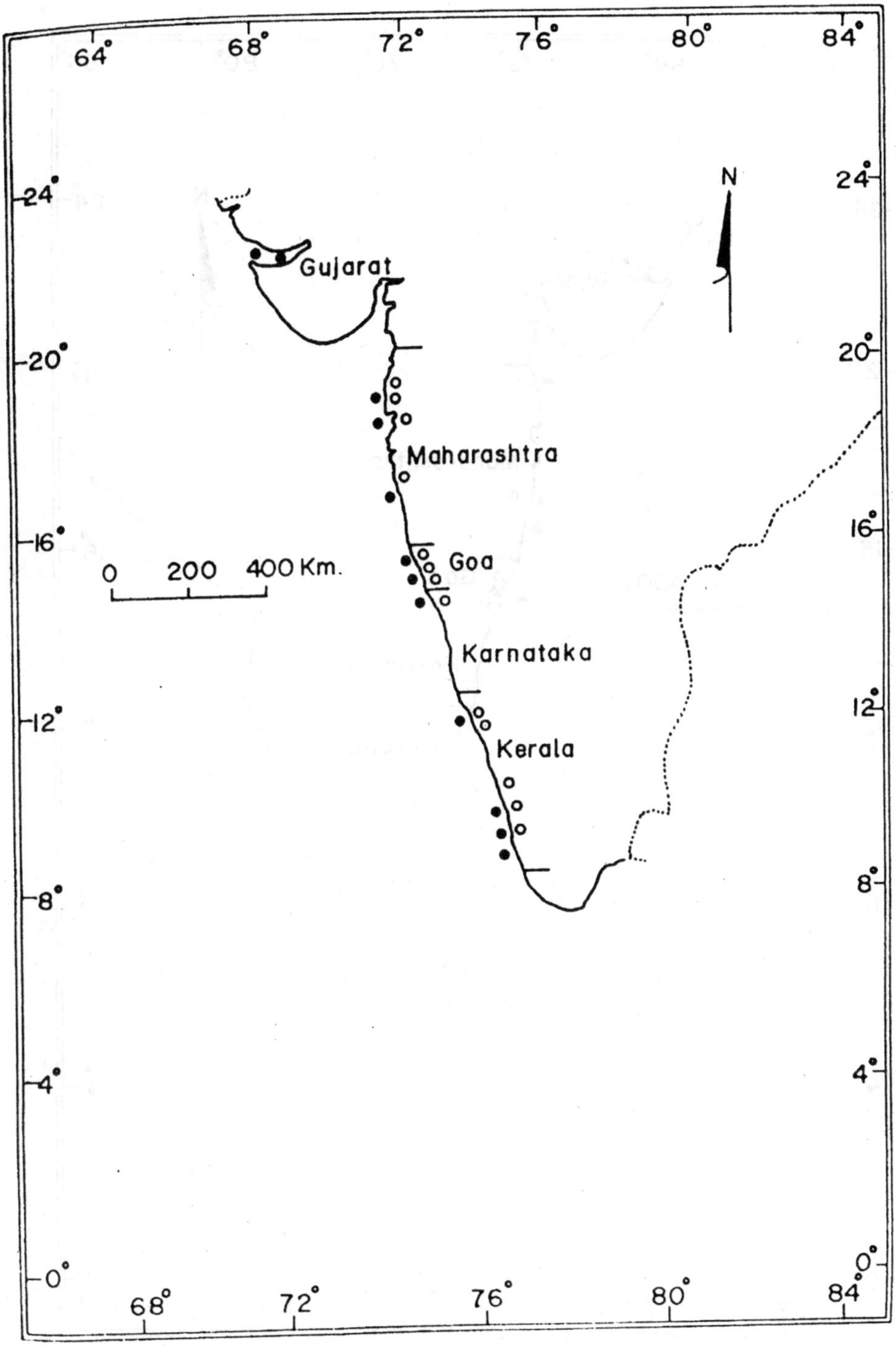

Fig. 5 Distribution of *B. gymnorrhiza* (•) & *B. cylindrica* (O) along the west coast of India.

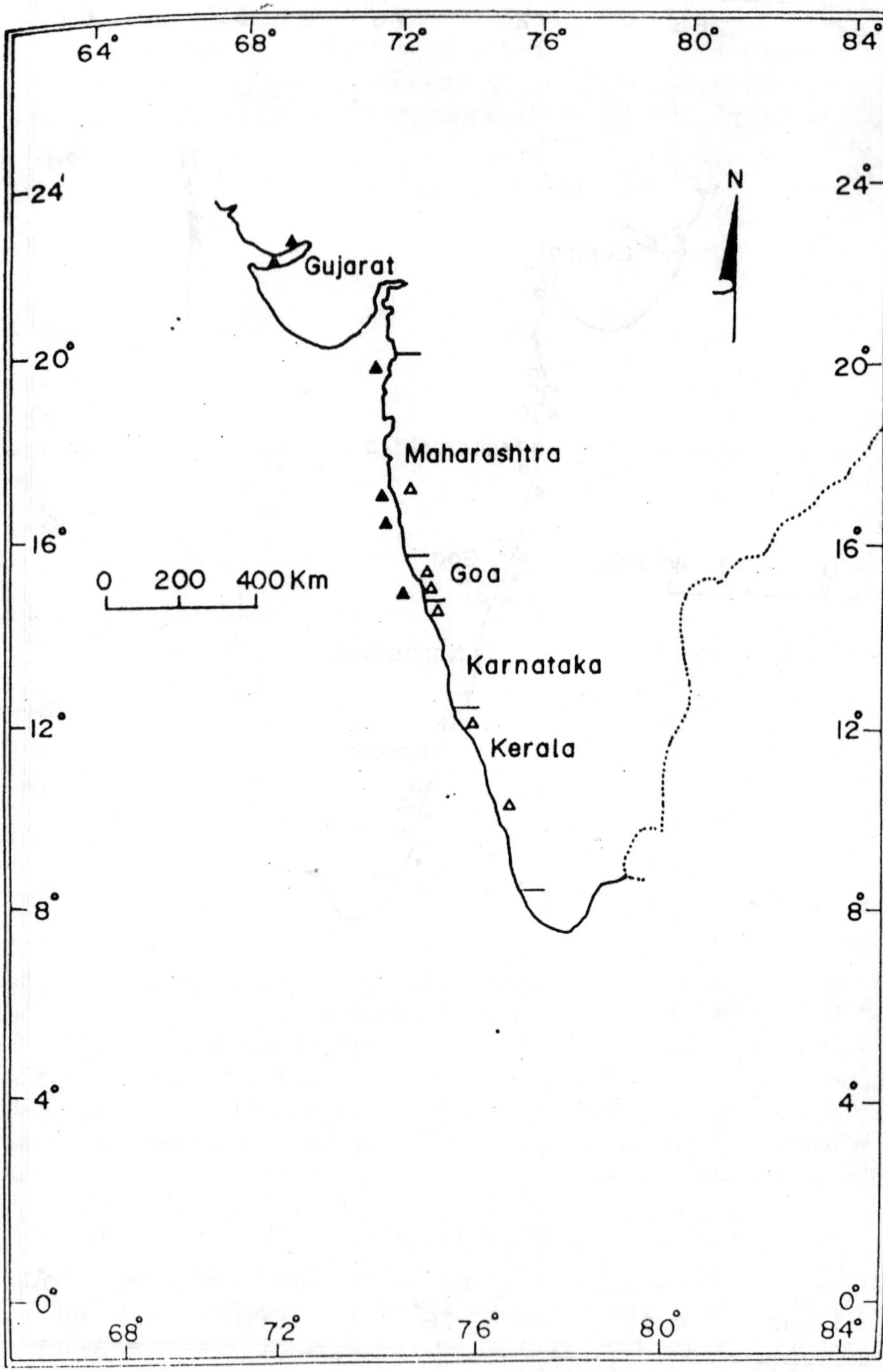

Fig. 6 Distribution of *C. tagal* (▲) & *K. candel* (Î) along the west coast of India.

distribution of *R. mucronata, R. apiculata C. tagal, B. gymnorhiza* and a very rare occurrence of *B. cylindrica and K. candel* from Maharashtra coast. Ratnagiri coast is reported to have a very common occurrence of *C. tagal* and a restricted distribution of *R. apiculata* from Bhatye creek, Ratnagiri. Joshi and Shinde (1979) surveyed Terekhol as well as Chiplun (Vashisti) and reported a dense Rhizophora zone along Terekhol inlet. The present study could not confirm the occurrence of *C. decandra* and *B. parviflora* from Maharashtra coast (Table 1) and found that members of Rhizophoraceae are very few in number and scattered (Figures 4,5,6).

Goa

Seven major rivers occur along the Goa coast of which the river Mandovi and Zuari along with Cumbarjua canal form the largest estuarine complex. It is estimated that about 2000 ha area is occupied by mangroves along the Goa coast (Untawale *et.al*, 1980). The mangrove flora of Goa consists of 15 species belonging to 10 genera and 7 families (Untawale *et. al* 1980) *R. mucronata* is one of the dominant species. *K. candel* exists as a co-dominant species along Mandovi and Zuari estuaries. *B. cylindrica, B. gymnorhiza, R. apiculata* have minimum range of distribution (Figs. 4,5,6 and Table 2). *C. tagal* is uncommon throughout these regions except few trees at Palolem, Canacona (Fig. 6). It is on the verge of disappearing from Goa Coast.

Karnataka

Untawale and Wafar (1988) reported about 14 species of mangroves belonging to 9 genera under 7 families from Karnataka coast. The dominant species of Rhizophoraceae are *R. mucronata* and *R. apiculata, R. mucronata* has been reported as the pioneer species here. Very old trees of *R. mucronata* are reported from Karki, Haldipur, Dareshwar and Honawar (Untawale & Wafar, 1988). *R. mucronata* and *R. apiculata* are found from Coondapur and estuaries like Sita and Kapil. A very scattered distribution of *B. gymnorhiza* and *B. cylindrica* was noticed (Table 2) from the banks of Kali river near the bridge at Karwar, (Figs. 4,5,6).

Kerala

Few centuries ago there were rich mangroves in Kerala, which is now reduced to isolated discrete strands of few species covering a few hundred hectares. From herbarium study it was clear that Quilon had housed a rich flora of mangrove including *R. mucronata, B. gymnorhiza,* and species of *Ceriops* and *Kandelia*. Today they all have been practically disappeared. *R. apiculata* attains a maximum height of 10 m or more in Kumarakam near Kottayam and *B. gymnorhiza* forms an important species here. At Edapally and Kannamali *R. mucronata* is very common and *K. candel* attains a maximum development near Pappinisseri and Baliapatanam railway bridge. *K. candel, B. gymnorhiza, B. cylindrica* and *R. apiculata* are common at Kungimangalam, Ezimala, Kavvy and Kotta (Figs. 4&5).

CONCLUSION

Tables 1-4 illustrate the distribution of common Rhizophoraceae along the east and west coast of India. It can be concluded that west coast house a fringing and degrading population of Rhizophoraceae because of various factors as reclamation, pollution, deforestation for firewood, cultivation, cattle grazing, and other developmental activities, It is found that the members of this family are not equally distributed along the coastal areas. Of the six species studied, *R.*

Table—3 Distribution of Rhizophoraceae along Kerala coast

Name of species	*Kerala*									
	*48	*49	*50	51	52	53	54	55	56	*57
R. mucronata	-	-	-	+	+	-	-	-	-	-
R. apiculata	+	+	+	+	+	-	+	-	+	+
B. gymnorhiza	+	+	+	+	-	-	-	+	-	-
B. cylindrica	-	+	+	+	-	+	+	+	+	-
C. tagal	-	-	-	-	-	-	-	-	-	-
K. candel	-	-	+	-	-	+	+	+	-	-

(48) Veli, (49) Quilon, (50) Kumarakom, (51) Kannamali,
(52) Chetwai, (53) Edakkad, (54) Pappinisseri, (55) Kungimangalam,
(56) Kotti, (57) Chileri (South of Kasargod).

1. * Locations of study area
2. +: Present, -: absent
3. Source: Ramachandran *et al* (1986)

Table—4 Distribution of Rhizophoraceae from East Coast of India

Species	*Tamilnadu: Cauvery delta*	*Andhra Pradesh: Godavari Krishna*	*Orissa: Mahanadi Delta*	*West Bengal: Gangetic delta*	*Andamans Nicobar Islands:*
B. cylindrica	C	*R/AB*	R	*R/AB*	NC
B. gymnorhiza	R	NC	NC	C	C
B. parviflora	AB	AB	R	R	C
B. sexangula	AB	AB	AB	NC	?
C. decandra	VC	VC	NC	VC	?
C. tagal	AB	AB	AB	VC	RC
K. candel	R	AB	R	R	AB
R. apiculata	VC	R	R	NC	C
R. mucronata	C	R	R	NC	NC

C - Common; NC - Not Common; VC - Very Common; R - Rare;
AB - Absent; RC - Rather Common
(Source: Blasco, 1974)

mucronata is the most dominant one and is of common occurrence in almost all places except in Kerala where its distribution is limited to certain places like Cochin and Kannamali. Though very scanty and scattered in occurrence *R. apiculata* and *B. cylindrica* occupies a second position in distribution. The number of trees of *B. gymnorhiza, C. tagal,* and *K. candel* are restricted to very few places. North Canara and Goa have moderate distribution of *K. candel*. A maximum distribution of C. tagal is noticed in Maharashtra at Tirakhol and Ratnagiri (Fig. 6). Kerala and Gujarat house only five species (Tables 1 & 3). *B. cylindrica* is not reported from Gujarat and *C. tagal* from Kerala coast. The varying distribution of Rhizophoraceae may be because of tidal, climatic, edaphic and human factors.

REFERENCES

Banijbatana, D. (1957). Mangrove forest in Thailand. In: Proc. 9th Pacific Sci. Conor., Bangkok, 1957, pp. 22-34.

Blasco, F. (1974). Main characteristics of Indian mangroves, In: Proc. Intern. Symp. On Biol. and Mgt. Mangroves (Eds. Gerald Walsh, Samuel Snedakar and Howard Teas), Vol. 1: 71-87.

Blasco, F. (1975). The Mangroves of India. Institute Francais de Pondicherry, Inde, Trav. Sec. Sc. Tech. No. XIV, pp. 175.

Breteler, F.J. (1977). Americas Pacific species of Rhizophora. Acta Bot. Neerl. 26: 225 –30.

Blatter, E. (1908). On the flora of Kutch. I. J. Bombay Nat. Hist. Soc., 16: 644-656.

Blatter, E (1909). On the flora of Kutch II. J. Bombay. Nat. Hist. Soc. 18: 756 – 777.

Ding Hou, (1960). A review of the genus *Rhizophora* with special reference to the Pacific species, Blumea 10: 625 – 634.

Duke, N.C. & J.S. Bunt, (1979). The genus *Rhizophora* (Rhizophoraceae) in North eastern Australia. Austr. J. Bot., 27: 657-678.

Fosberg, F.R. (1975). Phytogeography of Micronesian mangroves. In: Proceedings of the International Symposium on Biology and Management of mangroves (Eds. G.E. Walsh, S.C. Snedekar and H.J. Teas). Gainsville; Institute of Food and Agricultural Science, University of Florida, p. 23 – 42.

Joshi, G.V. and Shinde, S.D. (1979). Ecogeographical studies in Vashisti and Terekhol Rivers, Shivaji University, Publications, Kolhapur, pp. 2-56.

Jonker, E.P. (1959). The genus *Rhizophora* in Suriname, Acta Bot. Neerl. 8: 58-60.

Kuenzler, E.J. (1974). Mangrove swamp systems: In: Coastal Ecological systems of the United States. The conservation foundation, Washington, D.C.1: 346-371.

Kulkarni, D.H. (1957). Utilization of mangrove forest in Saurashtra and Kutch. Mangrove Symposium, Calcutta, 45-49.

Keay, J.W.R. (1953). *Rhizophora* in West Africa. New Bulletin, 1953: 121-127.

Linpeng and Weixinmen (1983). Ecological notes on the mangroves of Fujian China. In: Biology and Ecology of Mangroves (Eds. H.J. Teas): Dr. W. Junk Publishers, The Hauge, p. 31-36.

Lakshmanan, K.K. and Jayalakshmi, R. (1984). Floral biology in Ceriops (Rhizophoraceae). In: Marine plants, Proc. All India symposium on mangrove plants, (Eds. Krishnamurthy, V. and Untawale, A.G.), Dona Paula, 335-344.

Mathauda, G.S. (1957). The Mangrove forests of India, In: Proceedings of the Symposium on Mangrove forest. October 16-19, Calcutta, p. 66-87.

Navalkar, B.S. (1956). Geographical distribution of halophytic plants of Bombay and Salsette Islands, J. Bombay Natural Hist. Soc. 53: 335-343.

Patil, R.P. (1957). The mangroves in Salsette Island near Bombay. In: Proceedings of the Symposium on Mangrove Forest. Calcutta, p. 45-49.

Puri, G.S. and Jain, S.K. (1957). The mangrove vegetation of Western India. In: Proceedings of the Symposium on Mangrove Forest, Calcutta, pp. 50-57

Qureshi, J.M. (1957). Botanical silvicultural features of mangrove forest of Bombay state. In: Proceedings of the Symposium on the Mangrove Forest, Oct. 16-19, Calcutta (1957), P. 20-26.

Rosevear, D.R. (1947). Mangrove swamps, Farm and Forest, 8: 23-30.

Salvoza, F.M. (1936), *Rhizophora*. Nat. Appl. Sci. Bull., Un. Philip. 5:179-237.

Savory, H.J. (1953). A note on the ecology of *Rhizophora* in Nigeria. Kew Bull. 1953, 127-8.

Stearn, W.L. and G.K. Voigt (1959). Effect of salt concentration on growth of red mangrove in culture Bot. Gaz. 121(1): 36-39.

Snedaker, S.C. and Snedaker, J.G. (1977). The Mangrove Ecosystems: Research Methods, UNESCO, pp. 251.

Sidhu, S.S. (1963). Studies on mangroves of India. East Godavari region. Indian Forester, 89: 337-351.

Tomlinson, P.B. (1986). The Botany of Mangroves. Cambridge University Press, London, pp. 413.

Tomlinson, P.B. (1978). *Rhizophora* in Australia – some clarification of taxonomy and distribution. J. Arnold. Arbor. Harv. Uni., 59: 156-169.

Tissot, C. (1987). Recent evolution of mangrove vegetation in Cauvery delta: a palynological study. J. Mar. Biol. Ass. India. 29(1&2): 16-22.

Untawale A.G. and Parulekar, A.H. (1976). Some observations on the ecology of an estuarine mangrove of Goa, Mahasagar. Bull., N.I.O., Vol.19: No. 182, p. 57-62

Untawale, A.G. and Wafar, Sayeeda (1988). Distribution of mangroves along the Gulf of Kuchchh. In: Marine Archaeology of Indian Ocean Countries, pp. 135-138.

Untawale, A.G., Sayeeda Wafar and T.G. Jagtap (1980). Application of remote sensing techniques to study the distribution of mangroves along the estuaries of Goa. In: Wetlands Ecology and Management (edited by B. Gopal, R.E. Turner, R.G. Wetzel and D.F. Whigham), National Institute of Ecology and International Scientific Publications, India, p. 51-67.

Untawale, A.G. (1984). Present status of mangroves along the west coast of India. Proc. As. Sym. Mangr. Env. Res. of Manag. Kuala Lumpur, Malaysia. 57-74.

Wells, A.G. (1983). Distribution of mangroves spp. in Australia. In: Biology and Ecology of Mangroves (eds: H.J. Teas. Dr. Junk) Publishers Hague, pp. 57-76.

Walsh, G.E. (1974). Mangroves: A review: In Ecology of Halophytes. Academic Press, New York, pp. 605.

23

Application of Remote Sensing to Biological Oceanography: Indian Context

*Neera Chaturvedi

INTRODUCTION

The variability in the distribution of phytoplankton biomass and of primary production, in the ocean, in space and time is not very well understood due to lack of data. The perspective and sampling capabilities provided by remote sensing methods differ significantly from conventional oceanographic techniques. Remote sensing provides near instantaneous large-scale coverage of the oceans, thereby yielding information at scales inaccessible to ship or buoy. It is well known that the biological processes in the ocean are largely controlled by the presence of phytoplankton, which forms the base of food chain, and is responsible for CO_2^-fixation. Thus, phytoplanktons play an important role in the control of atmospheric carbon dioxide through absorption in the process of photosynthesis. This additional draw-down particularly in the case of oceanic waters is variable in space and time, but contributes significantly to the modification of anthropogenic atmospheric CO_2 input and may be involved with past climatic change as well (Bigg, 1998). Besides, the phytoplankton itself, affects the ocean surface heating and light penetration. Significant large-scale oceanic and atmospheric differences may sufficiently alter the circulation patterns and thermal structure of the region to produce variability in phytoplankton biomass. This is very well observed on satellite images. The penetrative power of solar radiation is a function of the water clarity and may vary at different locations. In the tropics there is sufficient light and heat throughout the year, so that there are continual small rises and falls in population of zooplankton, phytoplankton and nutrient levels. There are also changes from year to year in plankton population. These can be linked to changes in the climate, through winter

* Marine & Water Resource Group, Remote Sensing Applications Area, Space Applications Centre (ISRO), Ahmedabad 380 015— India

mixing variability, temperature and light availability. There may also be changes in the supply of nutrients from land run off. Disease and fishing practices may also alter the balances within the ocean and ultimately the phytoplankton stock. The variation in the wavelength characteristics of reflected light from the oceans due to changes in biological productivity will also have a climatic influence. Regions of greater productivity would reflect more incident radiation and allow less of it to warm the lower reaches of the upper ocean. A contrasting effect of phytoplankton is the warming of the upper ocean by the increased absorption of radiation as biomass increases (Sathyendranath et al 1991).

Bhattathiri et al (1996) observed that the seasonal fluctuations in the productivity regimes of the eastern and central Arabian Sea. The entire area becomes more or less oligotrophic during the April –May (intermonsoon period). During February-March (winter monsoon), the northern latitudes became more productive due to winter cooling and convective mixing (Madhupratap et al 1996). During July-August (summer monsoon), primary production increases along the eastern Arabian Sea as a result of upwelling.

At present, the understanding of the time and space variability of most biological properties is quite inadequate. Seasonality is one of the basic mode of variability and that is detectable by satellite sensors. Therefore, data on distribution of phytoplankton will be useful in climatological studies. Launch of SeaWiFS in August 1997 made the biological data available from 1997 onward. Weekly average data on chlorophyll distribution has been analysed for the period 1997-98, 1998-99. This was aimed to assess the temporal and spatial variability in the Arabian Sea on intra annual and inter annual basis.

The most common approach to remote sensing of biological oceanography is based on changes in ocean colour and its quantification. The signal for remote sensing of ocean colour is the upwelling radiance at the sea surface. Theoretical relationships are available that relate the spectral structure of water leaving radiance to ocean colour.

The first in the series of the Indian Ocean Satellite IRS- P4 (oceansat-1), was successfully launched on May 26,1999 using the indigenous Polar Satellite Launch Vehicle (PSLV) from Sriharikota. An eight band multispectral camera called Ocean Colour Monitor is flown in IRS P4. All the eight bands are in visible and near infrared region (table 1) having, spectral wavelengths spread between 400nm to 885 nm with 20/40 nm spectral resolution. The ground resolution is 360 m with swath about 1420 km and repetitivity is two days. Desai et al (2000) have discussed ocean research in India along with the ocean observation missions, which aim to provide operational ocean information, services in the country. Detailed studies of open ocean phenomena are best done with sensors designed for oceanographic observations, while coastal studies are often best done with remote sensing satellite designed for land applications.

The efforts have been made to use satellite based remote sensing approach for the identification and exploitation of potential fishing grounds to maximize fish catch. **Techniques have been developed to retrieve environmental parameters from various space borne sensors. Earlier in the absence of operational ocean colour data, efforts were made to provide operational satellite derived fishery forecast based only on sea surface temperature gradients derived from NOAA/ AVHRR data (Narain et al 1990). The validation results have indicated substantial increase in the catch per unit effort in the potential fishing zone (PFZ) areas compared to non- PFZ areas.**

Considering the limitations in the fishery forecast based on single parameter, efforts have been made to provide integrated fishery forecasting using ocean colour information along with SST to predict likely availability of fish more accurately (Solanki et al 2001).

Table—1 The OCM and SeaWiFS spectral bands

Band number	*Centre wavelength (nm)*	*Bandwidth (nm)*
1	412	20
2	443	20
3	490	20
4	510	20
5	555	20
6	670	20
7	765	40
8	865	40

GIFOV (Km)		Swath (Km)	Repetitivity (Days)
OCM	O.36	1420	2
SeaWiFS	1.1 LAC, 4.5 GAC	2801 LAC, 1502 GAC	2

OCM Ocean Colour Monitor, SeaWiFs Sea viewing Wide Field of View Sensor, LAC Local Area Coverage, GAC Global Area Coverage

The examples given here are based on the study of the oceanographic processes from Satellite data over the Arabian Sea and Bay of Bengal of the Indian Ocean. Being a tropical region it is observed that amongst the various parameters, the nutrient availability is a crucial factor in accounting for the seasonal variability in production at various levels. Besides an interrelationship between phytoplankton pigment (with optimized and nonoptimized sensors) and various other oceanographic parameters (sources: satellite and Climatological data) led to defining an approach for developing bio-optical coupled models.

TECHNIQUES DEVELOPED/STUDIES COMPLETED

The following is the theme wise details of the techniques developed or studies undertaken in the field of biological oceanography including attempts at defining the requirement of biophysical coupled models. This experience is gained while working on the data over the Arabian Sea and Bay of Bengal of the Indian Ocean.

REMOTE SENSING OF ALGAL BLOOM

1. Characterization of internal waves in the Andaman Sea
2. Interrelationship between oceanographic parameters
3. Intraannual and Interannual Chlorophyll variability
4. Remote sensing of Algal Bloom

Aim

Mapping and monitoring of algal bloom *Trichodesmium erythrium* (a blue green alga) using satellite data.

Objectives

To detect and characterize algal bloom,

To map area under bloom.

Background Information

The term "red tide" refers to bloom formation in massive proportions of certain dinoflagellates in the world oceans during certain periods, when physical and chemical conditions are conducive for their development. Red tide formation of toxic dinoflagellate such as *Gymnodinium* or *Gonyaulax* is a frequent and well-documented phenomenon in temperate seas. The process can be viewed as an interaction between the natural growth of the populations and concentrations both vertically, by buoyancy towards the top of the water column, and horizontally by converging water currents whereby surface material from a wide area accumulate in a discrete region where there are downward movements.

In tropical waters, the formation of red tide is very often associated with *Trichodesmium* (*Oscillatoria* according to Sournia 1968) blooms, though *Noctiluca* red tides also occur frequently. Red Tides due to *Trichodesmium* have been reported from almost all the tropical seas of the world. Wood (1965) observed that *Trichodesmium* Red Tide extended about 1600 Km and occupied an area of 52000 square Km. The ability of *Trichodesmium* to fix atmospheric nitrogen and the subsequent release of inorganic nutrients during their decay enriches the environment which leads to proliferation and succession of other planktonic organisms.

In the Arabian Sea the *Trichodesmium* red tides appear almost every year during February/ March and are known to persist till April/May (Devassy et al 1978, 1987). This period is characterized by fair weather, calm sea conditions and abundant sunlight (insolation greater than 10 h day^{-1} and incident solar radiation 500-600 cal cm^2 d^{-1}). The Red Tides are mostly limited to the top few centimeters of the sea.

Rationale

During bloom period, the chlorophyll concentration is reported to be as high as 600 mg m^{-3} and remains confined to the surface. This made it possible to be detected in spectral band 4 (500-600 nm) and 5 (600-700 nm) of Landsat MSS which is otherwise optimized for land applications.

- Study Area: **Off Goa (NW coast of India)**

Methodology

Images from US satellite Landsat were used for the bloom period during the year 1975. A comparative study was performed between the non-bloom images and those falling at the reported start and peak period of algal bloom. Landsat scenes of March 1 and 19 for the year 1975 covering coastal waters off Goa and approximately corresponding to reported start and peak bloom periods respectively were used. An image of December 2, 1972 for the study area was also used to assess the situation during non-bloom season. The study area reported to be under bloom lies between latitude 15° 15′ to 15° 35′ N and longitude 73° 40′ to 73° 55′ E.

Salient Results

A comparative study of MSS band 4 and 5 as diapositives could lead to map the area under algal bloom along with the suspended sediments. A ratio of band 4/5 could further improve upon the contrast of areas under suspended sediments due to differential absorption in band 4 and 5 by chlorophyll and suspended sediments. An enhanced FCC using a combination of band 4 (red), 6 (green) and ratio of 4/5 (blue) was found most useful in distinguishing area under algal bloom and suspended sediments. The bloom was approximately confined to 10m depth on March 1 and up to 20m depth on March 19 which closely matches the reported nature and occurrence of bloom by Devassy et al.1978, i.e. start on February 27 and reaching peak around March 10 during the year 1975.

Significance

The chlorophyll concentration is reported to be of the order of 600 mg m-3 during the peak. Thus, this may have a significant contribution in serving as carbon dioxide sink. In addition it plays an important role in the enrichment of water by fixing atmospheric nitrogen. The role played by the sudden appearance of bloom is significant from the point of view of biological productivity, carbon dioxide flux (acting as sink) and nutrient enrichment (nitrogen fixation). The mapping and monitoring can be done on a regular basis with the help of satellite data.

CHARACTERIZATION OF INTERNAL WAVES IN THE ANDAMAN SEA

Aim

To differentiate the initiation and maturation phase of internal waves by mapping phytoplankton pigment.

Objective

Detection of internal waves and the identification of development phase with optical sensors.

Background information

Due to temperature and salinity variations the surface region of marine waters is stratified and gives rise to internal waves. Disturbances created within the ocean are responsible for generation of these waves. Nutrients are brought to the surface by this process, which in turn lead to high phytoplankton growth. Internal waves are known to be present in Indian Seas and its characterization using SIR-A data has been studied in the Andaman Sea (Bhandari et al 1985). Mixing by internal waves is an important physical mechanism for maintaining upward flux of nutrients (Holligan et al 1985). Internal waves are reported to enrich the water (Sandstorm and Elliot 1984). In the Bay of Biscay where level of nitrate is considered to be a major factor regulating the standing stock of phytoplankton, the nutrient mixing takes place by large soliton like internal waves (Holligan et al 1985).

Rationale

It is known that there is a time lapse in phytoplankton growth from the time the nutrient become available. Thus it appears that the presence of phytoplankton pigment modulates the spectral behavior of the internal wave signatures which can readily be observed by satellite data in the optical region.

Methodology

Two IRS LISS I scenes of 31 August 1989 and 2 May, 1992 covering parts of the Andaman Sea have been used to generate phytoplankton pigment map representing the initiation and maturation phases respectively (Chaturvedi et al 1995).

Salient Results

The phytoplankton pigment map showed varying pattern in the two scenes. The chlorophyll map of May 2, 1992 shows distinct internal wave signatures. A high phytoplankton pigment concentration was observed along the crests. A distinct signature of the internal wave was seen in the image of August 31,1989 in the optical region (Band 1 of IRS LISS I). However, the chlorophyll map did not show any distinct variability in the pigment concentration. In view of this it was concluded that August image represents the initiation phase i.e. triggering of nutrient cycling and May 2,1992 image is of the maturation phase.

Significance

Internal waves are reported to enrich the water (Sandstorm and Elliot1984). In the Bay of Biscay where level of nitrate is considered to be a major factor regulating the standing stock of phytoplankton the nutrient mixing takes place by large soliton like internal waves (Holligan et al 1985). THUS, IRS LISS I data could be used to differentiate the initiation and maturation phase by mapping phytoplankton pigment, and thereby assessing the nutrient enrichment in the given area. The study demonstrates the use of seasonal data although of two different years in understanding the dynamics of internal waves.

INTERRELATIONSHIP BETWEEN OCEANOGRAPHIC PARAMETERS

A study using climatological data

Aim

Development of model to understand bio-physical coupling through relationship between parameters at temporal and spatial scales.

Objective

Inter-comparison of the features as seen on co-located CZCS derived phytoplankton pigment and climatological data on temperature profile, mixed layer depth and wind vector has been performed. Regression analysis has been performed to evaluate an understanding on region specific inter-relationship between parameters and its use as input towards developing biophysical coupled model.

Background Information

The spatial and temporal distribution pattern of phytoplankton pigment depends on various physical oceanographic parameters viz. temperature, wind, current, and mixed layer depth, which affect nutrient availability and direction of transport. In the ocean, the physical, chemical and biological processes are linked in an intimate manner. The impact of biological activity on surface heating and through carbon dioxide fixation on global warming make its role crucial in climate research.

Rationale

Physical variability tends to dominate the biological processes at the larger scale. The winds, which are hundreds of kilometers away, can disturb the local currents and thus affect the fertility. Other physical parameters like deepening and shallowing of MLD and SST also play a major role in biological production. Based on biomass/temperature relationship the water masses can be classified into mixed/stratified/frontal waters (Sathyendranath et al 1991). In the Arabian Sea, the pigment distribution reflects the introduction of nutrients into the photic zone. The spatial and temporal characteristics of the depth distribution of temperature and plant nutrients combine to yield latitudinally specific relationship in plant nutrients Vs temperature. The presence of phytoplankton is also known to increase the sea surface temperature by interception of short-wavelength radiation and thereby rise in the SST values (Sathyendranath et al 1991).

- **Study area: Northwestern Arabian Sea (Area bounds: 50°-80°E and 5°-25°N).**

METHODOLOGY

CZCS derived phytoplankton pigment maps for each month generated from using all available data for the particular month i.e. all scenes for the area from 1979 to 1986 which are cloud free have been used. This data has been obtained from DAAC, NASA, USA. These monthly phytoplankton pigment maps were compared with the climatological temperature information for each month. Inter-comparison of the features as seen on co-located points on CZCS derived phytoplankton pigment and climatological data (Climatological atlas Hastenrath and Greischar 1989) on temperature profile, mixed layer depth and the wind vector besides regression analysis has been performed.

SALIENT RESULTS

- The mean monthly pigment map derived from CZCS data showed seasonal change both qualitative as well as quantitative. The climatological temperature data shows variation with season and different patterns at different depths are observed.
- However, when compared with the pigment distribution pattern the temperature contour at 50 m depth shows a close match with the pigment concentration contours of 0.2 mg m^{-3} and 0.6 mg m^{-3} particularly during March, September and October. The MLD tends to be shallower during this period. The maximum annual pigment biomass in the Arabian Sea occurs when the mixed layer is relatively shallow. Therefore, at the time of sufficiently shallow MLD and less turbulent conditions, the pigment and temperature covary. The physical forcing is the primary factor influencing

the spatial pattern of mixed layer depth and pigment biomass during the monsoon seasons in the Arabian Sea (Bauer et al 1991).

- The vigorous and deep south west monsoon circulation is typically at peak in July and August. This is reflected on the CZCS derived pigment map, which shows a high value in August. However, for the rest of the period the pigment variation at different locations does not coincide with .the wind speed pattern indicating the regional specificity along with the seasonal variation. Moreover the influence of initial physical and chemical structures on the favourable wind conditions required for nutrient entrainment is highly determinant (Klein and Coste 1984). Nutrient entrainment is associated with strong current shear and strong temperature and nitrate gradient at the base of the mixed layer.
- The relationship between different oceanographic parameters at specific locations is shown in Figure 1. Mixed layer depth and wind speed showed a positive correlation (r^2= 0.63) in the offshore waters. In the nearshore waters the subsurface temperature and wind speed were found to be correlated (r^2= 0.60).
- At few locations reasonably good correlation is observed between phytoplankton pigment and subsurface temperature (r^2= 0.56, 0.72) in the nearshore waters while no significant correlation was observed with surface temperature. An inverse relationship between pigment concentration and temperature may be due to the negative correlation between temperature and nutrient (Kamykowski 1987) which ultimately affects phytoplankton growth. Since the satellite-derived information comes from one attenuation length rather than skin depth, the pigment concentration correlates better with temperature pattern at 50 m depth rather than at the surface.
- The phytoplankton pigment and temperature do not covary across the season. Surface and subsurface temperature may or may not covary depending upon the location and at times when sea surface temperature remains constant the subsurface temperature and phytoplankton pigment concentration show large variation.
- Locational differences in the relationship between different parameters are obvious.

Significance

This study shows the variation in relationship when a single parameter is considered on monthly basis. Therefore, more related parameters have to be considered and spatial variations have also to be observed before making any conclusive statement. However, one point comes out very clearly that the satellite derived pigment concentration cannot be directly compared with the satellite derived sea surface temperature unless one is sure as to how the temperature is distributed at different depths. Locational differences in the relationship between different parameters are obvious. It appears that besides season, location plays an important role in determining the relationship between various parameters (Chaturvedi et al, 1998). This study has demonstrated the need for developing region specific models.

A STUDY USING WEEKLY DATA

Aim

The present study aims at understanding the interrelationship between temperature and chlorophyll. Chlorophyll values (eight days average) derived from SeaWiFS images (DAAC: NASA) have been compared with the temperature profile data collected by the vessels of the Fishery Survey of India (FSI), Mumbai

Objective

The cruise data, on temperature at 1 and at 50m depth has been regressed with the chlorophyll value for the corresponding and the following week, at more or less the same location to understand the underlying relationship.

- **Study Area** North Western Arabian Sea(15° N – 22° N, 66° E - 72° E)

Background Information

Nutrient limitation interacting with changes in light intensity and temperature can lead to seasonality in primary and secondary production. The inverse relationship between temperature and nutrients concentration occurs seasonally in surface waters to varying degrees depending upon the geographical locations (Kamykowski 1987). It is also known that the spatial and temporal characteristics of the depth distribution of temperature and plant nutrients combine to yield latitudinally specific relationship in plant nutrient Vs temperature. The seasonal variation in mixed layer dynamics and upwelling is an important determinant of annual cycle of primary productivity in the Arabian Sea (Banse 1987).

Rationale

Variation in the absorption of solar irradiance in the upper ocean, is strongly influenced by variation in the concentration of phytoplankton. Variability in sea surface chlorophyll in time and space, will result in variation in the vertical radiative flux out of the base of the upper mixed layer. This variability may be important for tropical region, and overall heat budget on the global scale. Variability in vertical distribution of irradiance absorption is particularly important in tropical waters, because of the potential deep penetration of visible wavelengths, relative to shallow mixed layer. The presence of phytoplankton, is also known to affect the sea surface temperature (SST) by interception of short radiation absorption, and thereby increasing SST (Sathyendranath et al 1991).

Methodology

Eight days average chlorophyll Standard Mapped images (SMI) generated from SeaWiFS data (atmospherically, radiometrically and geometrically corrected data) covering Arabian Sea for the period November 1997 to April 1998 have been used. Variability within the data points is checked by taking chlorophyll value in the 3 by 3 pixel area and standard deviation was found to be less than 10 per cent in the open ocean. The chlorophyll values reported in the literature at certain locations for intermonsoon, winter monsoon and summer monsoon period is compared with satellite derived chlorophyll for the respective period and found to match reasonably well.

The temperature profile data collected by vessels of the Fishery Survey of India, Mumbai has been used. The cruise data points falling in the offshore waters lie between 15° N to 22° N and 66° E to 72° E. Seasonal coverage is from November 1997 to April 1998. The cruise data on temperature at 1 and 50m depth has been regressed with the chlorophyll value for the corresponding and following week for more or less at the same location.

Salient results

- This cruise data, on temperature at 1 and at 50m depth has been regressed with the chlorophyll value for the corresponding and the following week, at more or less the same location.
- The relationship is observed as season as well as region specific. Figure 2 shows the relationship during different months. During October an inverse relationship is observed for temperature at 1m (r^2=0.91) and 50m depth (r^2=0.82), with the chlorophyll for the following week. A narrow range of chlorophyll (0.21 to 0.29 mg m^{-3}) and temperature at 1m depth (28.7 to 29.2° C) was observed.
- For the month of November and December, no significant relationship was seen. February shows good correlation with temperatures, both at 1 and 50m depth for the corresponding as well as following week. Temperature observed was from 25.0 to 27.1°C for 1m and 23.9 to 25.1°C at 50m depth, where as the chlorophyll range was 0.13 to 0.36 mg m^{-3} and 0.16 to 0.53 mg m^{-3}. At low temperature (<25° C) a linear relationship is observed with chlorophyll till about 26° C and beyond this, it becomes inverse.
- March also shows a similar trend, but with the temperature at 1m depth only. The chlorophyll value increase up to 25.5°C, beyond which the trend is inverse. However, the temperature at 50m depth does not show any correlation. The range of temperature is also narrow around 1° C.
- During April, there is a declining trend in chlorophyll with increase in temperature values. Temperature at 1m and 50m depth shows correlation with the chlorophyll value of the following week and the relationship is inverse.
- The results indicate that during the months when temperature at 1m depth shows good correlation with the chlorophyll value, a good correlation with satellite derived temperature may also be observed. Very low temperature does not allow plankton growth as a physiological response whereas the waters with high temperature inhibit indirectly due to the lack of nutrients. Although the analysis is only for one season but variability in different months between chlorophyll and temperature distinctly brings out the specific nature of relationship.

Significance

The relationship between temperature and chlorophyll varies at different times of the year. During February and March, good correlation is observed between temperature at 1m depth and chlorophyll. Hence, it is expected that during these months Satellite derived temperature should as well be correlated with chlorophyll. During October, though the relationship is observed for the following week the chlorophyll range is very narrow. April showed relationship with the following week (Chaturvedi et al 2000). Such variations in phytoplankton pigment in space and time will increase or diminish the heat flux coming out of the upper mixed layer. Therefore, an understanding on the relationship between chlorophyll and temperature will be an important consideration for tropical and perhaps global heat budget.

INTRANNUAL AND INTERANNUAL CHLOROPHYLL VARIABILITY

Aim

To understand temporal and spatial variability in chlorophyll distribution pattern.

Objective

To examine the changes in chlorophyll values, at few locations, on weekly basis in order to understand the variability pattern. Interannual variability checked by analyzing the values at the same locations during the following year.

Background

Primary productivity and phytoplankton biomass are reported to show definite seasonal patterns in the Arabian Sea (Yentsch and Phinney 1992). They also show northward and eastward variability (Bhattathiri et al 1996) depending on physical events such as upwelling or winter cooling. Arabian Sea is an ocean basin where the strength of the physical forcing and biological responses vary seasonally. Banse and English (1994) have studied Geographic and temporal patterns of seasonality in the open sea using CZCS data.

Rationale

The space and time scales of biological processes are expected to reflect the scales of physical circulation elements as well as interactively induced scales i. e. from a competition between biological behaviour and physical transport, which may account for some scale of patchiness.

Methodology

Eight days average chlorophyll images have been analysed. This is level-3 global gridded product with 9 km resolution. The Standard Mapped Images (SMI) generated from SeaWiFS data {atmospherically, radiometrically and geometrically corrected) over the Arabian Sea for the period 1997-98 and 1998-99.

Chlorophyll images are generated using OC2 algorithm. Which takes the ratio of water leaving radiance in the spectral band 490nm to 555nm. Modified cubic polynomial is used and coefficients are applied. A detailed discussion of ocean colour algorithms for the SeaWiFS instrument is available in the work of O'Reilly et al (1998).

The chlorophyll values reported in the literature (Bhattathiri et al 1996) at certain locations for intermonsoon, winter-monsoon and summer-monsoon period is compared with satellite derived chlorophyll value for the respective period and found to match reasonably well. The chlorophyll values at all the stations were estimated in the same manner (JGOFS protocol). This was done to ascertain that the Satellite derived chlorophyll values can be used to understand the intraannual and interannual spatial and temporal variability in the Arabian Sea.

Nine locations have been selected; two near coast, five open ocean and two in offshore waters. These locations have been chosen to get a representative distribution pattern of chlorophyll in the near coast and open ocean waters in the Arabian Sea. Study area lie 20° N 70° E, 20° N 65° E, 20° N 60° E, 15° N 70° E, 15° N 65° E, 15° N 60° E, 10° N 70° E, 15° N 65° E and 10° N 60° E.

Variability within the data points is checked by taking the chlorophyll value in the 3 by 3 pixel area and standard deviation was found to be within 40 per cent (Fig.1). The variability may change with season and location and is more, in near coast regions. Open Ocean area is found to be more uniform.

Salient Results

- Intraannual comparison shows a definite pattern. The analysis of data for the period October to April shows that the highest chlorophyll concentration is during January and February. Figure 3 shows SeaWiFS derived chlorophyll image during November 1-8,1998 while Figure 4 shows the chlorophyll image during February 18-25,1999.
- Interannual comparison shows a distinct pattern of highs and lows at different locations.
- Inter-annual variability varies according to locations. Variability is more in nearcoast locations where chlorophyll concentration range is as well higher. In the offshore and open ocean waters year to year variation is not very significant.
- Seasonal periodicity is evident in interannual analysis. During the year 1997-98 as well as 1998-99 peak chlorophyll concentration observed during January and February at most of the locations (Chaturvedi and Narain, 2000).

Significance

Data is useful in understanding the distribution, magnitude and timing of phytoplankton variability. The study emphasizes the potential of satellite data for understanding chlorophyll distribution, which in turn forms an important part of the climate modelling.

Weekly average satellite data over few years gives a clearer picture of the underlying causes and associated oceanic phenomena. This along with cruise data will be extremely useful in drawing ultimate inferences on the oceanic processes and in its ultimate use in modelling. The data is useful in understanding the distribution, magnitude and timing of phytoplankton variability.

Acknowledgement

SeaWiFS data provided by DAAC, NASA, on a regular and timely basis is thankfully acknowledged. Thanks are due to Shri A K S Gopalan, Director, Space Applications Centre, Dr Aishwarya Narain, Remote Sensing Applications Area, Dr S R Nayak, Group Director, Marine

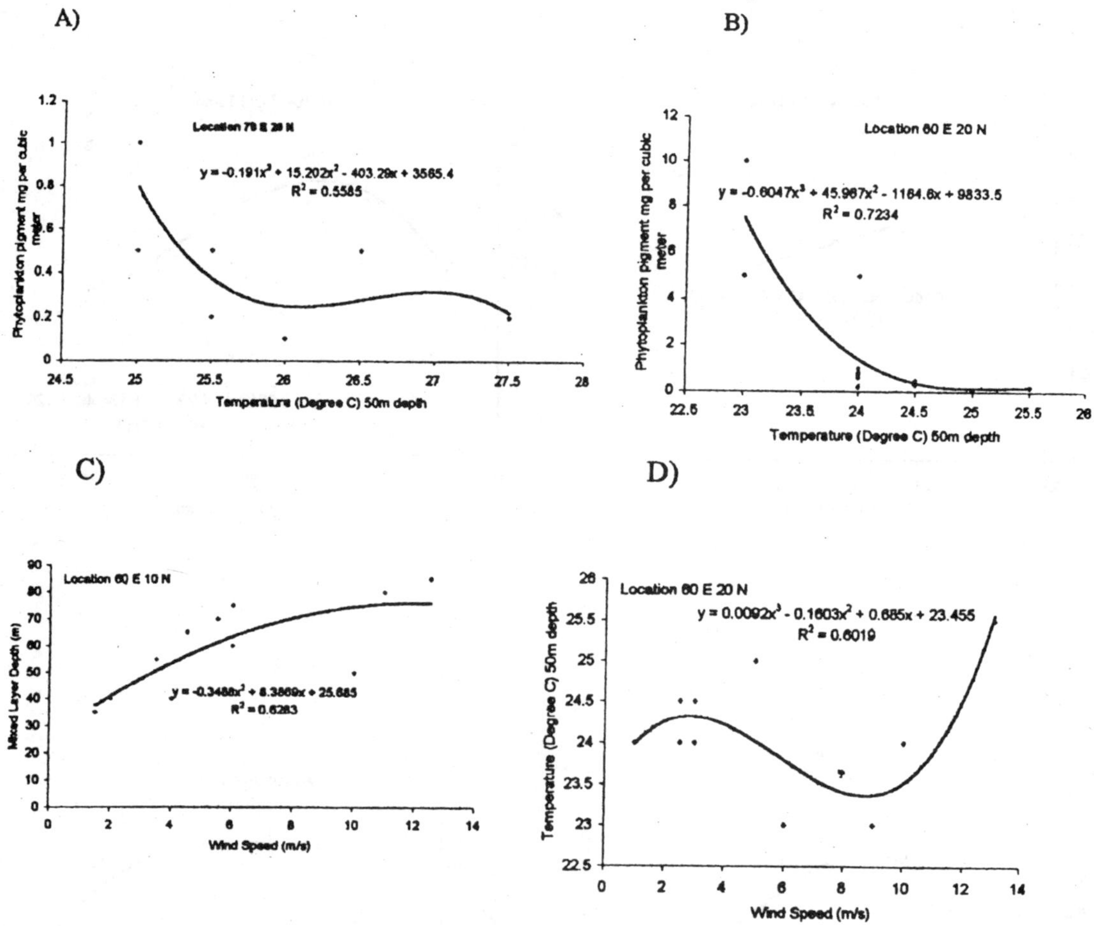

Fig. 1 **Interrelationship between different oceanographic parameters at specific locations i.e. phytoplankton pigment and subsurface temperature (A and B), between mixed layer depth and wind speed (C), and between subsurface temperature and wind speed (D), (Source: climatological atlas and CZCS data).**

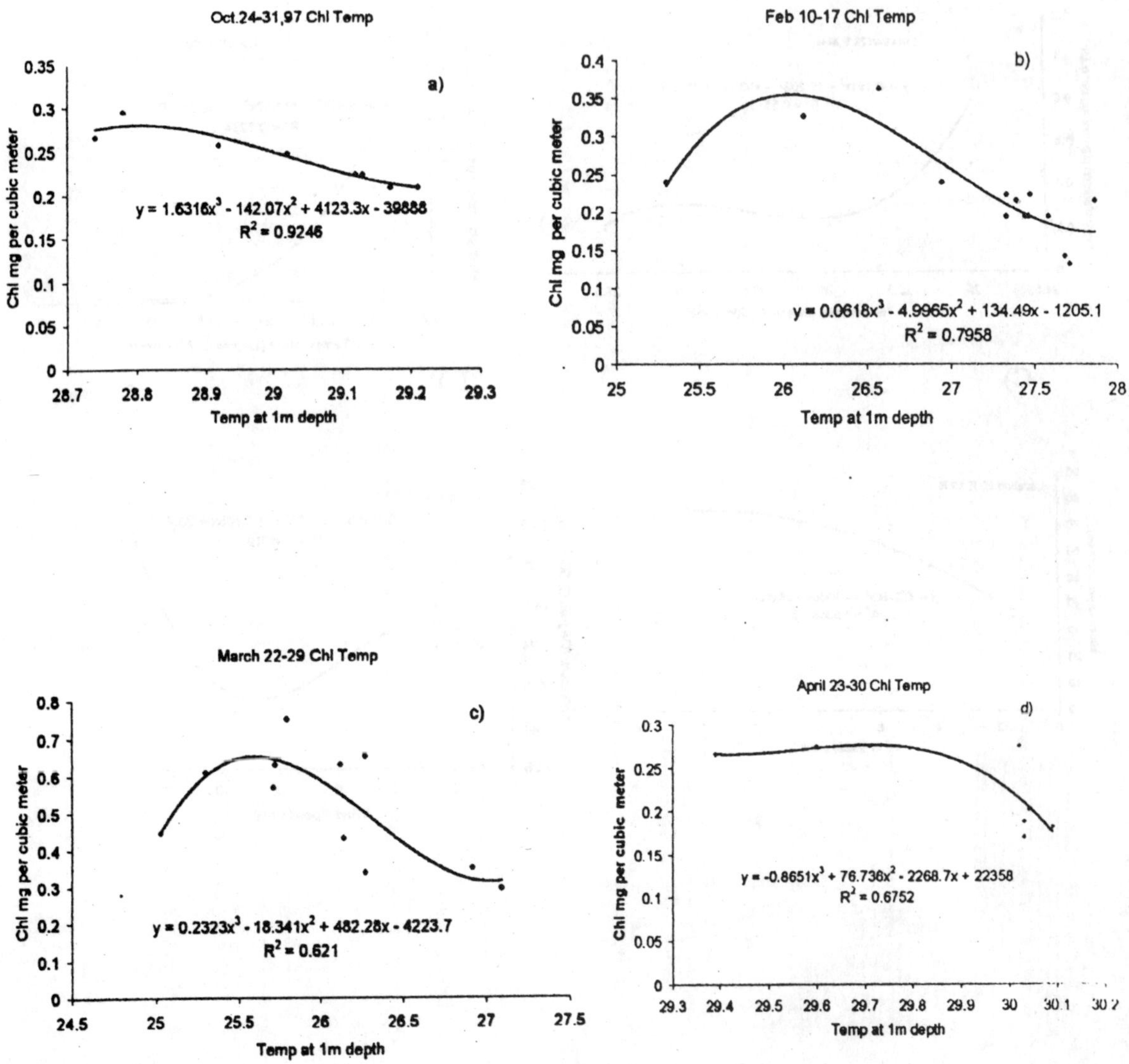

Fig. 2 Relationship between satellite derived chlorophyll and vessel based temperature for the period 1997-98: a) October, b) February, c) March, and d) April.

Fig. 3 SeaWiFS derived chlorophyll image for the period November 1 to 8, 1998.

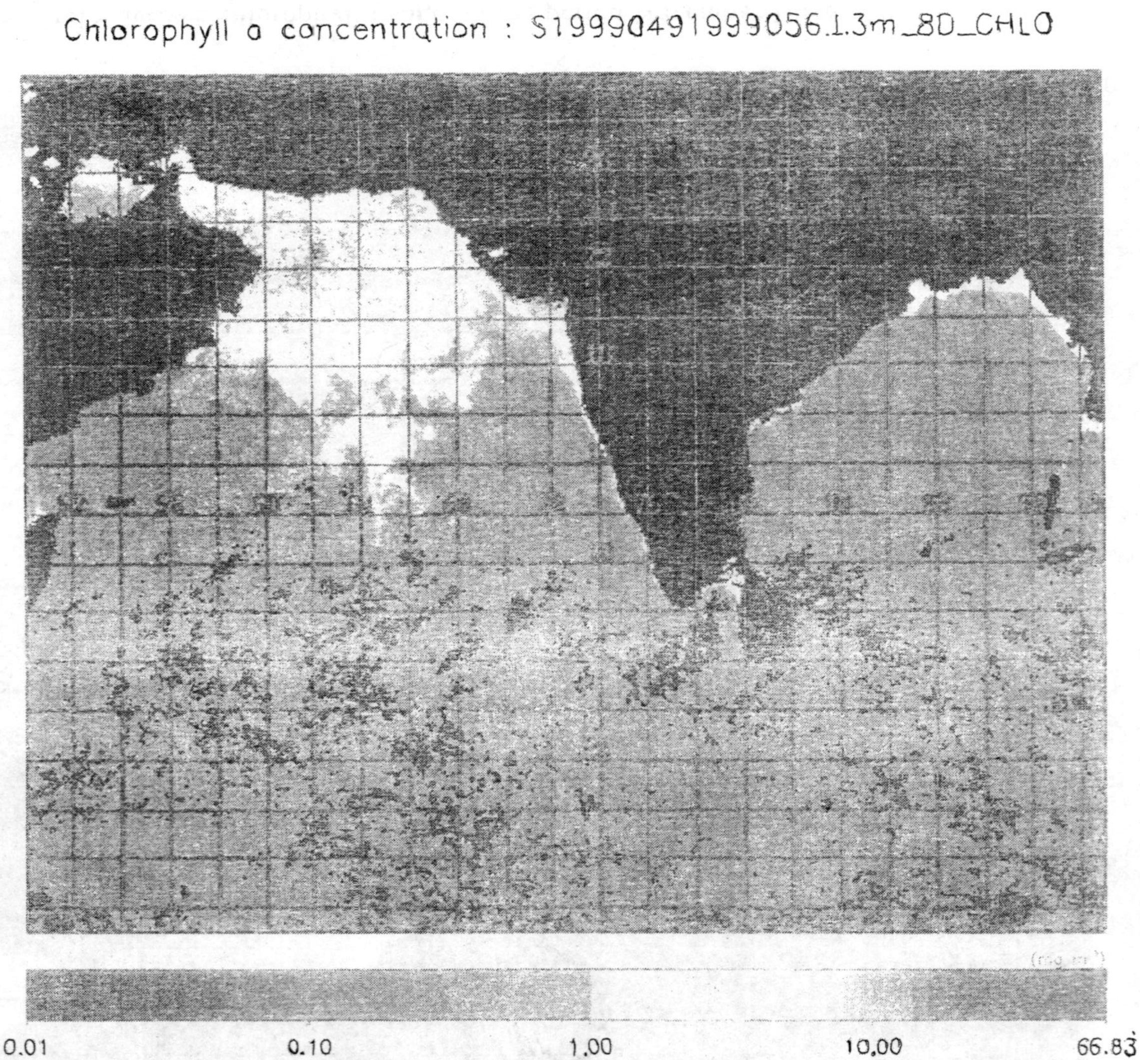

Fig. 4 SeaWiFS derived chlorophyll image for the period February 18 to 25,1999.

and Water Resource Group, and Dr M Mohan, Head, Marine Sciences and Coastal Environment Division for their constant support and encouragement.

REFERENCES

Banse K. 1987. Seasonality of phytoplankton chlorophyll in the central and southern Arabian Sea. Deep Sea Res., 34: 713-723

Banse, S. and D.C. English 1994. Seasonality of coastal zone color scanner phytoplankton pigment in the offshore oceans. Journal of Geophysical Research, 99: 7323-7345

Bauer Sonia, Gary Hetchock, L., Olson, and B. Donald 1991. Influence of monsoonally-forced Ekman dynamics upon surface layer depth and plankton biomass distribution in the Arabian Sea. Deep Sea Research 38: 531-553

Bigg G. R. 1998. The *Oceans and Climate*, Cambridge University Press.

Bhattathiri P. M. A. Aditi Pant, Surekha Sawant, M. Gauns, S. G. P. Matondkar and R. Mohanraju 1996. Phytoplankton production and chlorophyll distribution in the eastern and central Arabian Sea in 1994-95. Current Science. 71(11): 857-862

Bhandari S. M.; N. K. Vyas and H. I. Andharia 1985. Analysis and interpretation of SIR-A image of large internal waves in the Andaman Sea, In proceedings of Sixth Asian Conference on Remote Sensing, Hyderabad, India, 1985 p. 195.

Chaturvedi Neera, A Narayana and A Narain 1995. Proc. Third Thematic Conference Remote Sensing for Marine and Coastal Environments Technology and Applications, Vol II (Environment Research Institute of Michigan , USA), pp188.

Chaturvedi Neera, Aishwarya Narain and P. C. Pandey 1998. Remote sensing applications in oceanography with special reference to bio-physical coupling: A review Indian Journal of Marine Sciences. 27:1- 7

Chaturvedi Neera, and A. Narain 2000. Intra-annual and inter-annual variability in chlorophyll distribution pattern in the Arabian Sea: for the year 1997-98 and 1998-99 as observed from SeaWiFS data. Proceedings Sixth International Conference on Remote Sensing for Marine and Coastal Environments, South Carolina, pp I-386- I-391, USA 1-3 May 2000.

Chaturvedi Neera, A. Narain, D. K. Gulati, M. E. John and V. S. Somvanshi 2000. Interrelationship between satellite derived chlorophyll and temperature profile from cruise data: a study for Arabian Sea on seasonal basis. PORSEC Proceedings Vol 1 347-349, Goa, India, Dec.5-8, 2000.

Desai P. S., H. Honne Gowda and K. Kasturirangan 2000. Ocean research in India: Perspective from Space. Current Science. 78 (3): 268-278

Devassy V. P., P. M. A. Bhattathiri and S. Z. Qasim 1978. *Trichodesmium Phenomenon*. Indian J Marine Sciences. 7, 168-186

Devassy V. P. 1987. Trichodesmium Red tides in the Arabian Sea, Contributions in Marine Sciences pp61.

Hastenrath S. and Lawrence L Greischar 1989. Climatic atlas of the Indian Ocean part III: Upper ocean structure (University of Wisconsin Press, Madison, USA), 1989.

Holligan P. M., R. D. Pingree and G. T. Mardell 1985. Oceanic solitons, nutrient pulses and phytoplankton growth. Nature. 314: 348-350

Kamykowski Daniel 1987. A preliminary biophysical model of the relationship between temperature and plant nutrients in the upper ocean. Deep Sea Research. 34: 1067-1079.

Klein Patrice and Bernard Coste 1984. Effects of wind stress variability on nutrient transport into the mixed layer. Deep Sea Research. 31: 21-37

Madhupratap M., S. Prasann Kumar, P. M. A. Bhattathiri, M. Dileep Kumar, S. Raghukumar, K. K. C. Nair and N. Ramaiah 1996. Mechanism of the biological response to winter cooling in the northeastern Arabian Sea. Nature. 384: 549-552

A. Narain, R. M. Dwivedi, H. U. Solanki, B. Kumari, N. Chaturvedi, P. S. B. James, G. Subbaraju, D. Sudarsan and T. E. Sivaprakasam 1990. In Proceedings Seminar on remote Sensing applications for oceanography and fishery environment analysis , Beijing, China, (ESCAP/UNDP RAS/86/ 41 Bangkok, Thailand) 1990,226

O'Reilly, J. E., Maritorena, S. Mitchell, B. G. Siegel, D. A. Carder, K. L. Garver, S. Kahru A., and C. McClain 1998. Ocean color chlorophyll algorithms for SeaWiFS, Journal of Geophysical Research. 103: 24937-24953

Sandstorm H. and J. A. Elliot 1984. Internal tide and Solitons on the Scotian shelf: A nutrient pump at work. Journal of Geophysical Research. 89: 6415-6426

Sathyendranath S., T. Platt, E. P. W. Horne, W. G. Harrison, O. Ulloa, R. Outerbridge and N. Hoepffner 1991. Biological control of surface temperature in the Arabian Sea. Nature. 349: 54-56

Solanki H. U., R. M. Dwivedi, S. R. Nayak, J. V. Jadeja, D. B. Thakar and M. I. Patel 2001. Applications of Ocean Colour Monitor Chlorophyll and AVHRR SST for fishery forecast: Preliminary validation results off Gujarat coast, northwest coast of India Indian J Marine Sciences. 30: 132-138

Sournia, A. 1968 La Cynophyceae Oscillatoria (Trichodesmium, dans le Mozambique, Nova Hedwigia, 15: 1-12 cf Devassy V P 1987 Trichodesmium Red tides in the Arabian Sea, Contributions in Marine Sciences pp61.

Wood, E. J. F. 1965 Marine Microbial Ecology. Chapman and Hall, London 243pp cf Devassy V P 1987 Trichodesmium Red tides in the Arabian Sea, Contributions in Marine Sciences pp61.

Yentsch, C. S., and D. A. Phinney 1992, In Oceanography of the Indian Ocean (ed. Desai, B. N.), Oxford and IBH, New Delhi, pp.57-66.

24

Harmful Algal Bloom: Causes, Impacts and Concerns

*D. Monika Ram

INTRODUCTION

During the last two decades, increase in human population, coupled with technological advancement, industrial expansion and urbanization, has contributed to the greater abundance of toxic and noxious algae in water bodies all over the world. Consequently, there has been an escalating and worrisome trend in the incidence of problems associated with the harmful and toxic algae. Today, marine biotoxins and harmful algae represent a significant and expanding threat to human health and fisheries resources through out the world. Fortunately, there now seems to be a growing sense of both awareness as well as concern about this problematic situation, particularly in the coastal regions where the problem has been far more acute and has had major deleterious impacts. The present paper briefly discusses the major causes and impacts of harmful algal blooms (HAB), with particular emphasis on the serious environmental and human health hazards, which has evoked serious concerns to tackle this global menace.

HARMFUL ALGAL BLOOMS (HAB)

Harmful algae are microscopic, single—celled plants that live in aquatic habitats. Interestingly, most species of algae or phytoplankton are not harmful and serve as primary producers of the food web, without which higher life on this planet earth would not exist. Occasionally, however, some algal species grow very fast or "bloom" and accumulate into dense, visible patches near the surface of the water. "Red Tide" is a common name for such a phenomenon where phytoplankton species contain reddish pigments and "bloom" such that the water appears to be red. However, the term "red tides" is a misnomer because such species are not associated with tides and are not usually harmful, while those that are harmful may never

* Department of Botany, University of Delhi, Delhi–110 007.

reach the densities required to discolor the water!. Unfortunately, a small number of species produce neurotoxins that can be transferred through the food web where they affect and even kill the higher forms of life. The commonly affected include zooplanktons, shellfishes, fishes, birds, marine mammals and even humans that feed either directly or indirectly on them.

Today, scientists prefer to use the term "Harmful Algal Blooms" (HAB) to refer to the bloom phenomenon that contains toxins or that causes negative impacts. More significantly, HAB applies not only to microscopic algae, but also encompasses benthic or planktonic microalgae, which can also proliferate in response to anthropogenic nutrient enrichment and lead to serious ecological impacts. The development of dense canopies of microalgae in the benthos of shallow water bodies is an increasingly common phenomenon along virtually all the shorelines of the world. Human activities are increasing the concentration and fluxes of nitrogen and phosphorous in coastal waters, which in turn enhance seaweed productivity leading to high biomass levels.

CAUSES OF HAB

Most phytoplankton blooms result from a combination of physical, chemical and biological mechanisms that are, for the most part, poorly understood. Harmful algal blooms affect co-occurring organisms (Fig. 1) and alter food web function (Smayda, 1992).

Formerly, only a few regions were affected by HAB in scattered locations, but now virtually every coastal region of the world is threatened. In fact, the situation is grim and alarming with large geographical areas being adversely affected, and more so, by more than one harmful or toxic species. It is still a matter of debate as to the causes behind this expansion, with possible explanations ranging from natural mechanisms of species dispersal, to a host of anthropogenic activities. Some reasons that have been suggested for the expansion of HAB are as below:

(i) Species dispersal through currents, storms or other natural mechanisms.

(ii) Nutrient enrichment of coastal waters by human activities, leading to selection and proliferation of harmful algae.

(iii) Increased aquaculture operations, which enrich surrounding waters and stimulate algal growth.

(iv) Introduction of fisheries resources (through aquaculture development), which can reveal the presence of indigenous harmful algae.

(v) Transport and dispersal of exotic HAB species through ship ballast water or shellfish seeding activities.

(vi) Long term climatic trends in temperature, wind speed or isolation.

(vii) Increased scientific and regulatory scrutiny of coastal waters and fisheries products.

(viii) Improved chemical analytical capabilities that lead to the discovery of new toxins and toxic events.

Whatever may be the reasons for the expansion of HAB, today all nations throughout the world are faced with a bewildering array of toxic or harmful species, their adverse impacts and disturbing trends of increased incidence.

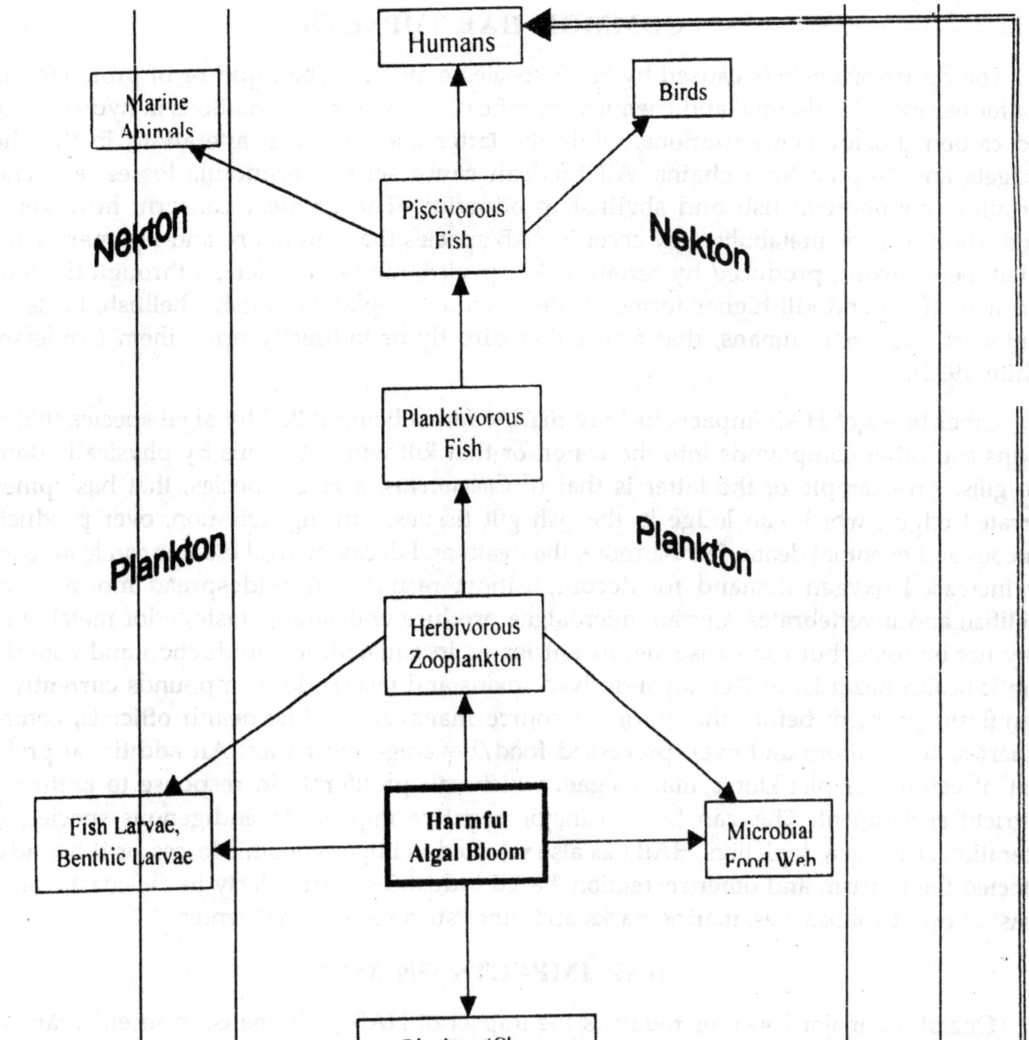

Fig. 1 Trophic linkages between harmful algal blooms and their ecosystems (modified after Smayda, 1992).

COMMON HAB IMPACTS

The disastrous effects caused by HAB species may either be rapid or of prolonged nature. The former includes thermal and chemical stratification, extreme fluctuations in hydrogen, oxygen and carbon dioxide concentrations, while the latter may result in alterations in the chemical budgets and trophic food chains. All this can cause serious economic losses, especially by curtailing commercial fish and shellfish production. The greatest concern, however, is the production of toxic metabolites by certain HAB species that can injure and kill marine life. The potent neurotoxins, produced by certain HAB species, can be transferred through the food web and may affect and kill higher forms of life, such as zooplankton, fish, shellfish, birds, marine mammals and even humans, that feed either directly or indirectly upon them (Anderson and White, 1992).

Other types of HAB impacts include marine fauna being killed by algal species that release toxins and other compounds into the water, or that kill without toxins by physically damaging the gills. An example of the latter is that of *Chaetoceros*, a HAB species, that has spines with serrated edges, which can lodge in the fish gill tissues, causing irritation, over production of mucous and eventual death. Furthermore, the death and decay of algal blooms can lead to annoxia by increased oxygen demand for decomposition, resulting in widespread mortality of fish, shellfish and invertebrates. Certain microalgae produce undesirable taste/odor metabolites that may not be toxic, but can cause significant losses in aquaculture production and contribute to public health hazards. In fact, algal-derived toxins and taste/odor compounds currently pose a significant problem before the marine resource managers, public health officials, commercial fisheries, aquaculture and even processed food/beverage industries. An additional problem is that of benthic or planktonic macroalgae, which can proliferate in response to anthropogenic nutrient enrichment. This can lead to major negative impacts on indigenous species, habitat alteration or oxygen depletion. HAB has also resulted in huge economic losses, as it has adversely affected the tourism and other recreation–based industries, particularly in the coastal areas that boast of beautiful beaches, marine parks and other such recreational centers.

HAB IMPACTS ON MAN

One of the major concerns today, is the impact of HAB on humans. In recent years, reports of harmful algal blooms associated with human illness (see Fig. 2) have received attention in the newspaper, electronic media and scientific literature. Man is principally exposed to the naturally occurring toxins produced by harmful algae through the consumption of contaminated seafood products. Very often, phytoplanktons are filtered from the water as food by organisms such as clams, mussels, oysters or scallops. These organisms accumulate the algal toxins to levels that can be lethal to humans or other consumers. In fact, a single clam can sometimes contain toxin sufficient to kill a human!

Generally, the consumption of contaminated seafood or its products is the main cause for human illness associated with HAB. An exception to this, however, is a type of toxin (brevetoxin), which causes aerosol formation of due to wave action.This aerosol causes asthma-like symptoms during blooms of the toxigenic organism. These diseases are caused by specific interaction of the toxins with tissues and organs responsible for carrying out vital cellular functions. Consequently, the toxins disrupt nerve conduction, uncouple communication between nerve and muscle and

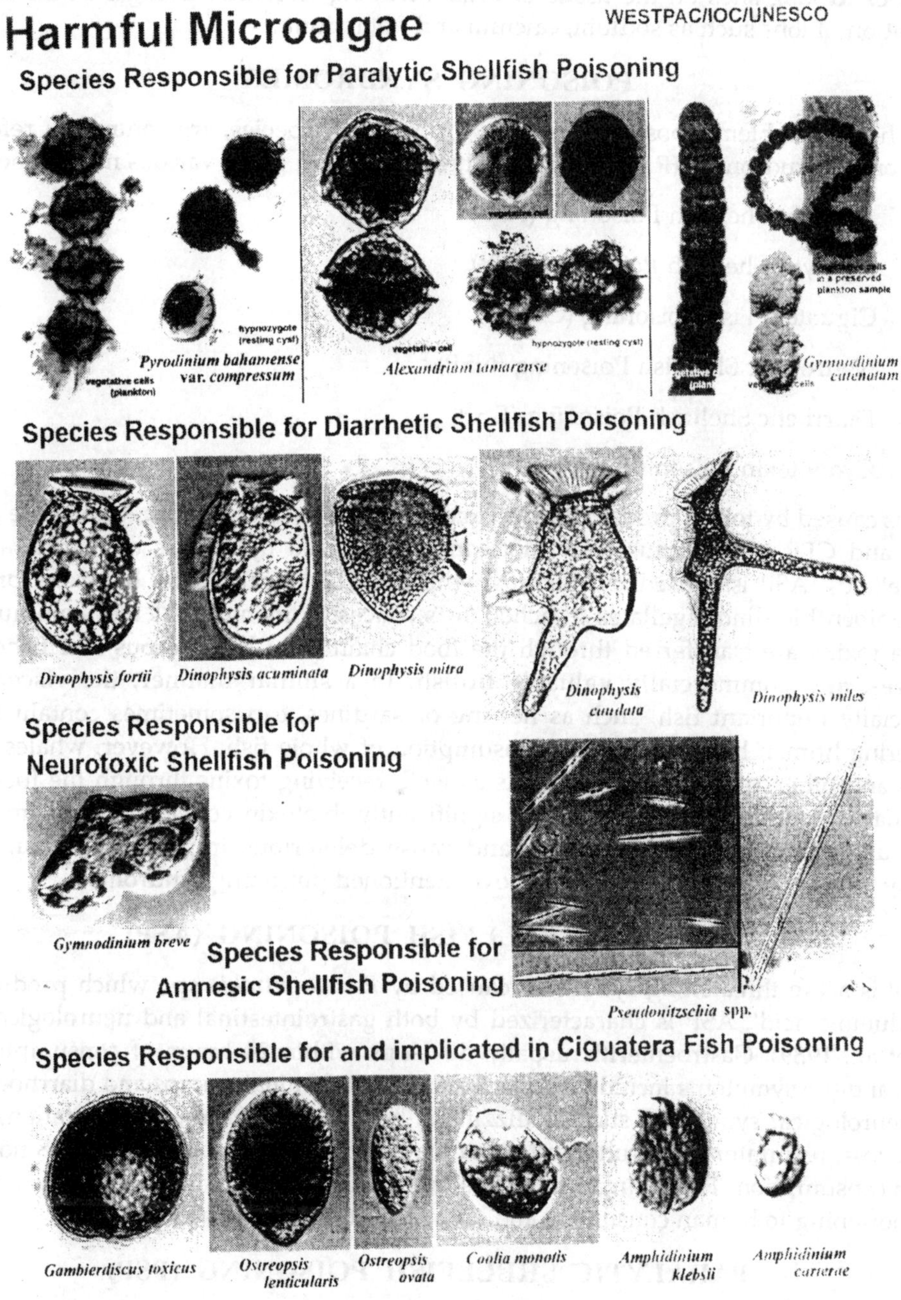

Fig. 2 **Harmful microalgae associated with human illness (Courtesy of Dr. Donald M. Anderson).**

affect critical physiological processes. Most of the toxins accomplish this by binding to specific receptors or locking sites on the tissue or organ, leading to critical changes in the intracellular concentration of ions such as sodium, calcium and potassium.

POISONING SYNDROMES

The health problems, posed to humans by the HAB species, are commonly referred to as the "Poisoning Syndromes"(Ram, 2000) and these have been given various names such as:

- Amnesic Shellfish Poisoning (ASF)
- Paralytic Shellfish Poisoning (PSP)
- Ciguatera Fish Poisoning (CFP)
- Neurotoxic Shellfish Poisoning (NSP)
- Diarrhetic Shellfish Poisoning (DSP)
- Cyanotoxin Poisoning (CP)

CP is caused by toxins produced by certain species of cyanobacteria. Among the rest, except for ASP and CFP, all are caused by biotoxins synthesized by a class of marine algae called dinoflagellates. ASP is caused by diatoms, whereas CFP is caused by biotoxins produced by certain epibenthic dinoflagellates attached to surfaces in many coral reef communities. The ciguatera toxins are transferred through the food chain from herbivorous reef fishes to larger carnivores, and commercially valuable finfish. In a similar manner, the viscera of other commercially important fish, such as herring or sardines, can sometimes contain PSP toxins, endangering human health following consumption of whole fish. However, whales, porpoises, seabirds and other animals can be victims as well, receiving toxins through the food chain via contaminated zooplankton or fish. More significantly, biotoxin conversions and magnification may occur during foodchain transfers and cause deleterious impacts (Shimizu, 1987). The following account briefly discusses the above mentioned poisoning syndromes.

AMNESIC SHELLFISII POISONING (ASP)

ASP is a life threatening syndrome caused by *Pseudo-nitzchia* sp., which produces a toxin called 'dumoic acid'. ASP is characterized by both gastrointestinal and neurological disorders (Bates et al., 1989). Gastroenteritis usually develops within 24 hours of consumption of toxic shellfish and the symptoms include nausea, vomiting, abdominal cramps and diarrhoea. In severe cases, neurological symptoms such as dizziness, headache, seizures, disorientation, short-term memory loss, respiratory difficulty and coma may also appear, usually within 48 hours of toxic shellfish consumption. The toxin accumulates in the fish and crab viscera, thereby increasing the risk of poisoning to human consumers and animals in the marine food chain.

PARALYTIC SHELLFISH POISONING (PSP)

PSP is caused by *Alexandrium* sp., *Gymnodium catenatum* and *Pyrodinium bahamense*, which produce toxins called 'saxitoxins'. PSP, like ASP, is a life threatening syndrome. Symptoms are purely neurological and their onset is rapid. The duration of effects is a few days in non-lethal cases. Symptoms include tingling, numbness and burning of the perioral region, ataxia, giddiness, drowsiness, fever, rash and staggering. The most severe cases result in respiratory arrest within 24 hours of consumption of toxic shellfish. If the patient is not breathing or if the pulse is not detected, then artificial respiration may be needed as a first aid. There is no antidote; supportive therapy is the rule and the survivors recover fully. PSP can be prevented by large scale proactive monitoring programs (e.g. assessing toxin levels in mussels, clams, scallops and oysters) and rapid closures to harvest of suspect or demonstrative toxic areas (Nishitani and Chew, 1988).

CIGUATERA FISH POISONING (CFP)

CFP is caused by *Gambierdiscus toxicus, Prorocentrum* sp., *Ostreopsis* sp. and *Coolia monotis*, which a produce a toxin called 'ciguatoxin'/ 'maitoxin'. CFP produces gastrointestinal, neurological and cardio-vascular symptoms (Withers, 1982). Generally, diarrhoea, vomiting, followed by neurological disfunction including reversal of temperature sensation, muscular aches, dizziness, anxiety, sweating, numbness, tingling of the mouth and digits are common. Paralysis and death may occur, but the symptoms are usually less severe, although debilitating. Recovery time is variable and may take weeks, months or years. Rapid treatment (within 24 hours) with mannitol is reported to relieve some symptoms. There is no antidote; support therapy is the rule and the survivors recover. Absolute prevention is possible by complete abstinence from eating any tropical reef fish, since there is currently no easy way to routinely measure 'ciguatoxin' or 'maitotoxin' in any seafood product before consumption.

NEUROTOXIN SHELLFISH POISONING (NSP)

NSP is caused by *Gymnodium breve*, which produces toxins called 'brevetoxins'. NSP produces syndrome nearly identical to that of CFP. In this case, gastrointestinal and neurological symptoms predominate. In addition, formation of toxic aerosols by wave action can produce asthma-like symptoms (Tester and Steindinger, 1997). However, this syndrome is less severe than CFP and no deaths have been reported so far. In addition, unlike CFP, recovery is generally complete in a few days. Monitoring programs generally suffice for preventing human consumption, except when officials are caught off-guard in previously unaffected areas.

DIARRHETIC SHELLFISH POISONING (DSP)

DSP is caused by *Dinophysis* sp., which produces a toxin called 'okadaic acid'. DSP produces gastrointestinal symptoms, usually beginning within 30 minutes to a few hours after the consumption of toxic shellfish (Yasumoto et al., 1978). The illness, which is not fatal, is characterized by incapacitating diarrhoea, nausea, vomiting, abdominal cramps and chills. Recovery occurs within 3 days, with or without medical treatment.

CYANOTOXIN POISONING (CP)

Some cyanobacteria produce toxins (cyanotoxins), which are physiologically of two types, namely 'neurotoxins' and 'hepatotoxins'. The two 'neurotoxins' (anatoxin and saxitoxin) are synthesized by species of *Anabaena, Aphanizomenon, Oscillatoria* and *Trichodesmium* (Negri et al., 1997). These toxins are alkaloids that block the signal transmission from neuron to muscle in animals and man. Symptoms include staggering, muscle twitching, gasping and convulsions. These toxins can be fatal at high concentrations due to respiratory arrest caused by failure of the muscular diaphragm. Among the two types of "hepatotoxins (microcystins and nodularins), microcystins are produced by species of *Microcystis, Anabaena, Nostoc, Nodularia* and *Oscillatoria,* while nodularins are produced specifically by species of *Nodularia* (Kotak et al., 1995). These toxins are inhibitors of certain protein phosphatases and affect the organism by causing bleeding in the liver. Clinical signs include weakness, vomiting, diarrhoea and cold extremities.

CONCERNS

HAB impacts include mass mortalities of wild and farmed fish, shellfish, death of seabirds, marine animals and alterations of marine habitats or even the trophic structure. These serious economic, public health and ecosystem impacts have evoked increased public awareness and concerns. It is being increasingly realized that " biological pollution", caused by HAB, unlike "chemical pollution", has the potential to reproduce and thereby increase in both intensity and incidence. Thus, the menace caused by problematic HAB, is much more severe and challenging and encompasses every aspect of the aquatic environment. Recent surveys have indicated that presently there are more toxic algal species and toxins, covering larger geographical areas, affecting fisheries resources and causing consequent higher economic losses, health hazards and ecological disasters (Hallegraeff, 1993). This is clearly indicative of the fact that the problem of HAB has had serious negative impacts in the past and poses potential threats for posterity as well.

CONCLUSIONS

Marine biotoxins and harmful algae represent a significant and expanding threat to human health and fisheries resources throughout the world. The impacts of HAB may be reflected in terms of heavy economic losses, serious human illnesses or alterations in the overall functioning of the affected aquatic ecosystems. However, the problems posed by HAB cannot be addressed by any single agency. Concerted efforts of all nations will be necessary to keep the lines of communication and coordination open. Today, HAB comprise a universal problem and therefore, it must be tackled globally.

Though the problem of HAB cannot be completely tackled, yet the losses and threats imposed by it can be minimized. For this, it is imperative to conduct periodic surveys of the coastal regions, identify the HAB species, assess and monitor their distribution and spread, identify factors that favor the occurrence and abundance of HAB and assess their potential economic, ecological and human impacts. This requires in-depth understanding of the bloom biology and ecology. Unfortunately, our understanding of the fundamental ecological, toxicological and oceanographic issues underlying the HAB phenomenon is woefully inadequate. In fact, our knowledge about each of the diverse harmful algal species varies significantly, and even the best studied, remain poorly characterized with respect to bloom biology or population dynamics. Perhaps, only combined efforts, public awareness, genuine concerns and an integrated approach to view the problem of HAB in totality, might just hold the key to address this global menace effectively.

ACKNOWLEDGEMENTS

I am extremely grateful to Dr. Donald M. Anderson of the Biology Department, Woods Hole Oceanographic Institution, Woods Hole, for sending the requested literature and photographs of harmful microalgae. I sincerely thank the Council of Scientific and Industrial Research (CSIR), New Delhi, for the award of Research Associateship.

REFERENCES

Anderson, D. M. and White, A. W. 1992. Marine toxins at the top of the food chain. *Oceanus*, Vol. 35: 55-61.

Bates, S. S., Bird C. J., deFreitas, A. S. W., Foxall, R., Gilgan, M., Hanic, L. A., Johnson, G. R. McCulloch, A. W., Odense, P., Pocklington, R., Quilliam, M. A., Sim, P. G., Smith, J. C., Subba Rao, D. V., Todd, E. C. D., Walter, J. A. and Wright, J. L. C. 1989. Pennate diatom *Nitzschia pungens* as the primary source of domoic acid, a toxin in shellfish from eastern Prince Edward Island, Canada. *Can. J. Fish. Aquat. Sci.*, Vol. 46: 1203-1215.

Hallegraeff, G. M. 1993. A review of harmful algal blooms and their apparent global increase. *Phycologia*, Vol. 32: 79-99.

Kotak, B. G., Lam, A. K-Y., Prepas, E. E., Kenefick, S. L. and Hurdey, S. E. 1995. Variability of the hepatotoxin microcystin-LR in hypertrophic drinking water lakes. *J. Phycol.*, Vol. 31: 248-263.

Negri, A. P., Jones, G. J., Blackburn, S. I., Oshima, Y. and Hideyuki, O. 1997. Effect of culture bloom development and of sample storage on paralytic shellfish poisons in the cyanobacterium *Anabaena circinalis*. *J. Phycol.*, Vol. 33: 26-35.

Nishitani, L. and Chew, K. K. 1988. PSP toxins in the Pacific Coast states: Monitoring programs and effects on bivalve industries. *J. Shell Res.*, Vol. 7: 653-669.

Ram, M. 2000. Poisoning syndromes: A major impact of harmful algal blooms. *J. Indian Ocean Studies*, Vol. 8: 264-272.

Shimizu, Y. 1987. Dinoflagellate toxin. In: Biology of Dinoflagellates, F. J. R. Taylor (ed.), pp. 282-315.

Smayda, T. 1992. Global epidemic of noxious phytoplankton blooms and food chain consequences in large ecosystems. In: *Food Chains, Yields, Models, and Management of Large Ecosystems*, K. L. M. A. Sherman and B. D. Gold (eds.), Westview Press, Boulder, CO, pp. 275-307.

Tester, P. and Steidinger, K. A. 1997. *Gymnodinium breve* red tide blooms: initiation, transport and consequences of surface circulation. *Limnol. Oceanogr.*, Vol. 45: 1039-1051.

Withers., N. W. 1982. Ciguatera fish poisoning. *Ann. Rev. Med.*, Vol. 33: 97-111

Yasumoto, T., Oshima, Y. and Sugawara, W. 1978. Identification of *Dinophysus fortii* as the causative organism of diarrhetic shellfish poisoning. *Bull. Jap. Soc. Sci.* Fisher., Vol. 46: 1405-1411.

25

Environmental Pollution Detection and Bioremediation by Marine Bacteria

*N. Ramaiah, Jaysankar De & S. R. Iyer

INTRODUCTION

Microorganisms, in particular bacteria, play far more important ecological roles in natural environments than their small sizes would suggest. The term environment for bacteria is everything that surrounds and sustains them including air, plant, animal surfaces and interiors, soil, water and even the micrometer sized organic particles. Thus, from an ecological perspective, bacteria are integral parts of the organismic communities that interact variously with each other and between other, higher -and lower- forms of life. These interactions in any ecosystem are of greater significance both in terms of ecological dynamics and stability. Their synthetic and degradative functions are pivotal to the continuum of life processes on earth (Brock et. al. 1994). Some microbes are autotrophic: capable of generating new organic matter. However, most microbes, in particular bacteria, are heterotrophic. Heterotrophic bacteria are largely responsible for biodegradation and recycling of organic matter. Thus, they are intimately involved in biogeochemical reactions of the key elements of living systems viz. C, H, N, S, P, O in addition to a few trace elements. In simple terms, biogeochemical reactions are all those changes an element undergoes in its oxidation state as it *moves* through an ecosystem, in and out of organisms. Significantly, heterotrophic microorganisms are the ONLY agents capable of regenerating biologically essential elements in forms usable particularly, by photosynthesizers.

Microbiology is a many faceted science. Its understanding is essential for a better appreciation of ecological and technological events to which microbes are intimately linked. Efficient and economically beneficial industrial applications of microbes have been possible by understanding the basic ecology, physiology, biology and biochemistry of microbes. Readers' attention is drawn

* National Institute of Oceanography, Dona Paula, Goa 403 004, India.

to several recent publications in arena of Microbiology and Biotechnology (Atlas et. al., 1984; Dewain and Solomon, 1986; Labeda, 1990; Glazer and Nikaido, 1995; Hunter and Belt, 1996 and Cooksy, 1998)

In this case study, we report the results of our studies on the usefulness of marine luminous bacteria in marine pollution pre-screening, the direct viable counts of bacteria in sensing marine environmental pollution stress. In addition we have furnished a few of our results on the use of marine bacteria for bioremediation of oil, pesticide and several other pollutants.

Bioluminescent Marine Bacteria

Bioluminescence, emission of light without heat, is a characteristic of many species of marine and estuarine bacteria whose taxonomy, ecology and phylogeny have been extensively studied for over several decades now (Velankar, 1955; Nealson and Hastings, 1979; Ruby, 1977; Ramaiah, 1989; Ramaiah, and Chandramohan, 1992; Ramaiah et. al., 2000). There are nine species of luminous marine bacteria (Nealson and Hastings, 1992) which are phylogenetically closely knit. Bacterial luminescence is extremely sensitive to toxicants, carcinogens, genotoxics at their nano or picomolar concentrations (Ulitzur, 1986; Bulich, 1982; Stebbing, 1985, Ramaiah and Chandramohan 1993, 1994). Many researchers have investigated on certain ecological (Ruby and Nealson, 1976; Ramaiah and Chandramohan, 1988; Lapota et. al., 1988 and Ruby et. al., 1998), physiological (Gupta et. al., 1985), biochemical (Kurfurst et. al., 1982), genetic (Engebrecht and Silverman 1983; Nealson et. al., 1993; Ruby, 1995 and Ramaiah et. al., 2000) and biotechnological (Lee et. al., 1982 and King et. al., 1990) facets of luminous bacteria. Thus, as the biology, biochemistry, physiology and genetics of these bacteria have been quite well understood, the research and commercial applications of marine luminous bacteria and their luminescence have been continuing.

Direct Viable Counts as Sensors of Environmental Pollution Stress

Various anthropogenic activities and some natural processes, bring about adverse environmental conditions that affect physiological responses such as viability (Byrd et. al., 1991; Colwell et. al., 1985; Oliver et. al., 1995 and Naganumo, 1996), metabolism, resting (dormant) stages and death (Amy and Morita, 1983) among microorganisms *in situ*. Through relevant bacteriological parameters, these responses can be understood and thus are useful for monitoring environmental quality. Among the many methods available for microbiologists, acridine orange direct counts (AODC) and direct viable counts (DVC) are widely employed to study total and viable fractions of bacteria in environmental samples. The technique of DVC, originally developed by Kogure et. al. (1979) is very useful for determining metabolically active bacterial cells and heterotrophic potential of microbial assemblages in the marine environment (Kogure et. al., 1987; Walker and Keevil, 1993; Joux and LeBaron, 1997 and Ramaiah et. al., 2002). Besides, it is easy to perform yielding highly reproducible results (Nilsson et. al., 1991; Wiechert et. al., 1992 and Ravel et. al., 1995). Its applications to study metabolic activities of bacteria in coastal marine regimes subjected to anthropogenic alterations were studied for the first time by Ramaiah et. al. (2002). The changes in DVC in ecological situations with gluts of extraneous organic matter input (e.g. through domestic sewage) on the one hand and, a plethora of xenobiotic toxicants (effluents of certain industries) on the other, were systematically analyzed and several laboratory experiments were done with an idea of evaluating the reliability of DVC for sensing the pollution stress in the marine environment. Actually these rationales evolved because the DVC recorded for the first time from the Indian coasts, were less than 20% of AODC from a polluted zone than those from a currently pollution free location in Gulf of Kuchch where DVC were ca. 35% of the AODC during the 1997 pre-monsoon months. Following this observation, it was planned to examine the fraction of DVC in samples originating from polluted, moderately polluted and relatively unpolluted coastal zones. From the extensive field and laboratory analyses, we report on the reliability of DVC for sensing risks in the marine environment caused by industrial activities.

Lipid Hydrolyzing Marine Bacteria

Oxidation and hydrolysis of lipids by heterotrophic bacteria lead to production of several intermediate molecules that are important in the formation of acetyl-CoA (Brock et. al., 1994) and in the nutrition and growth of many microbial communities (Harris et. al., 1991 and Chrost and Gajewski, 1995). Marine bacterial lipases are of great importance in the catabolic breakdown of varieties of lipid containing organic molecules and recycling of all biologically essential elements present in such complex compounds. Many recent studies highlight the involvement of bacterial lipases in the digestion of food in many marine animal guts (Myrnes et. al., 1994; Pollak and Montgomery, 1994 and Donachie et. al., 1995). Apparently, degradation of natural hydrocarbons by marine bacteria is dependent on their capacity to emulsify and solubilize lipids. In addition, most bacteria capable of breaking down lipids possess degradative properties useful in mitigating problems associated with marine oil spills (Ewell et. al., 1996 and Martinez et. al., 1996).

Although the growth of all bacteria including lipolytic ones is greatly influenced by environmental parameters (e.g., pH, temperature and salinity), physiological studies on growth of lipid degrading marine bacteria under varying conditions of these parameters are not available. A selection of bacterial isolates capable of high lipolytic activity would prove very useful in dealing with organic waste treatment including oil pollution, not only for their individual potential but also as suitable candidates in consortia to be constructed in marine pollution bioremediation (Ewell et. al., 1996). Further, from the available literature, it is also apparent that few studies have measured marine bacterial lipid hydrolysis rates. The present work was therefore aimed at finding out the optimal growth conditions for select strains of lipase positive bacteria, their rates of Tween-80 and tributyrin hydrolysis and also, their ability to breakdown and grow solely on a selection of crude and refined petroleum hydrocarbons.

Mercury Resistant Bacteria

Pollution of environment with heavy metals i.e. mercury, lead or cadmium has dramatically increased in many countries. Ground and wastewater are often considerably loaded with these unwanted toxic metals. Conventional methods to eliminate unwanted metals comprise precipitation, ion exchange, extraction and some other procedures are not always satisfactory under aspects of environment protection and with regard to costs, selectivity and sensitivity (Swain et. al., 1992). Chemicals including organo-mercury compounds, lead, arsenic and dichloro diphenyl trichloroethane (DDT) were historically used as antifoulants. Among the many heavy metals, mercury has an affinity to bind to organics forming generally recalcitrant and highly toxic organomercurial complexes in marine sediments (Gerlach, 1981; Misra, 1992) upon constant leaching and anthropogenic inputs.

The total amount of PCBs produced worldwide is estimated to be 1.2 million tons (Furukawa et. al., 2000) and in many cases exceed the permissible limit of 5 μg g^{-1} (w) in edible fishes according to the United States FDA guideline (Connolly and Thomann, 1992). Of these PCBs, almost the entire amount is retained in coastal sediments and open ocean waters, clearly indicating that the open ocean water serves as a vast reservoir and final source of PCBs (Sarkar, 1994).

Aquatic use of organotin compounds, which are toxic to both prokaryotes and eukaryotes, in particular their incorporation as tributyltin biocides in controlled release points on ships pose the greatest immediate impact upon harbour and coastal aquatic biota. Removal of toxic organotin compounds from the ocean is very important from a global environmental viewpoint (Fukagawa et. al., 1994).

Microorganisms in the marine environment are suggested to undergo selection pressure in the presence of toxic pollutants and develop resistance (Nies, 1999). In recent years a large number of microorganisms have been isolated, which are able to degrade compounds previously considered to be non-degradable. This suggests that under the selective pressure of environmental pollution, a microbial potential for the degradation of recalcitrant xenobiotics is developing and can be harnessed for pollutant removal by biotechnological processes. Taking into account that only a small portion of bacteria can presently be isolated by conventional methods, it is evident that nature harbours a significant resource useful in biotechnological applications. There is considerable evidence about mercury resistance among common microbial species (Pahan et. al., 1990; Barbieri et. al., 1996; Canstein et. al., 1999 and Macalady et. al. 2000). The toxicity of a compound depends ultimately on its concentration, exposure duration and its bioavailability. Thus the bioremediation of polluted groundwater and toxic waste sites requires that bacteria come into close physical contact with pollutants.

Biotic degradation of organomercury and inorganic mercury takes place in a number of forms. The most thoroughly researched one involves mercury resistance in bacteria possessing genes of the mer operon (Robinson, 1984 and Hobman and Brown, 1997). This capacity appears widespread in nature and has been found for both gram- negative and gram- positive bacteria and under aerobic and anaerobic conditions. Bacteria showing resistance to methyl mercury has been reported to posses resistance to TBT (Fukagawa et. al., 1994). Barkay (1987) had reported the inorganic mercury stress increased mercury resistant bacteria but they did not check the relationship between methyl mercury and other heavy metals. To examine the bioremediation potential of marine bacteria, we carried out several field surveys and obtained bacterial strains with unusual resistance to mercury.

MATERIALS AND METHODS

Assays with Luminous Marine Bacteria

In order to adapt tropical species of luminescent bacteria in pollution pre-screening efforts, we conducted various experiments with a variety of environmental strains of luminous bacterial species viz., *Vibrio harveyi*, *V. fischeri* and *Photobacterium leiognathi*. Details of methods followed are available in Ramaiah (1989) and in Ramaiah and Chandramohan (1993, 1994).

Experiments were carried out in order to observe the behaviour of bacterial luminescence when different chemicals were added in small volumes at very low concentrations available in Table 1.

Luminous bacterial strains of the species *V.harveyi, V. fischeri and P. leiognathi* were used for the study of light output kinetics with a view to find out whether there is any species specific behaviour of light output. These cultures were grown in seawater complete liquid medium containing 50 mM Tris and incubated at 28° C for 24 hours. Aliquots of 50, 100 or 200 µl of the cultures were pipetted out into sterile mini vials of luminometer. The contents in the vial were allowed for stabilization for one minute and constant light out put of ^{3}3500 relative light units were ascertained. On attaining such a state, known amounts of test solutions were dispensed into the vial and the decrease and/or increase in light out put was measured for over 300 seconds with a recording interval of 20 seconds.

Measurements of DVC and Related Parameters

Sampling was carried out from polluted (Ennore, Kulai, Padubidri and Ratnagiri), moderately polluted (Malvan) and non-polluted (Positra, Marmugao and Terekhol) coastal locations. Samples were collected during monsoon (June-September), post-monsoon (October- February) and pre-monsoon (March-May). Usually, water and sediment samples were collected from 5-9 stations located on 5, 10, 15 and 20m depth contours. Necessary care in handling the samples was always exercised and samples analysed following standard procedures of plate counts, total counts by epifluorescence microscopy following Hobbie et. al. (1977) and DVC by following Kogure et. al. (1979).

Additionally, the effect of cadmium on formation of DVC was by adding a solution of cadmium chloride (Qualigens, AR, Mumbai) to attain 5 mg Cd^{2+} l^{-1} in all water samples and sediment supernatants to understand how the DVC respond to toxic heavy metals. In addition, to check whether a higher concentration would completely inhibit colony formation as well as elongation of cells, $CdCl_2$ was added to a concentration of 100 mg Cd^{2+} l^{-1} to three each of water and sediment supernatants.

With a view to ascertain whether DVC as a sensitive method for detecting microbial metabolic activity was reliable, additional experiments were done. Measurements of ^{3}H-methyl thymidine incorporation by bacteria in water, sediment and by three bacterial cultures were carried out as this radio-isotope tracer technique is very powerful and useful, among others, for measuring microbial metabolic activities by following UNESCO Protocols (1994) as detailed in Ramaiah et. al. (1996a). These measurements were made also from metabolically stressed (by adding 25 mg Cd^{2+} l^{-1}) and unstressed (no added Cd^{2+}) conditions and, bacterial production rates (BPR) calculated.

Lipid Hydrolysis by Marine Bacterial Strains

Six isolates were studied for their growth under various physiological conditions, for estimating tween 80 (polyethylene sorbitan monooleate) and tributyrin (glycerin tributyrate) degradation rates and for understanding their potential to degrade various hydrocarbon moieties. Three isolates of luminous *Vibrio harveyi* (DP1, DP2, and DP3) and two isolates of *Pseudomonas* (CH03 and CH07) and one of *Aeromonas* sp. (SD15) were studied. These isolates originated from different locations along the Indian coast. Biochemical characteristics of these isolates are reported elsewhere (Ramaiah and Chandramohan, 1992 and De et. al., 2002). Amounts of lipid present in the tween 80 nutrient broth (TNB) inoculated individually with all six lipolytic isolates were quantified by a method described in Parsons et. al. (Parsons et. al., 1984). Detailed procedures have been presented elsewhere (Iyer and Ramaiah, 2002). To follow the hydrolysis rates of both Tween-80 and tributyrin, sub-samples from one luminous strain and one non-luminous strain were removed on day 2, 4, 6 and 8 and lipid contents determined. In addition, lipid estimations were made from 4 cultures incubated at 28° C for 8 days. Suitable blanks were kept for all sets of experiments.

The ability of all six strains to breakdown crude oils and other petroleum hydrocarbons and to use them as sole source of carbon and energy was studied in a mineral salt medium containing individual hydrocarbons listed in Table 3. The cultures were streaked on to the medium (composition l^{-1}: NH_4Cl, 3g; K_2PO_4, 0.1g; $FeSO_4$, 0.05 g; agar, 30g; seawater, 750 ml; distilled water, 250 ml and pH, 7.5). Approximately 15-25 mg crude oil or petroleum hydrocarbon was spread uniformly on the mineral salts-agar plates and different cultures streak inoculated. All plates were incubated for 21 days at 28° C and the growth was noted. Cultures were also grown on mineral salts agar plates without crude oil or petroleum hydrocarbon to check for the absence of growth (negative control). Four lipase negative marine strains were also checked for their ability to utilize these hydrocarbons.

MRB Isolation and Toxicity Assays

Water and sediment samples were collected off Chennai, Gopalpur, and from the coastal stretch between Velsao to Ratnagiri. The MRB were isolated from water samples either by membrane filtration technique or spread plate technique (De et. al. unpublished results). Suitable dilutions from sediment samples were spread plated for the isolation. Seawater nutrient agar (SWNA) with 10 ppm Hg was used as the medium for isolation of MRB.

Several isolates were checked for growth in seawater nutrient broth containing known concentration of different toxic chemicals including mercury, cadmium, phenol and many other toxicants. Seventeen of these strains isolated from samples collected off Goa and Gopalpur were grown in SWNB containing xenobiotics either individually or in combination. Cultures were incubated at 28 °C for 14 days and inspected regularly for growth. The absence of growth was confirmed by streaking a loopful of liquid culture on to SWNA in cases where visible growth was not evident or could not be discerned due to turbidity produced by some xenobiotics in SWNB. The ability of several isolates of MRB grown on SWNA containing 10 ppm Hg to tolerate higher concentrations of Hg was examined by inoculating them in to SWNB containing either 15, 25, 30, 40 50, 75 or 100 ppm Hg. 48 of these isolates could grow well at a mercury concentration of 25 ppm (and hereafter designated as bacteria highly resistant to mercury [BHRM]). The changes in growth of BHRM in the presence of different toxicants were monitored. A reference strain, *Photobacterium leiognathi* (ATCC 25521: PL3) [Ramaiah, 1989] unable to grow in medium containing $\geq$ 1.0 ppm Hg [20] was included to ascertain the Hg tolerance of environmental isolates.

Sensitivity to kanamycin, streptomycin, demechlocycline, neomycin, tetracycline and ampicillin by several strains capable of growth in the presence of mercury and different xenobiotics was examined on seawater nutrient agar medium by antibiotic disc sensitivity assay.

RESULTS

Effect of Various Chemicals on Light Output by Luminous Bacteria

The chemicals that decreased the light production by 50% or less within the assay period of 5 minutes (300 seconds) are listed in Table 1. The time at which the light level decreased to 50%

Table—1 Various chemicals and their concentrations in the assay vials that reduced the levels of light to below 50% in different strains of luminous bacteria

Chemical (Concentration in assay vials)		*V.harveyi*		*V. fischeri*		*P.leiognathi*	
		Time (Sec)	*Light (level %)*	*Time (Sec)*	*Light (level %)*	*Time (Sec)*	*Light (level%)*
Alcohols	Decanol (50%)	200	49.80	180	41.53	220	47.82
	Ethanol (0.5%)	20	23.04	-*	-	-	-
	Methanol (0.1%)	20	3.90	20	2.70	20	3.20
Aldehyde	Formalin (0.5%)	20	2.26	20	1.89	20	3.38
Amines	a-napthalamine (0.025%)	20	12.97	-	-	-	-
	Diaminonapthalamine (10nM)	20	3.11	20	3.7	20	3.89
Antibiotics	Chloramphenicol (50 µg)	20	21.61	20	26	120	45.27
	Penicillin (50 µg)	220	46.90	160	46	20	43.27
	Streptomycin (50 µg)	240	46.77	NR *	151	NR	55.90
	Nalidixic acid (50µg)	60	49.36	40	48.12	40	49
Chelaotors	Ammonium pyrolidine dithiocarbamate (10 nM)	20	0.64	20	0.73	20	0.39
	Tritriplex (10 nM)	20	6.20	20	3.62	20	4.80
Heavy metal salts	$CuSO_4$ (100 nM)	20	8.20	20	2.55	20	1.43
	$HgCl_2$ (100 nM)	20	2.71	20	1.35	20	5.60
	$HgNO_3$ (nM)	20	4.86	20	5.82	20	5.34
	$ZnCl_2$ (100 nM)	40	36.88	20	35.74	40	28.00
	$CdCl_2$ (100 nM)	40	48.85	60	32.62	80	39.33
	$AgNO_3$ (100 nM)	20	3.29	20	1.70	20	34.83
	Pb Ac (100 nM)	20	29.20	20	4.24	20	9.22
Pesticides	DDT (1 ppm)	20	44.67	-	-	-	-
	DDT (20 ppm)	240	48.14	20	1.64	20	1.93
	DDD (10 ppm)	300	49.41	220	44.17	260	45.27
Phenols	Phenol (0.0025%)	20	47.46	-	-	-	-
	DCP (8 µg)	20	5.56	20	1.64	20	1.93
	DCPAA (11 µg)	160	49.14	220	44.17	260	45.27
Others	Arsenous oxide (10µg)	NR	94.19	160	47.39	80	34.77
	Sodium arsenite (0.02 %)	60	30.94	-	-	-	-
	Acetone (0.5%)	20	15.22	-	-	-	-
	KOH (0.015%)	20	20.09	-	-	-	-
	NaOH (0.05 N)	20	26.67	-	-	-	-
	Hexane (50%)	180	48.38	-	-	-	-
	Surf (0.05%)	80	49.50	80	42.00	80	48.00
	Tris (20 nM)	20	34.40	20	28.38	20	21.43

* Not tested; NR= No reduction to below 50% within 300 seconds

was usually before 200 seconds. Twentyone chemicals brought down the luminescence to below 50% level in *V.harveyi* within the first 20 seconds. Ethyl alcohol (0.5%), methanol, formalin, α napthyl amine, DAN, chloramphenicol, streptomycin, chelators, all the heavy metal salts (except $ZnCl_2$ and $CdCl_2$), acetone, KOH, NaOH, phenol, DCP, DCPAA and Tris brought down the light to below 50% in the first 20 seconds. $ZnCl_2$, $CdCl_2$, and PPDDE, decrease the luminescence below 50% in 40 seconds. Hexane reduced the light to 48.38% in 180 seconds. However, 200 seconds or more was required for decanol, penicillin, streptomycin and DDD (10 and 20 ppm).

All the chemicals tested for *V.harveyi* strains were not included for *V. fischeri*, or *P. leiognathi*. But, those chemicals that were tested against all three species showed a similar trend as far as reduction of light levels is concerned. Only streptomycin did not bring down the levels of light to below 50% within 5 minutes in both *V. fischeri* and *P. leiognathi*.

Generally, all salts of heavy metals decreased the light levels to less than 10% within the first 40-60 seconds of addition. The results show that they are not only more toxic over the other chemical compounds but also at very low concentration, such as 100 nM. It was seen very clearly that all the heavy metal salts reduced the light out put in a very similar manner in all the three luminous bacterial species. $HgCl_2$, $HgNO_3$, $AgNO_3$ and $(CH_3COO)_2Pb$ showed almost the same behavior of reduction of light in all the three species. In the case of $CoCl_2$, peculiarly, there was an increase in the light levels of *V. harveyi* in the first 140 seconds. After that, a sharp decline to about 12.50% was seen at the end of 5 minutes.

Direct Viable Counts and Related Parameters

As the data on total direct (AODC), direct viable (DVC) and viable (TVC) counts from five to nine stations each off Ennore, Kulai, Padubidri and Positra during two seasons have been too numerous and reported elsewhere (Ramaiah et. al., 2002), we present here salient features. Also the analysis of variance of these data suggested no statistically significant variation between stations within any given sampling location. In the overall, their percentage contributions to TDC in water and sediment samples from various locations are presented in figures 1 & 2. It is easily discernible that the DVC were low during premonsoon and higher during monsoon and immediate post monsoon in locations subjected to pollution stress.

^3H thymidine incorporation rates within each subsample varied too widely particularly in water and sediment samples (Table 2) and ^3H incorporation rate was low in all samples and in bacterial suspensions treated with Cd^{2+}, hence, the low BPR. Unlike the generally low coefficients of variation (CV) in DVC, which ranged from 2.5 to 17% for the entire field and laboratory analyses, CV in TdR varied widely from 20 to 29% for water, 29 to 39% for sediment and from 17 to 22% for bacterial cultures.

Lipid Hydrolysis Rates

Tween-80: In the case of culture broths, the rate of Tween-80 hydrolysis increased steadily after the initial 2 days. The luminous strain DP1 (*Vibrio harveyi*), hydrolyzed 13.56% of the total tween -80 in the TNB medium in the first two days, 22.96% in four days, 27.52% by day six and 42 .48% the total Tween-80 by day eight. But in the case of non-luminous strain CH07 (*Pseudomonas* sp.), the hydrolysis rate was quite rapid after day six (*i.e.*, only 6.8% of the total by day two, 18% by day four, 26 % by day six). It increased rapidly between day six and eight, by which time 65.32% of the total Tween-80 in the TNB was hydrolyzed (Fig. 3).

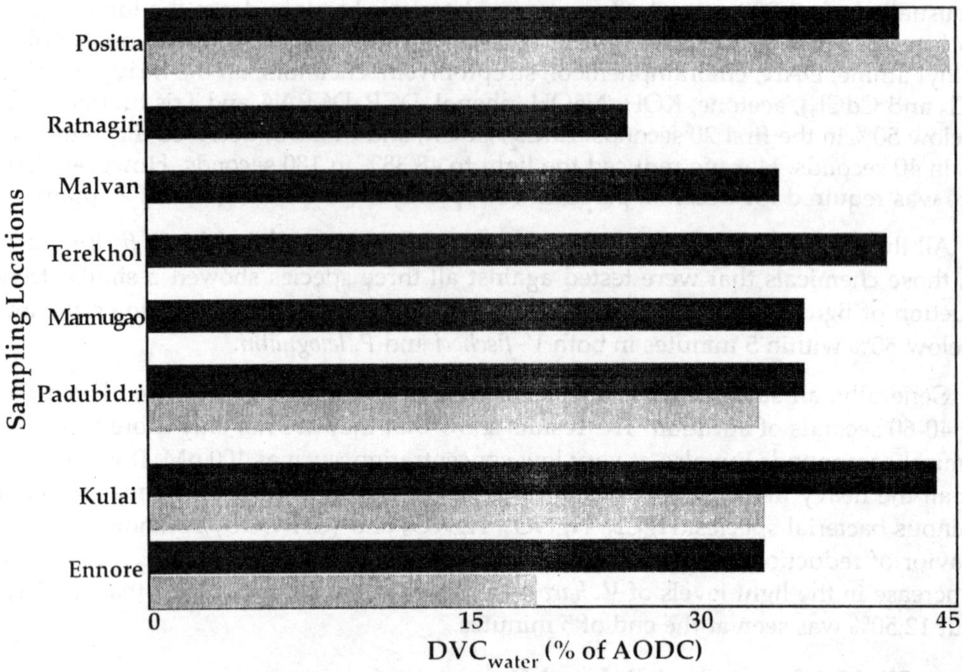

Fig. 1 Mean percentages of direct viable counts of bacteria in water samples collected from different sampling locations during premonsoon and postmonsoon (solid bars) seasons

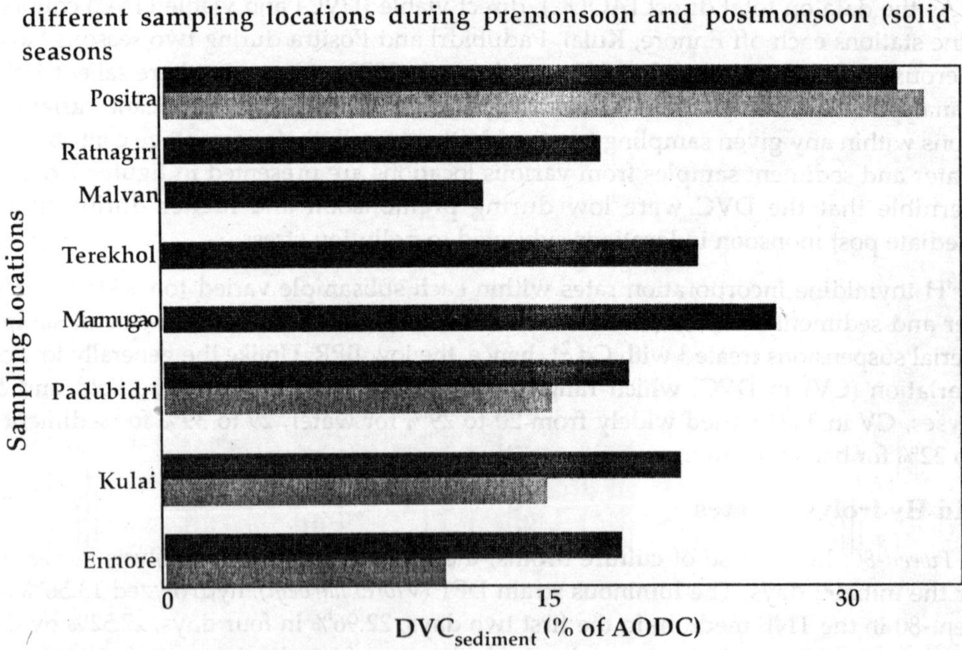

Fig. 2 Mean percentages of direct viable counts of bacteria in sediment samples collected from different sampling locations during premonsoon and postmonsoon (solid bars) seasons

Table—2 Bacterial production rates (BPR) calculated from ^3H methyl thymidine incorporation (TdR)[a] by bacteria in different samples (After Ramaiah et. al. 2002)

Sample	AODC[b]	BPR (number produced l^{-1} [or g^{-1}] h^{-1})	
		without Cd^{2+}	with 25 mg l^{-1} Cd^{2+}
water	(10^6 ml^{-1})		
Dona Paula Bay 1	6.91	$2.78 \pm 0.56 \times 10^5$	791.56[c]
Dona Paula Bay 2	7.82	$2.94 \pm 0.85 \times 10^5$	564.10
sediment	(10^9 g^{-1} dry wt)		
clay	4.92	$4.87 \pm 1.89 \times 10^6$	1840.34
silt	5.33	$6.42 \pm 1.87 \times 10^6$	2681.86
bacterial cultures (10^6 ml^{-1})			
Vibrio harveyi	56.5	$7.12 \pm 1.20 \times 10^5$	3112.67
Pseudomonas sp.	83.7	$7.37 \pm 1.69 \times 10^5$	2216.83

[a]calculated using formula: TdR (pM h^{-1} l^{-1}) = (DPMs-DPMb)/2200 * V/1000 * 60/T * 1/SA: where, DPM- disintegration per minute; s-sample (average of three replicates for all analyses); b-blank; V-sample volume (20 ml); T-incubation time in minutes (which was always 60 min); SA- specific activity of added ^3H thymidine (17000 mCi mmole^{-1}). A conversion factor of 2.17×10^{18} cells mole^{-1} thymidine incorporated was used for calculating bacterial production as per Ducklow, (1993) and, [b] as indicated in parentheses for water, sediment and bacterial cultures; [c] standard error was extremely small hence, not shown here.

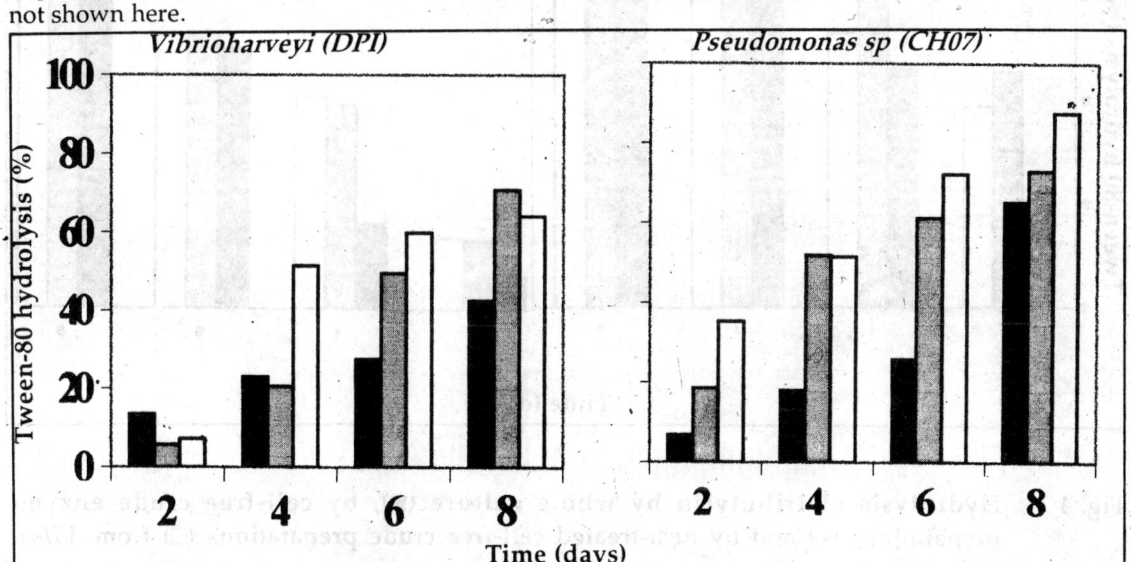

Fig. 3 **Hydrolysis of tween-80 by whole culture (■), by cell-free crude enzyme preparations (□) and by heat-treated cell-free crude preparations (□) from *Vibrio harveyi* and *Pseudomonas* sp.**

For the cell free crude enzyme preparations (CFE) the hydrolysis rate increased almost linearly. The CFE of DP1 hydrolyzed 5.84%, 20.68%, 49.44% and 70.84% of the total tween-80 by day 2, 4, 6 and 8 respectively. Where as, the CFE of the non-luminous strain CH07 hydrolyzed as much as 18.56% by day two that steadily increased to 73.12% by day eight. Generally there were differences in the rates of hydrolysis between the two strains.

For the heat-treated cell-free crude preparation, the tween-80 removal was almost similar to that of the cell free crude enzyme preparation. The heat-treating was mainly done to inactivate all the cells and to check if the extracelluar lipases were denatured at 65° C.

Tributyrin: In case of whole cell the rate of removal of tributyrin was similar to that of Tween-80 and found increased steadily after the first two days. The luminous strain, DP1, removed 7.28% of the total tributyrin in the TrNB medium on day 2, 23.68% on day 4, 30.44% on day 6 and finally it was 33.92% of the total tributyrin on day 8. But in the case of non-luminous strain CH07, its removal rate was quite rapid after the fourth day (Fig.4).

In the case of cell free crude enzyme preparations, tributyrin removal rate for the luminous culture increased steadily from the second day itself. The lipid removal rate by the cell-free extracts of this strain was 22.92%, 31.68 %, 34.36% and finally 35.88% on day 2, 4, 6 and on day 8 respectively. For the non-luminous strain the rate increased steadily from 6.52% to 30.92%.

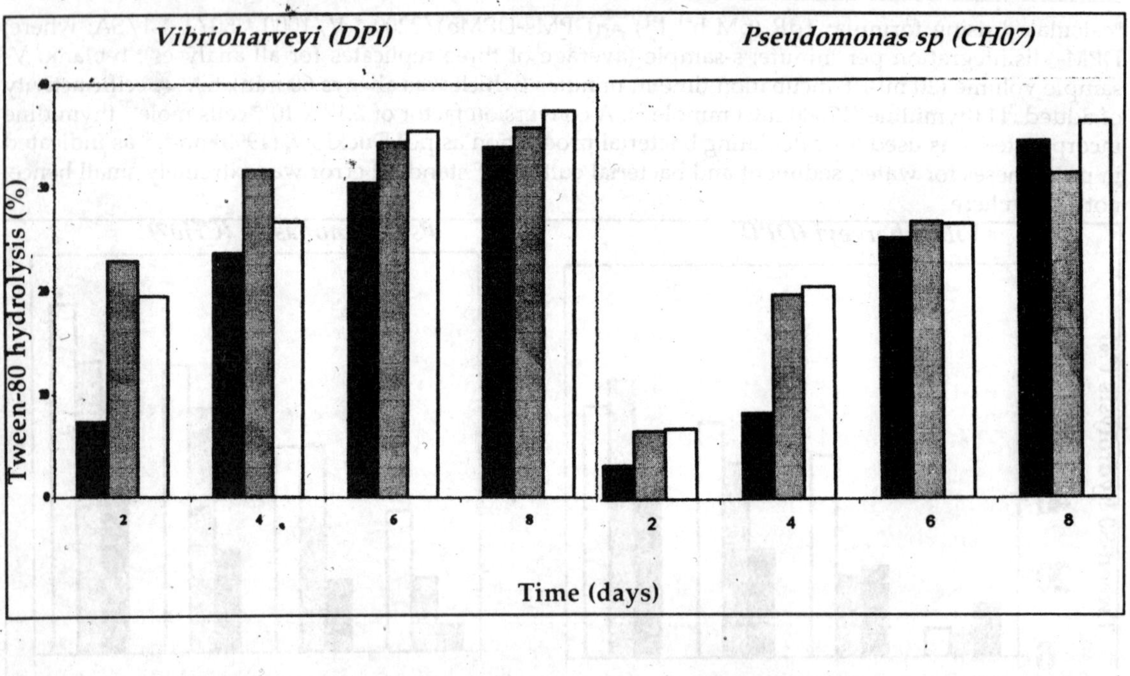

Fig. 4 Hydrolysis of tributyrin by whole culture (■), by cell-free crude enzyme preparations (□) and by heat-treated cell-free crude preparations (□) from *Vibrio harveyi* and *Pseudomonas* sp.

In the case of heat-treated cell free crude preparations, the tributyrin removal rate was 19.48% on the second, 26.72% on the fourth, 35.52% on the sixth and 37.44% on the eighth day by DP1. Whereas, the removal rates by CH07 were 6.68%, 20.24%, 26.16% and 35.88% on day 2, 4, 6 and 8.

Degradation of hydrocarbons: It was found that the *Vibrio harveyi* strains mainly degraded crude oils like Bombay high crude oil, Iranian crude oil much better than the petroleum hydrocarbons like naphthalene, xylene and benzene (Table 3). The degradation rates were poor for diesel and they did not degrade phenanthrene. The non-luminous bacteria grew well on the crude oils (Bombay High and Iranian crude) and also on petroleum hydrocarbons such as benzene and xylene but showed poor growth on phenanthrene and naphthalene. The growth of this strain was quite well evident from day 10 itself unlike the luminescent bacterial strains whose growth was moderate until the 15th day. None of the four lipase negative marine strains was able to grow on any hydrocarbon.

Table—3 Degradation of petroleum hydrocarbons by Vibrio harveyi and Pseudomonas sp.

Strain	*Petroleum hydrocarbons*						
	BMC	IRCO	DIES	NAPH	PHEN	BENZ	XYL
DP1	++	++	-	+	-	+	-
DP2	++	-	-	+	-	-	+
DP3	+	+++	+	++	-	-	+
SD15	++	+	+	++	++	+	++
CH03	+	+	+	-	-	+	++
CH07	++	+	++	++	-	+	+

BHC, Bombay-High crude; IRCO, Iranian crude oil; DIES, diesel; NAPH, naphthalene; PHEN, phenanthrene; BENZ, benzene; XYL, xylene. Isolates designated DP1, DP2 and DP3 belong to *Vibrio harveyi*; CH03 and CH07 to *Pseudomonas* and SD15 to *Aeromonas* sp.

MRB Identification and Toxicity Tolerance Potential

Based on the biochemical response of these BHRM isolates, 13 of the total 30 tested environmental isolates were found to belong to *Pseudomonas*, five each to *Proteus* and *Xanthomonas*. Four strains belonged to Enterobacteriacae. In addition, there was one strain each of *Aeromonas* and *Alteromonas* besides an unidentified one.

The ability of isolates of BHRM to grow in the presence of various toxic chemicals was highly variable. For example, not all isolates of Pseudomonas highly resistant to mercury were able to grow in SWNB containing 100 ppm DDT or 1000 ppm phenol or 10% (v/v) trichloroethane (TCE). Thirteen of the 25 isolates checked for growth in presence of 1500 ppm phenol in SWNB medium were found to be positive for growth. The ranges of concentrations of various toxic heavy metals and toxic xenobiotics tolerated by BHRM are listed in Table 4. Only one isolate out of thirteen was confirmed for growth in presence of 25 ppm phenyl mercury acetate and revealed the presence of both organomercurial lyase and mercuric reductase enzymes. When various toxicants were added together, growth of almost all isolates was very poor or absent (Table 5). None of the strains of BHRM tested during this study was able to tolerate and grow in the presence of formaldehyde (final concentration, 2% v/v) or formic acid (2% v/v). However, interestingly enough, three of the *Pseudomonas* isolates could grow in the presence of 10% TCE.

Table—4 Ranges of concentrations of different toxicants tolerated by bacteria highly resistant to mercury. Genera of bacteria studied; *Pseudomonas* (8 isolates), *Proteus* (5 isolates), *Xanthomonas* (1 isolate), *Aeromonas* (1 isolate), Enterobacteriaceae (1 isolate), unidentifiable (1 isolate)

Chemical	*Tolerance concentration in ppm*
Heavy metals	
Mercury	5-50
Cadmium	10-100
Copper	20-100
Zinc	20-100
Lead	10-100
Pesticides	
DDT	10-100
Penconazol	10-93
Propiconazol	10-95
Metolachlor	10-95
Pretilachlor	10-96
Profenofos	10-91
Others	
Phenol	50-1500
Tributyltin	1-25
Toluene	1-5
Trichloroethane	1-25 (% v/v)

All six strains (tested along with known sensitive strains) examined for antibiotic sensitivity were resistant to all antibiotics tested. Except for small inhibition by Kanamycin in the case of *Pseudomonas* and Enterobacteriaceae isolates, the other strains formed continuous growth on the antibiotic sensitivity assay agar suggesting their insensitivity to these antibiotics.

DISCUSSION

Luminous Bacteria as Reliable Aids for Pollution Pre-screening

One major advantage of bacterial bioluminescence over other currently used toxicity assays is that the instrumentation reduces the chance of human error. In bioluminescence assays, problems inherent in many of the cell systems in attempting to determine the extent of cellular alteration do not arise. The data output being the function of the instrument used, pipetting is the principal source of laboratory error, which can be eliminated easily by paying little more attention. For absolute values of toxicity, the monitoring time after adding the chemical solution to the bacterial suspension is critical. Hence, short-term assays adopted during this study certainly assure the lower limits of toxicity. Apart from being extremely suitable and far more advantageous, growing and handling these non-pathogens (to humans) is fairly simple and the live bacterial cells can be handled almost as chemical reagents.

Table—5 Growth response of mercury resistant bacterial strains to different combinations of toxic substances

Combination of toxic substances	*Pseudomonas (8)*[c]	*Xanthomonas (5)*	*Enterobacteriacene (4)*
M2[a]+D13+P10 (25 ppm)[b]	+	+	+
M5 + D10 + P10 (25 ppm)	+	+	+
M10+D5+P10 (25 ppm)	+	+	+
M20+D10+P20 (50 ppm)	- (1+)	-	-

[a]individual concentration (in ppm) of M, mercury; D, DDT; P, phenol in each combination; [b]cumulative concentration (in ppm) of M, D and P. For example, M2 means 2ppm Hg, D13, 13 ppm DDT and P10, 10 ppm phenol; [c]number of strains tested; +, positive growth; -, negative growth.

In the present study, the light output of *V. harveyi*, *V. fischeri* and *P. Leiognathi* has been studied to evaluate EC_{50} of various toxicants, most of which reach the marine environment through one or the other source. Information on the EC_{50} for different toxicants can very well serve to closely monitor and decide upon the disposal of wastes after treating the effluent in question. The speed of response of bacterial luminescence to the action of toxicants brings close to reality the prospect of online assays in the handling of industrial wastes.

Many currently employed short-term or long-term bioassays in marine pollution assessment studies have been found to be inadequate and unreliable owing to the adaptive responses shown by many assay organisms (Jensen et. al., 1974 and Simkiss and Mason, 1984), whereas such problems are avoided by the bacteria since the response in terms of light output in the cell suspensions are recorded for a very short time. The protocol we propose for pollution monitoring programmes in the tropical seas is briefly as follows. Any sample (e.g. seawater, sediment supernatant) suspected to be originating from the polluted region or any industrial effluent at various stages of treatment can be subjected to quick prescreening following a few simple steps. Small amounts (25 or 50 μl) of these samples are carefully transferred to a luminometer vial containing 25 or 50 μl luminous bacterial suspension and the response in terms of light output is measured in a luminometer. Once this step is completed, measures required for dilution of effluent can be worked out or any other measures can be decided upon to reduce the toxic effect of the effluent. In several earlier studies, environmental samples have been evaluated for pollution, based on the principle adopted in this study.

Marine Pollution Stress Detection Using DVC

One central, technical difficulty in marine microbial ecology has been of accurate enumeration of bacterial cells. Technical complexities with many methods of microbial abundance quantification as well as higher costs for analyzing certain microbial parameters often impede their inclusion for environment impact assessment studies. Plate counts, direct viable counts (Kogure et. al., 1979) and acridine orange direct counts (Hobbie et. al., 1977) are commonly employed methods to monitor cell viability and to measure the fraction of metabolically active bacterial cells. From many ecological studies it is now realized that plate counts do not generally reflect the metabolic potential of bacteria in nature. Thus, inferences based on plate counts are inadequate for an understanding of metabolic state of bacteria in environmental samples. Notwithstanding amendments that may be made to it from time to time, Kogure et. al. (1979) procedure for DVC enumeration is probably quicker, reliable and very useful in many situations. Despite the single antibiotic used, one can, with ease detect enlarged and brighter cells with certainty in 6h of incubation. Also, the fraction of substrate-responsive cells detectable by either procedure (i.e., Kogure et. al., 1979 and Joux and LeBaron, 1997) for DVC from natural and laboratory cultures during this study was greater than those recovered by traditional plating procedure. In addition, a comparative analysis of substrates that yield the best measure of DVC has been done earlier and, yeast extract was shown to be the best substrate that allows the growth of all active bacteria equal to that of a mixture of many amino acids (Kogure et. al., 1987).

Differences in fractions of DVC at different locations, and seasons make it possible to specify the locations that may not permit metabolic activity to the full potential of bacteria *in situ*. For instance, low percentages of DVC seen off locations with greater human interference such as Ennore, Padubidri, Kulai, Ratnagiri and, to some extent, Malvan suggest an imbalance in heterotrophic processes. These locations do experience an overload of xenobiotics that may not permit normal microbial activity. Such effect of toxic xenobiotics could be explained from the extremely low rates ^{3}H incorporations by bacteria - hence the negligible BPR- in environmental samples and laboratory cultures exposed to cadmium. Also, in areas affected by effluent discharges, very low uptake of ^{14}C glucose by bacterial assemblages *in situ* was found (Ramaiah et. al., 1996b). Low DVC in locations receiving higher quantities of industrial effluents allow to point out the reliability and usefulness of DVC in discerning metabolic stress on bacterial communities. It is important to take note of such lowered metabolic activities so as to plan regulatory and precautionary steps for containing indiscriminate outflow of toxic effluents into marine zones. On the other hand, it is discernible from the invariable fractions of DVC off Positra that the marine environment there is the least perturbed. This is despite the fact that low AODC were observed during both collections. Such low bacterial numbers with higher percentages of DVC imply a generally low abundance but enhanced activity of bacteria.

The effect of Cd^{2+}on DVC and plate counts was drastic as noticeable by reduced viability response of bacterial communities *in situ*. Plate counts and DVC reduced by 90-99% at 5 mg l^{-1}

Cd^{2+} and by over 99.5% at 25 mg l^{-1} Cd^{2+}. Toxicity measurements based on DVC and thymidine incorporation rates compared well and both were 3-4 orders of magnitude more sensitive than a method that utilized colony counts. Thus the simple tests with cadmium were helpful to realize reduced viability and BGR. From these observations it is proposed that impact(s) of toxic substances that retard/inhibit metabolic and ecological functioning of microbial communities can be detected by enumerating DVC. Through a set of easy -to- use and simple procedures (*i.e.*, plate counts, AODC and DVC), a highly reliable measure of bacterial activity potential is possible. Based on definitive results we affirmatively suggest DVC as very reliable not only to note ecological stress(es) native bacteria experience, but also to fix quantitative and qualitative threshold limits for effluent discharges, to establish carrying capacities for specific marine regions and, to identify risks to marine ecosystem by any given human activity causing habitat alteration/ retardation.

Usefulness of the Potential of Marine Bacterial Lipases in Oil Pollution Cleanup

All six strains of marine bacteria we studied elaborate extracelluâler lipases capable of rapid hydrolysis of both tween-80 and tributyrin. Probably, ours is the first study that has used a very sensitive chemical method to quantify the unutilized fatty acid and demonstrate that one can accurately quantify the lipids and get a reliable estimation of lipid hydrolysis rates. Although, the hydrolysis of tributyrin was generally slower, it is possible to suggest that the hydrolysis rates remain more or less the same until the lipids are completely utilized. We selected these cultures for determining lipid hydrolysis rates based on their rich growth in TNB. Similar to the observations made in this study, Sreekumari et. al. (Sreekumari et. al., 1988) observed that the lipolytic activity was elaborated by the members of various genera of marine bacteria, although their utilization of different lipid compounds (tributyrin, tween 80, tween 60, tween 40 and tween 20) varied with the strains.

The hydrolysis of both tween-80 and tributyrin increased with concomitant increase in cell numbers. For example from our OD measurements it is noticed that cell numbers of CH07 increased from 0.073×10^{10} cells ml^{-1} on day 2 to 0.101×10^{10} cells ml^{-1} on day 6 in the TNB (at pH 7, salinity 35 ppt and incubation at room temperature). However, the rate of hydrolysis on per cell bases appears to remain more or less constant. This is not only important for proving that the chemical method of determining the microbial lipid hydrolysis rates is reliable but also that these results are useful for an interpretation of bacterial lipid metabolism. Detailed investigations are necessary for qualitative characterization of lipolytic enzymes, their heat stability and many other chemical properties. Nonetheless, results from the study are useful in suggesting that the extracellular lipases from both luminous and nonluminous strains we studied are stable in PBS and active at 65° C. Also, the crude preparations from DP1 and CH07 when held at 100° C for 10 mins became inactive suggesting that they may function well around 65° C similar to the ones reported by Sigurgisladottir et. al. and Rath and Subramaniam.

From an ecological perspective, the ability to rapidly degrade lipids by marine bacteria is important. As the long chain carbon compounds reach the sea from various anthropogenic and natural processes, their rapid transformations by microbial communities are essential to mitigate problems associated with such complex organic molecules (Zilinskas et. al., 1995). Environmental pollution due to oil and petroleum hydrocarbon contamination requires attention in view of its complicated interaction with biological communities (Atlas, 1991). Many hydrocarbon degrading/modifying bacteria produce extracellular surfactants (Mohandass et. al., 1997) and render these hydrocarbons bioavailable (Harvey et. al., 1990). The potential of moderate to very good rates of lipid hydrolysis in the strains studied by us is useful to realize their capacity to emulsify all those hydrocarbon moieties they were able to utilize as sole source of carbon and energy. As all the four lipase negative strains examined for hydrocarbonoclastic activity in this study failed to grow on any of the tested hydrocarbons, it is possible to substantiate that the capacity to elaborate lipases is an important step in petroleum hydrocarbon degradation and hence in mitigation of oil pollution in the marine environment.

Importance of MRB in Marine Pollution Bioremediation

Microorganisms in the marine environment are suggested to undergo selection pressure in the presence of toxic pollutants and develop resistance (Nies, 1999). Such organisms become important in continuing the basic biological processes in contaminated habitats. It is possible that genes responsible for mercury resistance are linked to those that confer resistance to other heavy metals and xenobiotics under natural condition. Interest in bacterial resistance to metal salts, especially when associated with degradative activities, has been increasing since in industrial areas both metal and organic compound pollution is of environmental concern. Studies of these two bacterial abilities, namely the tolerance to toxic metals and the degradative capacities, have been extensively carried out, but the two features are often considered separately, not taking into account that in industrial pollution by organic compounds, i.e. fossil fuels or derivatives, is often accompanied by pollution by heavy metals, such as mercury. Factors affecting the chemical speciation, toxicity and bioavailability of mercury in contaminated environments are of a complex nature and require interdisciplinary approaches to understand the working of individual ecosystems (Zillioux et. al., 1993). Additionally, great care must be taken in determining the chemical abundance and speciation of mercury in environmental samples (Gill and Bruland, 1990). Because metals and metalloids, unlike organic compounds, cannot be transformed to harmless elements (CO_2, H_2O), their remediation consists of separation and volume reduction rather than elimination (Saouter et. al., 1995).

Environmental isolates that grew in the presence of 50 ppm Hg were also able to grow at very high concentrations of xenobiotic toxicants including TCE, DDT or other pesticides, heavy metals and phenols. However, in the combined presence of Hg, phenol and DDT, only a few isolates were able to grow. This suggests that most strains are unable to tolerate the additive toxicity of many toxicants in combination. It is interesting to note that all the examined strains were resistant to much higher concentration of mercury than the *in situ* concentrations. Environmental factors responsible for such high levels of mercury resistance need to be identified for a better understanding of the ecology of MRB including the BHRM recognized during this study.

On the whole, our data suggest that bacterial strains endowed with the capacity to degrade both aromatic compounds and tolerate toxic metals are not uncommon in polluted marine environments. Thus, such metal resistant and xenobiotic degrading isolates are suitable in any marine pollution bioremediation efforts. Therefore, basic studies such as the one reported here and those aimed at exploiting their characteristics in the treatment of polluted sites or wastes are very important. This is especially so in the cases of non-contained environmental applications, i.e. *in situ* bioremediation, where the use of genetically engineered bacteria is unadvisable.

In conclusion, results of the case studies presented here are a few leads useful in appreciating the 'role of the infinitely small' which is, as Pasteur had said "infinitely large". As can be discerned from the foregoing, many naturally occurring marine bacteria are very reliable agents useful to quickly and reliably pre-screen marine and other natural ecosystems for pollution, detect the prevalence of subtle stresses that may not allow full metabolic functioning of native microbial communities. That many marine bacterial genera possess innumerable potentials to degrade/biotransform some of the most deleterious chemicals is an important information from our studies. Marine bacteria, collectively as in the case of DVC, or individually as in the cases of luminous bacteria, MRB and lipolytic ones have proven that they come handy to assess the environmental health and are useful in even cleaning up the mess that is resulting from the anthropogenic causes. Small wonder that a great microbiologist, Dr. David Pearlman, termed bacteria as "bags of enzymes and ingenious than engineers".

ACKNOWLEDGEMENTS

We thank Drs.E. Desa, Director, NIO, D. Chandramohan, Dr. S. Y.S Singbal and, K. Sawkar for facilities and support. This is NIO contribution number 3759

REFERENCES

Amy P. S. and Morita R. Y. 1983. Starvation-survival patterns of sixteen freshly isolated open-ocean bacteria. **Appl. Environ. Microbiol.** 45:1109- 1115.

Atlas R. M., Brown A. E., Dobra K. W. and Miller L. 1984. **Experimental Microbiology- Fundamentals and application.** Macmillan, New York.

Atlas R. M. 1991 Bioremeidiation of oil spills promise and perils. International Marine Biotechnology Conference. Baltimore, MD.

Barbieri P., Bestetti G., Reniero D. and Galli E. 1996. Mercury resistance in aromatic compound degrading *Pseudomonas* strains. **FEMS Microbiol Ecol.** 20: 185-194.

Barkey T. 1987. Adaptation of aquatic microbial communities to Hg^{2+} stress. **Appl. Environ. Microbiol.** 53: 2725-2732.

Brock T. D.; Madigan M. T.; Martinko J. M.and Parker J. 1994. In **Biology of Microorganisms.** pp. 909. Prentice-Hall, New Jersey.

Bulich A. A. 1982. Practical and reliable method for monitoring toxicity of aquatic samples. **Prosses Biochem.** 17: 45-47.

Byrd J. J.; Xu H. S. and Colwell R. R. 1991. Viable but nonculturable bacteria in drinking water. **Appl. Environ. Microbiol.** 57: 875-878.

Canstein V. H., Li Y., Timmis K. N., Deckwer W. D. and Wagner-Döbler I. 1999. Removal of mercury from chloralkali electrolysis wastewater by a mercury-resistant *Pseudomonas putida* strain. **Appl. Environ. Microbiol.** 65: 5279-5284.

Chrost R. J.and Gajewski A. J. 1995. Microbial utilization of lipids in lake water. **FEMS Microbiol.Ecol.**18: 45-50.

Colwell R. R.; Brayton P. R.; Grimes D. J.; Roszak D. B. Huq S. A. and Palmer L. M. 1985. Viable, but non-culturable *Vibrio cholerae* and related pathogens in the environment: implications for release of genetically engineered microorganisms. **Biotechnology** 3: 817-820.

Connolly J. P. and Thomann R. V. 1992. Modelling the accumulation of organic chemicals in aquatic food chains. In J. L. Schnoor (Ed) Fate of pesticides and chemicals in the environment. pp. 385-405. John Wiley and Sons. Inc.

Cooksy E. K. (Ed.). 1998. **Molecular approaches to the study of the ocean.** pp. 549. Chapman & Hall. Lond, Weinheim, New York, Tokyo, Melbourne, Madras.

De J.; Ramaiah N.; Mesquita A. and Verlenkar X. N. 2002. Tolerance to various toxicants by marine bacteria highly resistant to mercury. **Mar Biotechnol.** (in press).

Dewain A. and Solomon N. A. (Eds.). 1986. **Manual of industrial microbiology & Biotechnology.** pp. 466. American Society for Microbiology. 1913, I Street. N.W. Washington DC 20006.

Donachie S. P.; Saborowski R.; Peters G. and Buchholz F. 1995. Bacterial digestive enzyme activity in the stomach and hepatopancreas of *Meganyctiphanes norvegica*. **J. Exp. Mar. Biol. Ecol.** 188: 151-165.

Ducklow H. W. 1993. Bacterioplankton distributions and production in the northwestern Indian Ocean and Gulf of Oman, September 1986. **Deep Sea Res.** 40: 753-771.

Engebrecht J.; Nealson K. and Silverman M. 1983. Bacterial bioluminescence: isolation and genetic analysis of functions from *Vibrio fischeri*. **Cell.** 32: 773-781.

Evans M. S., Birchenough A. C. and Brancato M. S. 2000. The TBT ban: out of the frying pan into the fire?. **Mar. Pollut. Bull.** 40 (3): 204-211.

Ewell M.; Hind J.; Washington D.; Fruchtman S.; Walch M.; Beard M.; Jones-Meehan, J. and Jones W. 1996 Applications of microbial consortia to the degradation of grease-containing waste. Abstr. Gen. Meet. Am. Soc. Microbiol. 96 Meet, 435.

Fukagawa T., Konno S., Takama K. and Suzuki S. 1994. Occurrence of tributyltin (TBT) and methyl mercury tolerant bacteria in natural seawater to which TBT was added. **J. Mar. Biotechnol.** 1: 211-214.

Furukawa K. 2000. Biochemical and genetic bases of microbial degradation of polychlorinated biphenyls (PCBs). **J. Gen. Appl. Microbiol.**, 46: 283-296.

Gill G. and Bruland K. W. 1990. Mercury speciation in surface freshwater systems in California and other areas. **Environ. Sci. Technol.** 24: 1392-1400.

Gerlach A. S. 1981. **Marine Pollution. Diagnosis and Therapy**. pp. 218. Springer-Verlag Berlin Heidelberg New York.

Glazer A. N. and Nikaido H. 1995. **Microbial Biotechnology: Fundamentals of applied microbiolgy.** pp. 622. W. H Freeman Company.

Harris J. M.; Seiderer L. J.and Lucas M. I. 1991 Gut microflora of two saltmarsh detritivore thalassinid prawns, *Upogebia africana* and *Callianassa kraussi*. **Microb. Ecol.**21: 277-296.

Hobbie J. E.; Daley R. J. and Jasper S. 1977. Use of Nuclepore filters for counting bacteria by fluorescence microscopy. **Appl. Environ. Microbiol.** 33: 1225-1228.

Hobman J. L and Brown N. L. 1997. In Siegel A. and Sigel H. (Eds) **Metal ions in Biological Systems**: vol. 34- Mercury and its effects on environment and biology. Marcel Dekker: New York. Chapter 19.

Hunter C. J. and Belt C. A. (Eds.). 1996. **Maintaining cultures for biotechnology and Industry.** pp. 263. Academic Press Inc.

Iyer S. R and Ramaiah N. 2002. Hydrolysis rates of Tween -80 and Tributyrin by marine bacteria. **In preparation.**

Izaki K. 1981. Enzymatic reduction of mercurous and mercuric in Bacillus cereus. **Can. J. Micrboiol.** 27: 192-197.

Jensen A., Rystad B. and Melson S. 1974. Heavy metal tolerance of marine phytoplankton. **J. Exp. Mar. Biol. Ecol.** 15: 145-157.

Joux F. and LeBaron P. 1997. Ecological implications of an improved direct viable count method for aquatic bacteria. **Appl. Environ. Microbiol.** 63: 3643-3647.

Kaiser K. L. E. and Ribo J. M. 1988. *Photobacterium phosphoreum* toxicity bioassay. II. Toxicity data compilation. **Tox. Assess. Internat.** 3: 195-238.

King J. M. H.; DiGrazia P. M.; Applegate B.; Burlage R.; Sanseverino J.; Dunbar P.; Larimer F. and Sayler G. S. 1990. Rapid, sensitive bioluminescent reporter technology for naphthalene exposure and biodegradation. **Science** 249: 778-781.

Kogure K.; Simudu U. and Taga N. 1979. A tentative direct microscopic method for counting living marine bacteria. **Can. J. Microbiol.** 25: 415-420.

Kogure K.; Simidu U.; Taga N. and Colwell R.R. 1987. Correlation of direct viable counts with heterotrophic activity for marine bacteria. **Appl. Environ. Microbiol.** 53: 2332-2337.

Kurfurst M., Ghisla S. and Hastings J. W. 1982. Bioluminescence emission of bacterial luciferase with FMN analogs in particular with 1-deaza-FMN. In Massey V. and Williams C. (Eds.) **Flavins and flavoproteins.** pp. 353-358. Elsevier, Amsterdam.

Labeda D. P. (Ed.). 1990. **Isolation of biotechnological organisms from nature.** pp. 321. McGraw Hill Publishing Company.

Lapota D., Galt C., Losee J. R., Huddel H. D., Orzech J. K. and Nealson K. H. 1988. Observations and measurements of planktonic bioluminescence in and around a milky sea. **J. Exp. Mar. Biol. Ecol.** 119: 55-81.

Larsen P. F., Huggett R. J., and Unger M. A. 1997. Assessment of organotin in waters of selected gulf of Maine estuaries. **Mar. Pollut Bul.** 34 (10): 802-804.

Lee J. 1982. Sensitization by lumazine protein bioluminescence emission from the reaction of bacterial luciferases. **Photochem. Photobiol.** 36: 689-697.

Macalady J. L., Mack E. E., Nelson D. C. and Scow K. M. 2000. Sediment microbial community structure and mercury methylation in mercury-polluted Clear Lake, California. **Appl. Environ. Microbiol.** 66: 1479-1488.

Makemson J. C. and Hastings J. H. 1986. Luciferase dependent growth of cytochrome deficient *Vibrio harveyi*. **FEMS Microbiol. Lett.** 38: 79-85.

Martinez J.; Smith D. C.; Steward G. F. and Azam F. 1996 Variability in ectohydrolytic enzyme activities of pelagic marine bacteria and its significance for substrate processing in the sea. **Aquat. Microb. Ecol.** 10, 223-230.

Misra T. K. 1992. Bacteria resistance to inorganic mercury salts and organomercurials. **Plasmid.** 27: 4-16.

Mohandass C., David J. J., Nair S., Loka Bharathi P. A. and Chandramohan D. 1997. Behavior of marine oil-degrading bacterial populations in a continuous culture system. **J. Mar. Biotechnol.** 5: 168-171.

Myrnes B.; Barstad H.; Olsen R. L and Elvevoll E. O. 1994. Enzymatic glycerolysis of minke whale oil. **Third Internat. Mar. Biotechnol. Conf. Prog. Abstracts.** 109.

Naganuma T. 1996 Differential enumeration of intact and damaged marine planktonic bacteria based on cell membrane integrity. **J. Aquat. Ecosys. Hlth.** 5: 217-222.

Nealson K. H. and Hastings J. W. 1979. Bacterial bioluminescence: its control and ecological significance. **Microbiol. Rev.** 43: 496-518.

Nealson K. H. and Hastings J. W. 1992. The luminous bacteria In: Balows, A., Trüper, H.G., Dworkin, M., Harder, W. and Schleifer, K.H. (Eds.). **The prokaryotes.** pp. 625-640. Springer-Verlag, New York.

Nealson K. H.; Wimpee B. and Wimpee C. 1993. Identification of *Vibrio splendidus* as a member of planktonic luminous bacteria from the Persian Gulf and Kuwait region with luxA probes. **Appl. Environ. Microbiol.** 59: 2684-2689.

Nies D. H. 1999. Microbial heavy metal resistance. **Appl. Microbiol. Biotechnol.** 51: 730-750.

Nilsson L.; Oliver J. D. and Kjelleberg S. 1991. Resuscitation *of Vibrio vulnificus* from the viable but nonculturable state. **J. Bact.** 173: 5054-5059.

Oliver J. D.; Hite F.; McDougald D.; Andon N. L. and Simpson L. M. 1995. Entry into, and resuscitation from, the viable but nonculturable state by *Vibrio vulnificus* in an estuarine environment. **Appl. Environ. Microbiol.** 61: 2624-2630.

Pahan K., Ray S., Gachhui R., Chaudhuri J. and Mondal A. 1990. Ecological and biochemical studies on MRB. **Ind. Env. Hlth.** 32 (3): 250-261.

Parsons T. R.; Maita Y. and Lalli C. M. 1984. In: **A manual of chemical and biological methods of seawater analysis.** pp. 75-80. Pergamon press.

Pollak P. E. and Montgomery W. L. 1994. Giant bacterium *(Epulopiscium fishelsoni)* influences digestive enzyme activity of a herbivorous surgeonfish *(Acanthurus nigrofuscus)*. **Comp. Biochem. Physiol.** 108A: 657-662.

Ramaiah N. and Chandramohan D. 1988. Distribution of luminous bacteria and bacterial luminescence in the equatorial region of the Indian Ocean. **Microbiologica.** 11: 243-254.

Ramaiah N. 1989. Ph.D thesis **Studies on Marine Luminous Bacteria.** pp 230.

Ramaiah N. and Chandramohan D. 1992. Ecology and biology of marine luminous bacteria in the Arabian Sea. In: Desai B. N. (Ed.) **Oceanography of the Indian Ocean.** pp. 11-23. Oxford IBH, New Delhi.

Ramaiah N. and Chandramohan D. 1993. Ecological and laboratory studies on the role of luminous bacteria and their luminescence in coastal pollution surveillance. **Mar. Pollut Bul.** 26: 190-201.

Ramaiah, N. and Chandramohan, D. (1994) Bacterial bioluminescence in marine pollution assessment In: Ocean technology perspectives, pp. 967-980 (Sushilkumar K., Agadi V. V., Das V. K. and Desai B. N., Eds.) Public Information Directorate, Council for Scientific and Industrial Research, New Delhi.

Ramaiah N., Raghukumar S. and Gauns M. (1996(a). Bacterial abundance and production in the central and eastern Arabian Sea; **Curr. Sci.** 71, 878-882.

Ramaiah N.; Raghukumar C.; Sheelu G. and Chandramohan D. 1996 (b). Bacterial abundance, communities and heterotrophic activities in the coastal waters off Tamil Nadu. **Ind. J. Mar. Sci.** 25: 234-239.

Ramaiah N.; Ravel J.; Straube W. L.; Matte M. and Colwell R. R. 2000. Detection of luciferase gene sequences in non-luminous bacteria from the Chesapeake Bay. **FEMS Microbiol. Ecol.** 34: 33-41.

Ramaiah N.; Kenkre V. D. and Verlekar X. N. 2002. Marine environmental pollution stress detection through direct viable counts of bacteria. **Wat. Res.** 36: 2383-2393.

Rath C. C. and Subramanyam V. R. 1996. Thermotolerant enzyme activities of Bacillus species isolated from the hot springs of Orissa. **Microbios** 86:157-161.

Ravel J.; Knight I. T.; Monahan C. E.; Hill R. T. and Colwell R. R. 1995. Temperature-induced recovery of *Vibrio cholerae* from the viable but nonculturable state: growth or resuscitation? **Microbiology** 141: 377-383.

Ribo J. M. and Kaiser K. L. E. 1987. *Photobacterium phosphoreum* toxicity bioassay. I. Test procedures and applications. **Tox. Assess. Internat.** 2: 305-324.

Robinson J. B. and Tuovinen O. H. 1984. Mechanisms of microbial resistance and detoxification of mercury and organomercury compounds: physiological, biological and genetic analyses. **Microbiol. Rev.** 48: 95-124.

Ruby E. G. and Nealson K. H. 1976. Symbiotic association of *Photobacterium fischeri* with the marine luminous fish *Monocentris japonica:* a model of symbiosis based on bacterial studies. **Biol. Bull.** 151: 574-586.

Ruby E. G. and Nealson K. H. 1977. A luminous bacteria that emits yellow light. **Science** 196: 432-434.

Ruby E. G. and Lee K. H. 1998. The *Vibrio fischeri-Euprymna scolopes* light organ association: current ecological paradigms. **Appl. Environ. Microbiol.** 64: 805-812.

Saouter E., Gillman M. and Barkay T. 1995. An evaluation of mer-specified reduction of ionic mercury as a remedial tool of a mercury-contaminated freshwater pond. **J. Ind.Microbiol.** 14: 343-348.

Sarkar A. 1994. Occurrence and distribution of persistent chlorinated hydrocarbons in the seas around India. In S. K. Majumdar, E. W. Miller, G. S. Forbes, R. F. Schamlz and Assad A Panah (Eds) **The Oceans: Physical-chemical Dynamics and human impact.** pp 444-458. The Pennsylvania Academy of Science.

Schlewe M. H., Hawk E. G., Actor D. J. and Krahn. 1985. Use of bacterial bioluminescence assay to assess toxicity of contaminated marine sediments. **Can. J. Fish. Aquat. Sci.** 42: 1244-1248.

Sigurgisladottir S.; Konraosdottir M.; Jonsson A.; Kristjansson J. K. and Matthiasson E. 1993. Lipase activity of thermophilic bacteria from Icelandic hot springs. **Biotechnol. Lett.** 15: 361-366.

Simkiss D. and Slawinski. 1985. Cellular responses of molluscan tissues to environmental metals. In: J. J. Steigman and G. W. Health (Eds.). **Responses marine organisms to pollutants.** pp. 103-118. Elsevier Applied Science Publishers, London.

Sreekumari K. R.; Chandrasekharan M. and Lakshmanaperumalsamy. 1988. Studies on lipolytic bacteria in stored fish *Etroplus suratensis* (Bloch). **Fish Technol.** 25: 64-68.

Stebbing A. R. D. 1985. Bioassay. In B. Bayne (Ed.) **The effects of stress and pollution in the marine animals.** pp. 133-140. Praeger Publishers, New York.

Stewart G. S. A. B. 1990. *In vivo* bioluminescence: New potentials for microbiology. **Lett. Appl. Microbiol.** 10: 1-8.

Swain E. B.; Engstrom D. R.; Brigham M. E.; Henning T. A. and Brezonik P. L. 1992. Increasing rates of atmospheric deposition in midcontinental North America. **Science** 257: 784-787.

Ulitzer S. 1986. Bioluminescence test for genotoxic agents. In S. P. Colowick and N. O. Klalpan (Eds.) **Methods in enzymology.** pp. 264-274. Vol. 133. Academic press, New York.

UNESCO. 1994. Protocols for the Joint Global Ocean Flux Study (JGOFS) Core Measurements, **IOC Manuals and Guides.** pp.170. 29, UNESCO, Paris.

Velankar N. K. 1955. Bacteria in the inshore environment at Mandapam. **Ind. J. Fish.** 2: 96-112.

Walker J. T. and Keevil C. W. 1993. Study of microbial biofilms using light microscope techniques. **Int. Biodet. and Biod.** 34: 223-236.

Weichart D.; Oliver J. D. and Kjelleberg S. 1992. Low temperature induced non-culturability and killing of *Vibrio vulnificus*. **FEMS Microbiol. Lett.** 100: 205-210.

Zilinskas; R.A., Colwell; R. R., Lipton; D. W. and Hill; R.T. (1995). In The Global Challenge of Marine biotechnology Status Report on the United States, Japan, Australia and Norway.224 -229 Maryland Sea Grant Publication.

Zillioux E. Z.; Porcella D. B. and Benoit J. M. 1993. Mercury cycling and effects in freshwater wetland ecosystems. **Environ. Toxicol. Chem.** 12: 2245-2264.

26

Conservation and Sustainable Management of Molluscan Resources Along Goa Coast, Central West Coast of India

***Dr. Yasmin Modassir & Sanitha Sivadas**

Native biodiversities a source of pride for each country, composing as it does a shining part of the national heritage. Anon

The rapid development and anthropogenic changes along the coastal regions call for measures to conserve the coastal regions, its fauna and flora. The South- West coast of India supports a wide diversity of niches and different landforms, each supporting different floral and faunal components. Biological productivity, which includes abundant, common, rare, endangered, vulnerable and threatened species along the coast line and in the estuaries is very important for coastal zone management.

Molluscs are one of the most conspicuous invertebrates of the animal kingdom and include such familiar forms as clams, mussels, oysters, squids, octopuses and gastropods. In abundance of species, mollusca is the second largest phylum after arthropods. It includes approximately 110,000 living species, mostly belonging to three major classes- Gastropoda, Pelecypoda and Cephalopoda. In addition, some 35,000 fossil species are known. Most molluscs are marine but many are found in brackish and fresh water and on land. The phylum name is derived from 'mollis' meaning soft, referring to the soft body within a hard calcareous shell, which is usually diagnostic. Aristotle in his famous *'Historia Animalium'* first coined the term to describe the various kinds of gastropods and cephalopods.

In India, molluscs occur in a wide range of habitats, characteristic of tropical zones. Molluscs occupy an important place in the affairs of mankind from time immemorial. Clams, mussels,

* Dept. of Zoology, Dhempe College of Arts & Science, Miramar, Panaji- Goa.—403 001 India

oysters and squids to mention a few among the molluscs, serve as food and are cheap source of protein. Most molluscan shells are used as ornament or shaped into a variety of household articles in all the maritime states. Their use as ornaments, utility articles and medicine has been widespread from ancient days. Dead shells washed ashore are collected for preparing lime and other multifarious purposes. Medicinal properties have been reported in some molluscs (Jones, 1970 Mukundan, 1968 and Chatterji personal communication). They also play destructive role in the marine environment causing damage to mud embankment with their burrows. Molluscan woodborers cause enormous ravage to wooden jetties, floating or submerged timber of the cargo boats or fishing crafts, and the rock-borers as the piddocks attack concrete piles in the harbour. The common men were so much inspired by these unique creatures that from ancient times molluscs have entered the folklore and legendary tales. The resources that can sustain regular and productive fisheries are abundant in the estuaries, backwaters and in coastal waters. Molluscs have already been identified as potential source for human consumption and along with fishes they can meet the country's quest for balanced and nutritious diet.

In Goa molluscan resources are abundant in the estuaries and coastal waters. Some of the economically important species sustain a regular fishery and contribute substantially to the economy of local fishermen. Very few species are commercially exploited in the state of Goa and several others are collected by the people for consumption. An estimated 887 tons of bivalves were reported to have been caught in 1991. However, the landing has been reduced to 632 tons in 1999 (Source: Directorate of Fisheries, Govt. of Goa). A serious problem in molluscan fishery has been over exploitation and mortality caused by man-made activities. Mining activities in Goa have also been reported to affect the bivalve population adversely (Paruleker *et al.* 1986). This has created a socio-economic problem for the local fishermen who depend upon molluscs for livelihood. From the view of molluscan biodiversity "conservation and management", a concept should be developed for sustainable utilization, which require following aspects to be looked into detail -

- Species composition and diversity.
- Faunal resources.
- Physico- chemical characteristics of the habitat.
- Public education by the way of holding training courses, preparing audio- video packages and publication on beach conservation by stopping over- exploitation of natural resources, especially protection of juveniles from being indiscriminately exploited.
- Identification of suitable sites along the coast for culture of molluscs.
- Developing indigenous technology for management of endemic species and provision for storage and marketing.
- Development of infrastructure to improve sustainability of the harvestable areas.

STATE OF GOA

Goa, is one of the smallest states in India located along the Central West Coast of India between the co-ordinate N 14° 54′ to N15° 48′ and E 73° 40′ to 74° 20′ E. It covers an area of about 3702-Km from north to south. The coastline stretches to a length of about 105-Km and from East to West, it is 65 Km. wide. Flanked between the continuous ranges of rugged hills constituting the Western Ghats on the east and the vast expanse of the Arabian Sea on the West, the region consists of a chain of high and prominent hills, characterized by deep gorges, steep valleys and ravines.

The state of Goa has one of the best waterways network. There are seven major rivers, namely Mandovi, Zuari, Terekhol, Chapora, Galgibag, Kumbarjua canal, Sal and Talpona (Fig 1). Mangrove vegetation is seen along the banks of all the rivers. All these rivers except the last two, which have local sources, rise on the slopes of the Western Ghats. Of all these rivers of Goa, Mandovi and Zuari are the two important rivers interspaced by Kumbarjua canal forming a major estuarine complex. Their basin covers 69% of the total geographical area of the state. The river Zuari after flowing through a stretch of 65 Km from the Dighi ghat joins the Marmugao Bay with wide opening between Marmugao and Cabo heads. Similarly, Mandovi River after traversing a stretch of 70 Km from the Parva ghat (Sahyadri hill) joins the Aguada Bay with wide funnel shaped opening in between Cabo and Aguada points. These rivers are fed by South-West monsoon rain and discharged from the catchement area. In the lower course of the Mandovi and the Zuari River several rich islands are formed, out of silt brought down by the rivers and deposited at their mouths.

Erosion and accretion features indicate that the coast is prograding along the beaches and retreating along the cliffs and headlands. Based on the disposition of landforms, the present coast is neither of emergent nature nor of submergent, but it is a combination of both. The continental shelf of Goa is relatively wide, with 50m contour depth occurring at 35 Km, 100 m depth contour at 80 Km and 200 m at 100 Km away from the coast. The seabed consists of silty clay up to 50 m water depth and sandy silt from 50 to 100 m water depths. Beach sediments mainly consist of quartz along with feldspar and other heavy minerals. They are represented by medium to fine sand, well to moderately sorted. The coastline of Goa is a scenic alteration of world famous beaches; headlands, bays, creeks, sea cliffs, promontories, pocket beaches, dunes, rocky headlands, or hard rock wave cut platforms, estuaries and river mouths.

The islands of Goa are of two types: the rocky islands like those of the Mormugao coast and the Piedade right in the Mandovi (and the Anjidiv too, belongs to this type), which are the protrusions of a drowned topography separated from the mainland. The other types are the alluvial islands like those of the Kumbarjua group created by the heavy deposition of silt by the river in their estuaries.

The oceanography of the Goa coast is influenced by three season's viz., Southwest monsoon (June - September), post monsoon period (October – January), and fair weather period (February – May). Sea would normally be very rough during monsoon season. On an average, Goa receives an annual rainfall of about 120 inches, over 90% of which is received in monsoon. The air temperature fluctuates between 22°– 38° C and the humidity is high at about 80%.

STUDY AREA

For the present study 37 beaches and seven major estuaries of Goa were surveyed (Fig 2). The study area is divided into sandy, rocky and muddy shore.

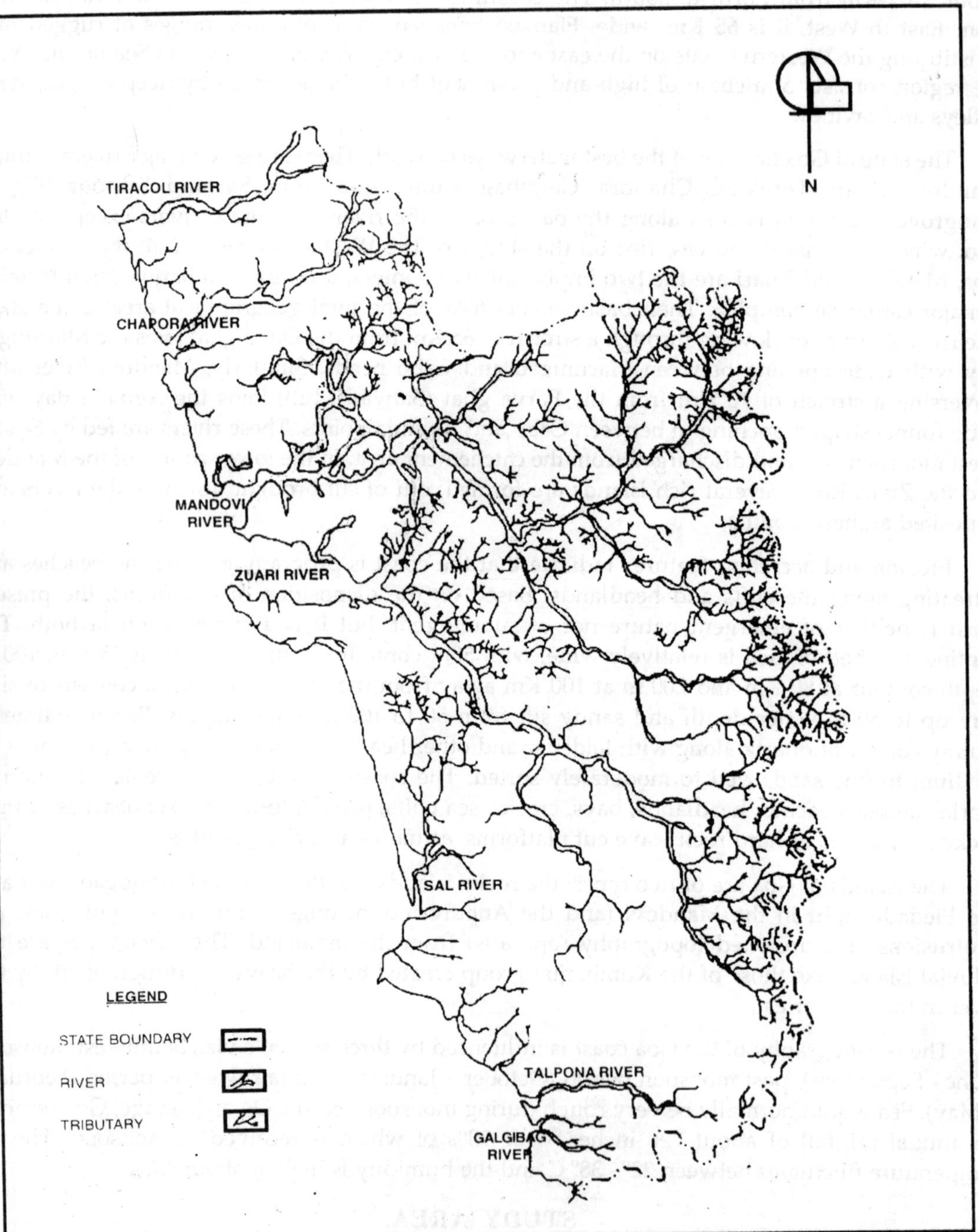

FIG. 1 : RIVER MAP OF GOA

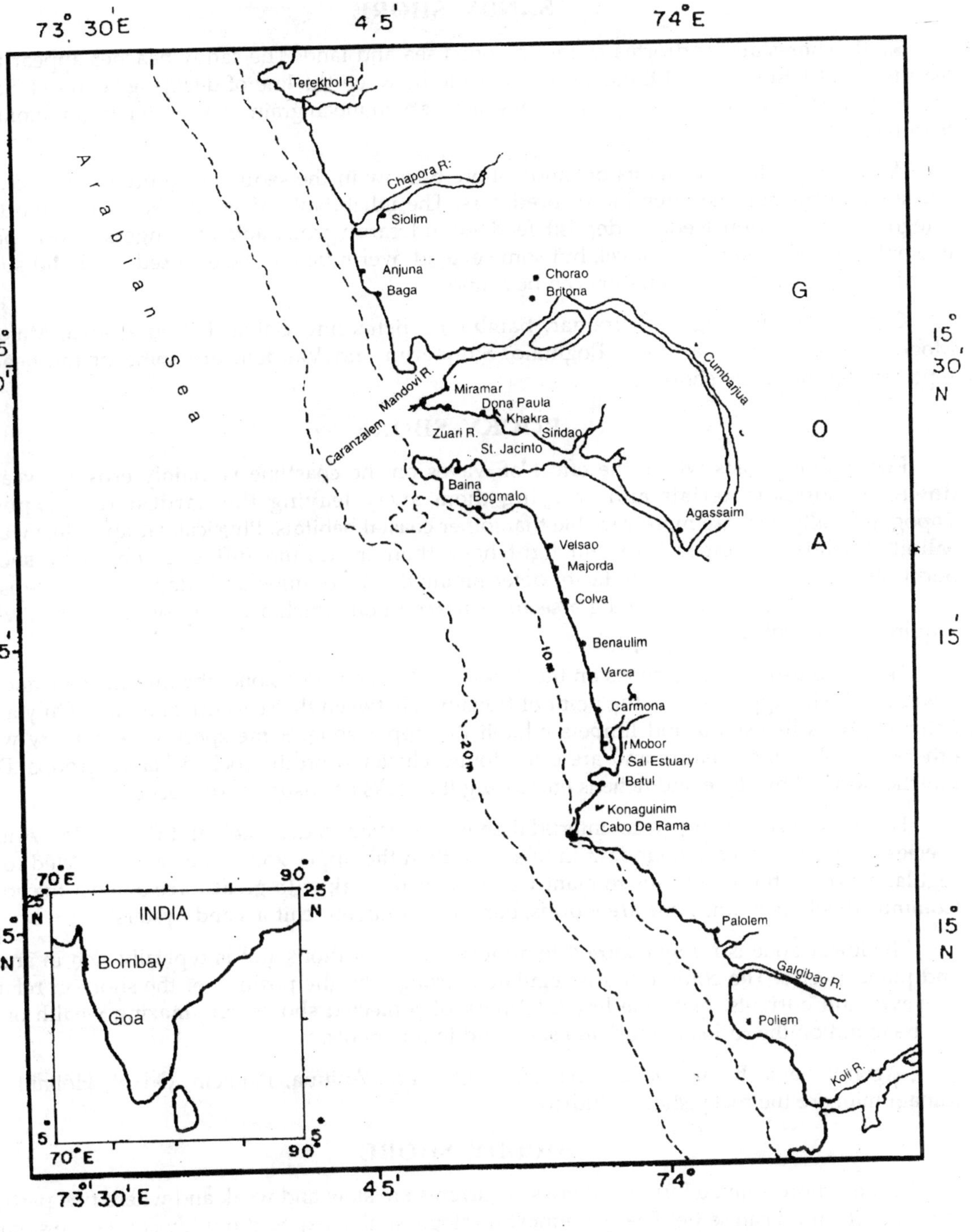

FIG. 2 : MAP OF GOA SHOWING THE AREAS UNDER STUDY

SANDY SHORE

Sandy shores are landforms facing the open sea and land. The sandy beaches appear to be featureless at first sight and there are no zones in between the line of decaying seaweed. Sandy shores do vary in slope aspect exposure; they also vary in mean grain size and other granulometric features.

At low tide, the inhabitants of sandy shore burrow in the sediment to prevent desiccation and avoid attracting the attention of predators. The inhabitants of sandy shores fall into three categories: suspension feeders, deposit feeders and carnivorous and scavengers. Most species live only near low tide water level, but some extend over much of the exposed sand and spread over in a way that distinct zonation can be found.

Caranzalem, Calangute, Miramar, Bambolim, Benaulim, Colva, Fatrade-Varca, Majorda, Mobor, Velsao, Zalor-Carmona, Bogmalo, Sinquerim and Vaddem are some of the beaches representing the sandy shore.

ROCKY SHORE

Rocky shore exists where the effect of waves on the coastline is mainly erosive, wearing down the softer materials and carrying them away leaving the hardest rock exposed. Topographically they are more variable than other coastal habitats. Physical variables like waves; salinity, isolation, temperature and light have their maximum influence on rocky shores. Substratum is the most important factor determining the distribution and adaptive characteristics. It is stable and permanent, forming a secure surface upon which a variety of organisms, which require attachment can grow.

There are usually three zones on the rocky coasts- the upper zone, the middle zone and the lower zone. The upper zone is the point of transition between the seashore and land. Only a few animal species like snails and limpets inhabit the upper zone. Some species are solitary while others are colonial by nature and are often found clustered on the rocks in large groups. Dark blotches formed by algae and lichens encrusting the rocks are usually conspicuous.

The middle zone is the 'true' intertidal zone, subjected to daily ebb and flood tides. Animal species are typically more abundant in this zone than the upper zone. They are subjected to the regular action of tides and thus to giant waves, which break with great force against the shore. Common dwellers in this zone are worms, barnacles, mussels, chitons and oysters.

The lower zone is rarely exposed to atmospheric conditions and is typically rich in animal and plant species. The community is regulated strongly by the position of the shore in relation to waves and currents. Near the low tide mark of protected shores, an amazing wealth of life can be found on the rocks, in mud and sand and in tide pools.

Agonda,Baga, Baina, Donapaula, Siridao,Khakra, Anjuna, Palolem, Odxel, Hollant and Kanaguinim are the rocky shores studied.

MUDDY SHORE

Muddy shores can be formed where sea currents are slow and weak and where fine particles that form the mud can settle. There is almost no slope; so that exposed muddy shores are suitably termed as mud flats because of relative lack of water movement. Such conditions are found in sheltered bays and in mouths of coastal streams. In these areas, fine organic particles are deposited to form muddy zone in which rich and varied animal life develops. As mud is more stable, animals that construct permanent tubes (Polychaeta) and burrows generally inhabit it. The members of the epifauna of mudflats are much the same as those of other intertidal zones. They are rather generalized types that tolerate a wide range of conditions occurring between low tide and high tide mark. A few groups of pelecypod molluscs are included among the epifauna of sandy mudflats. Several kinds of crustaceans inhabit the surface of mud and sandy mudflats. Some are highly valued as food. Mudflats are likely to contain starfishes, especially if oyster reefs are present.

Members of the infauna are somewhat specialized. They consist predominantly of Pelecypod molluscs, worms, few crustaceans and some fishes that spend at least part of their lives in under ground habitats. A number of researchers have found that particular group of burrowing animals are limited to various grades of soil, such as fine silt or coarse sand and to various proportions of sand and mud. Among the crustaceans adapted for burrowing in mud are few shrimps and crabs. Where certain animals construct tubes and burrows, other species are likely to find and occupy them. Thus, a high degree of commensalism develops among the various members of mudflat communities.

Muddy intertidal zones are difficult environments because of the fine texture of the substrate and the fact that oxygen, so necessary for life is likely to be in short supply. Food is also important in determining the number and kind of individuals present among the mud flat burrows. Nevertheless, the organic content does supply food to a great number of detritus feeders. Despite various difficulties, substantially more species inhabit mud flats than the intertidal zones of sandy shores.

In the state of Goa mangroves thrive along the banks of all the rivers. Total mangrove area is about 2,000 ha (Table 1). Mangrove vegetation consists of 15 species, belonging to 10 genera and 7 families. The dominant species are *Rhizophora mucronata, R. apiculata, Brugueira cylindrica, Kandelia candel, Avicennia officinalis, Avicennia alba, Sonneratia alba, Sonneratia caseolaris, Acanthus illicifolius* and *Excoecaria agallocha.*

Muddy shores include Brittona, Betul, Agacaim, Neura Nauxim Bay, St. Jacinto Island, Kumbarjua canal, River Chapora, River Terekhol, River Galgibag and River Sal.

Table—1 Rivers of Goa.

River	*Area (HA)*
Terekhol	30
Chapora	100
Mandovi	700
Sal	30
Zuari	900
Talpona	20
Galgibag	20
Kumbarjua Canal	200
Total Area	**2000**

PELECYPODA OF GOA

Bivalves constitute second largest and economically the most important class in the phylum Mollusca. Marine bivalves include oysters clams, mussels, scallops, ark shells and cockles, which form a regular fishery in the coastal region of India.

The present study showed that the state of Goa has a rich bivalve resource. 61 species of bivalves were identified during the present study. Of the three orders, Eulamellibranchiata was dominant over the other two orders and it includes nearly 66 % of the total number of species recorded in the present investigation. Bivalves were dominant on the sandy shores. Out of the total species collected 82 % have been identified up to species level.

The most widely distributed and abundant bivalves are *Meretrix casta, Paphia malabarica, Mactra violacea, Donax incarnatus, Donax cuneatus, Donax scortum* and *Sunetta scripta. Perna viridis, Crassostrea gryphoides* and *Saccostrea cucullata* are bivalves found on the rocky shore. *Polymesoda erosa* and *Anadara rhombea* represent bivalves of mangrove mud flats.

A list of bivlaves species identified (References sited) along with its distribution frequency and abundance is given in Table 2.

Table—2 Distribution Frequency of Bivalves Along the Goa Coast and A Comparison Showing Relative Abundance in different habits.

(- Absent; + Rare; + + Common; + + + Abundant)

Species	*Rocky*	*Sandy*	*Muddy*
1	2	3	4
ARCIDAE			
Arca gubernaculum (Reeve)	-	+	
Arca granosa (Lamarck)	-	+	-
Arca symmetrica (Reeve)	-	+	-
Arca tortuosa (Linn.)	-	+	-
Arca inaequivalvis	-	+	-
Arca complanta (Chemnitz)	-	+	-
Arca bistrigata (Dunkar)	-	+	-
Anadara rhombea (Born)	-	-	+ +
MYTILIDAE			
Perna viridis (Linn.)	+ + +	-	-
Modiolus metcalfei (Hanely)	+ +	-	-

(Contd...)

1	2	3	4
PECTINIDAE			
Pecten tranqebaricus (Gmelin)	-	+ +	-
Pecten splendiduls (Sowerby)	-	+	-
Pecten pyxidatus (Born)	-	+	-
Spondylus layardi (Chemnitz)	-	+	-
ANOMIIDAE			
Anomia achaeus (Gray)	-	+ +	-
Placuna placenta (Linn.)	-	-	+ + +
OSTERIDAE			
Crassostrea gryphoides (Scholotheim)	+ + +	-	-
Crassostrea madrasensis (Preston)	+ +	-	-
Saccostrea cucullata (Born)	+ + +	-	-
CARDITIDAE			
Cardita bicolor (Lamarck)	-	+ +	-
Beguina variegata (Bruguiere)	-	+	-
LUCINIDAE			
Codakia divergens (Philippi)	-	+	-
TRIDACNIDAE			
Tridacna elongata (Lamarck)	+	-	-
CARDIIDAE			
Cardium asiaticum (Bruguiere)	-	+ +	-
Cardium assimile (Reeve)	-	+ +	-
Cardium flavum (Linn.)	-	+ +	-
Cardium setosum (Redfern)	-	+ +	-
VENERIDAE			
Meretrix casta (Chemnitz)	-	+ + +	-
Merertix meretrix (Linn.)	-	+	-
Meretrix ovum (Deshayes)	-	+ + +	-
Paphia malabarica (Chemnitz)	-	+ + +	-
Paphia textile (Gmelin)	-	+ +	-
Sunetta scripta (Linn.)	-	+ +	-

(Contd...)

1	2	3	4
Sunetta donacina (Gmelin)	-	+	-
Gouldia divartica (Chemnitz)	-	+ +	-
Pitar erycina (Linn.)	-	+ +	-
Dosinia cretacea (Reeve)	-	+	-
Dosinia trigona (Reeve)	-	+	-
Chione scabra (Hanley)	-	+ + +	-
Chione tiara (Dillwyn)	-	+	-
Tapes radiatus (Chemnitz)	-	+ +	-
Catelysia opima (Gmelin)	-	+ +	-
MACTRIDAE			
Mactra turgida (Gmelin)	-	+ + +	-
Mactra violacea (Chemnitz)	-	+ + +	-
Mactra plicataria	-	+ +	-
DONACIDAE			
Donax scortum (Linn.)	-	+ +	-
Donax incarnatus (Chemnitz)	-	+ +	-
Donax cuneatus (Linn.)	-	+	-
Donax faba (Gmelin)	-	+	-
Donax spiculum (Reeve)	-	+	-
SEMELIDAE			
Semele crenulata (Sowerby)	-	+	-
TELLINIDAE			
Tellina ala (Hanley)	-	+	-
Tellina angulata (Gmelin)	-	+	-
Angulus sinuata (Melvill & Ambercromvie)	-	+	-
Gastrana polygona (Hanley)	-	+	-
SOLENIDAE			
Solen kempi (Preston)	-	-	+ + +
Solen truncatus (Wood)	-	+	-
Siliqua radiatus (Linn.)	-	+ +	-
PHOLIDAE			
Pholas orientalis (Gmelin)	-	+ + +	-
CORBICULIDAE			
Villorita cyprinoides (Gray)	-	+ + +	-
Polymesoda erosa (Lightfoot)	-	-	+ + +

GASTROPOD FAUNA

Of the six classes of Mollusca, the largest and the best known is the class Gastropoda, which contain species that have ventured permanently onto land. There are at least 30,000 species of univalves. Many of the gastropods like Turbo top shells, whelks, rock whelks, limpets, olives and chank are used as food by the people of Goa coastal region. Shells having bright colors and durable texture are used in the manufacture of buttons, brooches, buckles, and lampshades and in the interior decoration of houses.

Gastropoda was found to be the dominant class. 69 species belonging to three orders, 23 families and 40 genera have been identified (References sited) up to the species level. Both live and empty shells represent the collection. Gastropods show different pattern of distribution from the bivalves in being more dominant on the rocky shores. Of the three orders, Neogastropoda was the dominant with 28 species followed by Mesogastropoda with 25 species and Archeagastropod with 15 species. Among the Neogastropod, Muricidae was the dominant Family with 16 species.

Clypidina notata and *Cellana radiata* were the abundant on the rocky beaches. They were found in the upper zone of the rocky shore. Other species found include *Nerita albicella, Littorina undualta, Planaxis sulcatus, Thais bufo, Telescopium telescopium, Babylonia spirata, Umbonium vestiarium, Natica picta, Potamides cingulatus* and *Littorina scabra. Telescopium telescopium, Littorina scabra, Potamides cingulatus* and *Nerita crepidularia* are the gastropods of the muddy shore. *Oliva oliva, Umbonium vestiarium* and *Turritella duplicata* are gastropods representing the sandy shore.

A list of gastropods collected and identified up to the species level is given and described together with notes concerning their status and distribution. (Table 3).

Table—3 Distribution of Gastropod Fauna Along Goa Coast.

(- Absent; + rare; + + common + + + abundant)

Species	*Rocky*	*Sandy*	*Muddy*
1	**2**	**3**	**4**
TROCHIDAE			
Trochus radiatus (Linn.)	+ +	+ +	-
Trochus maculatus	+	-	-
Euchelus asper (Gmelin)	+ +	+	-
Umbonium vestiarium (Linn.)	-	+ + +	-
TURBINIDAE			
Turbo brunneus (Roding)	+ + +	-	-
NERITIDAE			
Nerita crepidularia (Lamarck)	-	-	+ +

(Contd...)

1	2	3	4
Nerita oryzarum (Recluz)	+	+	-
Nerita albicella (Linn.)	++	++	-
FISSURELLIDAE			
Diodora lima (Swerby)	+	-	-
Diodora bombyana (Swerby)	+	-	-
Scutus unguis (Linn.)	+	-	-
PATELLIDAE			
Cellana radiata (Born)	+++	-	-
Clypidina notata (Linn.)	+++	-	-
CYPRAEIDAE			
Cypraea arabica (Linn.)	+	-	-
Cypraea annulus ((Linn.)	+	-	-
C. moneta ((Linn.)	+	-	-
CYMATIIDAE			
Gyrineum natator (Roding)	+	-	-
STROMBIDAE			
Tibia curta (Sowerby)	-	-	+
BURSIDAE			
Bursa granularis (Roding)	+++	+++	+
Bufonaria rana (Linn.)	-	-	+
NATICIDAE			
Natica picta (Roding)	+	+	++
N. pulicaris	-	+	-
N. didyma (Roding)	-	+	+
Polinices mamilla (Linn.)	+	+	-
POTAMIDAE			
Cerithidae fluviatilis (Poteiz & Michaud)	-	+	-
Potamides cingulatus (Poteiz & Michaud)	-	-	++
Telescopium telescopium (Linn.)	+	-	+++
PLANAXIDAE			
Planaxis sulcatus (Born)	+++	-	

(Contd...)

1	2	3	4
TURRITELLIDAE			
Turritella terebra (Linn.)	+	-	-
T. duplicata (Linn.)	+	+ +	-
T. acutangula (Linn.)	+ + +	+ + +	-
T. atteunata (Reeve)	-	-	-
MITRIDAE			
Chrysame ambigua (Swanson)	+	-	-
LITTORINIDAE			
Littorina angulifera (Lamarck)	+	-	-
L. intermedia (Philippe)	+ +	-	-
L. scabra (Linn.)	+ +	-	+ +
MURICIDAE			
Murex trapa (Roding)	-	+ +	-
Murex virgenus (Sowerby)	-	+	-
Thais bufo (Lamarck)	+ +	+	-
T. rudolphi (Lamarck)	+	-	-
T. intermedia (Kiener)	+	-	-
T. tissoti (Petit)	+	-	-
T. carnifera (Lamarck)	+ +	+	+
Drupa margarticola (Broderip)	+	+	-
D. konkanesis (Blainville)	+	-	-
D. tuberculata (Blainville)	+ +	-	-
D. heptagonalis (Reeve)	+ +	-	-
Cronia contracta (Reeve)	+	-	-
C. subnodulosa (Melrill)	+	-	-
Morula granulata (Duclos)	+	-	-
Ocinebra bombayana (Melrill)	+	-	-
BUCCINIDAE			
Babylonia spirata (Linn.)	+	+ + +	+
Engina zea (Melrill)	-	+	-
Cantharus spiralis (Gray)	-	+	-

(Contd...)

1	2	3	4
TURRIDAE			
Surcula javana (Linn.)	-	+	-
S. amicta (Smith)	-	+	-
OLIVIDAE			
Oliva gibbosa (Born)	-	-	+
O. oliva (Linn.)	-	+ +	-
NASSARIUS			
Nassarius ornatus (Kiener)	+	+	+ +
Bullia lineolata (Wood)	-	+	-
PYRENIDAE			
Pyrene terpsichore (Sowerby)	-	+	-
VASIDAE	-	+	-
Xancus pyrenum (Linn.)	+	-	-
Hemifusus pugilinus (Born)	+	-	-

CEPHALOPOD FAUNA

They are so called because the foot is divided into a number of tentacle-like arms attached to the head. Many forms have an internal shell. It is blade-like and calcareous in the cuttlefish and thin, chitinous , feather-like in squids.

CLASS: CEPHALOPODA

SUBCLASS: DIBRANCHIATA

ORDER: DECAPODA

Family: Sepiidae

Genus: *Sepia*

Sepia aculeata (Orbigny)

Family: Loliginidae

Genus: *Loligo*

Loligo duvaucelii (Orbigny)

ORDER: OCTOPODA

Family: Octopodidae

Genus: *Octopus*

Octopus sp.

SCAPHOPOD FAUNA

The scaphopods are marine bilaterally symmetrical Molluscs, with the body and shell elongated and more or less cylindrical. They are commonly called the 'tusk shell'. Lees than 1,000 living species form the class. A single species–*Dentalium* sp. was collected during the present study.

CLASS: SCAPHOPODA

Family: Dentallidae

Genus: *Dentalium*

Dentalium sp.

PRESENT STATUS OF MOLLUSCAN FISHERY

Molluscs such as clams, oysters, mussels occur in appreciable quantities in the state of Goa and support a regular subsistence fishery. They are cheap source of protein, glycogen and other important minerals. Due to awareness created among the coastal people there is large consumption of molluscs. They are very much relished in Goa and are considered priority seafood item in Five Star hotels. The clam resources consisting species of *Meretrix, Paphia,* and *Villorita* are regularly exploited for food by the fisherfolk. Green mussel, *Perna viridis* forms a moderate resource and is under intense exploitation. Oysters are distributed throughout the state and are represented by *Crassostrea madrasensis, Crasssotrea gryphoides* and *Saccostrea cucullata*. An important observation made during the present study was that seven more species of bivalves, namely- *Mactra violacea, Sunetta scripta, Donax scortum, Polymesoda erosa, Anadara rhombea, Placuna placenta* and *Solen kempi* are consumed but are not commercially exploited. These species are found throughout the year except during the rainy season. Many people prefer these species to commercial ones as they find them tastier and are accessible to anybody. Cephalopods are caught as a by catch in fishing trawlers and it constitute 70 % of the total catch the remaining are taken by artisanal fishermen. The major part of the catch is exported. Gastropods are not very much exploited in Goa. A single species of fresh water snail, *Pila* sp. is the only gastropod marketed in the state. During the present study it was observed that two species of marine gastropods are collected by the people of Palolem. They are *Planaxis sulcatus* and *Trochus* sp.

Owing to overexploitation and other anthropogenic changes, the abundance of the molluscan resources especially those in the backwaters and estuaries have reduced and will be subjected to further damage, if left unchecked. Based on the data from the Directorate of Fisheries, Govt. of Goa, the bivalve production of the state for the year 1999 was about 632 tons. Table 4 shows the total production of bivalves in the last three years in state of Goa. Taking an overall view of these changes the future of the molluscs does not appear to be encouraging unless management and conservation measures are evolved and implemented. Therefore, proper management and utilization policy of the molluscs should be developed. Checking overexploitation, environmental monitoring, transplantation, culturing, legislation to govern shellfish export and market research are some of future management steps.

Table—4 Bivalve landing in tones in the state from 1997-1999 (Directorate of Fisheries)

Species	*1997*	*1998*	*1999*
M.casta	140	219	578
P.malabarica	69	322	48
Oysters	2	1	4
P.viridis	-	-	2

CULTURE PROSPECTS

The molluscs are mainly exploited for its meat. The present trend indicates very great interest in using the molluscan meat as diet. Accordingly the demand for bivalves is increasing. Being sedentary in nature, greater demand will lead to indiscriminate exploitation and depletion of natural resources. The exploitation of seed clams before attaining maturity is an important factor affecting the re-population of the clam's beds and the fishery during subsequent years. Another factor affecting the fishery is due to fluctuation in the natural beds. Therefore, the market demand can be met with by sustainable development of molluscan resources and by culturing bivalves.

Goa has rich bivalve resource. Sea mussels, edible oysters and clams are important among the cultivable species occurring in abundance in the varied water bodies of Goa (Ansari, 1978). There are a number of other species, which are collected and consumed by the local people.

The environmental conditions are suitable for molluscan culture in Goa waters. This can contribute to other objectives such as rural employment and improving socio-economic condition. Culture of the bivalves may prove to be a big step in regular production of the bivalves without affecting their natural beds.

CLAM CULTURE

Among the bivalves, clams form a regular fishery in the coastal region. Interest in culturing of clams species has grown in response to the decline in natural population. Intensive and semi-controlled clam culture has been attempted for almost three decades. In clam culture, the seed is generally collected from natural grounds and replanted in areas with suitable substratum, but where seed is not abundant. They are then allowed to grow to market size. In recent years, culture protocols have been established for hatchery, nursery and field growth of several clam species. Commercial site facilities have been developed to produce and supply seeds to the fisherman or to the grow seeds to marketable size. In general, they are amenable to high-density culture and species are suitable to the variety of sites available along the coast (Narasimham & Laxmilatha (1996).

MUSSEL CULTURE

In terms of production potential, mussel is one of the best resources of the sea and quantitatively it surpasses the production rate of any animal currently cultivated in land or sea. In addition, the mussels are known to grow at a faster rate in tropical waters.

In India, raft culture technique is practiced for mussel culture due to high yield rate of this culture system. In this technique the mussel seed is collected from natural beds and transplanted on a rope with the aid of mosquito netting. Mussels once submerged in seawater start growing and attach themselves to the rope with the help of byssus thread. The mussel grows at an average rate of 8-10 mm/month as against 3-5 mm in natural bed. The estimated production of mussel is 480 tons shell or (240 tons meat)/ha/year. The mussel production on rope is up to 23-kg/m. It is believed that mussel culture can replace shrimp farming in near future. Due to low investment and low recurring cost, mussel culture can be taken on large scale.

OYSTER CULTURE

Oysters are the most important among the cultured shellfish, as they are delicacy item. They are in great demand in foreign countries where they command high price.

From various studies carried out in institutes, a simple culture technique suitable for Indian conditions has been developed (Appukuttan. & Muthiah, 1996). The farming methods are broadly divided into-

1. On-bottom and
2. Off-bottom culture.

In on-bottom culture, the seed oysters are sown on the ground. This method is substrate-specific and the area sown should be free from silting and predators. The methods involved in off- bottom culture are-

1. Rack & tray
2. Rack & string
3. Stake and raft.

The technology of oyster farming followed in advanced countries is capital intensive. On the other hand, in India efforts have been directed towards development of low cost technology.

Advantages of Molluscs Culture

Molluscan culture has many benefits. They being sedentary in habit and feeding at lower level in food web they give high production per unit area. The culture techniques are very simple and use materials which are locally available hence can easily be adopted by fishermen. Culture techniques more over are eco- friendly, as fertilizers and artificial feeds are not required

MANAGEMENT STRATEGY FOR MOLLUSCAN FISHERY

The main objectives of fishery policy have been augmentation of fish production, increase in export of marine products and welfare of traditional fishermen. The development programme so far has not touched the molluscan resources. These resources have not been given any recognition except treating some of them like chank, pearl oyster and sub-soil shell deposits as they are revenue yielding items. Another factor affecting the management of molluscan fishery is that unlike other fish resources they are freely accessible to anybody who wants to exploit it. This has made it difficult to manage the resource because in the absence of property rights it is not possible to regulate and co-ordinate its use and excludes the free riders. Consequently, this resource has been over exploited in coastal region. Some of the management strategies for molluscs are presented below.

1. **Survey and estimation of total production of commercially important molluscan resources:**

Till date no serious scientific attempt has been made to estimate the production of molluscs, leaving them unaccounted. This situation arises from the fact that their exploitation is dispersed over a wide area and there is fluctuation of the resources in an area, the resource is accessible to anybody and production cannot be estimated in the absence of a system to cover such situation. The information available about the total production of bivalves in the state is only from the major fishing areas. Thus many areas, which have rich bivalves beds e.g. St. Jacinto Island where rich beds of *P. malabarica* are present has not been taken in to account since this is not a fishing area (Plate 4). Therefore, the data available about the total production of molluscs are not accurate.

A method, therefore is required to be developed to estimate the total bivalve production in Goa seasonally, as they are essential for formulating management policies. It is also felt that frequent surveys of all the estuaries should be conducted in different season, over the year to find out change in stock size in relation to environmental and biological factors. In order to stop the exploitation of these resources by everybody, rules of access and exploitation needs to be framed. The rules can be legitimized through legislative act. Co-operatives of molluscan fishermen can be set up for sustainable management of the molluscan resources. Such co-operatives will also help in improving the socio- economic condition of traditional fisher folk.

2. **Role of Eco-biological Parameters.**

It is a basic principle of fishery science that stock cannot be managed unless a solid base of biological information is available. Factors such as life span, age at maturity, fecundity, spawning, recruitment, population dynamics including mortality must be known. The harvest of mollusc species for shell collection and trade may lead to impoverishment or even extinction of population. Many species have become rare due to overexploitation. This information will be required for a sustainable development.

Environment largely influences the distribution and abundance of shellfish. In backwaters or estuarine ecosystem, the environmental parameters play a more decisive role than in an open bay or Gulf ecosystem. The rainfall, temperature, salinity, tide, discharge from industrial and agriculture all contribute to the quantity and quality production of shellfish. Methods have to be developed to check pollution.

3. **Marketing and Export of Molluscan Resources.**

Before encouraging production of bivalves on massive scale, the role and efficiency of various fish markets need to be studied. Marketing is a most important aspect in utilization of shellfish resources. Marketing survey has to be given priority to determine the turn over of shellfish marketed. Marketing of the molluscs are carried out using age old methods. Modern marketing methods have to be developed to improve the economic condition of fisher folk. This problem has to be solved and efforts are required to see that the fishermen get satisfactory price for their hard work. Market study is also important because it provides an overview of the domestic and potential export markets and identified some of the constraints.

The export sector has shown interest in exploitation in clam during the last five years purely from business motive. Table 5 gives the bivalve products exported from India in the last five years. In Goa, only the cephalopods are exported. In the year 1999 1256 t of cuttlefish and squids were exported from Goa (Directorate of Fishery, Govt. of Goa).

Table—5 Item wise export of bivalve products from India (MPEDA 1999)

Q: Quantity in tons V: Value in Rupees

Item	*1994*	*1995*	*1996*	*1997*	*1998*
1 Dried clam	Q: 16712	174626	0	0	0
Meat	V: 4406398	7007075	0	0	0
2 Dried Mussel	Q: 14880	83580	66196	14758	717
Meat	V: 246720	29421099	967058	671207	54198
3 Dried oyster	Q: 2336459	398000	0	1228032	511244
Shell powder	V: 4315336	637158	0	3146280	1848868
4 Sea Shells	Q: 83476	186620	744556	164069	363571
	V: 1126422	11817946	57195817	13030464	2015936
5 Boiled clam/	Q: 1272898	688389	657983	699686	16858
Whelk cockle	V:50848830	30186526	33334144	26446402	10907503
Meat					
6 Freeze Dried	Q: 0	0	0	500	410
Clam	V: 0	0	0	428640	596840
7 Frozen Mussel	Q: 58986	88015	12133	24773	20513
Meat	V: 2370384	533805	817938	1739989	1273227
8 Clam Meat	Q: 0	26820	0	0	0
Pickle	V: 0 1218191	0	0	0	
9 Canned Mussel	Q: 0	7250	0	0	0
Meat	V: 0	490157	0	0	0

4. Popularization of Post harvests Technology

Extension activities have special role in popularizing any food item. Training of farmers and entrepreneurs in mollusc culture and their popularization, educating the industry on post-harvest technology and promoting internal and external markets would go a long way to develop, establish and commercialize molluscan farming.

RECOMMENDATIONS

- It is important to create database about present status, biology and market demand.
- Method needs to be developed to survey the total production of molluscs in the state during different seasons. It will also help to find out changes in stock size in relation to environmental and biological factors.
- Conservation measures to protect the juveniles from being indiscriminately exploited.
- Steps to be taken to avoid degradation of the natural habitats.
- An extension service can be set up to popularize the molluscan products and the Government and financial institution may give financial support.
- Post- harvest technology has to be developed to suit local as well as international market.
- For culture on commercial scale, suitable sites have to be identified and made available to prospective farmers on lease. However, the allotments of coastal area for culturing should not at the same time come into conflict with other fishing activities.
- Prospect of exporting molluscs can be explored.
- Steps can be taken to commercialize the non commercial species
- Clam meat can be used as feed for shrimp farms, because of its high protein and low cost compared to artificial feed.
- Shells which are the by-product of culture can be used for preparing lime. The state at present depends on other state for its requirement.
- National Institute of Oceanography, Goa has successfully extracted hydrolysate from Green mussel, which has been found to have high anti -viral properties. The hydrolysate can be produced on commercial scale as the animal is found throughout the year in the state.
- Biological information of the bivalves found in the area is required for determining the scope of culture.
- Marketing surveys need to be conducted because it provides an overview of the domestic and potential export markets and identify some of the constraints.
- Workshops can be conducted to make the fishermen aware of the benefit of molluscan fishery.

SUMMARY

Molluscan fishery is an important part of fishery in the state and plays a major role in the economy of traditional fishermen. Pollution, over exploitation and other man made changes have caused decrease in the total production of molluscs in the last few years. This has created socio-economic problem for the local fishermen community who depend on molluscs for their livelihood. Until date, no effective measures have been taken for the management of the molluscan resource. No reliable data is available on their distribution, their total production and biology of the economically important molluscs, which are essentially for formulating management strategies. The objectives of the present investigation was to study provide data about occurrence of molluscan resources of Goa, identify the economically important molluscan species, their biology, biochemistry and propose future management strategies.

The demand for molluscs in Goa has led to exploitation of juveniles affecting their re-population. To solve this problem the bivalves can be cultured. Initially the culture can be carried out on small scale and if successful, it can be cultured on large scale. Export of the bivalves can also be taken up at a later stage.

Molluscan resources are abundant in the estuaries and coastal waters of the state. Seven species of bivalves are commercially exploited. An important observation during the present study was that seven more species of bivalves and two species of gastropods are collected by the local people for consumption. These species are found throughout the year barring rainy season. Many people prefer these species to commercial ones as they find them tastier and they are easily accessible to everybody.

In view of this, there is a need for greater awareness of these sedentary molluscs. Biological information of the oysters, clams and mussels found in the areas is required for determining the scope of culture. The base-line resource information will require to be periodically updated taking into account recruitment and exploitation.

It is hoped that with the technical know how, bivalve farming will be taken up so that the resources available along the coast are fruitfully utilized to augment marine fish production. This will also help in improving the socio-economic condition of molluscan fishermen.

REFERENCES

1. Ansari Z.A (1978): Fishery and culture possibilities of Molluscs of Goa. *Seafood Export Journ.* Vol. 10(4): 1-3.
2. Appukuttan K.K. & Muthiah P. (1996): Technology of edible oyster culture. *Bull. Cent. Mar. Fish. Res. Inst.* Vol 48: 64-69.
3. Apte Deepak (1998): The Book of Indian Shells, Bombay Natural History Society, Oxford University Press. (Identification)
4. Cephalopod Bionomics, fisheries & Resources of the exclusive economic zone of India (1985). Edited by E.G. Silas. *CMFRI Bull*: 37. (Suggested Reading)
5. Hornell J. (1921): Indian Molluscs, *J. Bombay nat. Hist. Soc.*Vol. 48: 543-569. (Identification)
6. Indo Pacific Mollusca (1965). Vol 1. The Delware Museum of Natural History Publication. (Identification)
7. Indo Pacific Mollusca (1969). Vol 2 (10). The Delware Museum of Natural History Publication. (Identification)
8. Indo Pacific Mollusca (1973). Vol 3 (13). The Delware Museum of Natural History Publication. (Identification)

9. Ingole B.S., Krishna Kumari L., Ansari. Z.A., Paruleker A.H (1994): New record of mangrove clam Geloina erosa (Solander, 1786) from the west coast of India, *J. Bombay. nat. Hist. Soc.* Vol. 91: 338-339. (Identification)

10. Jones S (1970): The Molluscan fishery resources of India. Proc. Symp. Mollusca. *Mar. Biol. Ass.* India, part 3: 906-918.

11. Kundu .H.L (1962 a), On the marine fauna of the Gulf of Kutch , Part-III Pelecypods, *J. Bombay. nat. Hist.* Soc. Vol. 61 . (Identification)

12. Kundu H.L (1962 b), On the marine fauna of the Gulf of Kutch, Part-III Pelecypoda, *J. Bombay. nat. Hist. Soc.* Vol.62(2): 210-235. (Identification)

13. Menon P.K.B, Gupta A.K. D, Gupta A.K.D, Gupta D.D: On The Marine Fauna of Gulf Kutch, PartII-Gastropoda. *J. Bombay nat. Hist. Soc.*, 58(2). (Identification)

14. Menon P.K.B., Dutta A.K., Gupta D.D: On the marine fauna of Gulf of Kutch Part II- Gastropoda. *J. Bombay. nat. Hist. Soc.* Vol. 8 (2): 475-494. (Identification)

15. Mookherjee. H.P. (1985): Contribution to the knowledge of molluscan fauna of India, part III, Marine molluscs of the Coramandel coast, Palk Bay & Gulf of Mannar- Gastropoda: Mesogastropoda (part 2). *Rec. Zool. Surv. India*, Occ. Papers no. 75. (Identification).

16. Mukundan C. (1968): Molluscs in Indian Tradition and Economy. Proc. Symp. Mollusca. *Mar. Biol. Ass.* India, Part 1.

17. Narasimham K. A., Laxmilatha P. (1996): Clam culture. *Bull. Cent. Mar. Fish. Inst.* Vol. 48: 76-87.

18. Oyster culture- Status & Prospects (1987). Ed. K. Nagappan Nayar & S. Mahadevan. *Bull. Cent. Mar. Fish. Inst.*:38. (Suggested reading).

19. Paruleker A.H., Ansari Z.A., Ingole B.S (1986): Effect of mining activities on the clam fisheries and bottom fauna of Goa estuaries. *Proc. Indian Acad. Sci. (Anim. Sci.)*, Vol. 95, No. 3: 325-339

20. Perera N.M.P.J (1975): Taxonomic study of the Cephalopods, particularly the Teuthoidea (Squids) and Sepoidea (Cuttlefish) in the waters around Sri Lanka. *Bull. Fish. Res. Stn., Sri Lanka*, Vol. 26. No. 1&2: 45-60. (Identification)

21. Pillai C.S.G & Appukutan K.K (1978): Distribution of molluscs in and around the coral reefs of the southern coast in India. *J. Bombay. nat. Hist. Soc.* Vol. 77: 26-48.(Suggested reading)

22. Rao K. V, Narasimham K. A & Algarswamy K (1962): Fishery of the Razor shell. Solen kempi (Preston) from Ratnagiri in Maharashtra state. *Ind. J. Fisheries*, 9(2) section A: 542-577.(Suggested reading)

23. Satyamurthi T. (1952): The mollusca of Krusdai Island I Amphineura & Gastropoda. *Bull. Madras. Govt. Mus.* (N.H.), I (2) 6. (Identification)

24. Satyamurti S. Thomas (1956): The Mollusca of Krusadai Island(in the Gulf of Mannar)II-Scaphopoda, Pelecypoda, and Cephalopoda., *Bull. of Madras government Museum.* Vol. I.(2). (Identification)

25. Silas E.G., Sarvasan R., Rao K.S., Prabhakaran K & Meiyappan M.M (1992): Identification of Common species of Cephalopods in India. *Bull. Cent. Mar. Fish. Inst.* Vol. 35. (Identification)

26. Subba Rao N.V. & Dey A. (1984): Contribution to the knowledge of Indian Marine molluscs I. Family Mitridae. *Rec. Zool. Surv. India*, Occ. Papers no. 61: 1-43. (Identification)

27. Subba Rao N.V. & Dey A. (1986): Contribution to the knowledge of Indian Marine molluscs I. Family Donacidae. *Rec. Zool. Surv. India*, Occ. Papers no. 91: 1-30. (Identification)

28. Subba Rao N.V. & Surya Rao K.V. (1993): Contribution to the knowledge of Indian Marine molluscs I. Family Muricidae. *Rec. Zool. Surv. India,* Occ. Papers no. 153: 1-233. (Identification)

29. Subba Rao N.V. (1971): On the collection of Stromboidae (Mollusca: Gastropoda) from Bay of Bengal, Arabian Sea & Western Indian Ocean with some new records. 1. Genus–Strombus. *J. mar. Biol. Ass. India.* Vol. 12 (182): 109-124. (Identification)

30. Witson B. (1993). Australian Marine Shells–Part 1 & 2. Odyssey Pub. Australia. (Identification)

31. Wye K. (1991). Illustrated Encyclopedia of Shells. Simon & Schuster Inc. N.Y. (Identification)

27

Response of Benthic Fauna to Different Pollutants: Some Case Studies and Relevance of Benthos to Environmental Impact Assessment

Z. A. Ansari, B.S. Ingole and Ramila Furtado

With rapid rate of urbanization, industrial growth and varieties of anthropogenic changes, the use of biological parameters as monitors of environment is being increasingly appreciated. As compared to plant, the animal communities have been used to a greater extent in toxicity tests and pollution monitoring programme. Biological monitoring thus precisely aims at locating adverse effects in terms of damage done to biota as a consequence to environmental deterioration caused by pollutants. The community level approach to the detection and monitoring of biological effects has been validated (Austen et al, 1994). Chemical analyses, although, valuable do not permit the assessment of combined impacts of several individual factors. Biological analyses on the other hand, permit realistic assessment and cover a wide range of possibilities (Agard et al., 1993).

Thus understanding of the mechanisms of ecological functions/parameters responsible for seasonal fluctuations in biological communities is an essential prerequisite in marine ecological monitoring programme (Christie 1980). It has been stated that monitoring fluctuation is easy, but without an understanding of causal mechanism only the most obvious effects of pollution will be observed (Lewis, 1976).

Once we realize and accept the importance of monitoring, it is necessary to decide about the suitable parameters to be monitored. If a large part of the variance of a parameter is found in the high frequencies of space and /or time, that parameter in general is unfit for monitoring. Benthos is a distinct entity among the biological components of marine ecosystem, directly related

National Institute of Oceanography, Dona Paula, Goa 403 004, India.

with bottom sediment. The distinction of benthic organisms lies in the sedentary and sessile life. Benthic fauna is represented by almost all invertebrate phyla. Seasonality in shallow water benthic fauna is a common feature (Parulekar et al, 1980), whereas the subtidal benthos show more uniformity and steady distribution unlike planktonic and nektonic forms. This is possible because of more or less stable subtidal benthic environment. Therefore, the qualitative and quantitative abundance in space and time will be suitable parameters to understand ecological functioning and monitoring of environments. Heip (1980) has proposed the use of meiofaunal density and biomass as tool in monitoring studies. Under similar considerations Warwick (1986) suggested the use of distribution of number of individuals among species (abundance) and their respective biomass in detecting pollution effects.

Benthos has a good history as bonafide community of environmentally disturbed ecosystems. They are the first casuality of any environmental disturbances. Benthic community structure is therefore being widely used in pollution monitoring programmes but often with equivocal results (Warwick, 1986). This paper summarizes results of some case studies of benthos under different stress conditions and suggest the distinctness of benthic data as tool in environmental impact assessment studies.

Municipal discharge and long – term changes in benthic community:

Discharges from Mumbai municipality find their way into the Arabian sea off Mumbai. Monitoring of the bottom communities for ten years in relation to water quality indicated major changes in the distribution, abundance and composition of macro invertebrates in the coastal waters of Mumbai. Salinity of bottom water (35.1 – 36.3 ppt) did not show great variations where as dissolved oxygen exhibited wide variation (0.76 to 5.37 ml/l) in the bottom waters. Dissolved oxygen decreased with passage of time. The values were higher in early 1970s as compared to 1980s. Such abnormally low DO distribution in Mumbai water has been reported (Zingde, et.al, 1979). Differences in both qualitative and quantitative distribution of benthos were observed (Table 1).

Table—1 Long-term changes in average population density (P, n/m²) and biomass (B, g/m²) of macrobenthos off Mumbai in different years.

Depth (m)	*1974*		*1976*		*1977*		*1981*		*1983*	
	P	**B**	**P**	**B**	**P**	**B**	**P**	**B**	**P**	**B**
0-20	117	12.5	169	21.65	28	0.24	89	0.18	nil	nil
21-40	297	2.94	1537	14.26	174	2.03	30	0.15	13	0.12
41-60	1094	22.30	1872	15.17	493	8.64	1302	11.65	236	4.04
61-80	406	6.94	1464	8.97	325	6.05	337	3.34	436	4.98
81-100	-	-	-	-	177	2.72	-	-	-	-
>100	-	-	-	-	174	11.02	-	-	-	-

(-) indicate no data

The macrobenthos showed highest population density of $1872/m^2$ in 40-60 m depth zone in 1976 and the minimum was $13/m^2$ during 1983 in the depth zone of 20-40 m (Table 1). A gradual decline in faunal abundance was noticed from 1974 through 1983. A decrease of more than 50% in the density was observed during these seven years particularly in the shallow depths. Major changes in the dominant macrofaunal group in the near shore region was also noticed. In 1974 the macrobenthic assemblage consisted of polychaetes, bivalves and crustaceans whereas in 1983 only polychaetes were encountered. Such differences in the faunal composition could be a direct consequence of the major environmental changes, including dumping of the municipal and industrial wastes, leading to extreme deterioration in the marine environment of the coastal waters of Mumbai (Anon., 1975). Shallow coastal regions receiving such high inputs of municipal sewage have been reported to undergo drastic changes in the composition of benthic fauna, as observed in the present study, due to significant increase in the sediment organics (Bozzini, 1975, Bhosle et al, 1983). This has resulted in the dominance of more resistant species of polychaete worms, especially deposit feeders, which are reported to be the biological indicators of organic pollution (Ansari et al, 1986) . The offshore regions, in contrast, showed no major time bound changes in benthic fauna.

In less than 10 years the abundance of macrobenthic fauna in the coastal waters off Mumbai has decreased both qualitatively and quantitatively by more than 50%. Such an impoverishment in the distribution and relative abundance of benthic fauna indicates drastic deterioration of the environment and could be explained with reference to high magnitude of sewage and industrial pollution in the coastal waters of Mumbai (Zingde et al 1979, Gajbhiye et al 1981). In such environmental conditions the more resistant and opportunistic species such as deposit feeding polychaetes will thrive and predominate the benthic population. These opportunistic species can be used as biological indicators of organic pollution (Mearns and Greene, 1979).

It can reasonably be surmised that environmental disturbances arising out of increasing anthropogenic perturbations are the important causative factors responsible for such impoverishment in benthic fauna of the coastal waters at Bombay. While assessing the pollution effect on ecological changes seasonality in the occurrence of benthic organisms particularly in the shallow coastal waters should be taken into account. A detailed knowledge about response of individual species to varying environmental stress is important to quantify the effect and could be used for monitoring the specific cases (Grovaere et al, 1980). Thus it is suggested that benthic biomass and density can effectively and pragmatically be used as monitoring parameters of environmental impact assessment.

ORGANIC ENRICHMENT AND BENTHIC FAUNA

The introduction of organic rich wastes such as domestic sewage, trash fish and washings of the fishing fleets causes significant changes in benthic, meio and macrofaunal community structure in the intertidal and subtidal habitats (Goving & Huling, 1976; Ansari et al, 1986). Response of benthos appears to be related to distance of the source of organic pollution (Ansari et al, 1984). Generally three zones are recognized: grossly polluted, polluted and unpolluted. However, a fourth zone called enriched zone was recognized by Pearson and Rosenberg (1976). An enriched zone is that zone located at short distance from the polluted zone where the faunal abundance rise dramatically. The increase in number is caused by extremely abundant population of one or two opportunistic species. It has a spatially narrow distribution and the number rapidly decrease with increasing distance from pollution point.

The effect of the impact of organic enrichment on meio and macrobenthos was studied in the Mandovi estuary. The meiofaunal population of the sandy beach receiving untreated domestic sewage responded in a fashion (Fig. 1) similar to that reported earlier (Goving & Hulings, 1976). The effect was most pronounced in the vicinity of discharge and decreased progressively with increasing distance from the discharge point, as expected. The population showed stability in the density at about 400 m away from the discharge point. The moderate quantity of untreated sewage discharged into Mandovi estuary, was found to enhance the production of distinct category of organisms called "opportunistic species" among meiobenthic groups, in the vicinity of the discharge point as reported earlier (Gray and Ventilla , 1971; Pearson and Rosenberg, 1978).

Fig. 1 Effect of sewage pollution on beach meiofauna

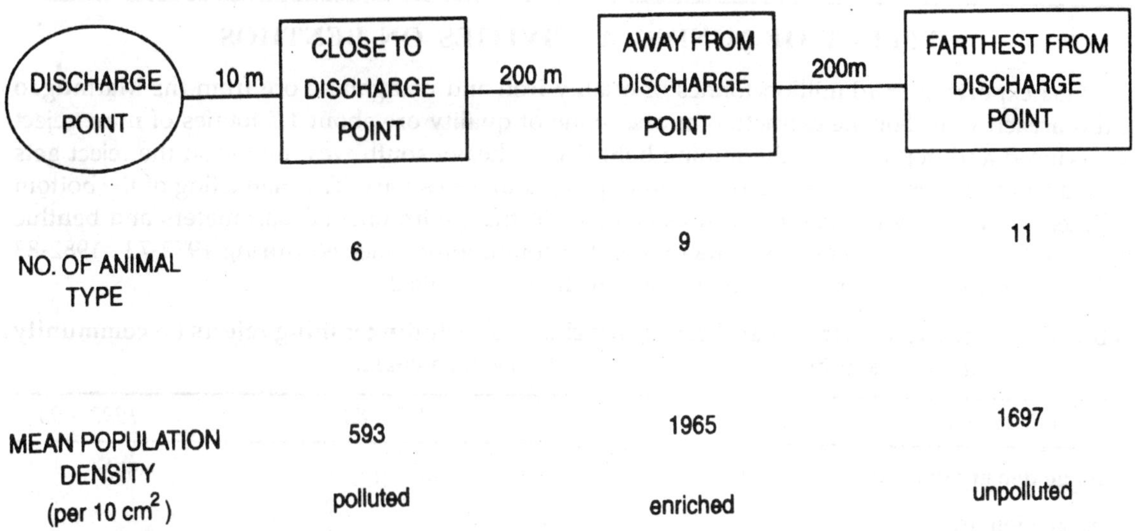

The effect of organic rich effluent whether harmful or beneficial to benthic community is determined by the circumstances of the specific situation. It has been reported that gradual but consistent input of organics to the sediment may increase the rate of carbon assimilation and influence the benthic productivity (McIntyre, 1977).

A comparison of the macrobenthic community data and its distinctness at two sites, one receiving moderate input of organic enrichment (polluted) and the other without any input (unpolluted) was made. The biological community data is given in Table 2 which re-emphasized that low organic enrichment enhances the benthic productivity at taxa level. In such cases some species take advantage of the changing condition and proliferate abundantly and form large population. *Dendronereis aestuarina, Prionospio polybranchiata,Perinereis nuntia* and *Lumbriconereis polydesma* were four opportunistic polychaetes which recorded very high density in organically rich sediment at polluted site and they can be considered as opportunistic species in this case.

This corroborates the earlier findings (Dauer & Conner,1980). The distinct benthic level organization while undergoing major changes in community parameters such as density and biomass, shows richness of the opportunistic species (Mearns & Greene, 1979). Benthic taxonomic distinctness encompassing genetic diversity is high in species which characterize unstable stressed environment, as opposed to their counterparts from more stable conditions (Nevo et al, 1984). It has been demonstrated therefore that benthic community in organically enriched area are kept in a successional stage from the source of pollution (Pearson & Rosenberg, 1978).

Table—2 Comparison of data on macrobenthos from a polluted and unpolluted area in Mandovi estuary.

	Biomass g /m²	*Density No /m²*	*No of species*	*Diversity*	*Evenness*
Polluted	1.19 ± 0.95	947 ± 284	9.82 ± 1.80	2.79 ± 0.6	2.8 ± 0.3
Unpolluted	0.35 ± 0.27	476± 162	7.25 ± 1.04	2.25 ± 0.6	2.27 ± 0.2

EFFECT OF MINING ACTIVITIES ON BENTHOS

Goa exports over 10 million tonnes of quality iron and manganese ore from the Marmugao harbour each year. For the extraction of one tonne of quality ore about 1.5 tonnes of mine reject is produced and deposited in the mining belt. Due to heavy south-west monsoon the reject gets eroded and transported down the river and deposited in the estuary. The blanketing of the bottom sediment by this eroded material causes changes in the environmental parameters and benthic community. A comparison of the data on benthic community collected during 1972-73 , 1982-83 and again in 1992-93 has revealed major faunal changes (Table 3).

Table—3 Long term effect of anthropogenic changes including mining rejects on community data of macrobenthos in Mandovi estuary, Goa coast.

Community parameters	*1972 - 73*	*1982 – 83*	*1992 – 93*
Average density (No /m²)	1604	722	572
Average biomass (g /m²)	17.79	9.37	5.71
Number of species	111	69	47
Diversity index	4.49	3.82	2.29
Dominant faunal group	Polychaeta, Bivalvia	Polychaeta, Crustacea	Polychaeta, Crustacea

There was clear cut reduction in both the qualitative and quantitative distribution of benthos. The effect of iron and manganese on benthos vary greatly. The macrofaunal density of 1604 recorded in 1972-73 had reduced to 572/m² in 1992-93 and the biomass showed over 70% reduction in the same period. The reduction was mainly attributed in the disappearance of clam, *Meretrix casta* in the estuary. The biomass of this clam recorded in 1972-73 was more than 0.5 kg while in 1992 it was less than 100 g. The high suspended load due to washing of the mining reject could clog the gills of the clams, resulting in heavy mortality (Parulekar et al, 1986). Out of 111 species recorded in 1972-73 only 47 species were observed in 1992-93. It has been reported that macro-invertebrates are sensitive to changes in the environment as they integrate the combined effect of many stresses through changes in their ecology, morphology and physiology (Ketchum, 1972). As a consequence of the major changes in water quality the dominant filter feeding and detritus feeding species of 1972-73 have been replaced, by scavenger or migrant species comprising of opportunists such as amphipod, alpheids and errant polychaetes.

Table—4 Percentage composition of macrobenthic feeding types

Year	*Filter feeders*	*Detritus feeders*	*Carnivores*	*Scavengers*
1972-73	33.5	28.9	12.5	25.1
1982-83	26.5	14.3	8.6	50.6
1992-93	13.8	8.3	6.6	71.3

From the foregoing account it is concluded that pollution greatly affect benthos leading to either general impoverishment under severe condition or enhancement in fauna under moderate enrichment. It is well recognized that marine organisms integrate the combined effect of pollutant and therefore biological monitoring particularly with distinct benthic assemblages has its merit. The manifestation of marine pollution can best be studied in a generalized stress response among benthic communities using density and biomass criteria and this can be recorded through reduction in number of sensitive species and increase in the number of tolerant and opportunistic species. Hence changes in species composition of benthic community may be conveniently used as sensitive sentinels of pollution.

ACKNOWLEDGEMENT

The authors are thankful to Dr. E. Desa, Director of N.I.O., Goa for his encouragements. This is contribution No. 3731 from NIO.

REFERENCES

Agard, J.B.R., J. Gobin and R.M. Warwick, 1993. Analysis of marine macrobenthic community structure in relation to oil pollution, natural oil seepage and seasonal disturbance in a tropical environment. Mar. Ecol. Prog. Ser., 92:233-243.

Ansari Z.A., A. Chatterji and A.H. Parulekar, 1984. Effect of domestic sewage on sand beach meiofauna at Goa, India. Hydrobiologia, 111, 229-233.

Ansari, Z.A. , B.S. Ingole and A.H. Parulekar, 1986. Effect of high organic enrichment on benthic polychaete population in an estuary. Mar. Poll. Bull., 17, 361-365.

Anonymous, 1975. Oceanographic and waste water disposal studies around Bombay. NIO Project Report, 23 pp (mimeo).

Austen, M.C., A.J. McEvoy and R.M. Warwick, 1994. The specificity of meiobenthic community responses to different pollutants: Results from microcosm experiments. Mar. Poll. Bull. 28:557-563.

Bhosle, N.B., C. D'Silva and C.V.G. Reddy, 1983. Lipid content in surface water of Arabian Sea. Indian J. Mar. Sci., 12:242.

Bozzini, G., 1975. Enrichment phenomena and danger of eutrophication in marine environment. In: E.A. Pearson and De Fraja Franjipane (ed.) Marine pollution and waste disposal. Pergamon Press New york. pp.193-196.

Christie, H. 1980. Methods for ecological monitoring: Biological interactions in a rocky sub-tidal community. Helgolande Meeresunters, 33: 367-374.

Dauer, D.M. and W.G. Conner, 1980. Effect of moderate sewage input on benthic polychaete population. Estuar. Coast. Mar. Sci., 10, 335-346.

Gajbhiye, S.N., Jiyalal Ram and B.N. Desai, 1981. distribution of copepods from the polluted and unpolluted regions off Bombay. Indian j. Mar. Sci., 10:346-349.

Goving, M.M. and N.C. Hulings, 1976. A spatial study of meiofauna on a sewage polluted Lebanese sand beach. Acta Adriatica, 18, 341-362.

Gray, J.S. & R.S. Ventilla, 1971. Pollution effects on micro and meiofauna of sand. Mar. Poll. Bull., 2:39-43.

Grovaere, J.C.R., d. Van Damme, C. Heip and L.A. De Coninck, 1980. Benthic communities in the southern Bight of the North Sea and their use in ecological monitoring. Helgolander Wiss Meeresunters., 33:507-521.

Heip, C. 1980. Meiobenthos as tool in the assessment of marine environmental quality. Rapp. P.v. Reun. Cons. Int. Explor. Mer., 179:182-187.

Ketchum, B.H., 1972. In: Special publication dedicated to Dr. N.K. Pannikar. Mar. Biol. Asso. of India, 7-28.

Lewis, J.R., 1976. Long term ecological surveillance: practical realities in the rocky littoral. Oceanogr. Mar. Biol., 14:371-390.

McIntyre A.D., 1977. Effect of pollution on inshore benthos. In B.C. Coull (ed.) Ecology of marine benthos (University of south California Press.) pp. 301-312.

Mearns, A.J. and C.S. Greene, 1979. Composition of benthos at several waste discharge point. In coastal water research project, Southern California, pp. 211-218.

Nevo, E. A. Beiles and R. Ben-Shlomo, 1984. The evolutionary significance of genetic diversity: ecological demographic and life history correlates. In: Mani G (ed.) Evolutionary dynamic of genetic diversity. Lecture notes in biomathematics 53, Springer Verlag Berlin, pp. 13-213.

Parulekar, A.H. V.K. Dhargalkar and S.Y.S. Singbal, 1980. Benthic studies in Goa, estuaries. 3. Annual cycle of macrofaunal distribution, production and trophic relation. Indian J. Mar. Sci., 9:189-200.

Parulekar, A.H., Z.A. Ansari & B.S. Ingole, 1986. Effect of mining activities on the clam fisheries and bottom fauna of Goa estuaries. Proc. Indian Acad. Sci. (Animal Sci.), 95:325-339.

Pearson T.H. and R. Rosenberg, 1976. A comparative study of the effect on the marine environment of waste from cellulose industries in Scotland and Sweden. Ambio, 5:77-79.

Pearson T.H. and R. Rosenberg, 1978. Macrobenthic succession in relation to organic enrichment and pollution of the marine environment. Oceanogr. Mar. Bol. Ann. Rev , 16:229-311.

Warwick, R.M. 1986. A new method of detecting pollution effects on marine macrobenthic communities. Mar. Bol., 92:557-562.

Zingde, M.D., S.K. Trivedi and B.N. Desai, 1979. Physico-chemical studies on coastal pollution off Bombay, Indian. J. Mar. Sci., 8:271-277.

28

Environmental Studies for Deep Seabed Mining

Rahul Sharma*

ABSTRACT

Minerals from the deep-sea are potential sources of metals such as Cu, Ni, Co, Mn, Fe, that could be mined in future by developing suitable technologies for mining as well as extracting metals from them. Several groups of environment scientists and mining engineers have been conducting studies for working out these minerals. These studies have revealed unknown physical, chemical, biological and geological conditions under which the mining system will have to operate, and also the potential impacts that may occur.

The data generated through environmental studies would be useful not only in understanding the potential effects of large scale mining in marine ecosystem, but also in the development of various mining subsystems. However, owing to the uncertainty of the design of mining system and the difference in the scale of mining compared to the experiments conducted, many questions remain unanswered, which need to be considered in future.

INTRODUCTION

Deepsea mining has been a subject of interest around the world for over three decades, due to its potential for the economical recovery of large reserves of minerals, such as manganese nodules and crusts, which are considered as alternative source of strategic metals, such as Cu, Ni, Co, Pb, Zn, Cd besides Fe and Mn, for industrial development. The value of imports of ores and minerals during 1994-1995 accounted for ~21.5% of the total imports value of all merchandise to India and amounted to about 190 billion Indian rupees (Desa, 1999). The imports include the minerals and metals that also have a marine source. Demand for these minerals and metals has increased (Table 1), which are met either from domestic production or from imports. In case of some of the metals and minerals, such as zinc, phosphorite, ilmenite and rutile, the domestic

* National Institute of Oceanography, Goa 403004, India.

Table—1 Production and Imports of Selected Minerals and Metals by India

	Domestic production (tonnes)			Imports (tonnes)			
Metal	1992-93	1993-94	1994-95	1992-93	1993-94	1994-95	% increase
Cobalt	—	—	—	137	204	260	31
Copper	5,209,584	5009,050	4767,628	142,973	232,849	233,361	22
Nickel	—	—	—	6,022	7,650	10,907	27
Zinc	126,600	143,961	149,213	20,455	25,398	35,192	27
Lead	38,382	25,299	34,476	15,538	49,058	80,950	67
Rock phosphate	—	—	—	2230,859	1922,739	2540,138	7
TiO_2	—	—	—	2,226	3,454	2,118	-2
Phosphorite	689,774	1035,517	1086,291	—	—	—	
Apatite	16,912	12,342	11,066	—	—	—	
Ilmenite	245,322	270,546	297,444	—	—	—	
Rutile	11,757	11,928	13,457	—	—	—	

*Average annual increase in imports (based on 1992-93, 1993-94, 1994-95 figures).
Source: IBM (1996) (Desa, 1999)

production has steadily grown. On the other hand, in case of metals such as cobalt and nickel, the entire demand is met from imports. Irrespective of the trend in domestic production, the imports have grown with years for all the metals, except for TiO_2, which is supplied from placer minerals. Due to growing demand of metals, there is an increasing need to look at the deepsea minerals as an alternative source, for the future.

A deep-sea mining operation would offer variety of challenges, owing to distant locations (thousands of miles from the coast), deep-sea occurrences (5-6 km of water depths), extreme physico-chemical conditions (high pressure, low temperature) and unknown environmental settings. Moreover, due to the location of most of the potential mining areas in the international waters, it is a collective responsibility to ensure that no damage is caused to the marine environment. As no deepsea mining activity has been started on a commercial scale, it is a favourable situation to foresee the likely effects and work out the means to minimize such effects by designing 'environment friendly' mining system in advance and plan mining operations accordingly.

This paper describes the environmental studies conducted by various groups around the world to understand the potential environmental impact of deepsea mining, evolve methods to conserve marine environment and plan the activities to be undertaken in future.

STATUS OF DEEP SEABED MINING AND RELATED STUDIES

Considerable progress by different groups of scientists and engineers has been made in all the 4 areas of deep seabed mining.

Survey and Exploration

The survey and exploration of deep-sea minerals has led to detailed studies on distribution and grades of these minerals, their genesis and mineralogy, petrology and sedimentology of associated substrates and bathymetry maps of the seafloor. In the quest for faster and more accurate survey and data collection, various survey equipment ranging from free-fall devices, mechanical equipment, remotely operated vehicles, tethered online cameras to remotely sensed backscatter and subbottom profiling data, have been developed and used. As a result, India, France, Japan, Russia, China, a consortium countries of eastern European countries (IOM) and Korea have claimed large areas in the Indian and Pacific Oceans.

Mining System Development

International consortia had conducted tests of prototype mining systems and components in the Pacific Ocean in the seventies (DOMES, 1976). The subsequent progress in the development of mining technology for deep-sea minerals has been slow since 1980 due to unfavorable metal market condition. However, research groups have been working at designing different subsystems of the mining system and presently the scientists have narrowed down to basically two miner/ collector concepts: self-propelled seafloor miner vehicles (Brink and Chung, 1980;) and tow-sled collectors (Chung and Tsurusaki, 1994).

Mineral Processing

Some international consortia have developed manganese nodule processing techniques including pilot processing tests. However, the results were not made available to non-members. Recently, some of the research groups have conducted laboratory-scale tests (Zhong et al., 1999; Premchand and Jana, 1999), which may be scaled up to pilot-plant-scale in future. However, the unwanted tailings, depending on the type of processing, will amount to between 72% and 97% of the original ore (Black, 1982), and their safe disposal into the terrestrial or marine environment, or their alternative uses have not been studied to sufficient extent.

Environmental Studies

Owing to growing concern for the environmental impact of deep-sea mining, multi-disciplinary studies (physical & chemical oceanography, geology, geochemistry, ecology and geotechnical engineering) have been undertaken in different parts of the world oceans. Initial environmental impact studies were conducted during pilot scale mining tests for manganese nodules, in the seventies (DOMES, 1976, Burns et al., 1980; Ozturgut et al., 1980), followed by several experiments in the last decade, creating benthic disturbances in the Pacific and the Indian Oceans (Foell et al., 1990; Trueblood, 1993; Fukushima, 1995; Tkatchenko et al., 1996; Sharma and Nath, 1997). These studies include collection of baseline environmental data in the proposed mining areas, experimentally disturbing the area and monitoring its effects over a period of time. These aim at prediction of the intensity of mining impacts on different environmental parameters as well as evaluating the processes of restoration and re-colonization of deep-sea environment and animal communities. A comparison of the sediment resuspended during these experiments, shows that these were much smaller in scale than a commercial mining system (Yamazaki and Sharma 2001).

POTENTIAL ENVIRONMENTAL EFFECTS OF DEEP SEABED MINING

The impact of offshore mining is expected to be in the form of 'plume' at the seafloor, turbidity in the water column and addition of bottom sediments to the surface resulting in change in the marine ecosystem (Fig. 1). According to an estimate, an area of 300-600 sqkm will be disturbed every year for mining of 3 million metric tones of nodules per year (Sharma, 1993). With every tonne of manganese nodule mined from seabed, 2.5-5.5 tonnes of sediment will be resuspended (Amos and Roels, 1977). Hence, for an average 10,000 metric tonnes of nodules mined per day, about 40,000 metric tonnes of sediment will be disturbed (at 1 tonne nodule: 4 tonne sediment). The adjacent areas will not only have higher sedimentation rates, but the suspended loads may remain over long periods and also travel laterally, clogging filter feading apparatus of benthic organisms in the area. Sudden increase in the amount of suspended matter due to sediment plume and mining discharge, will increase the turbidity of these waters, and may also affect pelagic organisms (Morgan, 1981).

The debris and sediments mixed with water, which are lifted and transported with the minerals, will be discharged at the surface, and create turbidity. Their dispersion by currents will decrease the available sunlight for photosynthesis causing long term effects on biological productivity. At the same time introduction of bottom water, with its higher nutrient values, would result in artificial upwelling, increasing the surface productivity (Pearson, 1979).

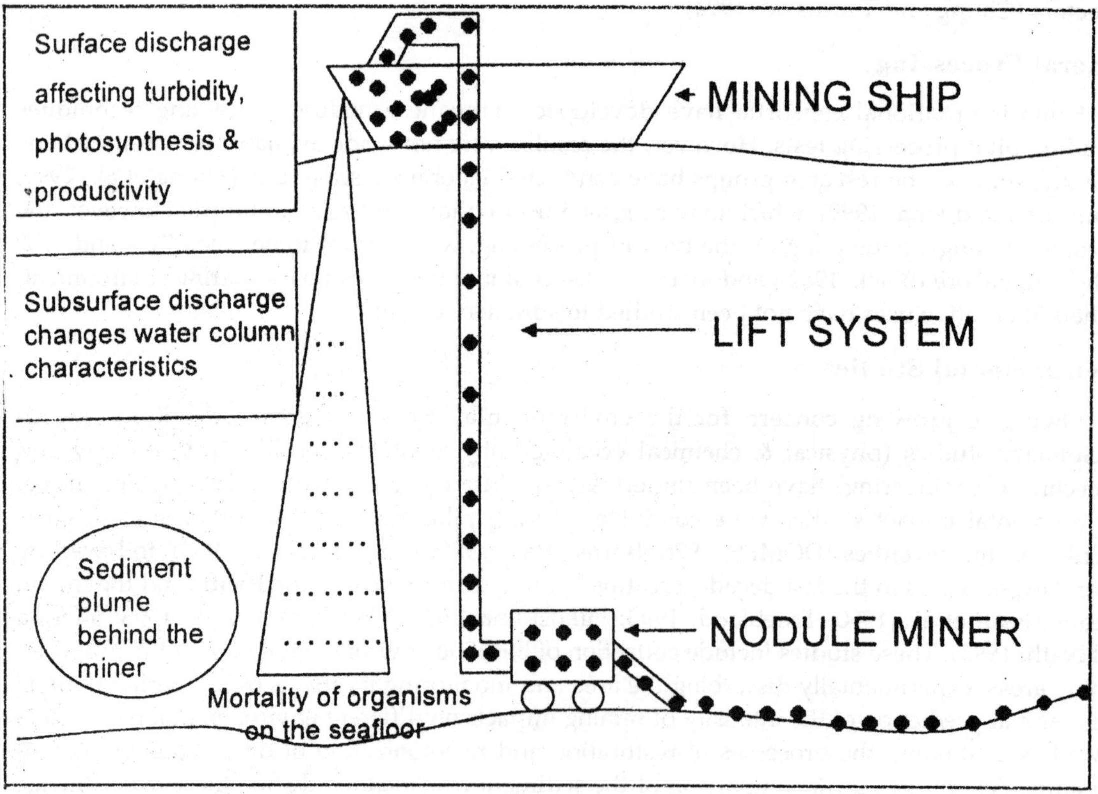

Fig. 1. Schematic for environmental impact of deepsea mining

A summary of the probable impacts at various levels in the water column have been given in the following section (from ISA, 1998):

Potential Benthic Impacts

It is anticipated that the primary benthic impacts caused by mining will be:

- direct impacts along the track of the nodule collector, where the sediments and associated fauna will be crushed or dispersed in a plume and the nodules removed
- smothering or entombment of the benthic fauna away from the site of nodule removal, where the sediment plume settles; and
- clogging of suspension feeders and dilution of deposit-feeders food resources.

Potential Water-column Impacts

Discharge of tailings and effluent below the oxygen-minimum zone may cause some environmental harm to the pelagic fauna, such as:

- mortality of zooplankton species
- effects on meso-and bathypelagic fishes and other nekton, impacts on deep-diving marine mammals, on bacterioplankton and depletion of oxygen by bacterial growth on suspended particles.
- effects on fish behaviour and mortality caused by the sediments or trace metals
- mortality of and changes in zooplankton species composition caused by discharges
- dissolution of heavy metals (e.g. copper and lead) within the oxygen-minimum zone and
- the possible clogging of zooplankton by filtering particles in the plume.

Potential Upper-water Column Impacts

If tailings consisting of sediments (including clay) and effluent are discharged in near-surface waters then there are additional impacts to those listed above, which are:

- the potential for trace-metal bioaccumulation
- reduction in primary productivity due to shading of phytoplankton
- effects on phytoplankton from trace metals
- behaviour of marine mammals.

ENVIRONMENTAL STUDIES CONDUCTED BY DIFFERENT GROUPS

A brief description of studies conducted so far is given in the following paragraphs.

DOMES

The DOMES (1976) (Deep Ocean Mining Environment Study: 1972-1981) conducted by National Oceanographic and Atmospheric Administration (NOAA, USA) monitored environmental impacts during two of the pilot scale mining tests conducted by the Ocean Mining

Inc. and the Ocean Mining Associates (OMA) in 1978 in the Pacific Ocean (Burns et al., 1980; Ozturgut et al., 1980). During the study the concentration of particulates was measured in the discharge and the biological impacts in the surface as well as benthic plumes were assessed.

DISCOL

The DISturbance and Re-COLonisation (DISCOL) experiment was conducted by the scientists of the Hamburg University, Germany in the Peru Basin in the Pacific Ocean from 1988 to 1998. Collection of pre-disturbance baseline environmental data was followed by the disturbance caused by a plow harrow in a circular area of 10.8 km^2 (Foell et al., 1990). Post disturbance studies were carried out to monitor the impact and recolonisation after 6 months, 3 years and 7 years using sediment cores, deeptowed photographic systems, current meters, and nephelometers. The results have shown that over a period of time, although certain groups of benthic organisms may show a quantitative recovery, the faunal composition is not the same as the undisturbed one (Schriever et al., 1997). This implies that complete re-colonisation is a slow process and some of the benthic organisms may need more time to re-establish themselves in the disturbed areas in terms of density and diversity.

NOAA-BIE

The benthic impact experiment (NOAA-BIE) by the National Oceanographic and Atmospheric Administration (NOAA, USA) was conducted in the Clarion Clipperton Fracture Zone (CCFZ) of the Pacific Ocean (1991 to 1993). After baseline studies in a pre-selected area, the Deep-Sea Sediment Resuspension System - DSSRS (Brockett and Richards, 1994) was used for 49 times in an area of 150x3000 m and the post-disturbance sampling with CTD, sediment traps and core samples indicated changes in the faunal distribution in the area (Trueblood, 1993). The impact assessment after 9 months indicated that, whereas some of the meiobenthos showed a decrease in abundance, the macrobenthos showed an increase in their numbers probably due to increased food availability (Trueblood et al., 1997). Hence, mixed effects of resedimentation on benthic organisms could be expected and the effects cannot be generalized. The resedimentation in turn would depend on the total volume of sediment resuspended and the prevailing current patterns in the area.

JET

The Japan deep-sea impact experiment (JET, 1994 - 1997) was conducted by MMAJ (Metal Mining Agency of Japan) in the CCFZ of the Pacific Ocean in 1994 with the DSSRS. Disturbance was created during 19 transects over two parallel tracks of 1600 m length (Fukushima, 1995). The impact was assessed from sediment samples, deep-sea camera operations, sediment traps, and current meters and the results show that the abundance of meiobenthos decreased in deposition areas immediately after the experiment and returned to originals levels 2 years later, but the species composition was not the same; whereas the abundance of certain groups of mega and macro-benthos was still lower than the undisturbed area (Shirayama, 1999). This observation shows that certain groups are more susceptible than others to adapt to the changed conditions on the seafloor.

IOM-BIE

A benthic impact experiment (IOM-BIE) was conducted by Interoceanmetal (IOM) Joint Organisation, in CCFZ in Pacific Ocean in 1995, using DSSRS. In all, 14 tows were carried out on a site of 200x2500 m and the impact was observed from deep-sea camera tows and sediment samples (Tkatchenko et al., 1996). The results have shown that whereas no significant change was observed in meiobenthos abundance and community structure in the re-sedimented area, alteration in meiobenthos assemblages within the disturbed zone was observed (Radziejewska). This indicates that the intensity of disturbance and the proximity to the site could have variable effects on composition of the faunal assemblages.

INDEX

The Indian Deep-sea Environment Experiment (INDEX) was conducted by National Institute of Oceanography (Goa, India) in 1997 in a pre-selected area in the Central Indian Ocean Basin after a detailed baseline study from 1995-97. The DSSRS was used 26 times in an area of 200x3000 m, during which about 6000 m^3 of sediment was resuspended (Sharma and Nath, 1997). Photographic surveys showed reduction in benthic organisms due to the movement of the disturber on the seafloor as compared to that before the experiment (Fig. 2, 3). The post disturbance impact assessment studies have indicated that the effects of disturbance are the

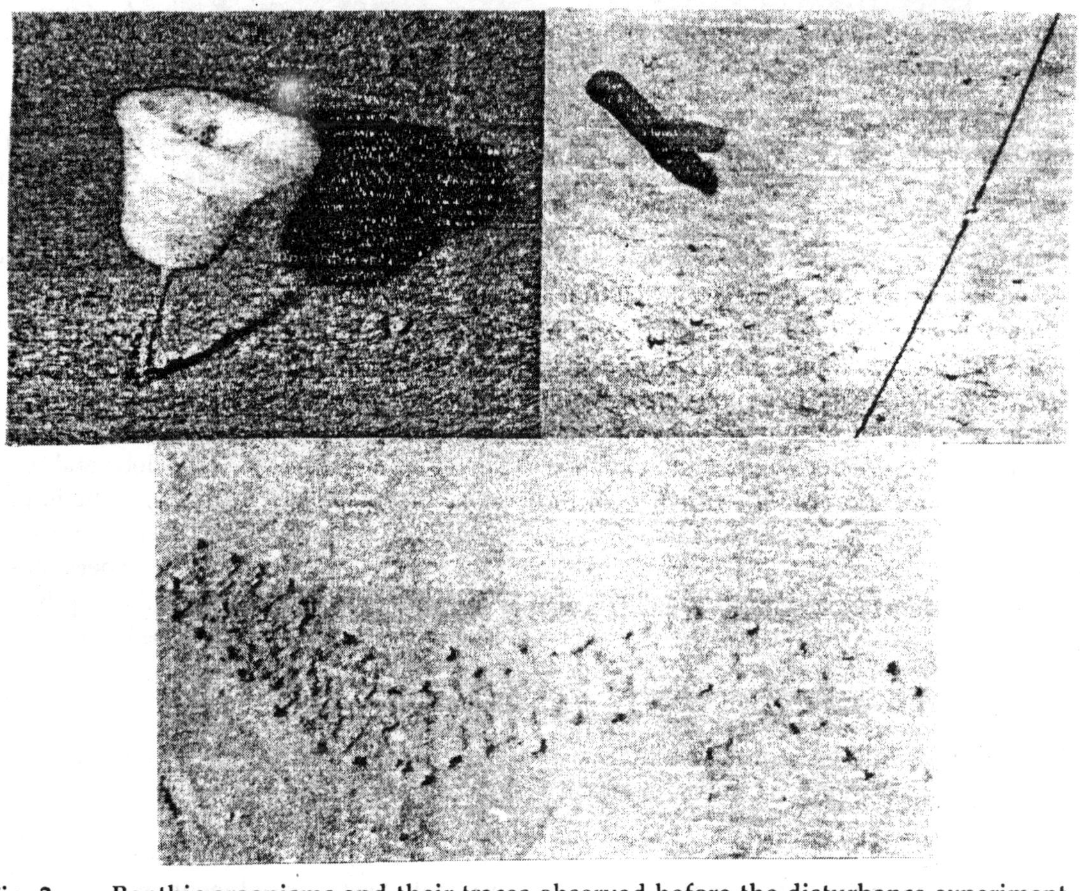

Fig. 2. Benthic organisms and their traces observed before the disturbance experiment.

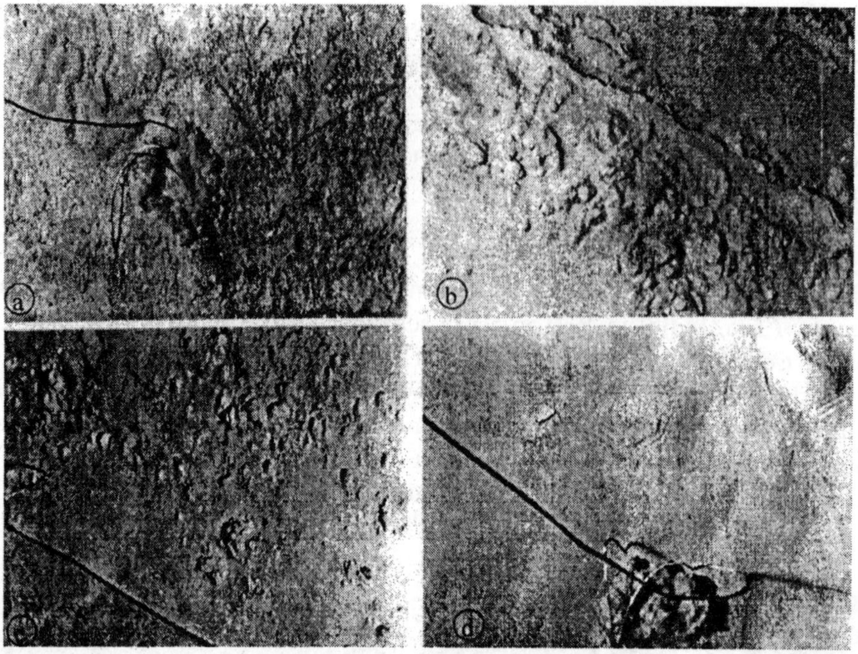

Fig. 3. (a) Seafloor with animal tracks before the disturbance, (b) 'disturber' track on the seafloor, sediment lumps close to the disturber track, (d) resedimented surface after disturbance experiment.

highest in and around the disturbed area (Sharma et al., 2001) and there are differential effects on microbial, meio and macrobenthos within and outside the disturbance track (Ingole et al., 1999, Nair et al, 2000). Longterm monitoring indicates that favourable conditions have developed for restoration of benthic environment.

ENVIRONMENTAL CONSIDERATIONS FOR DEEP SEABED MINING

It is anticipated that the environmental impact will be greatest at the seafloor and at the depth zones of discharge of mine tailings and effluent (Thiel et al., 1997). Although, the impacts at the seafloor and in the water column cannot be avoided, they must be taken into account during the development of the mining equipment and in designing the mining operations to minimize the effects. As there is a higher potential for environmental harm in the depth zone from the surface to 1,000 m, it is advisable that discharges should be released below the depth of the oxygen-minimum layer (e.g. in 1,000 m in many parts of the Pacific Ocean), which has to be evaluated in advance of commercial mining activities.

In order to limit the impacts to minimum levels, the following measures need to be taken in the design considerations of the deep-sea mining system (ISA, 1998)

- minimizing sediment penetration of collector and mining vehicle,
- avoiding the disturbance of the more consolidated suboxic sediment layer,

- reducing the mass of sediment swirled up into the near bottom water layer,
- inducing a high rate of resedimentation from the plume behind the miner,
- minimizing the transport of sediment and abraded nodule fines to the ocean surface
- reducing the discharge of tailings into bathyal or abyssal depth, and
- reducing the drift of tailings by increasing their sedimentation rate.

Certain environmental factors that must be considered to predict the impact of offshore mining (Sharma and Rao, 1992) are:

(i) The proportion of total area that will be affected, with respect to total area of the waterbody.

(ii) Influence of surface and subsurface currents in different seasons

(iii) Distance of coastal belts and inhabited areas from the area of influence.

(iv) Existence of fishing potential or any other commercial activity in the area of influence.

SOME OF THE UNANSWERED QUESTIONS

In view of small scale of experiments conducted and calculations based on limited data for the possible effects of future mining activities, many questions remain unanswered, due to the limited scope of these studies (Chung et al, 2001). Some of the questions are:

1. What would be the effects of surface discharge of effluents or debris mixed with bottom waters?
2. What would be the effects of movement of a collector mechanism or miner/collector vehicle tracking on the seafloor?
3. What should be the acceptable depth of discharge of effluents in the water column below the free surface?
4. What would be impacts of discharges from nodule-transport pipe and other sub-systems (for example, buffer) suspended in the water column?
5. How would the resettlement or redistribution of suspended sediments occur close to the seafloor?
6. What would be the net effect on the marine food chain of different subsystems / activities of deep-sea mining at various depth zones?
7. Could alternative uses be found for the large quantities of debris from deep-sea minerals?
8. What would be an ideal size of each mining area to allow re-colonisation and restoration of the benthic communities?
9. What alterations in the mining system design could be used to minimize the effects on the marine ecosystem? What engineering data should the environment experiment provide?

10. Which key parameters would be useful as indicators of environmental impacts and how do we prioritize them?
11. What results, data and statements each contractor would be required to deliver in order to claim mining regions?
12. What will be a likely mining system or systems in next 20 years to determine a basis of environmental tests? How do we categorize and define potentially real mining system and miner or collector systems, and narrow down to one or two systems?
13. Which other environmental parameters do we need to study in case of at-sea processing?
14. How do we categorize and identify pollutant or effluent discharges from the ship, pipe, buffer and miner/collector for land processing and at-sea processing?
15. Upon generalizing the 13 and 14 above, how do we approach to develop a disturbance flow simulation model or component disturbance tests, and verify the simulation model?
16. How do we prevent disturbance and fine discharges and generalize the test parameters from the tow-sled collector and miner/collector vehicle?
17. How much room do we leave in environment test planning for unidentified questions?
18. How do we integrate the test data of bottom disturbance with other effects?

CONCLUSIONS

Since, offshore mining may become essential in order to replenish the mineral resources for industrial development in future, steps should be taken to arrive at such design parameters that the mining operation will cause least possible damage to the marine environment. Marine biological communities survive in a fragile ecosystem by maintaining a very delicate balance with their environment. The marine environment supports optimum life on every square meter and any change due to mining operations, could upset the whole marine ecosystem. Time series data on various oceanographic parameters (geological, biological, chemical, physical, meteorological etc.) before, during and after the offshore mining activities will have to be collected to assess the effects of mining as well as to modify the design of the mining system.

However, the scale of the experiments conducted so far is too small compared to the potential disturbance caused by a nodule mining system. The time scales for the re-establishment of geochemical and ecological conditions after commercial mining disturbance will be very long, may be in the order of decades. Hence all the results of these experiments cannot be extrapolated to future mining activities. There are many more questions that must be answered and more problems to be solved prior to the commencement of a commercial mining operation. Environment scientists and engineers should form a team in their efforts so as to achieve the ultimate target of recovering minerals from deep ocean with acceptable environmental impacts.

REFERENCES

Amos, A. F. and Roels, O.A. (1977). Environmental aspects of maganese nodule mining. Mar. Pol. Int. Bull, 1, pp. 160-162.

Black, JRH. (1982). "Metals Recovery from Manganese Nodules: A Review of Processing Schemes," in Humphrey, PB, ed, *Marine Mining: A New Beginning*, pp 93-107, Department of Planning and Economic Development, State of Hawaii (USA).

Brink, AW, and Chung, Jin S. (1980). "Automatic Position Control of a 300,000 Ton Ship Mining Systems," *Proc Offshore Technology Conf*, Houston, 1980; also in ASME Trans. *J Energy Resources Technology*, 1982.

Brockett, T, and Richards, CZ (1994). "Deep-Sea Mining Simulator for Environmental Impact Studies," *Sea Technology*, August, pp 77-82.

Burns, RE Erickson, BH Lavelle, JW, and Ozturgut, E (1980). "Observation and Measurements During the Monitoring of Deep Ocean Manganese Nodule Mining Tests in The North Pacific, March-May 1978," NOAA Technical Memo ERL MESA-47, National Oceanic and Atmospheric Administration, Colorado (USA), 63.

Chung, JS, Schriever, G, Sharma R, Yamazaki, T (2001). Deep seabed mining environment: preliminary engineering and environmental assessment. Proceedings of Int. Symposium on Ocean Mining, Szczecin (Poland), 8-13.

Chung, S.S. and Tsurusaki, K., 1994. Advance in Deep-Ocean Mining Systems Research. Proc. of the Fourth International Offshore and Polar Engineering Conference, pp. 18-31.

Desa, E. (1999). Opportunities for offshore minerals exploration in Indian Ocean. Proceedings of Int. Symposium on Ocean Mining, Goa (India), 6-13.

DOMES (1976). "Summary of the Minutes of the DOMES Mining Industry Meeting," NOAA Pacific Marine Environmental Lab., Seattle, USA, 12 pp.

Foell, EJ, Thiel, H, and Schriever, G (1990). "DISCOL: A Longterm Largescale Disturbance – Recolonisation Experiment in the Abyssal Eastern Tropical Pacific Ocean," Proc of Offshore Technology Conference, Houston, USA, Paper No. 6328, pp 497-503.

Fukushima, T (1995). "Overview "Japan Deep-sea Impact Experiment = JET," Proc. of 1st ISOPE Ocean Mining Symp, Tsukuba, Japan, ISOPE, pp 47-53.

Ingole, B.S; Ansari, Z.A; Matondkar, S.G.P; Rodrigues, N. (1999). Immediate response of meio and macrobenthos to disturbance caused by benthic disturber. Proc. of the 3rd ISOPE – Ocean Mining Symp., Goa (India), Int. Soc. Offshore and Polar Eng., 191-197.

ISA (1998). "Draft Guidelines for the Assessment of the Possible Environmental Impacts Arising from Exploration of Deep-Seabed Polymetallic Nodules from the Area," Workshop by the International Seabed Authority ISA), Sanya, Hainan, China, 1-5 June 1998.

Morgan, C. (1981). "Environmental impacts of deep-ocean mining – the importance of manganese". In: R.A. Geyer (Editor) Marine Environmental Pollution (2), Dumping and Mining, Elsevier Oceanography Series, Elsevier, New York, 415-435.

Nair. S, Mohandass, C.Loka Bharati, P.A, Sheelu, G, Raghukumar, C. (2000). Micro-scale response of sediment variables to benthic disturbance in the central Indian Ocean Basin. Marine Georesources and Geotechnology, 18 (3):273-284.

NOAA (1981). "Deep Seabed Mining: Final Programmatic Environmental Impact Statement," National Oceanic and Atmospheric Administration, Washington, DC, Vol. 1.

Ozturgut, E, Lavelle, JW, Steffin, O, and Swift, SA (1980). "Environmental Investigation During Manganese Nodule Mining Tests in the North Equatorial Pacific, in November 1978," NOAA Technical Memorandum ERL MESA-48, National Oceanic and Atmospheric Administration, Colorado (USA), 50.

Pearson, J.S., 1979. Ocean Floor Mining, Noyes Data Coprn., N.J. 201.

Premchand; Jana, R. (1999). "Processing of Polymetallic Sea Nodules - An Overview," *Proc. of 3rd ISOPE Ocean Mining Symp*, Goa, India, ISOPE, pp 237-245.

Radziejewska, T (1997). "Immediate Responses of Benthic Meio- and Megafauna to Disturbance Caused by Polymetallic Nodule Miner Simulator," *Proc Int Symp Environmental Studies for Deep-sea Mining, Metal Mining* Agency of Japan, Tokyo, Japan, pp 223-236.

Schriever, G, Ahnert, A; Borowski, C, and Thiel, H (1997). "Results of the Large Scale Deep-sea Impact Study DISCOL during Eight Years of Investigation," *Proc Int Symp Environmental Studies for Deep-sea Mining*, Metal Mining Agency of Japan, Tokyo, Japan, pp 197-208.

Sharma, R., (1993). Quantitative Estimation of Seafloor Features from Photographs and Their Application to Nodule Mining. Proc. of Marine Georesources and Geotechnology, Vol. 11, pp. 311-331.

Sharma, R, and Nath, BN (1997). "Benthic Disturbance and Monitoring Experiment in Central Indian Ocean Basin," *Proc 2nd ISOPE Ocean Mining Symp*, Seoul, Korea, ISOPE, pp 146-153..

Sharma, R, Nath, B.N., Parthiban, G., Sankar, S.J. (2001). Sediment redistribution during simulated benthic disturbance and its implications on deep seabed mining. Deep-sea Research II, 48, 3363-3380.

Sharma, R. and Rao, A.S. (1992). Geological factors associated with megabenthic activity in the Central Indian Basin. Prod. of Deep-Sea Research, Vol. 39, No. ¾, pp. 705-713.

Shirayama, Y (1999). "Biological Results of JET Project: an Overview," *Proc 3rd ISOPE Ocean Mining Symp*, Goa, India, ISOPE, pp 185-190.

Thiel, H (1991). "Environmental Impact Resulting from Deep-Sea Mining and Risk Assessment," pp. 87-96 in *Law of the Sea at the Crossroads*, R. Wolfrum, Ed. "The Continuing Search for a Universally Accepted Regime," *Proc an Interdisciplinary Symp of the Kiel Institute of International Law*. Veroeffentlichungen des Instituts fuer Internationales Recht an der Universitaet Kiel.

Thiel, H, Angel, MV, Foell, EJ; Rice, AL, and Schriever, G. (1997). "Environmental Risks from Largescale Ecological Research in the Deep-Sea," EC Report, Bremerhaven, Germany, pp 210.

Tkatchenko, G, Radziejewska, T, Stoyanova, V, Modlitba, I, Parizek, A (1996). "Benthic Impact Experiment in the IOM Pioneer Area: Testing for Effects of Deep-sea Disturbance," *Int Seminar on Deep Sea-bed Mining Tech*, China Ocean Mineral Resources R&D Assoc., Beijing, China, C55-C68.

Trueblood, DD (1993). "US Cruise Report for BIE¾ II Cruise. NOAA Technical Memo OCRS 4, National Oceanic and Atmospheric Administration, Colorado, USA, pp 51.

Trueblood, DD, Ozturgut, E, Pilipchuk, M, and Gloumov, IF (1997). "The Echological Impacts of the Joint U.S.-Russian Benthic Impact Experiment," *Proc. Int. Symp. Environmental Studies for Deep-sea Mining*, Metal Mining Agency of Japan, Tokyo, Japan, pp 237-243.

Yamazaki, T, and Sharma, R (2001). "Estimation of Sediment Properties during Benthic Impact Experiments, " *Marine Georesources and Geotechnology*', 19: 269-289.

Zhong, X, He, Z, Shen, Y, Mao, Y, Qu, S, and Li, X (1999). "The Smelting-Rusting Solvent Extracting Process to Recover Valuable Metals from Polymetallic Nodules," *Proc 3rd ISOPE Ocean Mining Symp*, Goa, India, Int Soc Offshore and Polar Eng, pp 227-231.

29

Biofilm Development on Metal Surfaces in Tropical Marine Waters

*Fraddry D'souza & Naryan B. Bhosle

The immersion of solid surfaces in aquatic environment results in the rapid adsorption of dissolved organic matter, thereby conditioning the surfaces (Loeb and Neihof, 1977; Characklis and Escher, 1988; Zutic and Tomaic, 1988). A number of compounds including glycoproteins (Baier, 1980), humic material (Loeb and Neihof, 1975) and/or unspecified macromolecules (Zaidi *et al.*, 1984) have been implicated as conditioning agents. Such adsorption of organic matter onto surface results in the concentration of useful nutrients at the surface. Conditioned surfaces are then colonized by microorganisms like bacteria, diatom, fungi and protozoa. Marshall *et al.*, (1971) reported that very small bacteria selectively colonize surfaces in the first 24h following immersion of surfaces in seawater. These early colonizing copiotrophic bacteria were predominantly small Gram-negative rods of which 50 to 90% were *Pseudomonas* species and 10 to 49% were pigmented *Flavobacterium* species or non-pigmented non-motile *Achromobacter* species. Further it has been observed that oligotrophic bacteria *Caulobacter*, *Hypomicrobium* and *Saprospira* also adhere to surfaces immersed in marine water for 24h or more (Marshall *et al.*, 1971; Corpe, 1973; Fletcher and Marshall, 1982). Thus the biomass, the number and diversity of microorganisms increase over the period of immersion (Jordan and Staley, 1976; Gerchakov *et al.*, 1977; Sonak and Bhosle, 1995). Biofilm may influence the settlement of larvae of macrofouling organisms through physical and chemical modification of the surface (Crisp, 1974; Dean and Hurd, 1980; Wieczorek *et al.*, 1995).

There are many studies on the biofilm of surfaces immersed in aquatic environments (Corpe, 1973; Jordan and Staley, 1976; Gerchakov *et al.*, 1977; Baier, 1984; Little, 1984; Bhosle *et al.*, 1989, 1990). However, little is known about biofilm bacteria developed on metal surfaces, especially

* Marine Corrosion and Materials Research Division, National Institute of Oceanography, Dona Paula 403004, Goa, India.

immersed in tropical marine waters (Sonak and Bhosle, 1995). Similarly, not much is known about the nature of organic matter deposited on the surfaces over the period of immersion. Therefore, the aims of the present study was to measure the changes in the biofilm carbon, nitrogen, and C/N ratio and to study the development of biofilm microorganisms including bacteria and microalgae on metal surfaces immersed in the surface waters of the Dona Paula Bay over a period of 28 d.

MATERIALS AND METHODS

Test panels

Test panels (100x150x2mm) of mild steel (MS), stainless steel (SS) and copper (Cu) were cleaned with 20%(v/v) HCl, washed with tap water followed by distilled water. Panels were dried at 60^0C and kept covered until marine immersion.

Deployments of panels

Test panels were fixed to a fiber glass assembly using PVC nut and bolts. Fiber glass assembly with attached test panels was deployed in the surface seawater (~1m) of the Dona Paula Bay ($15^o 31'$ N, $73^o 59'$) in November/December, 1995, January/February, 1997 and April/May, 1997. During each deployment, the panels were covered with polythylene bags in order to avoid the contamination from the surface microlayer (Bhosle *et al.*, 1989). Plastic bags were removed when the panels were in seawater. The assembly was maintained at 1m depth using a float. Replicate panels were removed after 24h and thereafter at weekly intervals over a period of a 28 d.

Estimation of biofilm biomass

After retrieval biofilm material from the test panels was scraped in laminar flow chamber using a nylon brush and filter sterilized (0.22mm) seawater (White and Benson 1984; Bhosle *et al.*, 1989; Sharma *et al.*, 1990). The scraped material was adjusted to a known volume. A known quantity was filtered through GF/F filters which were preashed (450°C, 3h) before use. The replicate filters were then used for the analysis of biofilm biomass measured as organic carbon (OC), organic nitrogen (ON) and chlorophyll *a* (Chl *a*). The POC and Chl *a* were analysed following the methods described by Parsons *et al.* (1984). The ON was estimated following the method of Smart *et al.* (1983). The coefficient of variation of the analytical methods used for OC, ON and Chl *a* was generally less than 8%.

Estimation of viable count of bacteria

The biofilm material (1ml) obtained as above was serially diluted and 0.1ml aliquot was plated on ZoBell Marine Agar (ZMA) by the surface spread plate method (Buck and Cleverdon, 1960). The Petri dishes were incubated at room temperature (28 ±2°C) for 24h and the total viable count was noted. During each sampling fifteen to twenty colonies (~ 24% of the total) were picked up from the greatest dilution showing growth. Thus the colonies in this study represent the numerically most dominant aerobic strains from the test panels. Colonies were purified by repeatedly streak plating on ZMA plates, routinely subcultured and maintained on ZMA slopes at 4°C. The coefficient of variation of the analytical method was generally less than 20%.

Isolation and characterization of bacteria developed on MS, SS and Cu panels.

Two hundred and eighty three bacterial colonies were isolated during the period of study. Some of the colonies did not survive. Rest of the colonies were studied for morphological, physiological and biochemical characteristics using standard methods as described earlier (Sonak and Bhosle, 1995). Based on these characteristics the isolates were identified upto the genus.

Scanning Electron Microscopy

Experimental panels (0.3 x 1 mm) of MS, SS, and CU were deployed in the surface waters as described above and were retrieved after 8 d following immersion. The panels were fixed with 2% glutraldehyde in filtered (0.2 mm pore size) seawater for 2 hours. They were dehydrated through an acetone: water series beginning with 20 % acetone and ending with 100 % acetone. Samples were then critical point dried, sputter coated with gold using a Sputter Coater, and then examined with a Jeol SEM.

Environmental Parameters

During each biofilm sampling surface seawater samples were collected and analyzed for temperature, salinity, dissolved oxygen and nutrients like nitrate, nitrite and phosphate. Temperature was measured using a thermometer. Salinity, dissolved oxygen, nitrate, nitrite, phosphate, were measured using the methods described by Parsons *et al.* (1984). Seawater samples were filtered through preignited GF/F (450°C, 3h) filters for the analysis of particulate organic carbon (POC), particulate organic nitrogen (PON) and chlorophyll *a* (Chl *a*). The POC, PON and Chl *a* were analysed following the methods described above for the analysis of biofilm material.

RESULTS

Environmental parameters

Changes in temperature, salinity, dissolved oxygen, POC, PON and Chl *a* of the surface seawater samples collected during the period of biofilm sampling are shown in Table 1 to 3. Surface seawater temperature was low initially at day 1 of the immersion period in December/ November, 1995. As the immersion period increased there was a small increase in the temperature of the surface water (Table 1). Temperature did not vary much during the sampling period in January/February 1997, whereas, it was relatively higher for the period April/May, 1997 (Table 2 and 3). A similar trend was recorded for the salinity of the seawater samples collected during all the sampling periods. Dissolved oxygen concentration showed small variation (2.83 to 3.81mg l^{-1}) over the period of study in November/December 95. The concentration of dissolved oxygen increased (5 to 6 mg l^{-1}) during the period of sampling in January/February and April/ May97. Nitrate concentrations varied from 2 to 3.64 m mol l^{-1} during first 8 days and decreased to ~0.94 m mol l^{-1} at day 21 followed by a small increase again at day 28 of the immersion period in November/December 95. Nitrate concentrations were generally low for the subsequent sampling periods. Nitrite distribution showed a pattern similar to that recorded for nitrate. Phosphate concentration showed small changes without exhibiting any particular trend throughout the period of study. POC did not show any strong short term variation, while PON content varied and was relatively higher during the latter period (21 to 28 d) of immersion in November/December 95. Concentration of POC was low during the first 8 d and then increased

Table—1 Variation in some physical, chemical and biological parameters in the surface sea waters of Dona Paula Bay during the period November/December 1995

Immersion period (days)	*T* °C	*S* ‰	*D.O* *mg.l*$^{-1}$	*Nitrate* *μmol.l*$^{-1}$	*Nitrate* *μmol.l*$^{-1}$	*Phosphate* *μmol.l*$^{-1}$	*POC* *μg C.l*$^{-1}$	*PON* *μg N.l*$^{-1}$	*Chl a* *μg.l*$^{-1}$
1	29 ± 0.82	27.42 ± 0.85	3.81 ± 0.46	2.03 ± 0.46	0.68 ± 0.19	0.44 ± 0.01	1498 ± 190	56.8 ± 11.6	5.42 ± 0.56
8	28 ± 0.47	29.90 ± 1.25	3.41 ± 0.31	3.64 ± 0.64	0.56 ± 0.11	1.69 ± 0.24	1166 ± 119	35.5 ± 0.24	1.01 ± 0.15
15	29 ± 0.85	30.82 ± 2.51	3.54 ± 0.14	2.19 ± 0.81	1.33 ± 0.17	1.54 ± 0.39	1189 ± 36	84.5 ± 17.9	1.35 ± 0.38
21	30 ± 0.82	31.51 ± 1.47	2.83 ± 0.57	0.94 ± 0.14	0.35 ± 0.09	0.73 ± 0.54	703 ± 95	102 ±	1.84 ± 2.18
28	31 ± 1.41	30.83 ± 0.82	3.28 ± 0.23	1.46 ± 0.58	0.39 ± 0.08	2.71 ± 0.16	1493 ± 23	75.3 ± 9.8	1.94 ± 1.83

± = standared deviation; T = temperature; S = salinity; D.O = dissolved oxygen; POC = particulate organic carbon; PON = particular organic nitrogen; Chl a = chlorophyll a.

Table—2 Variation in some physical, chemical and biological parameters in the surface sea waters of Dona Paula Bay during the period January/February 1997.

Immersion period (days)	*T* °C	*S* ‰	*D.O* *mg.l⁻¹*	*Nitrate* *μmol.l⁻¹*	*Nitrate* *μmol.l⁻¹*	*Phosphate* *μmol.l⁻¹*	*POC* *μg C.l⁻¹*	*PON* *μg N.l⁻¹*	*Chl a* *μg.l⁻¹*
1	30.67 ± 0.47	31.82 ± 1.24	3.95 ± 1.19	0.87 ± 0.69	0.28 ± 0.16	0.37 ± 0.15	277.21 ± 12.60	53.63 ± 1.23	1.36 ± 0.29
8	30.00 ± 0	32.54 ± 1.68	4.83 ± 0.25	1.36 ± 0.62	0.48 ± 0.03	0.46 ± 0.11	528.0 ± 354.19	49.78 ± 6.13	1.56 ± 0.60
15	30.00 ± 0	33.54 ± 0.34	4.78 ± 0.39	1.13 ± 0.63	0.48 ± 0.03	0.57 ± 0.25	1129.59 ± 103.95	47.72 ± 4.82	1.68 ± 0.51
21	30.33 ± 0.47	33.90 ± 0.56	5.32 ± 0.94	1.20 ± 0.55	0.52 ± 0.02	0.58 ± 0.24	1080.42 ± 109.72	45.29 ± 3.23	1.92 ± 0.85
28	30.33 ± 0.47	34.81 ± 1.45	5.16 ± 0.38	0.65	0.40 ± 0.19	0.57 ± 0.26	983.4 ± 203.7	48.17 ± 1.91	2.67 ± 0.94

± = standard deviation; T = temperature; S = salinity; D.O = dissolved oxygen; POC = particulate organic carbon; PON = particulate organic nitrogen; Chl a = chlorophyll a.

Table—3 Variation in some physical, chemical and biological parameters in the surface sea waters of Dona Paula Bay during the period April/May 1997.

Immersion period (days)	*T*	*S*	*D.O*	*Nitrate*	*Nitrate*	*Phosphate*	*POC*	*PON*	*Chl a*
	°C	‰	$mg.l^{-1}$	$\mu mol.l^{-1}$	$\mu mol.l^{-1}$	$\mu mol.l^{-1}$	$\mu g\ C.l^{-1}$	$\mu g\ N.l^{-1}$	$\mu g.l^{-1}$
1D	32	36.35	5.54	0.08	0.12	0.43	918.72	106.68	5.85
	± 0.82	± 0.51	± 1.22	± 0.04	± 0.04	± 0.24	± 359.74	± 46.62	± 0.15
8D	33.67	36.78	6.55	0.11	0.17	0.36	670.56	23.42	5.76
	± 1.88	± 0.82	± 2.79	± 0.07	± 0.09	± 0.10	± 158.61	± 1.40	± 2.16
15D	35.0	36.62	4.52	0.12	0.26	0.69	677.19	25.93	11.09
	± 1.63	± 0.78	± 0.37	± 0.05	± 0.06	± 0.14	± 263.75	± 2.18	± 5.94
21D	35.67	36.99	4.13	0.05	0.22	0.68	798.77	18.77	16.59
	± 0.94	± 0.78	0.11	± 0.01	± 0.09	± 0.25	± 177.39	± 8.95	± 2.75
28D	34.67	36.34	4.39	0.11	0.37	0.91	647.89	20.74	12.77
	± 0.47	± 0.85	± 0.47	± 0.08	± 0.09	± 0.34	± 137.53	± 7.93	± 3.32

± = standard deviation; T = temperature; S = salinity; D.O = dissolved oxygen; POC = particulate organic carbon; PON = particulate organic nitrogen; Chl a = chlorophyll a.

during 15 to 28 d following immersion in January/February 97. Small decrease in POC was recorded for the period April/May 97. The concentration of PON did not vary much in January/February while it decreased over the period of sampling in April/May 97. Small changes in the Chl *a* concentration of the surface seawater was recorded during the periods November/December 95 and January/February 97. Conversely, the Chl *a* concentration increased during the period of observation in April/May 97.

Biofilm biomass

Biofilm biomass developed on MS, SS and Cu over the period of immersion was measured as OC, ON, Chl *a* and bacterial abundance. The data are shown in Fig. 1 to 3. Changes in C/N (atomic) are given in Fig. 4. The SEM photographs of the biofilm on MS, SS and CU are given in Fig. 5.

Mild steel

OC of the biofilm material developed on MS was very low on day 1 following immersion and showed a gradual increase during the first 15 day period of immersion in November/December 95 [Fig. 1(a)]. Thereafter, OC of the biofilm did not vary much over the period of study. ON of the biofilm also showed an increasing trend until day 21 followed by a small decrease at 28 d of the immersion in November/December 95 [Fig. 1(b)]. For the subsequent sampling periods of January/February and April/May 97, biofilm OC and ON showed consistent increase over the period of immersion [Fig. 2(a), 2(b) and 3(a), 3(b)]. Bacterial abundance on MS increased and was highest on day 15 following immersion and decreased thereafter [Fig. 1(c)]. A similar trend was evident for the bacterial abundance in the biofilm material developed on MS during the deployment periods of January/February and April/May 97. Chl *a* content of the biofilm showed increase over the period of immersion during all the three deployment periods [Fig. 1(d), 2(d) and 3(d))]. It was interesting to note the change in the abundance of microorganisms developed on MS. Bacteria were more abundant during the first 15 day period of deployment of test panels in the surface waters. Thereafter bacterial numbers generally decreased and the abundance of microalgae increased [Fig.1 to 3]

Stainless steel

Biofilm OC increased as a function of the immersion period during the first 15 d and thereafter it showed a small decrease at day 21 followed by a small increase again at day 28 of the immersion [Fig. 1(a)]. As observed for OC, ON also increased during first 15 d following immersion and then showed small decrease during the remaining period of immersion [Fig. 1(b)]. When the sampling was repeated again in January/February and April/May 97, OC and ON of the biofilm material developed on SS showed consistent increase over the period of deployment [Fig. 2a, b and 3(a), (b)]. Bacterial cell numbers increased initially for 15 days and decreased thereafter at day 21 and 28 following immersion [Fig.1(c)]. A similar trend in the abundance of biofilm bacteria developed on SS was also recorded for the subsequent sampling periods of January/February and April/May 97 [Fig. 2(c) and 3[c)]. Chl *a* of the biofilm material was low during the first 15 d period of immersion and increased thereafter and the highest concentration was generally recorded at day 21 following immersion [Fig. (1) to [3]). At day 28 of the immersion the Chl *a* concentration either decreased [Fig. 1(d) and 3(d)] or increased (Fig. 2d).

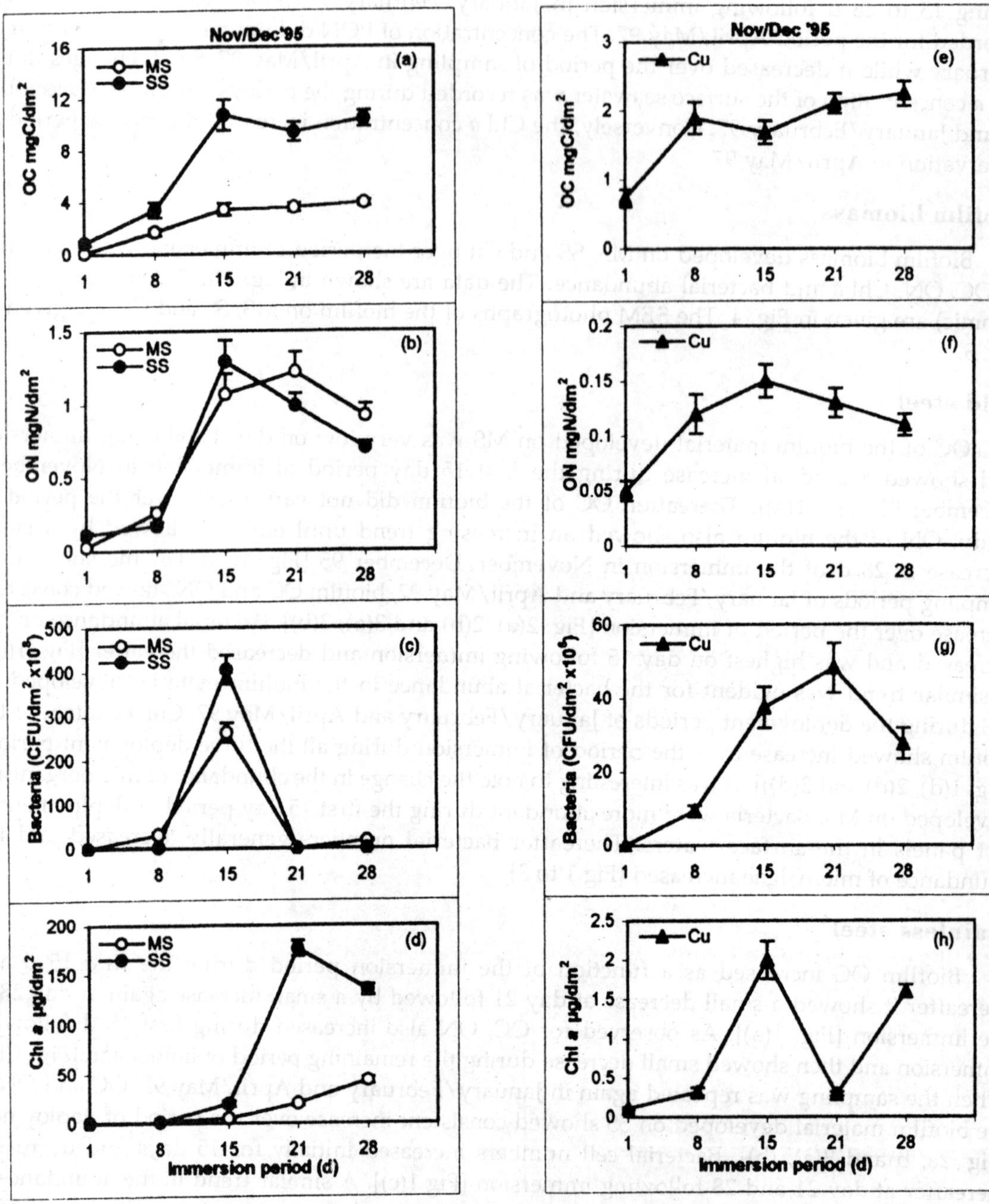

Fig. 1 **Changes in biofilm biomass as organic carbon, nitrogen, bacterial abundance and algal biomass as Chl a on MS (a-d), SS (a-d) and Cu (e-h) panels immersed in the surface waters of Dona Paula Bay for a period of 28 d in November/December 1995. Bar indicates mean ± standard deviation.**

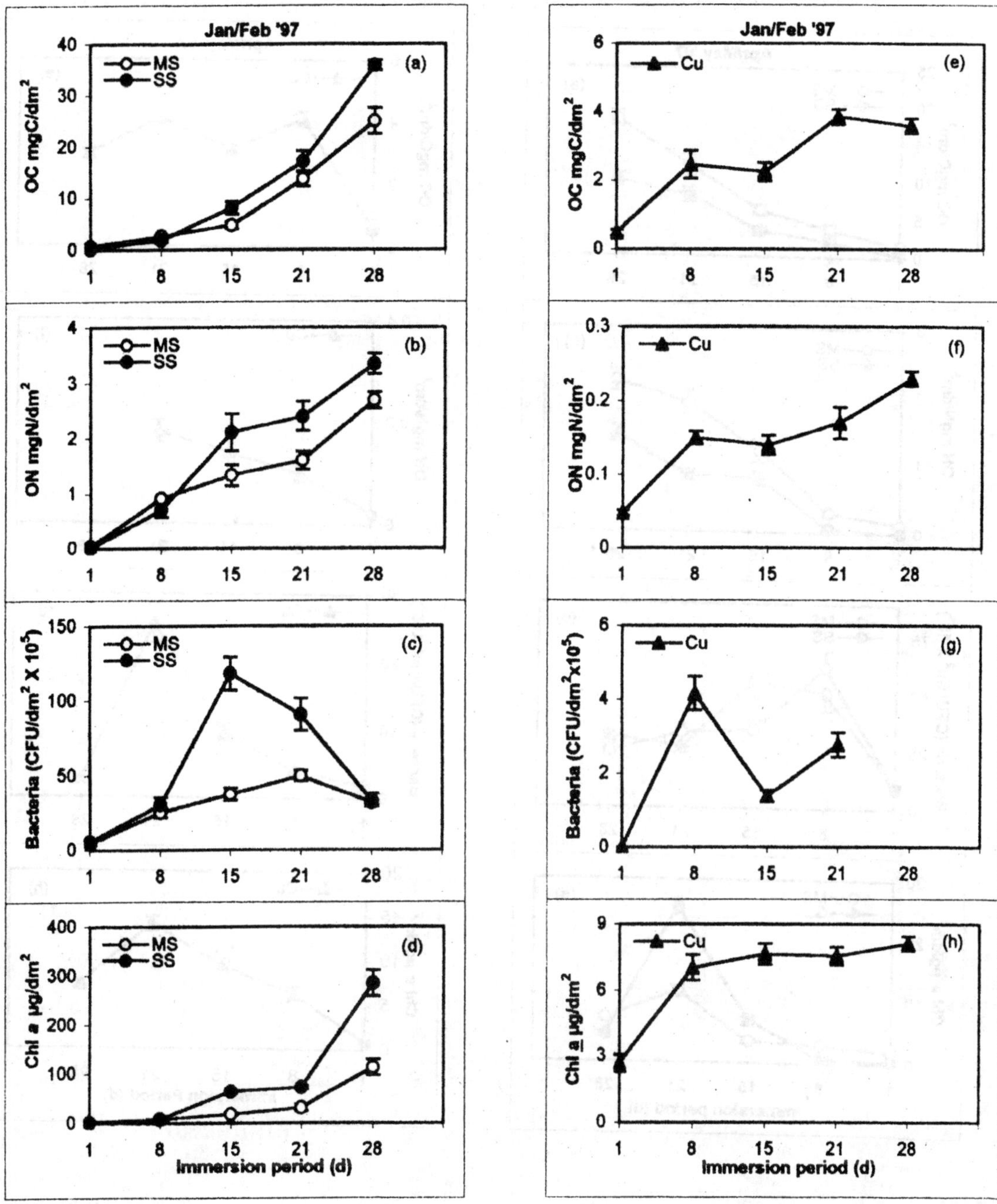

Fig. 2 Changes in biofilm biomass as organic carbon, nitrogen, bacterial abundance and algal biomass as Chl a on MS (a-d), SS (a-d) and Cu (e-h) panels immersed in the surface waters of Dona Paula Bay for a period of 28 d in January/February 1997. Bar indicates mean ± standard deviation.

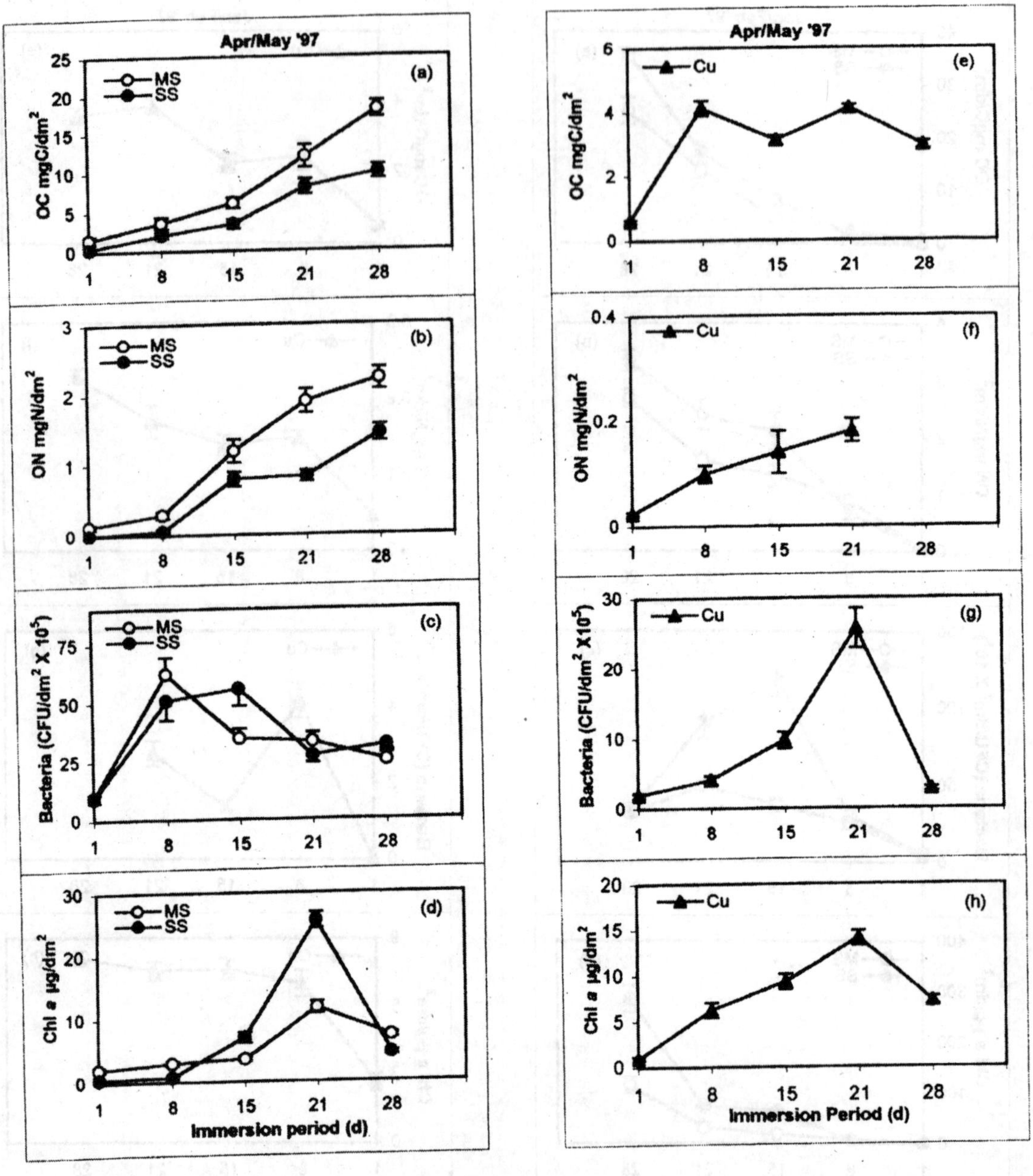

Fig. 3 Changes in biofilm biomass as organic carbon, nitrogen, bacterial abundance and algal biomass as Chl a on MS (a-d), SS (a-d) and Cu (e-h) panels immersed in the surface waters of Dona Paula Bay for a period of 28 d in April/May 1997. Bar indicates mean ± standard deviation.

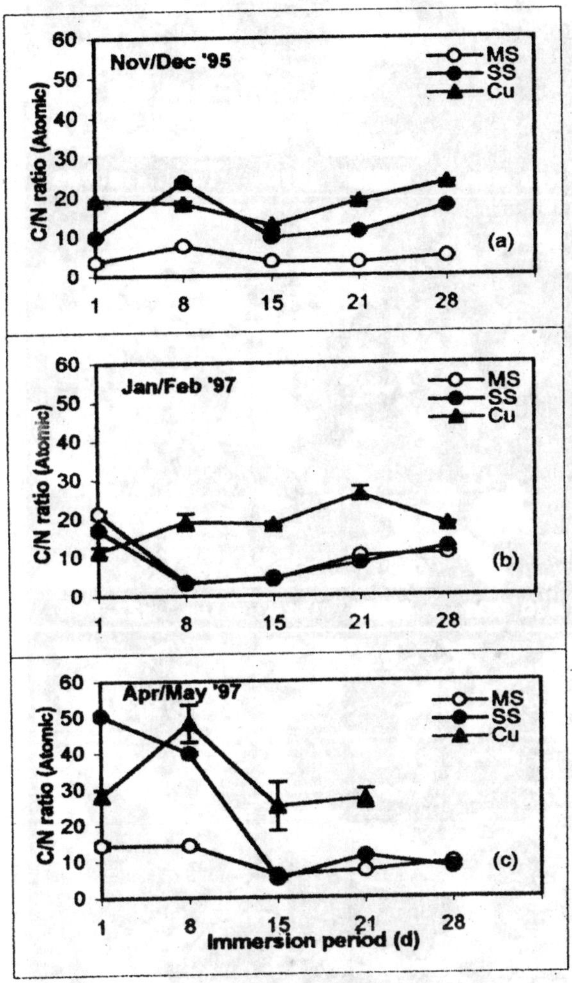

Fig. 4 Changes in C/N ratio of the biofilm material on MS, SS and Cu panels immersed in the surface waters of Dona Paula Bay for a period of 28 d in Nov/Dec' 95 (a), Jan/Feb'97 (b) and Apr/May'97 (c). Bar indicate mean ± standard deviation.

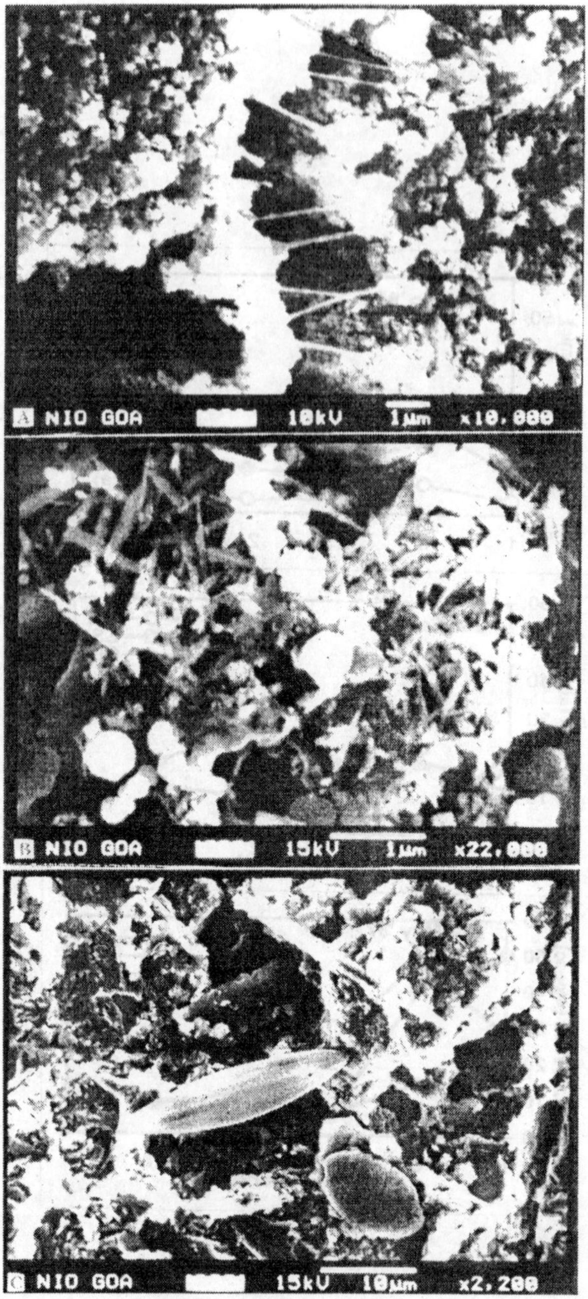

Fig. 5 Scanning Electron Microscopic (SEM) observation of the biofilm developmed on mild steel (A), stainless steel (B) and copper (C) immersed for 8 days in the surface sea water of the Dona Paula Bay.

Copper

OC and ON of the biofilm material developed on Cu panels was very low as compared to MS and SS [Fig. 1(e), 1(f), 2(e), 2(f) and (3(e), 3(f)]. OC and ON of the biofilm generally increased over the period of immersion during all the sampling periods. Bacterial numbers and Chl *a* were also relatively low on Cu as compared to those observed on MS and SS [Fig. 1(g), 1(h), 2(g), 2(h) and 3 (g), 3(h)]. Both Chl *a* and bacteria generally increased during the period of immersion. Biofilm bacteria and microalgae on Cu did not show any succession, during the period of immersion as observed for the biofilm on MS and SS.

C/N ratio (atomic) of the biofilm material

C/N ratio of the biofilm material developed on MS, SS and Cu varied during the 28 day period of immersion [Fig. 4]. The ratio was relatively low on MS and SS as compared to (Fig. 4). C/N ratio of the biofilm on MS, SS and Cu showed large variations during the first 15d period of immersion and thereafter showed increase with the immersion period especially during November/December 95 and January/February 97.

SEM

The SEM photographs of the biofilm on MS, SS and Cu showed the presence of bacteria and microalgae (Fig. 5). Bacteria appear to be more abundant on MS and SS. Bacteria interconnected with exopolymeric thread like structures are clearly seen in the biofilm on the MS panels. Cu panels also showed the presence of bacteria and a few diatoms.

Isolation and characterization of bacterial cultures

A total of 283 colonies were isolated from the biofilm developed on MS (85), SS (113) and Cu (85). Of these 30, 32 and 24 colonies from MS, SS and Cu respectively could not be maintained. The remaining 197 colonies were identified using standard morphological, physiological and biochemical characteristics of the isolates (Table 4). Most isolates were Gram-negative rods, oxidase positive, facultative anaerobes and oxidative and fermentative in nature. Many were motile and did not reduce nitrate to nitrite. The isolates in general produced catalase and utilized some sugars including glucose, xylose, fructose and sucrose. On the basis of these characteristics the cultures were tentatively identified upto the generic level (Table 4).

The generic composition of the bacterial biofilm developed on MS, SS and Cu panels appear to vary over the period of immersion. However no particular trend in the distribution of various bacteria was evident with some exception. For example, Pseudomonas was the only most abundant group (27.33 ± 4.02 %) on MS, and SS during the initial 24h immersion period. Subsequently, as the immersion period increased large variation in the bacterial composition was observed on all the surfaces during the immersion period.

DISCUSSION

Several parameters such as dry weight, carbon, nitrogen, ATP and carbonate contents of fouling material have been employed to quantify the extent of biofouling (Aftring and Taylor, 1979; Mayack *et al.*, 1984; Bhosle *et al.*, 1990; Bhosle and Wagh, 1997). In this study organic carbon and nitrogen were used to estimate biofilm biomass developed on MS, SS and Cu. Biofilm biomass

Table—4 Morphology and biochemical characteristics of bacteria on mildasteel, stainless steel and copper panels immersed in the surface sea waters of Dona Paula Bay for a period of 28d.

Bacteria	*Gm stain & morph.*	*Mot.*	*O*	*C*	*I*	*NO_3 red.*	*U*	*Hugh Oxd.*	*eifson Fermt.*	*Citrate slant*	*H_2S prod.*	*Gela.*	*G*	*X*	*F*	*S*	*MR*	*Pigment*	*VP*
Acinetobacter (4)	Gm -ve rods	-	-	+	-	-	-	+	-	-	-	2	+	-	+	+	+	-	-
Aeromonas (8)	Gm -ve rods	+	2	+	-	+	-	+	+	-	-	+	+	+	+	+	-	-	-
Bacillus (28)	Gm +ve rods	12	+	+	-	-	-	+	-	-	-	-	+	8	+	4	3	-	+
Chromobacterium (5)	Gm -ve rods	+	+	-	-	+	-	+	+	-	-	-	+	+	+	+	-	4	+
Corynebacterium (4)	Gm +ve rods	-	-	-	-	-	+	-	+	-	-	-	+	-	-	-	+	-	-
Enterobactereacea (12)	Gm -ve rods	+	-	-	-	-	-	+	-	-	-	+	+	-	-	-	+	-	-
Flavobacterium (9)	Gm -ve rods	-	-	-	-	+	-	+	+	-	-	2	+	-	+	+	-	-	+
Halobacter (11)	Gm - ve rods	+	+	-	-	-	-	+	-	-	-	+	+	-	+	-	-	-	-
Kingella (3)	Gm -ve rods	+	+	-	-	-	-	-	-	-	-	-	-	-	-	-	-	-	-
Micrococcus (2)	Gm +ve cocci	-	+	-	-	+	-	+	-	-	-	+	+	+	+	-	+	-	-
Moraxella (7)	Gm -ve rods	-	+	+	+	-	-	+	-	-	-	-	-	-	-	-	-	-	-
Morganella (1)	Gm -ve rods	+	-	+	-	+	-	+	-	-	-	-	-	-	-	-	-	-	-
Neisseria (16)	Gm -ve cocci	-	+	3	-	4	-	3	3	4	-	3	+	+	+	-	2	-	3
Pastuerella (19)	Gm -ve rods	-	+	+	-	+	-	+	-	-	-	-	+	-	-	-	-	-	-
Photobacter (2)	Gm -ve rods	+	+	+	-	-	-	+	-	-	-	-	+	+	+	+	-	-	+
Pseudomonas (47)	Gm -ve rods	44	5	45	-	3	-	+	-	-	-	43	+	5	+	+	2	-	13
Staphylococcus (3)	Gm +ve cocci	-	-	2	-	+	-	+	-	-	-	-	+	+	+	-	-	+	-
Vibrio (2)	Gm -ve rods	+	+	+	-	+		+	+	-	-	-	+	+	+	+	-	-	-
Xanthomonas (2)	Gm -ve rods	+	+	+	-	+	-	+	-	-	-	+	+	-	-	-	+	-	-
Yersinia (1)	Gm -ve rods	+	-	+	-	+	-	+	-	-	-	+	+	+	-	-	-	-	-

Gm stain & morph = Gram stain & morphology, Mot = motility; O = oxidase; C = catalase; I = Indole; U = urease; gela = gelatine; G = glucose; X = xylose; F = ructose; S = sucrose; MR = Methyl Red; VP = Voges Proskauer; ± = All isolates giving positive results; - = All isolates riving negative results; (n) = Number of isolates giving positive results.

(measured as OC and ON) developed on MS, SS and Cu generally increased over the period of immersion at the study site. This agrees well with earlier studies suggesting a non-linear increase in biofilm biomass as a function of the immersion period (Martinez *et al.*, 1984; White and Benson, 1984; Yanshun *et al.*, 1984; Chamberlain and Graner, 1988; Sonak and Bhosle, 1995; Bhosle and Wagh, 1997). The physical, chemical and biological parameters of the surface seawater showed small variation. The observed increase in the biofilm biomass developed on MS, SS and Cu over the period of study was therefore due to increase in settlement and/or due to growth of attached microorganisms. The latter seems likely because the abundance of bacteria and/or diatoms generally increased during the period of immersion. As compared to MS and SS, biofilm biomass as OC, ON, Chl *a* and bacterial numbers was relatively low on Cu. The observed low biomass on copper panels was probably due to the toxic effect of the copper ions.

The C/N ratio (atomic) is a useful parameter to understand the nature of organic material deposited on the experimental panels immersed in marine waters (Sharma and Wagh, 1990; Bhosle and Wagh, 1997). C/N ratios of the biofilm material on these surfaces at day 1 following immersion were generally high (>10). Bhosle and Wagh (1997) reported high C/N ratio (> 15) of the biofilm material on aluminium panels immersed (1 to 4 days) in surface waters of the Dona Paula Bay. These results suggest that irrespective of the surface, the biofilm material developed on metal surfaces was generally enriched in degraded material during the initial period of immersion (1 to 8 d). The high C/N ratio also indicates that bacteria were not the major contributor to the organic matter of the biofilm deposited on metal surfaces during the initial period of immersion.

The C/N ratio of the biofilm material showed large variations during the first 15 d following immersion. Subsequently, as the period of immersion increased, C/N ratio of the biofilm material developed on MS, SS and Cu increased, especially during the period November/December 95 and January/February 97. As compared to Cu, C/N ratio of the biofilm material developed on MS and SS was relatively low. The observed low C/N ratio of the biofilm on MS and SS is similar to that observed for bacteria and diatoms (Hedges *et al.*, 1997). This implies that the biofilm material on MS and SS was relatively rich in bacteria and diatoms. This was also evident from the data on bacterial and microalgal abundance in the biofilm material on these surfaces.

Conversely, relatively higher C/N ratio (>15) of the biofilm on MS, SS and Cu were also observed. It could be argued that the observed high C/N ratio of the biofilm material was due to production of exopolysaccharides by the microorganisms attached to these surfaces. Although biofilm on Cu panels had relatively higher exopolysaccharides as compared to the biofilm on MS and SS, the contribution of exopolysaccharides to the total carbon was not more than 2 to 3 % (data not shown). This suggests that exopolysaccharide was not the source for high C/N ratio of the biofilm material developed on these surfaces. Alternatively, it is well documented that the predominance of nitrogen-free biomolecules (e.g. lignin, hemicellulose, tannins, cellulose, cutin etc.) over proteins makes higher plant tissues characteristically carbon-rich (C/N ~20 to 500) (Hedges *et al.*, 1997). Similarly, degraded organic matter is also enriched in carbon as compared to nitrogen. Thus high C/N ratio (> 15) indicates the abundance of terrestrial organic matter and/or degraded organic matter in the biofilm material developed on these surfaces.

Viable cell count of the biofilm bacteria showed an increasing trend over the period of immersion during the first 15 d period of immersion. Therefore, the observed increase in the viable cell count was due to the enhanced settlement and/or growth of bacteria on these panels. Two hundred and eighty-three colonies were isolated from the microfouling material on MS, SS and Cu over the period of immersion. These were randomly selected from the greatest dilution showing growth, therefore the colonies represent the numerically most dominant aerobic strains from the test panels. Most isolates were Gram-negative rods. This observation is in agreement with several other studies. For example, Corpe (1973) isolated several strains of bacteria developed on glass that were generally Gram negative. In another study Sonak and Bhosle (1995) isolated 100 bacterial colonies from aluminum panels immersed in Mandovi estuary, which were also Gram-negative rods.

We isolated and characterized a large number (283 isolates) of bacteria from these surfaces. However with exception of *Pseudomonas*, we did not observe the abundance of any particular bacterial genera in the biofilm on these surfaces during the period of immersion. *Pseudomonas* species were the most abundant (c = 27.33 ± 4.02 %) initial colonizers on MS and SS panels during the initial 24h following immersion. Marshall *et al.* (1971) reported that very small bacteria selectively colonize surfaces in the first 24h following immersion of surfaces in seawater. These early colonizing bacteria were predominantly small Gram-negative rods of which 50 to 90% were *Pseudomonas* species. Similarly, Sonak and Bhosle (1995) reported the abundance of *Pseudomonas* species on aluminum panels during the 24h period of immersion. From this it is evident that irrespective of the type of metal used generally, *Pseudomonas* are the most abundant bacteria colonizing the surfaces during initial 24h period of immersion in various environments. With increase in the immersion period bacterial composition changed, however no particular organism showed very high abundance. The observed changes in the composition of bacteria suggest the changes in the biofilm community over the period of immersion.

Bacteria were generally most abundant during 15 day period of immersion of all the three surfaces in marine waters. This agrees well with the earlier studies suggesting the abundance of bacteria in microfouling on metal surfaces during the initial period of immersion (Marshall *et al*, 1971; Sharma and Wagh, 1990; Sonak and Bhosle, 1995). As the period of immersion increased the abundance of bacteria decreased. This decrease was followed by an increase in the abundance of microalgae (measured as Chl *a*). It was also interesting to note that these changes were consistent for the three observations carried out at different immersion periods. These successional changes compared well with the microfouling electron transport system activity (ETS) studies reported by Bhosle *et al.* (1994). These authors measured the ETS activity and microalgal biomass (as Chl *a*) of the biofilm material developed on aluminum panels immersed in the surface waters of the Dona Paula Bay. A regression equation between ETS and Chl *a* was employed to estimate the contribution of autotrophic and heterotrophic microorganisms. Heterotrophic contribution to the total ETS was higher (79 %) during the initial period of immersion. With subsequent increase in the immersion period microalgal contribution to the total ETS increased from 21% to 61%.

In conclusion, development of biofilm on MS, SS and Cu panels immersed in surface waters of the Dona Paula Bay was monitored with respect to carbon, nitrogen, C/N ratio, microalgal biomass, viable cell count and generic composition of bacteria. Biofilm showed temporal and seasonal variations. Biofilm development on MS, SS and Cu suggests that bacteria were the most abundant microorganisms initially followed by microalgae. *Pseudomonas species* were generally more abundant during the initial period of fouling. High C/N ratio suggests the abundance of terrestrially derived and/or degraded organic matter in the biofilm material developed on these surfaces. Further studies are needed to understand the role of the degraded and/or terrestrial organic matter in subsequent fouling.

ACKNOWLEDGEMENTS

Authors thank Dr E Desa, the Director of the institute for encouragement and facilities. We also thank the Department of Ocean Development, Govt. of India, New Delhi for financial support to carry out this work. This is NIO contribution No. 3736.

REFERENCES

Aftring, R. P. & Taylor, B. F. 1979 Assessment of microbial fouling in an ocean energy conversion experiment. *Applied Environmental Microbiology* **38**, 734-739.

Baier, R. E. 1980 Substrate influences on adhesion of microorganisms and their resultant new surface properties. In *Adsorption of microorganisms to surfaces* (Edited by Bitton G. and Marshall K. C.) pp 59-104. John Wiley & Sons, Inc, New York.

Baier, R. E. 1984 Initial events in microbial film formation. In *Marine Biodeterioration: An interdisciplinary study* (Edited by Costlow J. D. and Tipper R. C.), pp 63-67. US Naval Institute, Annapolis M D.

Bhosle, N. B., Nandakumar, K., Venkat, K., Dhople, V. M., Sawant, S. S. & Wagh, A. B. 1989 Biological and biochemical characterization of microfouling on aluminium panels placed in the Arabian Sea. *Proceeding of Indian National Science Academy* **55**, 51-56.

Bhosle, N. B., Nandakumar, K. & Wagh, A. B. 1990 Influence of particulate material on microfouling biomass in the Arabian Sea. *Biofouling* **2**, 65-75.

Bhosle, N. B., Avinash, S., Tulaskar & Wagh, A. B. (1994) Electron transport system activity of microfouling material: Relationships with biomass parameter. *Biofouling* **8**, 1-11.

Bhosle, N. B. & Wagh, A. B. 1997 Amino acids in biofilm material on aluminum panels immersed in Marine waters. *Biofouling* **11**, 149-166.

Buck, J. D. & Cleverdon, R. C. 1960 The spread plate as a method for the enumeration of marine bacteria. *Limnology and Oceanograplhy* **5**, 78-80.

Chamberlain, A. H. L. & Graner, B. J. 1988 The influence of iron content on biofouling resistance of 90/10 copper-nickel alloy. *Biofouling* **1**, 79-96.

Characklis, W. G. & Escher, A. R. 1988 Microfouling in initial events. In *Marine Biodeterioration: Advanced Techniques applicable to Indian Ocean* (Edited by Thompson, M. F., Sarojini, R., Nagabushanam, R.), pp 249-260. Oxford & IBH Publ Co. Pvt. Ltd, New Delhi, Bombay, Calcutta,

Corpe, W. A. 1973 Microfouling: The role of primary film forming bacteria. In *Proceeding of 3rd International Congress of Marine Corrosion and fouling* (Edited by Acker, R. F., Brown, B. F., De Palma, J. R. & Inversion W. P.), pp. .598-601. North Weston University Press Evanston III.

Crisp, D. J. 1974 Factors influencing the settlement of marine organisms. In *Chemoreception in marine organisms* (Edited by Grant, P. T. & Mackie, A. M.), pp. 107-265. Academic press. New York.

Dean, T. A. & Hurd, L. E. 1980 Development in an esturain fouling community: The influence of earlier colonists on latter arrivals *Oceanologia* **46**, 295-301.

Fletcher, M. & Marshall, K. C. 1982 Are solid surfaces of ecological significance to aquatic bacteria. In *Advances in Microbial Ecology* (Edited by Mrashall, K. C.), Vol.6, pp. 191-336. Plenum Press, New York.

Gerchakov, S. M., Marzalek, O. S., Ruth, I. G, & Udey, L. R. 1977 Succession of periphytic microorganisms on metal and glass surfaces. In *Proceeding of 4th International Congress of Marine Corrosion and Fouling* (Edited by Ramanovsky V.), pp. 203-221. 1 Centrede Recherches et dEtudes Oceanographiques, Boulogne, France.

Hedges, J. I., Keil, R. G. & Benner R. 1997 What happens to terrestrial organic matter in the ocean? *Organic Geochemistry* **27**, 195-212.

Jordan, T. L. & Staley, J. T. 1976 Electron microscopic study of succession in the periphyton community of lake Washington. *Microbial Ecology* **2**, 241-251.

Little, B. J. 1984 Succession in microfouling. In *Marine Biodeterioration: An interdisciplinary study* (Edited by Costlow J. D. andTipper R. C.), pp. 63-67. US Naval Institute, Annapolis MD.

Loeb, G .I. & Neihof, R. A. 1975 Marine conditioning film. *Advance Chemistry series* **145**, 319-335.

Loeb, G. I. & Neihof, R. A. 1977 Adsorption of an organic film at Platinum sea water interface. *Journal of Marine Research* **35**, 283-291.

Marshall, K. C., Stout, R. & Mitchell, K. 1971 Selective sorption of bacteria from sea water. *Canadian Journal of Microbiology* **17**, 1413-1416.

Martinez, D., Bastida, R. & Brancevich, G. 1984 Ecological aspects of marine fouling at the port of Ingeniero white (Argentina). In *Sixth International Congress of Marine Corrosion and Fouling*, Athens 5-8 sept 1984, Greece, pp. 521-537.

Mayack, K. A., Sorraco, R. J., Wilde, E. W. & Pope, D. H. 1989 Cooperative aspects of chlorine and chlorine dioxide biocide regimes for biofouling control. *Water Research* **18**, 595-599.

Parsons, T. R., Matai, Y. & Lalli, C. M. (eds) 1984 *A manual of chemical and biological methods of sea water analysis*. Pergamon Press, New York, pp. 63.

Sonak, S. & Bhosle, N. B. 1995 Observation on biofouling bacteria isolated from aluminum panels immersed in estuarine waters. *Biofouling* **8**, 243-254.

Sharma, M. O., Bhosle, N. B. & Wagh, A. B. 1990 Methods for removal and estimation of microfouling biomass. *Indian Journal of Marine Science* **19**, 174-–176.

Sharma, M. O. & Wagh, A. B. 1990 The nature of particulate matter settled on solid substrata. *Oceanologica Acta* **13**, 471-474.

Smart, M. M., Kada, R. G. & Donnermeyer, G. N. 1983 Determination of total nitrogen in sediments and plants using persulphate digestion. *Water Research* **17**, 1207-1211.

White, D. C. & Benson, P. H. 1984 Determination of biomass, physiological status, community structures and extracellular plaque of the microfouling film. In *Marine Biodeterioration: An interdisciplinary study* (Edited by In Costlow J. D. and Tipper R. C.), pp. 68-74.US Naval Institute, Annapolis M D.

Wieczorek, S. K., Clare, A. S. & Todd, C. D. 1995 Inhibitory and faciltatory effects of microbial films on settelment of *Balanus amphitrite* larve. *Marine Ecology Progress Series* **119**, 221-228.

Yánshun, L., Dezan, Y., Siyan, L. & Ruimei, Y. 1984 Characteristics of microfouling organism in Xiamin Harbour. In *Sixth International Congress of Marine Corrosion and Fouling*, Athens, Greece, pp. 9-21.

Zaidi, B. R., Bard, R. F. & Tosteson, T. R. 1984 Microbial specificity of metallic surfaces exposed to ambient sea water. *Applied Environmental Microbiology* **48**, 519-524.

Zutic, V. & Tomaic, J. 1988 On the formation of organic coating on marine particles: Interactions of organic matter at hydrous alumna/seawater interfaces. *Marine Chemistry* **23**, 51-68

30

Benthic Life on the Tropical Sandy Shore: Miramar Beach a Case Study

Baban Ingole

INTRODUCTION

Sandy beaches are a dynamic environment in equilibrium, where sand, water and air are in motion, absorbing the energy of the waves and constituting one of the most important aspects of the coastline (McIntyre, 1977;Brown & McLachlan, 1990; Ingole, 1992; Ingole & Parulekar, 1998). Though sandy beaches have been studied for ages, investigators are yet to completely understand the different processes that take place as a result of the dynamic nature of the ecosystems. Coastal ecosystems are exposed to increasing urbanization and industrialization worldwide. Sandy beaches are integral components of these ecosystems and are greatly influenced, usually negatively, by increasing anthropogenic activities.

After viewing the rocky shore at low tide, the sandy shore seems barren. The reason for this is that the rocky shores are flourished with fascinating prospect to both layman as well as scientists (Brown & McLachlan, 1990). Sandy shore has no lush growth of macro-algae and apart from sea-grasses; there is no obvious plant life, except for the odd rock or pebble, which act as a suitable substratum for small plants. Few animals live on the surface, so the explorer has to look for telltale signs of life below the sand, such as tube openings and burrow mouths. Most organisms colonizing the sand live in it rather than on it. This creates a need for a suitable burrowing mechanisms, which may take the form of the proboscis and parapodia of polychaete worms such as the catworm *Nephtys*, and the lugworm *Arenicola* or the foot of shore-dwelling bivalve molluscs. The bivalve body is particularly well suited for this life-style because it is entirely enclosed in a streamlined shell. The foot of intertidal and subtidal clams such as *Mactra violacea*

* Biological Oceanography Division, National Institute of Oceanography, Dona Paula, Goa: 403004, INDIA

and *Meretrix spp.* is relatively large and powerful. It can pull the shell below the sand surface very quickly symbolising the burrowing life style. Other animals that have evolved for life in sand are the burrowing starfishes and brittle stars. Sea cucumbers and anemones employ their soft bodies to burrow, but they have no special burrowing appendages, unlike the burrowing sea urchins with their highly specialized spines and tube feet.

Burrowing is only one of the problems that have to be overcome before life in the sand can be established. Once in position under the sand, a stream of fresh oxygen-bearing seawater has to be supplied to the gills. In polychaete like the lugworm, this water is pumped through the burrows by peristaltic activity of the body wall. Others extend out from their burrows, withdrawing, if danger threatens. In the case of bivalve molluscs the body wall is extended to form two siphons, one bringing clean water to the gills and one carrying oxygen-depleted water away from them. This system combines well with filter feeding mechanism, as the inhalent water stream carries food particles, which can be sieved by the gills. Other animals feed by ingesting the substrate and digesting any suitable organic material. Larger forms such as eel and flat fishes are efficient predators, detecting their prey at a distance by smell and moving swiftly over the surface to engulf it; they may also capture by burrowing.

Some animals do not burrow completely. The common shrimp nestles in the ripple marks of the sand and scavenges for carrion. Various crabs live on the surface sediment and uses its two antennae like a snorkel to draw water below the sand for respiration. The cuttle-fish Sepia is one of the larger predators of the sandy shore, moving in at high tide to prey on small fishes. It has astonishing powers of camouflage and can adjust its coloration to match the background.

MIRAMAR BEACH-A CASE STUDY

Located along the central west coast of India, the sandy beach at Miramar, Goa is special - a city beach (Figure 1 and Plate 1). High-energy waves and variable amplitudes of spring and neap tides influence the intertidal zone, which is inhabited by intricate life of different communities. These communities inhabit the beach sand and play a critical role in maintaining ecological balance. This estuarine sandy beach is rich in benthic fauna with over 47 species of macrobenthos (Table 1) as well as high abundance of micro-and meiobenthic organisms having been collected till date (Dwivedi, et al., 1973; Eleftheriou & Jones, 1976; Bhosle, 2000, Modassir, 2000, Fernandes, 2001; Ingole, 2002).

COLLECTION AND ANALYSIS OF INTERTIDAL MACROFAUNA

To study the abundance and diversity of intertidal macrofauna, benthic sampling was conducted along a fixed transect as shown in figure 1. Sediment samples were collected with the help of metallic quadrant (25x25x10 cm size) and minimum 6 quadrants were taken at each location. Macrofaunal samples were washed through a 0.5-mm mesh sieve and preserved in 10% seawater formalin - Rose Bengal solution. All the samples were again washed through 0.5-mm mesh in running water to clear adhering sediment. All stained organisms were sorted from the samples, preserved in 90% alcohol, identified up to species level and enumerated. Species level identification was based primarily on keys of Fauvel (1953); Gosner (1971), Parulekar (1972), Day (1973) and Vongpanich (2000).

Plate 1. Miramar beach during the low tide.

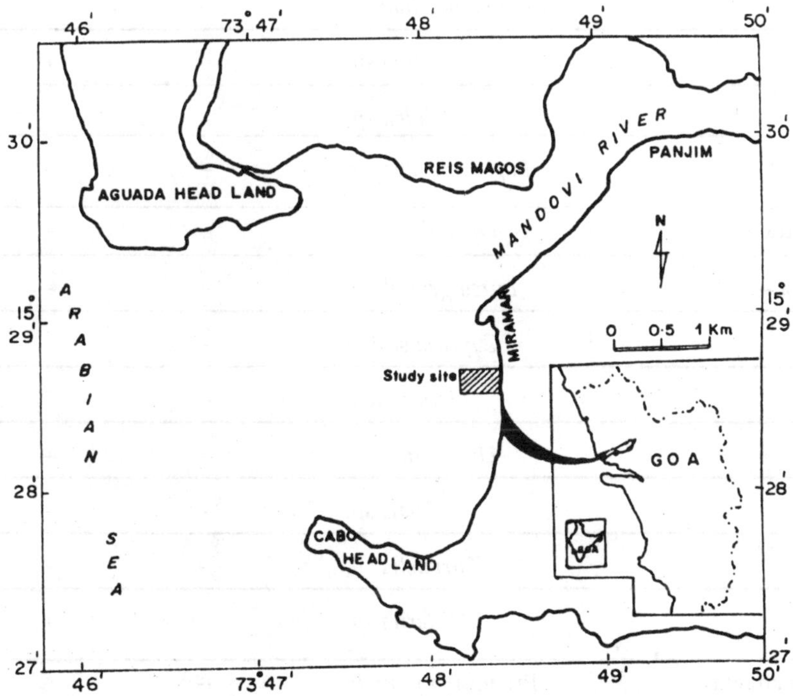

Fig. 1. Location of the sampling site.

Table—1 Macrofaunal species collected during October 2000 from the intertidal area of Miramar beach.

Taxon		High tide	Mide tide	Low tide
Nematoda:	*Viscosia sp.*	–	+	+
Mollusca: Bivalvia –	*Mactra violacea*	–	+	+
	Meretrix casta	–	–	+
	Donax incarnatues	–	+	+
	D. scortum	–	+	–
	D. faba	+	+	+
	Sunetta scripta	–	+	+
	Sunetta sp. 1	–	–	+
	Sunetta sp. 2	–	–	+
	Silliqua radiata	–	+	+
	Gouldia pectinatum	–	–	+
	Tellina sp.	–	+	+
	Solen sp.	–	+	+
	Paphia malabarica	–	–	+
Scaphopoda –	*Dentallium sp.*	–	–	+
Gastropoda –	*Bursa granularis*	–	–	+
	Cypraea scabra	–	–	+
	Natica picta	–	–	+
	Umbonium sp.	–	+	+
	Thais sp.	–	–	+
	Turritella sp.	–	+	+
	Nerita sp.	–	+	+
Annelida: Polychaeta	*Prionospio pinnata*	–	–	+
	P. steenstrupi	–	–	+

	Pisionereis indica	–	–	+
	Streblospio benedicti	–	+	+
	Nephthys oligobranchia	–	–	+
	Nephthys gravieri	–	–	+
	Nephthys inermis	–	–	+
	Glycera alba	–	–	+
	Diopatra neapolitana	–	–	+
	Tomopteris elegans	–	–	+
	Lacydonia sp.	–	–	+
	Polydora sp.	–	+	–
	Onuphis sp.	–	+	+
	Lumbriconeries sp.	–	–	+
Oligochaeta		–	–	+
Echiuroida		–	+	+
Arthropoda: Crustacea—Decapoda:				
	Ocypoda ceratopthalma	+	+	–
	Dotilla myctiroides	–	+	+
	Cyathura sp.	–	+	–
	Cirolana borealis	+	+	–
	Emerita holthuisi	–	+	+
	Eurydice pulchara	–	+	+
	Matuta sp.	–	+	+
	Scottolana longipes	–	+	+
	Squilla sp.	–	+	+
	Gastrosaccus sp.	–	–	+
Ophichthidae—(Snake eel)	*Pisodonophis boro*	–	–	+

RESULTS AND DISCUSSION

Since the beach remains exposed (to sunlight, air, predation and humans) during low tide, the faunal communities in the intertidal habitats are comprised of ten different major groups, dominated by burrowing forms such as polychaetes, bivalves and crustaceans. Polychaeta was by far the most dominant group in term of density and diversity which constitute almost 40 - 72% of the macrobenthic population at low tide level (Table 2). Bivalve was second in the order of dominance with 12 species, followed by Crustacea with 10 species (Table 1). Out of the 12 bivalve species recorded from Miramar beach, 8 species are edible and 3 surf clam species (*viz, Mactra violacea; Paphia malabarica; Sunetta scripta* Plate 2) are exploited discretely for human consumption (Ingole, 2002). Among the crustaceans, sand crab *Dotilla myctiroides* (Plate 3) and mole crab *Emerita holthuisi* were the most abundant in mid and low tide area respectively.

Table—2 Abundance (individuals m^{-2}; n = 6). of macrofauna on the intertidal beach at Miramar observed at low tide level on 2 February 1998.

	Range	Mean	±SD
Annelida: Polychaeta			
Glycera alba	16–488	138	177.3
Nereis spp.	16–122	51.3	61.2
Nephthys spp.	368–1220	603.3	317.5
Maldanidae	0–122	122	0
Pisionidae	122–1098	699	538.2
Unidentified polychaetes	0–244	244	0
Arthopoda: Crustacea			
Cyathura sp.	16–32	24	11.3
Cirolana borealis	16–122	69	74.9
Scottolana sp.	48–336	186.7	144.3
Tanaidacae	32–48	40	11.3
Thallasiniidae	112–128	101.3	33.3
Dotilla myctiroides	0–16	16	0
Mollusca: Bivalvia-*Donax* spp	100–1200	822.4	332.3
Mactra spp.	16–244	130	161.2
Echuiroida	48–366	171.2	130.9
Nemertina	0–16	16	0
Nematoda	16–122	69	74.9
Total faunal density (no. m^{-2})	1342–3782	2615	926.9
Polychaete density (no. m^{-2})	656–3050	1527	934.9
% of Polychaetes	30–74	56.1	19.9

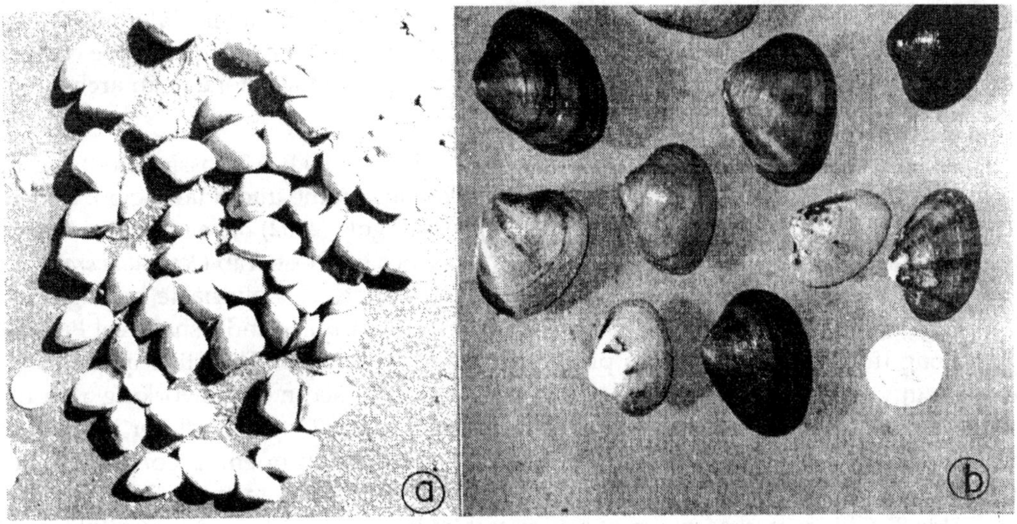

Plate. 2 Edible clams (a) (*Mactra violacea*; (b) *Paphia Malabarica*

Burrowing sand crab *Dotilla myctiroides*, edible clam *Mactra violacea* and polychaete *Nephthys* spp., were the most dominant in terms of population density. These burrowing organisms spend most of their life span inside the burrow (except for a brief larval planktonic stage during their early life in the water column). For obvious reason they do not make themselves easily visible, thus giving a wrong impression that the intertidal beach is less productive. Meiobenthos [the largest of which are smaller than one half millimeter] are made up of juveniles of clams, shrimps, worms and other crustaceans with densities as high as 38-1553 individuals /10 cm^2 (mean: 592 ±561 SD). Almost 75% of this population lives concentrated in the top 2 cms of the sediment layer (Ingole 2002). This fine top layer is thus crucial in the recruitment of the benthic population. Similarly the population density of the larger macrobenthic forms [ranging in size upto 3-cms] ranged from 1342 - 3782 individuals/m^2 (mean: 2615±846 SD; n=6, Table, 2).

The majority of these macrobenthic organisms live in under-ground burrows emerging to release their eggs in the tidal water. Eggs are hatch in the near shore waters and after a brief planktonic stage, the newly hatched young recruits return to the beach with the incoming tide where they settle into the beach sediment. Being small and not well developed, they cannot bury themselves immediately in the sediment and are constrained to spend some time in the top sediment layer before they are big enough to construct their own burrows. This is a critical time for them, and a first estimate is that about 90% of the juveniles perish or are eaten by the predators. The fortunate few are those that serendipitously stumble on already available burrows.

Intertidal Zonation: Zonation of organisms occurs on the sandy shore as well as on the rocky shore, but the effects of the ebb and flow of the tides are less marked below the sand. The sun does not warm up the sand beyond 10 centimeters depth as much as on the surface and water between the sand grains is not completely drained, so the risk of desiccation and overheating is less at low water level.

As shown in table 1, 47 macrobenthic species have been identified on this intertidal beach. Of this *Dotilla myctiroides*, *Nephthys* sp. *Donax incarnatus*, *Mactra violacea*, and *Pisodonophis boro* can be termed as important keystone species as they occupy distinct intertidal zones. Keystone species are those whose existence within a community is critical to the stability of the community structure. Removal of a keystone species has an effect much like a keystone in an arch – the arch comes crumbling down.

While studying the macrofauna of Miramar beach Bhosle (2000), Modassir (2000) and Ingole (2002) have reported distinct zonation pattern of macrofauna on the intertidal region. Depending upon the distribution and abundance of dominant species Ingole (2002) have identified 7 keystone species. Accordingly, the high tide was marked by presence of ghost crab *Ocypoda ceratopthalma*, whereas the midtide by the sand crab *Dotilla myctiroides* (Plate 1). Polychaete belonging to the genus *Nephthys* and *Diopatra;* bivalves *Donax incarnatus; Mactra violacea* and a snake eel *Pisodonophis boro* are among the major bioturbators present in the low tide area. They live in underground burrows, occupying a large part of the low tide area. The sediment-reworking rate of two polychaete species (*Nephthys* sp. and *Diopatra* sp.) is very high (27 to 133 ml per individual per pumping; mean = 74.4 ± 36.1; n=44). Considering one burrow opening correspond to one individual, the total amount of reworked sediment was calculated to be 1.57 kg per day per burrow (Table 3). This reworked material is very rich in bacterial flora, organic matter, dissolved nutrients and meiofauna. Therefore, the sediment reworking is very helpful in augmenting the biological productivity of the area.

Plate. 3 Sand crab *Dotilla myctiroides* at Miramar beach.

During 5 years of collection on this beach, the author has found that 3 of these 7 species [sand crab *Dotilla myctiroides*, surf clam *Mactra violacea* and polychaete *Nephthys sp.*] live in underground burrows and continually rework the intertidal sediment of Miramar beach (Plate 1). According to Bhosle (2000); Fernandes (2001) and Ingole (2002), the structure of polychaete and snake eel burrows observed on Miramar beach can be far more complicated and could play very important role in nutrient recycling and sedimentation.

These burrowers discard sediment onto the surface, producing tiny volcano like mounds that all of us as kids have wondered at. The sediment-reworking process of the burrowers not only recycles the dissolved nutrients into the sediment-water interface, but through their pumping action also supplies dissolved oxygen to the deeper sediment layers. This sediment reworking [also known as bioturbation] and pumping of water from burrow depth to surface [also known as bioirrigation] work simultaneously, transforming the simple mechanical one-dimensional exchange into a biogeochemical interaction between sediment pore-water and the overlying water column. Thus, the nature and rate of bioturbation can act as major factor in influencing the biogeochemical processes at the beach and the organisms responsible for this process are therefore key elements of the beach ecosystem.

Figure 2, shows the drawing of a resin cast of a ghost shrimp (*Callianassa japonica)* burrow found in the Amakusa beach in Japan (Tamaki & Ingole, 1993). Although only one small burrow opening is seen on the sediment surface, the burrow infact occupies a large space below the sediment layer, acting like a multi-chambered "house" for the animal. In this "house", one chamber is used to store freshly arriving organic material, which is allowed to breakdown along

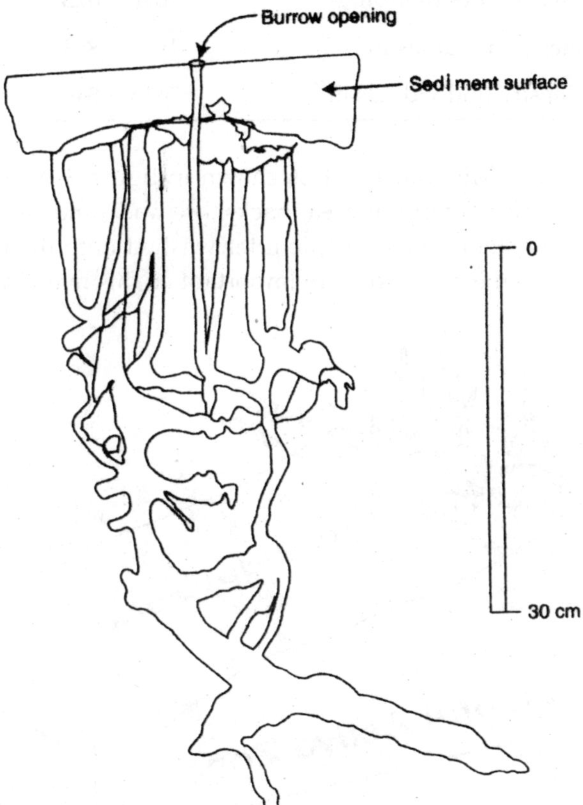

Fig. 2 Resin cast of a typical burrowing animal.

with the animals fecal material. This mixture develops into a rich bacterial flora deep inside the burrow chamber. We could call this the "*kitchen*" of the burrowing organism. Another part of the "house" is used for storing unwanted detrital material. Once this chamber is full the waste material then is pumped back to the sediment surface, creating those delicate star like patterns that we spoke about earlier. These are like the many signposts that we see in Miramar adverting different housing colonies.

Table—3 Sediment reworking rate of polychaete (*Nepthys* spp.) studied during February 1998 on the intertidal beach at Miramar, Goa.

Burrow density (Nos. m^{-2})	1.33 ± 0.56	(n=65)
Burrow diameter (mm)	6.61 ± 1.57,	(n=50)
Burrow mound diameter (mm)	96.1 ± 13.9	(n=37)
Irrigation rate (l^{-d} per burrow)	8.932	(n=17)
Sand ejection rate (kg^{-d} per burrow)	1.574	
Pumping duration (second per pumping)	20.3 ± 10.3	(n=13)
Density of polychaetes (individuals m^{-2})	1527 ± 853	
Total macrofaunal density (individuals m^{-2})	2615 ± 846	

A typical section of Miramar beach is shown in figure 3. This littoral active zone includes the near shore, surf zone, intertidal area, backshore and sand dunes. All this form an integral part of the beach ecosystem. Even organic litter washed onto the beach along with the detritus and algal deposition on the backshore are important as this material actually stabilizes the beach

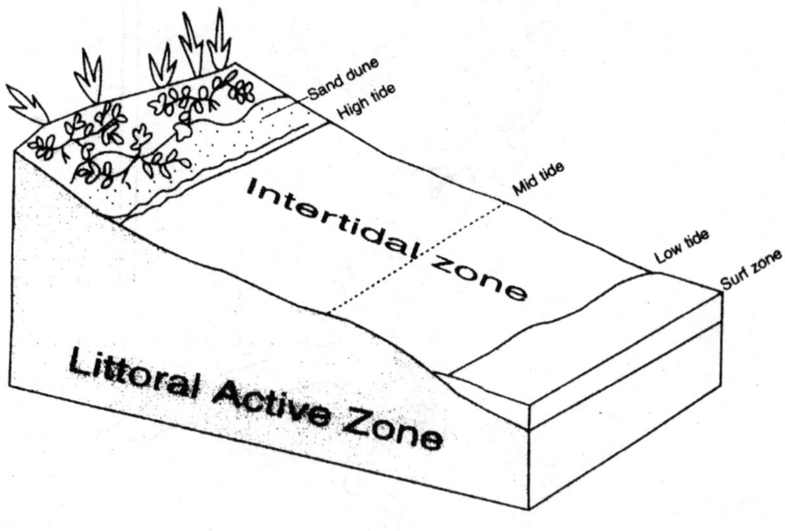

Fig. 3: Section of an intertidal beach.

sand and provides food and shelter for beach fauna. Larger semi-terrestrial animals such as the ghost crab and talitrid amphipods occupy the backshore areas and depend on intertidal organisms that the author has described for food. The upper littoral zone and the intertidal section are therefore essentially interdependent and interacting. Because beach ecosystems are so fragile, even the slightest man-induced alteration to them may result in significant effects. Therefore, monitoring the impact of anthropogenic activities on the beach ecosystem is very essential to maintain natural ecological balance.

ACKNOWLEDGEMENTS

I wish to thank Dr. Ehrlich Desa, Director, NIO, Goa for encouraging me to write this manuscript and correcting the initial draft. I wish to thank my colleagues Ms. Critabell Fernades and Ms. Hema Bhosle for field and laboratory help and to Dr. S. Raghukumar, Scientist NIO, Goa for providing helpful comments on the manuscript. This is the contribution No. 3756 of NIO, Goa.

REFERENCES

Bhosle, H., 2000. Sediment reworking by macrobenthic organisms on an intertidal sandy beach. M.Sc. (Marine Science) Thesis, Goa University, Goa, (unpublished). 58 pp.

Brown, A.C. and McLachlan, A., 1990. *Ecology of Sandy Shores* (eds), Elsevier Scientific publication, Amsterdam, 321 pp.

Day, J. H., 1973. Polychaeta collected by U.D. Gaikwad at Ratnagiri, south of Bombay, *Zoological Journal of Linneous Society*, 52: 337-361.

Dwivedi, S. N., Rahim, A. and Ayyappan Nair, S., 1973. Ecology and Production of intertidal sand fauna during southwest monsoon along Mandovi estuary (Goa). In: *Recent Researches in Estuarine Biology*, R. Natarajan (ed.) Hindustan Publishing Corporation New Delhi (India), pp: 1-20.

Eleftheriou, A., and Jones, D. A., 1976. The genus *Eurydice* on the west coast of India. *Journal of Zoology London*, 178: 385-394.

Fauvel, P., 1953.*The fauna of India, including Pakistan, Ceylon, Burma and Malaya* (The Indian Press, Ltd, Allahabad), 507 pp.

Fernandes, C. E. G., 2001. Influence of sediment reworking on the nutrient recycling and microbiota. M.Sc. (Marine Science) Thesis, Goa University, Goa (unpublished), 72 pp.

Gosner, K. L., 1971. *Guide to Identification of Marine and Estuarine Invertebrates*. John Wiley & Sons, 693 pp.

Ingole, B. S., 1992. Culture and Population Dynamics of Benthic intertidal harpacticoid copepods, *Amphiascoides subdebilis WILLEY*, 1935 & *Scottolana longipes* THOMPSON & SCOTT, 1903. - Ph.D., thesis University of Bombay, Bombay, pp.130.

Ingole, B. S., 2002. Miramar beach is no more habitat of exotic species. *The Navhind Times*, 13th February 2002, pp.1 &14.

Ingole, B. S., and Parulekar, A. H., 1998.The Role of Seasonal Fluctuations in Salinity in structuring the intertidal meiofauna of a tropical estuarine beach: Field Evidence. *Indian Journal of Marine Science*, 27: 356-361.

McIntyre, A.D., 1977. Sandy Foreshore, in: *The Coastline*, (ed.) by R.S.K. Barnes and J. Wiley & Sons London.

Modassir, Y., 2000.The intertidal fauna inhabiting an exposed sandy beach at Miramar, Goa. *Advance in Biosciences*, 19(1): 19-32.

Parulekar A. H., 1972. Polychaetes from Maharashtra and Goa. *Journal of Bombay Natural Histor Society*, 68 (3): 726-749.

Tamaki, A. and Ingole, B. S.,1993. Distribution of juvenile and adult ghost shrimps, *Callianassa japonic* Ortmann (Thalassinidea), on an intertidal sandflat:Intraspecific facilitation as a possible pattern generating factor. *Journal of Crustacean Biology*, 13 (1): 175-183.

Vongpanich, V., 2000. Family Mactridae (Mollusca: Bivalvia) in Thai waters. *Puket Marine Biologic Center Special Publication*, 21(2): 483-498.